ACOUSTICS OF DUCTS AND MUFFLERS

ACOUSTICS OF DUCTS AND MUFFLERS

Second Edition

M. L. Munjal
Department of Mechanical Engineering
Indian Institute of Science
Bangalore, India

WILEY

This edition first published 2014
© 2014 John Wiley & Sons Ltd

Registered office

John Wiley & Sons Ltd, The Atrium, Southern Gate, Chichester, West Sussex, PO19 8SQ, United Kingdom

For details of our global editorial offices, for customer services and for information about how to apply for permission to reuse the copyright material in this book please see our website at www.wiley.com.

The right of the author to be identified as the author of this work has been asserted in accordance with the Copyright, Designs and Patents Act 1988.

All rights reserved. No part of this publication may be reproduced, stored in a retrieval system, or transmitted, in any form or by any means, electronic, mechanical, photocopying, recording or otherwise, except as permitted by the UK Copyright, Designs and Patents Act 1988, without the prior permission of the publisher.

Wiley also publishes its books in a variety of electronic formats. Some content that appears in print may not be available in electronic books.

Designations used by companies to distinguish their products are often claimed as trademarks. All brand names and product names used in this book are trade names, service marks, trademarks or registered trademarks of their respective owners. The publisher is not associated with any product or vendor mentioned in this book.

Limit of Liability/Disclaimer of Warranty: While the publisher and author have used their best efforts in preparing this book, they make no representations or warranties with respect to the accuracy or completeness of the contents of this book and specifically disclaim any implied warranties of merchantability or fitness for a particular purpose. It is sold on the understanding that the publisher is not engaged in rendering professional services and neither the publisher nor the author shall be liable for damages arising herefrom. If professional advice or other expert assistance is required, the services of a competent professional should be sought.

Library of Congress Cataloging-in-Publication Data

Munjal, M. L. (Manohar Lal), 1945-
 [Acoustics of ducts and mufflers with application to exhaust and
ventilation system design]
   Acoustics of ducts and mufflers / M. L. Munjal. – Second edition.
       pages cm
   Revised edition of: Acoustics of ducts and mufflers with application to
exhaust and ventilation system design / M. L. Munjal. 1987.
   Includes bibliographical references and index.
   ISBN 978-1-118-44312-5 (hardback)
 1. Air ducts–Acoustic properties. 2. Automobiles–Motors–Mufflers–Acoustic properties.
 3. Ventilation–Noise. 4. Automobiles–Motors–Exhaust systems–Noise. I. Title.
   TH7683.D8M86 2014
   629.25′2–dc23
                                                                       2013041131

A catalogue record for this book is available from the British Library.

ISBN   9781118443125

Set in 10/12pt, Times-Roman by Thomson Digital, Noida, India

1   2014

To the memory of my parents

Contents

Preface — xiii

1 Propagation of Waves in Ducts — 1
   1.1 Plane Waves in an Inviscid Stationary Medium — 2
   1.2 Three-Dimensional Waves in an Inviscid Stationary Medium — 5
        1.2.1 Rectangular Ducts — 5
        1.2.2 Circular Ducts — 8
   1.3 Waves in a Viscous Stationary Medium — 12
   1.4 Plane Waves in an Inviscid Moving Medium — 16
   1.5 Three-Dimensional Waves in an Inviscid Moving Medium — 18
   1.6 One-Dimensional Waves in a Viscous Moving Medium — 20
   1.7 Waves in Ducts with Compliant Walls (Dissipative Ducts) — 23
        1.7.1 Rectangular Duct with Locally Reacting Lining — 24
        1.7.2 Circular Duct with Locally Reacting Lining — 27
        1.7.3 Rectangular Duct with Bulk Reacting Lining — 29
        1.7.4 Circular Duct with Bulk Reacting Lining — 32
   1.8 Three-Dimensional Waves along Elliptical Ducts — 34
   References — 39

2 Theory of Acoustic Filters — 41
   2.1 Units for the Measurement of Sound — 41
   2.2 Uniform Tube — 43
   2.3 Radiation Impedance — 46
   2.4 Reflection Coefficient at an Open End — 48
   2.5 A Lumped Inertance — 49
   2.6 A Lumped Compliance — 50
   2.7 End Correction — 51
   2.8 Electroacoustic Analogies — 51
   2.9 Electrical Circuit Representation of an Acoustic System — 52
   2.10 Acoustical Filter Performance Parameters — 53
        2.10.1 Insertion Loss, IL — 54
        2.10.2 Transmission Loss, TL — 55

	2.10.3	Level Difference, LD	56
	2.10.4	Comparison of the Three Performance Parameters	58
2.11		Lumped-Element Representations of a Tube	58
2.12		Simple Area Discontinuities	60
2.13		Gradual Area Changes	62
	2.13.1	Conical Tube	62
	2.13.2	Exponential Tube	64
	2.13.3	Elliptical Tube	65
2.14		Extended-Tube Resonators	67
2.15		Helmholtz Resonator	69
2.16		Concentric Hole-Cavity Resonator	70
2.17		An Illustration of the Classical Method of Filter Evaluation	71
2.18		The Transfer Matrix Method	74
	2.18.1	Definition of Transfer Matrix	74
	2.18.2	Transfer Matrix of a Uniform Tube	76
	2.18.3	A General Method for Derivation of Transfer Matrix	76
	2.18.4	Transfer Matrices of Lumped Elements	77
	2.18.5	Transfer Matrices of Variable Area Tubes	78
	2.18.6	Overall Transfer Matrix of the System	81
	2.18.7	Evaluation of TL in Terms of the Four-Pole Parameters	83
2.19		TL of a Simple Expansion Chamber Muffler	85
2.20		An Algebraic Algorithm for Tubular Mufflers	88
	2.20.1	Development of the Algorithm	88
	2.20.2	Formal Enunciation and Illustration of the Algorithm	89
2.21		Synthesis Criteria for Low-Pass Acoustic Filters	91
References			94

3 Flow-Acoustic Analysis of Cascaded-Element Mufflers — 97

3.1		The Exhaust Process	97
3.2		Finite Amplitude Wave Effects	101
3.3		Mean Flow and Acoustic Energy Flux	102
3.4		Aeroacoustic State Variables	105
3.5		Aeroacoustic Radiation	108
3.6		Insertion Loss	111
3.7		Transfer Matrices for Tubular Elements	112
	3.7.1	Uniform Tube	113
	3.7.2	Extended-Tube Elements	114
	3.7.3	Simple Area Discontinuities	118
	3.7.4	Physical Behaviour of Area Discontinuities	118
3.8		Perforated Elements with Two Interacting Ducts	119
	3.8.1	Concentric-Tube Resonator	124
	3.8.2	Cross-Flow Expansion Element	124
	3.8.3	Cross-Flow Contraction Element	125
	3.8.4	Some Remarks	125
3.9		Acoustic Impedance of Perforates	126
3.10		Matrizant Approach	129

	3.11	Perforated Elements with Three Interacting Ducts	131
		3.11.1 *Three-Duct Cross-Flow Expansion Chamber Element*	135
		3.11.2 *Three-Duct Reverse Flow Expansion Chamber Element*	136
	3.12	Other Elements Constituting Cascaded-Element Mufflers	137
	References		143
4	**Flow-Acoustic Analysis of Multiply-Connected Perforated Element Mufflers**		**147**
	4.1	Herschel-Quincke Tube Phenomenon	147
	4.2	Perforated Element with Several Interacting Ducts	151
	4.3	Three-Pass Double-Reversal Muffler	154
	4.4	Flow-Reversal End Chambers	158
	4.5	Meanflow Lumped Resistance Network Theory	163
	4.6	Meanflow Distribution and Back Pressure Estimation	169
		4.6.1 *A Chamber with Three Interacting Ducts*	169
		4.6.2 *Three-Pass Double-Reversal Chamber*	171
		4.6.3 *A Complex Muffler Configuration*	173
	4.7	Integrated Transfer Matrix Approach	175
		4.7.1 *A Muffler with Non-Overlapping Perforated Ducts and a Baffle*	176
		4.7.2 *Muffler with Non-Overlapping Perforated Elements, Baffles and Area Discontinuities*	180
		4.7.3 *A Combination Muffler*	184
	References		186
5	**Flow-Acoustic Measurements**		**187**
	5.1	Impedance of a Passive Subsystem or Termination	187
		5.1.1 *The Probe-Tube Method*	188
		5.1.2 *The Two-Microphone Method*	193
		5.1.3 *Transfer Function Method*	197
		5.1.4 *Comparison of the Various Methods for a Passive Subsystem*	202
	5.2	Four-Pole Parameters of a Flow-Acoustic Element or Subsystem	203
		5.2.1 *Theory of the Two Source-Location Method*	204
		5.2.2 *Theory of the Two-Load Method*	207
		5.2.3 *Comparison of the Two Methods*	208
		5.2.4 *Experimental Validation*	208
	5.3	An Active Termination – Aeroacoustic Characteristics of a Source	210
		5.3.1 *Direct Measurement of Source Impedance*	211
		5.3.2 *Indirect Measurement of Source Characteristics*	214
		5.3.3 *Numerical Evaluation of the Engine Source Characteristics*	224
		5.3.4 *A Comparison of the Various Methods for Measuring Source Characteristics*	228
	References		229
6	**Dissipative Ducts and Parallel Baffle Mufflers**		**233**
	6.1	Acoustically Lined Rectangular Duct with Moving Medium	234
	6.2	Acoustically Lined Circular Duct with Moving Medium	239
	6.3	Transfer Matrix Relation for a Dissipative Duct	241

	6.4	Transverse Wave Numbers for a Stationary Medium	244
	6.5	Normal Impedance of the Lining	245
	6.6	Transmission Loss	249
	6.7	Effect of Protective Layer	251
	6.8	Parallel Baffle Muffler	257
	6.9	The Effect of Mean Flow	259
	6.10	The Effect of Terminations on the Performance of Dissipative Ducts	260
	6.11	Lined Bends	261
	6.12	Plenum Chambers	261
	6.13	Flow-Generated Noise	262
	6.14	Insertion Loss of Parallel Baffle Mufflers	263
	References		264
7	**Three-Dimensional Analysis of Mufflers**		**267**
	7.1	Collocation Method for Simple Expansion Chambers	268
		7.1.1 Compatibility Conditions at Area Discontinuities	269
		7.1.2 Extending the Frequency Range	272
		7.1.3 Extension to Other Muffler Configurations	275
	7.2	Finite Element Methods for Mufflers	275
		7.2.1 A Single Element	276
		7.2.2 Variational Formulation of Finite Element Equations	281
		7.2.3 The Galerkin Formulation of Finite Element Equations	285
		7.2.4 Evaluation of Overall Performance of a Muffler	288
		7.2.5 Numerical Computation	290
	7.3	Green's Function Method for a Rectangular Cavity	292
		7.3.1 Derivation of the Green's Function	292
		7.3.2 Derivation of the Velocity Potential	295
		7.3.3 Derivation of the Transfer Matrix	297
		7.3.4 Validation Against FEM	299
	7.4	Green's Function Method for Circular Cylindrical Chambers	301
	7.5	Green's Function Method for Elliptical Cylindrical Chambers	303
	7.6	Breakout Noise	306
		7.6.1 Breakout Noise from Hoses	306
		7.6.2 Breakout Noise from Rectangular Ducts	310
		7.6.3 Breakout Noise from Elliptical Ducts	312
		7.6.4 Breakout Noise from Mufflers	315
	References		316
8	**Design of Mufflers**		**321**
	8.1	Requirements of an Engine Exhaust Muffler	321
	8.2	Simple Expansion Chamber	322
	8.3	Double-Tuned Extended-Tube Expansion Chamber	324
	8.4	Tuned Concentric Tube Resonator	326
	8.5	Plug Mufflers	327
	8.6	Side-Inlet Side-Outlet Mufflers	329
	8.7	Designing for Insertion Loss	331

8.8	Three-Pass Double-Reversal Chamber Mufflers	338
8.9	Perforated Baffle Muffler	347
8.10	Forked Dual Muffler System	349
8.11	Design of Short Elliptical and Circular Chambers	353
	8.11.1 Short Chamber Configurations	354
	8.11.2 Theory	355
	8.11.3 Results and Discussion	356
	8.11.4 Acoustical Equivalence of Short Expansion and Flow-reversal End Chambers	361
8.12	Back-Pressure Considerations	362
8.13	Practical Considerations	365
8.14	Design of Mufflers for Ventilation Systems	367
8.15	Active Sound Attenuation	369
References		375

Appendix A: Bessel Functions and Some of Their Properties — 377

Appendix B: Entropy Changes in Adiabatic Flows — 381
- B.1 Stagnation Pressure and Entropy — 381
- B.2 Pressure, Density, and Entropy — 382

Appendix C: Nomenclature — 385

Index — 389

Preface

Engine exhaust noise pollutes the street environment and ventilation fan noise enters dwellings along with the fresh air. Work on the analysis and design of mufflers for both these applications has been going on since the early 1920s. However, it gained momentum in the 1970s as people became more and more conscious of their working and living environment. Governments of many countries have responded to popular demand with mandatory restrictions on sound emitted by automotive engines. The exhaust system being the primary source of engine noise (combustion-induced structural vibration sound is the next in importance for diesel engines), exhaust mufflers have received most attention from researchers.

The text deals with the theory of exhaust mufflers for internal combustion engines of the reciprocating type, air-conditioning and ventilation ducts, ventilation and access openings of acoustic enclosures, and so on. The common feature of all these systems is wave propagation in a moving medium. The function of a muffler is to muffle the sound through impedance mismatch or to dissipate the incoming sound into heat, while allowing the mean flow to go through almost unimpeded. The former type of muffler is called reflective, reactive, or nondissipative mufflers, while the latter are called dissipative or absorptive mufflers, or simply silencers. However, every muffler contains some impedance mismatch and some acoustic dissipation. Therefore, this book deals with mufflers that are primarily reflective as well as with those that are primarily dissipative. However, combination mufflers have also been discussed.

The monograph 'Acoustics of Ducts and Mufflers', the original edition of which was published in 1987, has continued to be the only monograph on this subject. Elsewhere in the literature, the theory of exhaust mufflers is generally covered in a chapter or two in books on industrial noise control, a treatment which is too superficial for a student, researcher, or prospective designer; the actual design of practical exhaust and ventilation systems is treated in a simplistic, textbook style approach.

Over the last 26 years, since publication of the original edition of this monograph, there have been several significant developments in this important area. The mainframe computers, desktops, personal computers and laptops have got much faster with the random access memory (RAM) being several orders larger. Very efficient algorithms have been developed for analysis of one-dimensional (1D) as well as three-dimensional (3D) analysis of mufflers. Commercially available softwares with user-friendly graphic user interface (GUI) have made the analysis (and thence design) of mufflers much more convenient as well as faster.

Concurrently, efficient algorithms have been developed for plane wave analysis of multiply-connected perforated-element mufflers as well as the cascaded-element mufflers. Methods have been developed to estimate the meanflow pressure drop across complex muffler configurations making use of the lumped-element flow resistance networks or the flow network models, without resorting to tedious and time-consuming computational fluid dynamics (CFD) modeling. It is now possible to estimate the unmuffled exhaust as well as intake noise spectrum of internal combustion reciprocating engines making use of empirical expressions of acoustic source characteristics. Recently 3D as well as 1D models have been developed for analysis of multiport elements as well as two-port elements.

The break-out noise radiating from the shell as well as end plates of a muffler can now be predicted reasonably well, thereby enabling prediction of net transmission loss (TL) and insertion loss (IL) of mufflers. All these developments are reflected in this second edition of the monograph.

Like the original edition, this second edition of the monograph is an outcome of my research in the analysis and design of mufflers for over 40 years, and of a course that I have been offering to graduate students all along. My experience in industrial consultancy is amply reflected in the application orientated treatment of the subject. Although a bias in favor of the methods developed by me and my students over the years is unavoidable, every effort has been made to offer the best to the reader. Emphasis is on the latest and/or the best methods available, and not on the coverage of all the methods available in the current literature on any particular topic or aspect. A substantial portion of the book represents recent unpublished material. References have been cited from all over the English-language literature, but no effort has been made to make the lists (at the end of every chapter) exhaustive.

The symbols of parameters used throughout the book are presented in an appendix. Otherwise, every chapter has been prepared so that it is complete in itself. Generally, symbols are accompanied by the names of the parameters they represent in order to make the reading as smooth as possible.

The text starts with the propagation of waves in ducts, which forms the base for subsequent chapters. Exhaustive treatment of the one-dimensional theory of acoustic filters is followed by aeroacoustics of exhaust mufflers where the convective as well as dissipative effects of moving medium are incorporated.

The time-domain approach (method of characteristics) in Chapter 4 of the original edition has been replaced with flow-acoustic analysis of multiply-connected perforated-element mufflers, where estimation of back pressure making use of the flow resistance network approach has been discussed apart from the 1D frequency domain acoustic analysis.

Experimental methods needed for supplementing analysis corroborating analytical results, and verifying the efficacy of a muffler configuration, are discussed in Chapter 5. Dissipative ducts or mufflers are dealt with in Chapter 6. These days, a variety of muffler configurations have come into commercial use in which three-dimensional effects predominate. These configurations can be analyzed best by means of the finite element methods. In this second edition, 3D analytical methods have also been introduced in Chapter 7 in order to account for the higher-order modes for prediction of the transverse TL (relevant to break-out noise) as well as axial TL, and thence the net TL.

The last chapter is devoted exclusively to the design of mufflers for various applications. Active noise control in a duct has also been touched upon.

Preface

This book is addressed to designers and graduate students specializing in technical acoustics or engineering acoustics. Researchers will find in it a state-of-the-art account of muffler theory. An effort has been made to make the book complete in itself, that is, independently readable. Engine exhaust systems and ventilation systems are the primary targets. However, methods discussed here can be applied to the inlet and discharge systems of reciprocating compressors as well.

I owe my first interest in vibrations and dynamical systems to my former teacher, M. V. Narasimhan who has continued to be my friend, philosopher, and guide over the last four decades. I benefited greatly from my association and discussions with A. V. Sreenath, B.S. Ramakrishna, M. Heckl, B. V. A. Rao, Colin H. Hansen, Mats Abom, Hans Boden, Ahmet Selamet, J. E. Sneckenberger, M. G. Prasad, Istvan L. Ver, Larry J. Eriksson, S. Soundranayagam, V. R. Sonti, C. S. Jog, Rudra Pratap, Joseph W. Sullivan, and S. Anantharamu, among others.

I have drawn heavily from the published work of my former students, Prakash T. Thawani, M. L. Kathuriya, V. B. Panicker, Mohan D. Rao, K. Narayana Rao, U. S. Shirahatti, A. D. Sahasrabudhe, V. H. Gupta, V. Easwaran, G. R. Gogate, T. S. S. Narayana, V. Bhujanga Rao, S. N. Panigrahi, Trinath Kar, R. N. Hota, B. Venkatesham, N. K. Mukherjee, P. Chaitanya, M. Harikrishna Reddy, N. K. Vijayasree, Akhilesh Mimani, D. Veerababu, Ramya Teja and Abhishek Verma. Some of them, who used the original version of the monograph, pointed out several typos and made useful suggestions that have been gratefully addressed in this second edition of the monograph.

I wish to particularly thank R. Mangala and C. Srinivasa for their help with typing drawings and formatting of the manuscript through several revisions and corrections.

This book took nearly two years of preparation, writing, and processing. If it is completed today, it is largely due to the constant forbearance, understanding, and cooperation of my wife, Vandana alias Bhuvnesh.

This manuscript has been catalyzed and supported by the Department of Science and Technology (DST), under its Utilization of Scientific Expertise of Retired Scientists (USERS) Scheme.

M. L. Munjal
Bangalore, India
July 2013

1

Propagation of Waves in Ducts

Exhaust noise of internal combustion engines is known to be the biggest pollutant of the present-day urban environment. Fortunately, however, this noise can be reduced sufficiently (to the level of the noise from other automotive sources, or even lower) by means of a well-designed muffler (also called a silencer). Mufflers are conventionally classified as dissipative or reflective, depending on whether the acoustic energy is dissipated into heat or is reflected back by area discontinuities.

However, no practical muffler or silencer is completely reactive or completely dissipative. Every muffler contains some elements with impedance mismatch and some with acoustic dissipation. In fact, combination mufflers are getting increasingly popular with designers.

Dissipative mufflers consist of ducts lined on the inside with an acoustically absorptive material. When used on an engine, such mufflers lose their performance with time because the acoustic lining gets clogged with unburnt carbon particles or undergoes thermal cracking. Recently, however, better fibrous materials such as sintered metal composites have been developed that resist clogging and thermal cracking and are not so costly. Besides, long strand unglued glass fibers can stand high temperatures. Nevertheless, no such problems are encountered in ventilation ducts, which conduct clean and cool air. The fan noise that would propagate through these ducts can well be reduced during propagation if the walls of the conducting duct are acoustically treated. For these reasons the use of dissipative mufflers is much more common in air-conditioning systems.

Reflective mufflers, being nondissipative, are also called reactive mufflers. A reflective muffler consists of a number of tubular elements of different transverse dimensions joined together so as to cause, at every junction, impedance mismatch and hence reflection of a substantial part of the incident acoustic energy back to the source. Most of the mufflers currently used on internal combustion engines, where the exhaust mass flux varies strongly, though periodically, with time, are of the reflective or reactive type. In fact, even the muffler of an air-conditioning system is generally provided with a couple of reflective elements at one or both ends of the acoustically dissipative duct.

Clearly, a tube or pipe or duct is the most basic and essential element of either type of muffler. A study of the propagation of waves in ducts is therefore central to the analysis of a muffler for its acoustic performance (transmission characteristics). This chapter is devoted to

---

Acoustics of Ducts and Mufflers, Second Edition. M. L. Munjal.
© 2014 John Wiley & Sons, Ltd. Published 2014 by John Wiley & Sons, Ltd.

the derivation and solution of equations for plane waves and three-dimensional waves along rectangular ducts, circular tubes and elliptical shells without and with mean flow, without and with viscous friction, with rigid unlined walls and compliant or acoustically lined walls. We start with the simplest case and move gradually to the more general and involved cases.

1.1 Plane Waves in an Inviscid Stationary Medium

In the ideal case of a rigid-walled tube with sufficiently small cross dimensions[*] filled with a stationary ideal (nonviscous) fluid, small-amplitude waves travel as plane waves. The acoustic pressure perturbation (on the ambient static pressure) p and particle velocity u at all points of a cross-section are the same. The wave front or phase surface, defined as a surface at all points of which p and u have the same amplitude and phase, is a plane normal to the direction of wave propagation, which in the case of a tube is the longitudinal axis.

The basic linearized equations for the case are:
Mass continuity

$$\rho_o \frac{\partial u}{\partial z} + \frac{\partial \rho}{\partial t} = 0; \qquad (1.1)$$

Dynamical equilibrium

$$\rho_o \frac{\partial u}{\partial t} + \frac{\partial p}{\partial z} = 0; \qquad (1.2)$$

Energy equation (isentropicity)

$$\left(\frac{\partial p}{\partial \rho}\right)_s = \frac{\gamma(p_o + p)}{\rho_o + \rho} \cong \frac{\gamma p_o}{\rho_o} = c_o^2 (say) \qquad (1.3)$$

where

z is the axial or longitudinal coordinate,
p and ρ are acoustic perturbations on pressure and density,
p_o and ρ_o are ambient pressure and density of the medium,
s is the entropy,
$p/p_o \ll 1, \quad \rho/\rho_o \ll 1$

Equation 1.3 implies that

$$d\rho = \frac{dp}{c_o^2}; \quad \frac{\partial \rho}{\partial t} = \frac{1}{c_o^2} \frac{\partial p}{\partial t}; \quad \frac{\partial \rho}{\partial z} = \frac{1}{c_o^2} \frac{\partial p}{\partial z}. \qquad (1.4)$$

The equation of dynamical equilibrium is also referred to as momentum balance equation, or simply, momentum equation. Similarly, the equation for mass continuity is commonly called continuity equation.

---

[*] The specific limits on the cross dimensions as a function of wave length are given in the next section.

Substituting Equation 1.4 in Equation 1.1 and eliminating u from Equations 1.1 and 1.2 by differentiating the first with respect to (w.r.t.) t, the second with respect to z, and subtracting, yields

$$\left[\frac{\partial^2}{\partial t^2} - c_o^2 \frac{\partial^2}{\partial z^2}\right] p = 0. \tag{1.5}$$

This linear, one-dimensional (that is, involving one space coordinate), homogeneous partial differential equation with constant coefficients (c_o is independent of z and t) admits a general solution:

$$p(z,t) = C_1 f(z - c_o t) + C_2 g(z + c_o t). \tag{1.6}$$

If the time dependence is assumed to be of the exponential form $e^{j\omega t}$, then the solution (1.6) becomes

$$p(z,t) = C_1 e^{j\omega(t - z/c_o)} + C_2 e^{j\omega(t + z/c_o)}. \tag{1.7}$$

The first part of this solution equals C_1 at $z = t = 0$ and also at $z = c_o t$. Therefore, it represents a progressive wave moving forward unattenuated and unaugmented with a velocity c_o. Similarly, it can be readily observed that the second part of the solution represents a progressive wave moving in the opposite direction with the same velocity, c_o.

Thus, c_o is the velocity of wave propagation, Equation 1.5 is a wave equation, and solution (1.7) represents a standing wave defined as superposition of two progressive waves with amplitudes C_1 and C_2 moving in opposite directions.

Equation 1.5 is called the classical one-dimensional wave equation, and the velocity of wave propagation c_o is also called phase velocity or sound speed. As acoustic pressure p is linearly related to particle velocity u or, for that matter, velocity potential ϕ defined by the relations

$$u = \frac{\partial \phi}{\partial z}; \quad p = -\rho_o \frac{\partial \phi}{\partial t}, \tag{1.8}$$

the dependent variable in Equation 1.5 could as well be u or ϕ. In view of this generality, the wave character of Equation 1.5 lies in the differential operator

$$L \equiv \frac{\partial^2}{\partial t^2} - c_o^2 \frac{\partial^2}{\partial z^2}, \tag{1.9}$$

which is called the classical one-dimensional wave operator.

Upon factorizing this wave operator as

$$\frac{\partial^2}{\partial t^2} - c_o^2 \frac{\partial^2}{\partial z^2} = \left(\frac{\partial}{\partial t} + c_o \frac{\partial}{\partial z}\right)\left(\frac{\partial}{\partial t} - c_o \frac{\partial}{\partial z}\right), \tag{1.10}$$

one may realize that the forward-moving wave [the first part of solution (1.6) or (1.7)] is the solution of the equation

$$\frac{\partial p}{\partial t} + c_o \frac{\partial p}{\partial z} = 0, \tag{1.11}$$

and the backward-moving wave [the second part of solution (1.6) or (1.7)] is the solution of the equation

$$\frac{\partial p}{\partial t} - c_o \frac{\partial p}{\partial z} = 0, \quad (1.12)$$

Equation 1.7 can be rearranged as

$$p(z,t) = \left[C_1 e^{-jkz} + C_2 e^{+jkz}\right] e^{j\omega t}, \quad (1.13)$$

where
$k = \omega/c_o = 2\pi/\lambda$,
k is called the wave number or propagation constant, and λ is the wavelength.

As particle velocity u also satisfies the same wave equation, one can write

$$u(z,t) = \left[C_3 e^{-jkz} + C_4 e^{+jkz}\right] e^{j\omega t}. \quad (1.14)$$

Substituting Equations 1.13 and 1.14 in the dynamical equilibrium equation (1.2) yields

$$C_3 = C_1/\rho_o c_o, \quad C_4 = -C_2/\rho_o c_o,$$

and therefore

$$u(z,t) = \frac{1}{Z_o}\left(C_1 e^{-jkz} - C_2 e^{+jkz}\right) e^{j\omega t}, \quad (1.15)$$

where $Z_o = \rho_o c_o$ is the characteristic impedance of the medium, defined as the ratio of the acoustic pressure and particle velocity of a plane progressive wave.

For a plane wave moving along a tube, one could also define a volume velocity v_v ($= Su$) and mass velocity

$$v = \rho_o S u, \quad (1.16)$$

where S is the area of cross-section of the tube. The corresponding values of characteristic impedance (defined now as the ratio of the acoustic pressure and the said velocity of a plane progressive wave) would then be as follows:

$$\text{For particle velocity u, characteristic impedance} = p/u = \rho_o c_o; \quad (1.17a)$$

$$\text{For volume velocity } v_v, \text{ characteristic impedance} = p/v_v = \rho_o c_o/S; \quad (1.17b)$$

$$\text{For mass velocity } v, \text{ characteristic impedance} = p/v = c_o/S. \quad (1.17c)$$

For the latter two cases, the characteristic impedance involves the tube area S. As it is not a property of the medium alone, it would be more appropriate to call it characteristic impedance of the tube. For tubes conducting hot exhaust gases, it is more appropriate to deal with acoustic mass velocity v. The corresponding characteristic impedance is denoted in these pages by the symbol Y for convenience:

$$Y_o = c_o/S. \tag{1.18}$$

Equations 1.15, 1.16 and 1.18 yield the following expression for acoustic mass velocity:

$$v(z,t) = \frac{1}{Y_o}\left(C_1 e^{-jkz} - C_2 e^{+jkz}\right)e^{j\omega t} \tag{1.19}$$

Subscript 0 with Y and k indicates nonviscous conditions. Constants C_1 and C_2 in Equations 1.13 and 1.19 are to be determined by the boundary conditions imposed by the elements that precede and follow the particular tubular element under investigation. This has to be deferred to the next chapter, where we deal with a system of elements or an acoustic filter.

1.2 Three-Dimensional Waves in an Inviscid Stationary Medium

In order to appreciate the limitations of the plane wave theory, it is necessary to consider the general 3D (three-dimensional) wave propagation in tubes. The basic linearized equations corresponding to Equations 1.1 and 1.2 for waves in stationary nonviscous medium are obtained by replacing $\partial/\partial z$ with the 3D gradient operator ∇. Thus,

$$\text{Mass continuity:} \quad \rho_o \nabla \cdot u + \frac{\partial \rho}{\partial t} = 0; \tag{1.20}$$

$$\text{Dynamical equilibrium:} \quad \rho_o \frac{\partial u}{\partial t} + \nabla p = 0. \tag{1.21}$$

The third equation is the same as Equations 1.3 or 1.4. On making use of this equation in Equation 1.20, differentiating Equation 1.20 w.r.t. to t, taking divergence of Equation 1.21 and subtracting, one gets the required 3D wave equation,

$$\left[\frac{\partial^2}{\partial t^2} - c_o^2 \nabla^2\right] p = 0, \tag{1.22}$$

where the Laplacian ∇^2 is given as follows.
Cartesian coordinate system (for rectangular ducts)

$$\nabla^2 = \frac{\partial^2}{\partial x^2} + \frac{\partial^2}{\partial y^2} + \frac{\partial^2}{\partial z^2}; \tag{1.23}$$

Cylindrical polar coordinate system (for circular tubes)

$$\nabla^2 = \frac{\partial^2}{\partial r^2} + \frac{1}{r}\frac{\partial}{\partial r} + \frac{1}{r^2}\frac{\partial^2}{\partial \theta^2} + \frac{\partial^2}{\partial z^2}. \tag{1.24}$$

1.2.1 Rectangular Ducts

For harmonic time dependence, making use of separation of variables, the general solution of the 3D wave equation (1.22) with the Laplacian given by Equation 1.23 can be seen to be

$$p(x,y,z,t) = \left(C_1 e^{-jk_z z} + C_2 e^{+jk_z z}\right)\left(e^{-jk_x x} + C_3 e^{+jk_x x}\right)\left(e^{-jk_y y} + C_4 e^{+jk_y y}\right) e^{j\omega t}, \quad (1.25)$$

with the compatibility condition

$$k_x^2 + k_y^2 + k_z^2 = k_o^2. \quad (1.26)$$

Here, k_x, k_y and k_z are wave numbers in the x, y and z direction, respectively. In the limiting case of plane waves, $k_x = k_y = o$. Then, Equation 1.26 yields $k_z = k_o$ and Equation 1.25 reduces to Equation 1.13.

It may be noted from Equation 1.25 that x-dependent factor involves two unknowns k_x and c_3, and the y-dependent factor involves the unknowns k_y and C_4. These may be evaluated from the relevant boundary conditions as follows.

For a rigid-walled duct of breadth b and height h (Figure 1.1), the boundary conditions are

$$\frac{\partial p}{\partial x} = 0 \quad at \quad x = 0 \quad and \quad x = b \quad (1.27a)$$

and

$$\frac{\partial p}{\partial y} = 0 \quad at \quad y = 0 \quad and \quad y = h. \quad (1.27b)$$

Substituting these boundary conditions in Equation 1.25 yields, respectively,

$$C_3 = 1; \quad k_x = \frac{m\pi}{b}, \quad m = 0, 1, 2, \ldots\ldots \quad (1.28a)$$

and

$$C_4 = 1; \quad k_y = \frac{n\pi}{h}, \quad n = 0, 1, 2, \ldots\ldots, \quad (1.28b)$$

and Equation 1.25 then becomes

$$p(x,y,z,t) = \sum_{m=0}^{\infty} \sum_{n=0}^{\infty} \cos\left(\frac{m\pi x}{b}\right) \cos\left(\frac{n\pi y}{h}\right) \left(C_{1,m,n} e^{-jk_{z,m,n} z} + C_{2,m,n} e^{+jk_{z,m,n} z}\right) e^{j\omega t}, \quad (1.29)$$

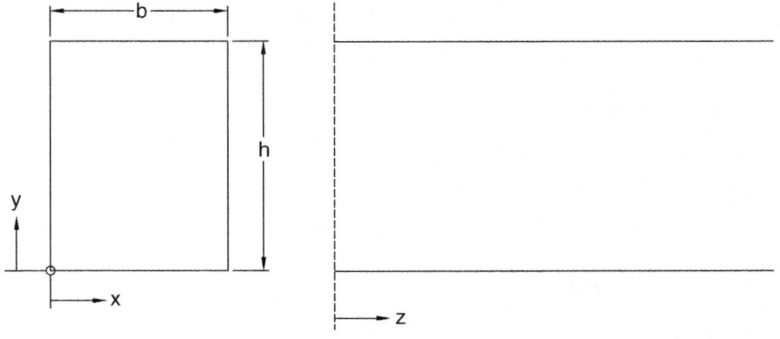

Figure 1.1 A rectangular duct and the Cartesian coordinate system (x, y, z)

where, as per Equation 1.26, the transmission wave number for the (m, n) mode, $k_{z,m,n}$ is given by the relation

$$k_{z,m,n} = \left[k_o^2 - (m\pi/b)^2 - (n\pi/h)^2\right]^{1/2}. \tag{1.30}$$

In order to evaluate axial particle velocity corresponding to the (m, n) mode, we make use of the z-component of the momentum equation (1.21)

$$\rho_o \frac{\partial u_{z,m,n}}{\partial t} + \frac{\partial p}{\partial z} = 0, \tag{1.31}$$

which yields

$$\begin{aligned} u_{z,m,n} &= \frac{-\partial p/\partial z}{j\omega \rho_o} \\ &= \frac{k_{z,m,n}}{k_o \rho_o c_o} \{C_{1,m,n} e^{-jk_{z,m,n}z} - C_{2,m,n} e^{+jk_{z,m,n}z}\} \cos\left(\frac{m\pi x}{b}\right) \cos\left(\frac{n\pi y}{h}\right) e^{j\omega t}. \end{aligned} \tag{1.32}$$

Now, mass velocity can be evaluated by integration over the area of cross-section in Figure 1.1:

$$\begin{aligned} v_{z,m,n} &= \rho_o \int_o^h \int_o^b u_{z,m,n} dx dy \\ &= \int_o^b \cos\left(\frac{m\pi x}{b}\right) dx \int_o^h \cos\left(\frac{n\pi y}{h}\right) dy \frac{k_{z,m,n}}{\omega} \{C_{1,m,n} e^{-jk_{z,m,n}z} - C_{2,m,n} e^{+jk_{z,m,n}z}\} e^{j\omega t}, \end{aligned} \tag{1.33}$$

which yields

$$\begin{aligned} v_{z,m,n} &= 0 \quad \text{for} \quad m \neq 0 \quad \text{and/or} \quad n \neq 0 \\ &= \frac{bh}{c_o} \{C_1 e^{-jk_o z} - C_2 e^{+jk_o z}\} e^{j\omega t} \quad \text{for} \quad m = n = 0. \end{aligned} \tag{1.34}$$

Thus, acoustic mass velocity is nonzero only for the plane wave or (0, 0) mode for which Equation 1.19 is recovered. Incidentally, it shows that the concept of acoustic volume velocity or mass velocity does not have any significance for higher-order modes. Equation 1.32 shows that for the same acoustic pressure, amplitude of the particle velocity for the (m, n) mode is less than $(k_{z,m,n}/k_o$ times) that for the plane wave. It can be noted that for the (0, 0) mode, $k_{z,m,n} = k_o$ and Equation 1.29 reduces to Equation 1.13. Thus, plane wave corresponds to the (0, 0) mode solution in Equation 1.29.

Any particular mode (m, n) would propagate unattenuated if $k_{z,m,n}$ is greater than zero. Then, use of Equation 1.30 yields

$$k_o^2 - \left(\frac{m\pi}{b}\right)^2 - \left(\frac{n\pi}{h}\right)^2 > 0 \tag{1.35a}$$

or

$$\lambda < \frac{2}{\left\{\left(\frac{m}{b}\right)^2 + \left(\frac{n}{h}\right)^2\right\}^{1/2}} \qquad (1.35b)$$

Obviously, a plane wave of any wavelength can propagate unattenuated, whereas a higher mode can propagate only insofar as inequality (1.35b) is satisfied. Thus, if $h > b$, the first higher mode (0, 1) would get cut-on (that is, it would start propagating) if

$$\lambda < 2h \quad \text{or} \quad f > \frac{c_o}{2h}. \qquad (1.36)$$

In other words, only a plane wave would propagate (all higher modes, even if present, being cut-off, that is, attenuated exponentially) if the frequency is small enough so that

$$\lambda > 2h \quad \text{or} \quad f < \frac{c_o}{2h}, \qquad (1.37)$$

Thus, the cut-off frequency of a rectangular duct (Figure 1.1) is given by

$$f_{co} = \frac{c_o}{2h}, \qquad (1.38)$$

where h is the larger of the two transverse dimensions of the rectangular duct.

1.2.2 Circular Ducts

The wave equation (1.22), with the Laplacian given by Equation 1.24 governs wave propagation in circular tubes (see Figure 1.2). Upon making use of the method of separation of variables, and writing time dependence as $e^{j\omega t}$ and θ dependence as $e^{jm\theta}$, one gets

$$p(r, \theta, z, t) = \sum_m R_m(r) e^{jm\theta} Z(z) e^{j\omega t}. \qquad (1.39)$$

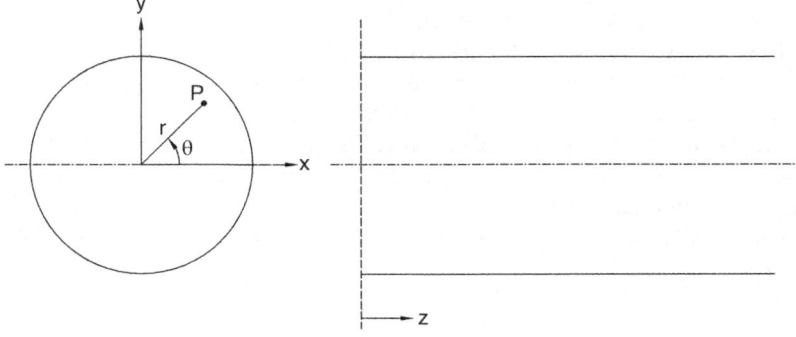

Figure 1.2 A cylindrical duct/tube and the cylindrical polar coordinate system (r, θ, z)

Assuming the z-dependence function $Z(z)$ as in Equation 1.25 with

$$\frac{d^2Z}{dz^2} = -k_z^2 Z \qquad (1.40)$$

and substituting Equations 1.39 and 1.40 in the wave equation, one gets a Bessel equation for $R(r)$:

$$\frac{d^2R_m}{dr^2} + \frac{1}{r}\frac{dR_m}{dr} + \left(k_o^2 - k_z^2 - \frac{m^2}{r^2}\right)R_m = 0. \qquad (1.41)$$

As indicated in Appendix A, Equation 1.41 has a general solution

$$R_m = C_3 J_m(k_r r) + C_4 N_m(k_r r), \qquad (1.42)$$

where the radial wave number k_r is given by

$$k_r^2 = k_o^2 - k_z^2, \qquad (1.43)$$

and $J_m(\cdot)$ and $N_m(\cdot)$ are Bessel function and Neumann function, respectively.

$N_m(k_r r)$ tends to infinity at $r = 0$ (the axis). But acoustic pressure everywhere has got to be finite. Therefore, the constant C_4 must be zero.

Again, the radial velocity at the walls $(r = r_o)$ must be zero. Therefore,

$$\frac{dJ_m(k_r r)}{dr} = 0 \text{ at } r = r_o. \qquad (1.44)$$

Thus, k_r takes only such discrete values as satisfy the equation

$$J'_m(k_r r_o) = 0. \qquad (1.45)$$

Upon denoting the value of k_r corresponding to the nth root of this equation as $k_{r,m,n}$, one gets

$$p(r,\theta,t) = \sum_{m=0}^{\infty}\sum_{n=1}^{\infty} J_m(k_{r,m,n}r)e^{jm\theta}e^{j\omega t} \times \left(C_{1,m,n}e^{-jk_{z,m,n}z} + C_{2,m,n}e^{+jk_{z,m,n}z}\right), \qquad (1.46)$$

where

$$k_{z,m,n} = \left(k_o^2 - k_{r,m,n}^2\right)^{1/2}. \quad \text{(cf. Eq. 1.30)} \qquad (1.47)$$

As the first zero of J'_o (or that of J_1) is zero, $k_{r,0,1} = 0$ and $k_{z,0,1} = k_o$. Thus, for the (0, 1) mode, Equation 1.46 reduces to Equation 1.13, the equation for the plane wave propagation. Hence, the plane wave corresponds to the (0, 1) mode of Equation 1.40 and propagates unattenuated.

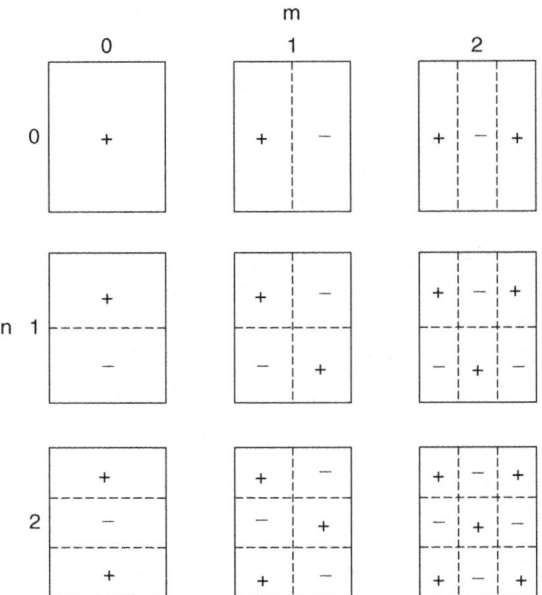

Figure 1.3 Nodal lines for transverse pressure distribution in a rectangular duct up to m = 2, n = 2 (Reproduced with permission from [5])

In most of the literature [1–3], n represents the number of the zero of the derivative $J'_m(k_r r_o)$ as per Equation 1.45. This introduces a dissimilarity between the notation for rectangular ducts and circular ducts. In rectangular ducts, m and n represent the number of nodes in the transverse pressure distribution as shown in Figure 1.3. A similar picture could emerge for circular ducts if n were to denote the number of circular nodes in the transverse pressure distribution. This is shown in Figure 1.4. With this notation [4,5], the plane mode would have the (0, 0) label in circular as well as rectangular ducts, and m and n would have the same connotation, that is, the number of nodes (in respective directions) in the transverse pressure distribution.

This new notation is adopted here henceforth. According to this, $n = 0$ would represent the first root of Equation 1.45 and n would represent the $(n + 1)^{st}$ root thereof. In Equation 1.46, the summation $n = 1$ to ∞ would read $n = 0$ to ∞ as in Equation 1.29 for rectangular ducts.

The first two higher-order modes (1, 0) and (0, 1) will get cut-on if $k_{z,1,0}$ and $k_{z,0,1}$ are real, that is, if $k_o > k_{r,1,0}$ and $k_{r,0,1}$. The first zero of J'_1 occurs at 1.84 and the second zero of J'_o occurs at 3.83. Thus, the cut-on wave numbers would be $1.84/r_o$ and $3.83/r_o$, respectively. In other words, the first azimuthal or diametral mode starts propagating at $k_o r_o = 1.84$ and the first axisymmetric mode at $k_o r_o = 3.83$. If the frequency is small enough (or wave length is large enough) such that

$$k_o r_o < 1.84, \quad \text{or} \quad \lambda > \frac{\pi}{1.84} D, \quad \text{or} \quad f < \frac{1.84}{\pi D} c_o, \qquad (1.48)$$

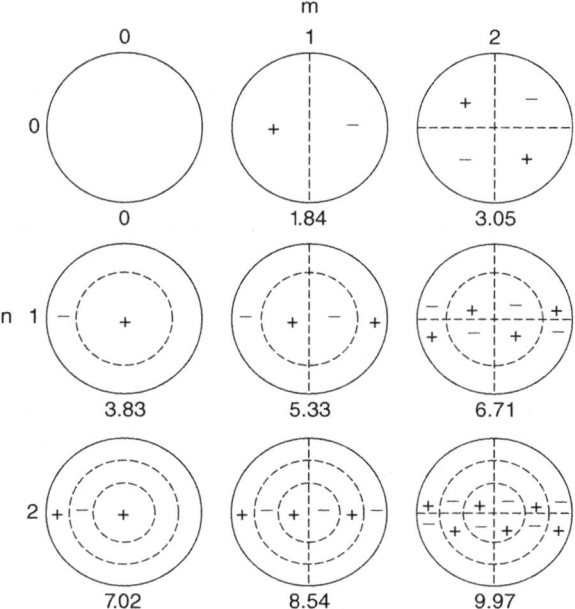

Figure 1.4 Nodal lines for transverse pressure distribution in a circular duct up to m = 2, n = 2 (Reproduced with permission from [5])

where D is the diameter $2r_o$, then only the plane waves would propagate. Thus the cut-off frequency of a circular tube is given by

$$f_{co} = \frac{1.84\, c_o}{\pi\, D} = 0.5857 \frac{c_o}{D} \qquad \text{(cf. Eq. 1.38)} \tag{1.49}$$

Fortunately, the frequencies of interest in exhaust noise of internal combustion engines are low enough so that for typical maximum transverse dimensions of exhaust mufflers Equation 1.49 is generally satisfied. Therefore, plane wave analysis has proved generally adequate. In the following pages, as indeed in most of the current literature on exhaust mufflers, one-dimensional wave propagation has been used throughout, with only a passing reference to the existence of higher modes or three-dimensional effects. In practice, muffler configurations are designed making use of the 1D analysis, and 3D analysis is used for a final check.

Substituting the (m, n) mode component of Equation 1.46 in the equation of dynamical equilibrium for the axial direction, that is,

$$\rho_o \frac{\partial u_z}{\partial t} + \frac{\partial p}{\partial z} = 0, \tag{1.50}$$

yields

$$u_{z,m,n} = -\frac{\partial p/\partial z}{j\omega \rho_o}$$

or

$$u_{z,m,n} = J_m(k_{r,m,n}r)e^{jm\theta}e^{j\omega t}\frac{k_{z,m,n}}{k_o\rho_o c_o} \times \left\{C_{1,m,n}e^{-jk_{z,m,n}z} - C_{2,m,n}e^{+jk_{z,m,n}z}\right\}. \quad (1.51)$$

Thus, as compared to the plane wave, acoustic particle velocity for the (m, n) mode is $k_{z,m,n}/k_o$ times, for the same acoustic pressure. Of course, as just shown for rectangular ducts, volume or mass velocity does not have a meaning for higher order modes.

1.3 Waves in a Viscous Stationary Medium

The analysis of wave propagation in a real (viscous) fluid with heat conduction from the walls of the tube is originally due to Kirchhoff [6,7]. The presence of viscosity brings into play a coupling between the axial and radial motions of the particle in a circular tube. Even if one were to assume axisymmetry (freedom from θ dependence), the wave propagation in a circular tube would be two-dimensional.

Neglecting heat conduction in the first instance, the basic equations governing axisymmetric wave propagation in stationary medium are [8]:

Mass continuity

$$\frac{\partial \rho}{\partial t} + \rho_0 \left(\frac{u_r}{r} + \frac{\partial u_r}{\partial r} + \frac{\partial u_z}{\partial z}\right) = 0; \quad (1.52)$$

Dynamical equilibrium (Navier–Stokes equations)

$$\rho_0 \frac{\partial u_z}{\partial t} + \frac{\partial p}{\partial z} = \mu\left(\frac{\partial^2 u_z}{\partial r^2} + \frac{1}{r}\frac{\partial u_z}{\partial r} + \frac{\partial^2 u_z}{\partial z^2}\right) + \frac{\mu}{3}\left(\frac{\partial^2 u_r}{\partial r\partial z} + \frac{1}{r}\frac{\partial u_r}{\partial z} + \frac{\partial^2 u_z}{\partial z^2}\right); \quad (1.53)$$

$$\rho_0 \frac{\partial u_r}{\partial t} + \frac{\partial p}{\partial r} = \mu\left(\frac{\partial^2 u_z}{\partial r^2} + \frac{1}{r}\frac{\partial u_r}{\partial r} - \frac{u_r}{r^2} + \frac{\partial^2 u_r}{\partial z^2}\right) + \frac{\mu}{3}\left(\frac{\partial^2 u_r}{\partial r^2} + \frac{1}{r}\frac{\partial u_r}{\partial r} - \frac{u_r}{r^2} + \frac{\partial^2 u_z}{\partial z\partial r}\right). \quad (1.54)$$

The thermodynamic process being isentropic for small-amplitude waves, Equation 1.3 is the third equation.

Eliminating ρ from Equation 1.52 with the help of Equation 1.3, and using the resulting equation to eliminate p from Equations 1.53 and 1.54 yields

$$\frac{\partial^2 u_z}{\partial t^2} - c_0^2\left(\frac{\partial^2 u_z}{\partial z^2} + \frac{1}{r}\frac{\partial u_r}{\partial z} + \frac{\partial^2 u_r}{\partial z\partial r}\right) = \frac{\partial}{\partial t}\left[\frac{\mu}{\rho_0}\left(\frac{\partial^2 u_z}{\partial r^2} + \frac{1}{r}\frac{\partial u_z}{\partial r} + \frac{1}{3}\frac{\partial^2 u_r}{\partial r\partial z} + \frac{1}{3}\frac{1}{r}\frac{\partial u_r}{\partial z} + \frac{4}{3}\frac{\partial^2 u_z}{\partial z^2}\right)\right]; \quad (1.55)$$

$$\frac{\partial^2 u_r}{\partial t^2} - c_0^2\left(\frac{\partial^2 u_r}{\partial r^2} + \frac{1}{r}\frac{\partial u_r}{\partial r} - \frac{u_r}{r^2} + \frac{\partial^2 u_z}{\partial r\partial z}\right) = \frac{\partial}{\partial t}\left[\frac{\mu}{\rho_0}\left(\frac{\partial^2 u_r}{\partial z^2} + \frac{1}{3}\frac{\partial^2 u_z}{\partial z\partial r} + \frac{4}{3}\frac{\partial^2 u_r}{\partial r^2} + \frac{4}{3r}\frac{\partial u_r}{\partial r} - \frac{4u_r}{3r^2}\right)\right]. \quad (1.56)$$

For a sinusoidal forward progressive wave, if the input is only axial, the steady-state solution would be of the form

$$u_z = U_z(r)e^{j\omega t}e^{-j\beta z}; \tag{1.57}$$

$$u_r = U_r(r)e^{j\omega t}e^{-j\beta z}. \tag{1.58}$$

Upon substituting these in Equations 1.55 and 1.56, decoupling the equations for U_z and U_r, using the order-of-magnitude relation

$$\frac{\mu\omega}{\rho_0 c_0^2} \ll 1, \tag{1.59}$$

which is true for most of the gases (and liquids), and applying the rigid-wall boundary condition, one gets, after considerable algebra [9],

$$U_z(r) = A\{J_0(Cr) - J_0(Cr_0)\}, \tag{1.60}$$

$$U_z(r) = \frac{j\beta A}{C}J_1(Cr), \tag{1.61}$$

where amplitude A is a constant, and

$$C = -\frac{1}{1+j}\left(\frac{2\rho_0\omega}{\mu}\right)^{1/2} = (-1+j)\left(\frac{\rho_0\omega}{2\mu}\right)^{1/2} \tag{1.62}$$

Substituting Equations 1.57, 1.58 and 1.62 in the continuity equation (1.52) gives

$$p = -\frac{\rho_0 c_0^2 \beta}{\omega}A_1 J_0(Cr_o)e^{j\omega t}e^{-j\beta z}, \tag{1.63}$$

which indicates that acoustic pressure p is independent of the radius, where U_r and U_z are not. Figure 1.5 shows typical profiles of the axial velocity v_z, radial velocity u_r and pressure p.

Upon integrating u_z over the cross-section of the tube to calculate volume velocity, multiplying it with ρ_0 to get mass velocity v, dividing p by v, and noting that

$$\frac{J_1(Cr_0)}{J_0(Cr_0)} = -j \quad \text{for} \quad |Cr_0| > 10, \tag{1.64}$$

one gets for characteristic impedance Y:

$$Y = \frac{p}{v} = \pm\frac{c_0}{\pi r_0^2}\left\{1 - \frac{1}{r_0}\left(\frac{\mu}{2\rho_0\omega}\right)^{1/2} + \frac{j}{r_0}\left(\frac{\mu}{2\rho_0\omega}\right)^{1/2}\right\}. \tag{1.65}$$

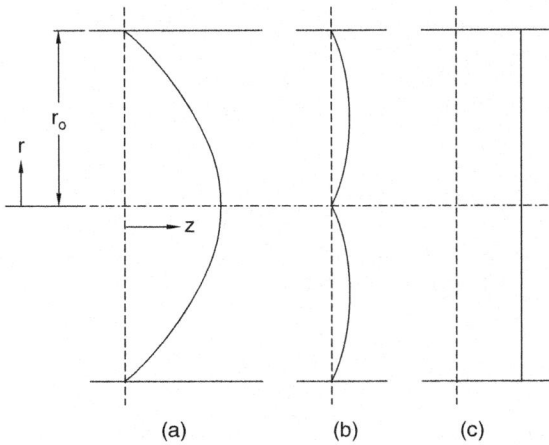

Figure 1.5 Profiles of (a) axial velocity, (b) radial velocity and (c) pressure, at some cross-section of the pipe

Writing Y as c/S [cf. Equation 1.18] gives the velocity of wave propagation in the tube c:

$$c = \pm c_0 \left\{ 1 - \frac{1}{r_0}\left(\frac{\mu}{2\rho_0\omega}\right)^{1/2} + \frac{j}{r_0}\left(\frac{\mu}{2\rho_0\omega}\right)^{1/2} = \pm c_0 \left\{ 1 - \frac{\alpha}{k_0} + j\frac{\alpha}{k_0} \right\} \right\}. \quad (1.66)$$

The corresponding expressions for β become

$$\beta = \pm k_0 \left\{ 1 + \frac{1}{r_0}\left(\frac{\mu}{2\rho_0\omega}\right)^{1/2} - \frac{j}{r_0}\left(\frac{\mu}{2\rho_0\omega}\right)^{1/2} \right\} = \pm\{(k_0 + \alpha) - j\alpha\}$$

$$= \pm k_0 \left\{ 1 + \frac{\alpha}{k_0} - j\frac{\alpha}{k_0} \right\} \quad (1.67)$$

where α is the attenuation constant

$$\alpha = \frac{1}{r_0 c_0}\left(\frac{\omega\mu}{2\rho_0}\right)^{1/2}. \quad (1.68)$$

Thus, wave number k for a progressive wave in the tube is

$$k = k_0 + \alpha = k_0\left(1 + \frac{\alpha}{k_0}\right). \quad (1.69)$$

Notably, k is slightly higher than k_0, the wave number in the free medium.

The standing wave solution (1.13) becomes

$$p(z, t) = \left\{ C_1 e^{-\alpha z - jkz} + C_2 e^{\alpha z + jkz} \right\} e^{j\omega t}. \quad (1.70)$$

The acoustic mass velocity v can be got from Equations 1.70 and 1.65:

$$v(z,t) = \frac{1}{Y}\{C_1 e^{-\alpha z - jkz} - C_2 e^{\alpha z + jkz}\} e^{j\omega t}, \qquad (1.71)$$

where Y is the characteristic impedance for the forward wave, corresponding to the positive sign of Equation 1.65; that is,

$$Y = Y_0 \left\{ 1 - \frac{\alpha}{k_0} + j\frac{\alpha}{k_0} \right\}, \qquad (1.72)$$

Y_0 being the characteristic impedance for the inviscid medium, given by Equation 1.18:

$$Y_0 = \frac{c_0}{S}, \quad S = \pi r_0^2.$$

Kirchhoff [6,7] takes into account heat conduction as well. Following a slightly different but more general analysis, he gets expressions that are identical to Equations 1.67 and 1.68 with μ being replaced by μ_e, an effective coefficient of viscothermal friction, given by

$$\mu_e = \mu \left\{ 1 + \left(\gamma^{1/2} - \frac{1}{\gamma^{1/2}} \right) \left(\frac{K}{\mu C_p} \right)^{1/2} \right\}^2, \qquad (1.73)$$

where C_p is the specific heat at constant pressure, and K is the coefficient of thermal conductivity. It may be noted that $\mu C_p / K$ is the Prandtl number. Incidentally, for air at normal temperature and pressure (NTP), Prandtl number is 0.7 and the specific heat ratio γ is 1.4. Thus, for air, Equation 1.73 yields $\mu_e = 1.65\mu$.

Experimental measurements of α by several investigators [2] show disagreement with theoretical values, the discrepancy ranging from 15 to 50%. However, almost all of them confirm the functional dependence of α on $\omega^{1/2}$ and r_0 implied in Equation 1.68. Of course, the attenuation constant α is also a function of surface roughness, flexibility of the tube wall, humidity of the medium, and so on.

In the foregoing analysis, it has been observed that the axial component of acoustic velocity u_z is a function of radius, and its radial dependence remains the same along the axis. This latter property enabled us to define an acoustic mass velocity v, and we got Equation 1.71 to go with Equation 1.70. These two equations correspond to Equations 1.13 and 1.19 for undamped plane waves. This formal similarity of the standing wave solutions suggests strongly that one could perhaps write the basic equation in terms of a mean axial particle velocity u defined as

$$u \equiv \frac{v}{\rho_0 S}, \qquad (1.74)$$

taking into account the effect of α in the equation of dynamical equilibrium as an additional pressure-drop term, looking at the velocity of wave propagation c as a real number equal to the real part of Equation 1.66, the corresponding k as in Equation 1.69, and dropping the radial component of acoustic particle velocity altogether. These basic equations would then lead to a one-dimensional damped wave equation with essentially the same solutions as given. Such a representation would make conceptualization as well as analysis considerably easier, and would admit useful generalizations for damped wave propagation in a moving medium, as shown later in Section 1.6.

Thus, two of the basic equations for damped plane waves, the equation of mass continuity and thermodynamic (isentropic) process, are the same as Equations 1.1 and 1.3, whereas the equation for dynamical equilibrium (1.2) becomes

$$\rho_0 \frac{\partial u}{\partial t} + \frac{\partial p}{\partial z} + 2\alpha \rho_0 c u = 0, \qquad (1.75)$$

where $2\alpha\rho_0 cu$ is the pressure drop per unit length due to viscothermal friction as given by Rschevkin [10].

These three basic equations lead to the one-dimensional damped wave equation [cf. Equation 1.5]

$$\left[\frac{\partial^2}{\partial t^2} - c^2 \frac{\partial^2}{\partial z^2} + 2c\alpha \frac{\partial}{\partial t}\right] p = 0. \qquad (1.76)$$

Looking for a propagating solution of the type

$$p = C e^{j\omega t} e^{\beta z} \qquad (1.77)$$

one gets, on substituting Equation 1.77 in Equation 1.76,

$$\begin{aligned} \beta &= \pm \left(-k^2 + 2jk\alpha\right)^{1/2} \\ &\cong \pm jk\left(1 - j\frac{\alpha}{k}\right) \\ &= \pm(jk + \alpha), \end{aligned} \qquad (1.78)$$

where the following inequality has been assumed:

$$\alpha^2/k^2 \simeq \alpha^2 \quad k_0^2 \ll 1. \qquad (1.79)$$

Thus, we recover Equation 1.70 for acoustic pressure p and, hence, Equation 1.71 for the acoustic mass velocity.

It is important to note here that Equation 1.75 is not an exact equation and therefore should not be used to find the values of c, k, α and Y, which are to be adopted from the foregoing relatively rigorous analysis.

1.4 Plane Waves in an Inviscid Moving Medium

Wave propagation is due to the combined effect of inertia (mass) and elasticity of the medium, and therefore a wave moves relative to the particles of the medium. When the medium itself is moving with a uniform velocity U, the velocity of wave propagation relative to the medium

Propagation of Waves in Ducts

remains c. Therefore, relative to a stationary frame of reference (that is, as seen by a stationary observer), the forward wave would move at an absolute velocity of $U + c$ and the backward moving wave at $U - c$. The waves are said to be convected downstream by mean flow. This is borne out by the following analysis.

Let the medium be moving with a velocity U, the gradients of which in the r direction as well as z direction are negligible. The basic linearized equations for this case are the same as for stationary medium [Equations 1.1–1.3] except that the local time derivative $\partial/\partial t$ is replaced by substantive derivative D/Dt, where

$$\frac{D}{Dt} = \frac{\partial}{\partial t} + U\frac{\partial}{\partial z}. \tag{1.80}$$

Thus, the mass continuity and momentum equations are

$$\rho_0 \frac{\partial u}{\partial z} + \frac{D\rho}{Dt} = 0 \tag{1.81}$$

and

$$\rho_0 \frac{Du}{Dt} + \frac{\partial p}{\partial z} = 0, \tag{1.82}$$

respectively. The third equation is, of course, the isentropicity relation (1.3).

Eliminating ρ and u from these three equations yields the convective one-dimensional wave equation

$$\left(\frac{D^2}{Dt^2} - c_0^2 \frac{\partial^2}{\partial z^2}\right) p = 0 \tag{1.83}$$

or

$$\frac{\partial^2 p}{\partial t^2} + 2U\frac{\partial^2 p}{\partial z \partial t} + (U^2 - c_0^2)\frac{\partial^2 p}{\partial z^2} = 0. \tag{1.84}$$

Making use of the separation of variables and assuming again a time-dependence function $e^{j\omega t}$, the wave equation (1.84) may be seen to admit the following general solution:

$$p(z,t) = \left(C_1 e^{-j\omega/(c_0+U)z} + C_2 e^{+j\omega/(c_0-U)z}\right) e^{j\omega t}, \tag{1.85}$$

or

$$p(z,t) = \left(C_1 e^{-jk_0 z/(1+M)} + C_2 e^{+jk_0 z/(1-M)}\right) e^{j\omega t}. \tag{1.86}$$

Writing

$$u(z,t) = \left(C_3 e^{-jk_0 z/(1+M)} + C_4 e^{+jk_0 z/(1-M)}\right) e^{j\omega t}, \tag{1.87}$$

substituting Equations 1.86 and 1.87 in convective wave Equation 1.82 and equating the coefficients of $e^{-jk_0 z/(1+M)}$ and $e^{+jk_0 z/(1-M)}$ separately to zero yields.

$$C_3 = \frac{C_1}{\rho c_0} \quad \text{and} \quad C_4 = -\frac{C_2}{\rho c_0}$$

Thus, acoustic mass velocity $v(z,t)$ is given by

$$v(z,t) = \rho_0 S u(z,t) = \frac{1}{Y_0}\left(C_1 e^{-jk_0 z/(1+M)} - C_2 e^{+jk_0 z/(1-M)}\right)e^{j\omega t}, \tag{1.88}$$

where the characteristic impedance Y_0 is the same as for stationary medium—Equation 1.18.

Equation 1.85 indicates (symbolically) the convective effect of mean flow on the two components of the standing waves, as mentioned in the opening paragraph of this section.

1.5 Three-Dimensional Waves in an Inviscid Moving Medium

As indicated earlier in Section 1.2, analysis of three-dimensional waves in a flow duct is needed for understanding the propagation of higher-order modes and for evaluating the limiting frequency below which only the plane wave would propagate unattenuated.

Combining the arguments presented in Sections 1.2 and 1.4 yields the following basic relations:

$$\text{Mass continuity:} \quad \rho_0 \nabla \cdot u + \frac{D\rho}{Dt} = 0; \tag{1.89}$$

$$\text{Dynamical equilibrium:} \quad \rho_0 \frac{Du}{Dt} + \nabla p = 0; \tag{1.90}$$

$$\text{The convected 3D wave equation:} \quad \left(\frac{D^2}{Dt^2} - c_0^2 \nabla^2\right)p = 0. \tag{1.91}$$

Here, the mean-flow velocity is assumed to be constant in space and time, that is, independent of all coordinates.

For a rectangular duct (Figure 1.1), the solution to Equation 1.91 would be

$$p(x,y,z,t) = \sum_{m=0}^{\infty}\sum_{n=0}^{\infty} \cos\frac{m\pi x}{b}\cos\frac{n\pi y}{h} \times \left\{C_{1,m,n}e^{-jk_{z,m,n}^+ z} + C_{2,m,n}e^{+jk_{z,m,n}^- z}\right\}e^{j\omega t}, \tag{1.92}$$

where $k_{z,m,n}^+$ and $k_{z,m,n}^-$ are governed by the equation [cf. Equation 1.30]

$$k_{z,m,n}^2 + \left(\frac{m\pi}{b}\right)^2 + \left(\frac{n\pi}{h}\right)^2 = (k_0 \pm M k_{z,m,n})^2 \tag{1.93}$$

or

$$k_{z,m,n}^{\pm} = \frac{\mp M k_0 + \left[k_0^2 - (1-M^2)\left\{\left(\frac{m\pi}{b}\right)^2 + \left(\frac{n\pi}{h}\right)^2\right\}\right]^{1/2}}{1-M^2}. \tag{1.94}$$

Thus, the condition for higher-order modes ($m, n > 0$) to propagate unattenuated is given by the condition that the sum under the radical sign is not negative, or

$$k_0^2 - (1 - M^2)\left\{\left(\frac{m\pi}{b}\right)^2 + \left(\frac{n\pi}{h}\right)^2\right\} \geq 0. \tag{1.95}$$

In other words, only a plane wave would propagate if the frequency is small enough so that

$$\lambda > \frac{2h}{(1 - M^2)^{1/2}} \quad \text{or} \quad f < \frac{c_0}{2h}(1 - M^2)^{1/2} \tag{1.96}$$

[cf. inequality (1.37)], where h is the larger of the two transverse dimensions of the rectangular duct.

Clearly, the cut-off frequency for the first higher mode (0, 1) for a flow duct is lower than that of a stationary-medium duct by a factor $(1 - M^2)^{1/2}$, where M is the average Mach number of the mean flow.

It is worth noting here that the cut-off frequency is the same for downstream as well as upstream propagation.

The same remarks hold for propagation of higher-order modes in a circular duct, the solution for which can readily be seen to be (following the algebra of Section 1.2.2)

$$p(r, \theta, z, t) = \sum_{m=0}^{\infty}\sum_{n=0}^{\infty} J_m(k_{r,m,n}r)e^{jm\theta}e^{j\omega t} \times \left\{C_{1,m,n}e^{-jk_{z,m,n}^+ z} + C_{2,m,n}e^{+jk_{z,m,n}^- z}\right\}, \tag{1.97}$$

where $k_{z,m,n}^+$ and $k_{z,m,n}^-$ are governed by the equation

$$k_{z,m,n}^2 + k_{r,m,n}^2 = (k_0 + Mk_{z,m,n})^2 \tag{1.98}$$

or

$$k_{z,m,n}^{\pm} = \frac{\mp Mk_0 + \left[k_0^2 - (1 - M^2)k_{r,m,n}^2\right]^{1/2}}{1 - M^2}. \tag{1.99}$$

Thus, the condition for higher-order modes (m and/or $n > 0$) to propagate unattenuated is given by

$$k_0^2 - (1 - M^2)k_{r,m,n}^2 \geq 0. \tag{1.100}$$

In other words, only a plane wave would propagate if the frequency is small enough so that

$$k_0 r_0 < 1.84(1 - M^2)^{1/2},$$

or

$$\lambda > \frac{\pi D}{1.84(1 - M^2)^{1/2}},$$

or

$$f < \frac{1.84 c_0}{\pi D}(1-M^2)^{1/2} = 0.5857\frac{c_0}{D}(1-M^2)^{1/2} \qquad (1.101)$$

The lowering of the cut-off frequency by mean flow has been demonstrated experimentally by Mason [11,12]. In particular, he has shown that the cut-off frequency for circular tubes with flow is indeed lowered by a factor $(1-M^2)^{1/2}$ for low Mach numbers ($M < 0.2$) that are typical of exhaust mufflers.

Now the particle velocity $u(x, y, z, t)$ can be determined by assuming for it a form similar to that of pressure [i.e. Equation 1.92], with constant $C_{1,m,n}$ and $C_{2,m,n}$ replaced by new constants $C_{3,m,n}$ and $C_{4,m,n}$, and summing $u_{z,m,n}$ so obtained over m and n. Thus,

$$\begin{aligned} u_z(x, y, z, t) = \frac{1}{\rho_0 c_0} \sum_{m=0}^{\infty}\sum_{n=0}^{\infty} \cos\frac{m\pi x}{b}\cos\frac{n\pi y}{h} \\ \times \left\{ \frac{k_{z,m,n}^+}{k_0 - Mk_{z,m,n}^+} C_{1,m,n} e^{-jk_{z,m,n}^+ z} - \frac{k_{z,m,n}^-}{k_0 + Mk_{z,m,n}^-} C_{2,m,n} e^{+jk_{z,m,n}^- z} \right\}. \end{aligned} \qquad (1.102)$$

Similarly, the particle velocity $u(r, \theta, z, t)$ for 3D waves in a circular tube with mean flow can be readily proved to be given by the equation

$$\begin{aligned} u_z(r, \theta, z, t) = \frac{1}{\rho_0 c_0} \sum_{m=0}^{\infty}\sum_{n=0}^{\infty} J_m(k_{r,m,n} r) e^{jm\theta} e^{j\omega t} \\ \times \left\{ \frac{k_{z,m,n}^+}{k_0 - Mk_{z,m,n}^+} C_{1,m,n} e^{-jk_{z,m,n}^+ z} - \frac{k_{z,m,n}^- z}{k_0 + Mk_{z,m,n}^-} C_{2,m,n} e^{+jk_{z,m,n}^- z} \right\}. \end{aligned} \qquad (1.103)$$

1.6 One-Dimensional Waves in a Viscous Moving Medium

As has been shown in Section 1.3, a wave front in a tube containing a viscous fluid is not plane inasmuch as axial particle velocity is not the same all over the cross-section, although acoustic pressure is constant for most of the common gases for which inequality (1.59) is satisfied. Nevertheless, as shown later in that section, one could write the equivalent one-dimensional equations following Rschevkin [10]. These equations are extended here to account for the additional aeroacoustic losses due to turbulent friction, and also the convective effect of mean flow. They imply use of a quasi-static approach [13] wherein it is assumed that the steady flow relations apply with acoustic perturbations as well. On subtracting one from the other and linearizing in terms of acoustic perturbations ρ and u, we get the required aeroacoustic equation for propagation of one-dimensional waves in a moving medium with friction. This principle or approach is indeed the very basis of aeroacoustics, and is used extensively in Chapter 3.

With subscripts 0 and T denoting mean and total (perturbed) states, we can write

$$\rho_T = \rho_0 + \rho; \qquad p_T = p_0 + p; \qquad u_T = U + u; \qquad (1.104)$$

where, for the linear case,

$$\left(\frac{\rho}{\rho_0}\right)^2 \ll 1, \quad \left(\frac{p}{p_0}\right)^2 \ll 1, \quad \left(\frac{u}{c_0}\right)^2 \ll 1, \tag{1.105}$$

so that terms involving quadratic terms in the acoustic perturbation variables p, ρ and u can be neglected.

Substituting these relations in the mass continuity equation

$$\frac{D\rho_T}{Dt} + \rho_T \frac{\partial u_T}{\partial z} = \frac{\partial \rho_T}{\partial t} + \frac{\partial}{\partial z}(\rho_T u_T) = 0, \tag{1.106}$$

Subtracting from it the corresponding unperturbed steady flow equation, and noting that both the time derivative as well as space derivative of the mean quantities p_0, ρ_0 and U are zero by definition, yields the equation

$$\frac{\partial \rho}{\partial t} + U\frac{\partial \rho}{\partial z} + \rho_0 \frac{\partial u}{\partial z} = 0 \tag{1.107}$$

which, of course, is identical to Equation 1.81 when one notes that

$$\frac{D}{Dt} = \frac{\partial}{\partial t} + u_T \frac{\partial}{\partial z} \cong \frac{\partial}{\partial t} + U\frac{\partial}{\partial z}. \tag{1.108}$$

The one-dimensional equation for dynamical equilibrium with viscothermal dissipation and turbulent friction loss can be written as [10,13]

$$\rho_0 \frac{Du_T}{Dt} + \frac{\partial p_T}{\partial z} + 2\alpha\rho_0 c u_T + \xi\rho_0 u_T^2 = 0, \tag{1.109}$$

where $2\alpha\rho_0 c u_T$ is the pressure drop per unit length due to viscothermal friction, $\xi = F/2d$, $F =$ Froude's friction factor, defined as ratio of the pressure drop in an axial length equal to one diameter divided by the dynamic head $1/2\rho_0 u_T^2$, and $d =$ diameter of the tube, or hydraulic diameter (four times the ratio of area and perimeter) if the tube is not circular.

Thus, $\xi\rho_0 u_T^2$ is the pressure drop per unit length due to boundary-layer friction or wall friction. Froude's friction factor F can be obtained as a function of Reynolds number from textbooks on fluid mechanics (see, for example, [14,15]).

For the typical flow velocities in exhaust mufflers, F is given by Lees formula

$$F = 0.0072 + \frac{0.612}{R_e^{0.35}}, \quad R_e < 4 \times 10^5, \tag{1.110}$$

where R_e is the Reynolds number $Ud\rho_0/\mu$ and μ is the coefficient of dynamic viscosity.

Substituting Equation 1.104 in Equation 1.109, subtracting from it the corresponding unperturbed steady flow equation, and making use of the order-of-magnitude relations for the small-amplitude (i.e. linear) waves gives

$$\rho_0 \frac{\partial u}{\partial t} + \rho_0 U \frac{\partial u}{\partial z} + \frac{\partial p}{\partial z} + 2\rho_0 \alpha c u + 2\xi\rho_0 U_0 u = 0$$

or

$$\rho_0 \frac{Du_T}{Dt} + \frac{\partial p}{\partial z} + 2\rho_0(\alpha c + \xi U)u = 0, \quad (1.111)$$

For small-amplitude wave propagation in a moving medium with no transverse gradients, the thermodynamic process is still almost isentropic and therefore Equation 1.3 holds.

Eliminating ρ and u from Equations 1.3, 1.107 and 1.111 yields the desired wave equation

$$\left[\frac{D^2}{Dt^2} - c_0^2 \frac{\partial^2}{\partial z^2} + 2(\xi U + c\alpha)\frac{D}{Dt}\right]p = 0. \quad (1.112)$$

This equation is very similar to Equation 1.76 except that local time derivative operator $\partial/\partial t$ is replaced by the substantive derivative D/Dt, thereby incorporating the convective effect of mean flow, and a flow-acoustic friction term has been added.

On assuming a solution of the form

$$p(z,t) = Ce^{j\omega t}e^{\beta z}, \quad (1.113)$$

substituting it in Equation 1.112, making use of the order-of-magnitude considerations

$$\begin{aligned} M^2\alpha^2 &< \alpha^2 \ll k^2, \\ \xi^2 M^4 &< \xi^2 M^2 \ll k^2, \\ 2\xi M^3 \alpha &< 2\xi M\alpha \ll k^2, \end{aligned} \quad (1.114)$$

and some algebraic manipulations [16], one gets two values of β:

$$\beta^\pm \cong \mp \left(\frac{\alpha + \xi M + jk}{1 \pm M}\right). \quad (1.115)$$

Thus,

$$p(z,t) = \left[C_1 \exp\left(-\frac{\alpha + \xi M + jk}{1 + M}z\right) + C_2 \exp\left(+\frac{\alpha + \xi M + jk}{1 - M}z\right)\right]\exp(j\omega t). \quad (1.116)$$

This solution shows clearly that

i. total aeroacoustic attenuation in a moving medium $\alpha(M)$ is a sum of the contributions of the viscothermal effects and turbulent flow friction, and
ii. the factors $1 \pm M$ that represent the convective effect of mean flow apply to the attenuation constants as well as to the wave numbers.

The attenuation constants are

$$\alpha^+ = \frac{\alpha + \xi M}{1 + M} = \frac{\alpha(M)}{1 + M}, \quad (1.117)$$

$$\alpha^- = \frac{\alpha + \xi M}{1 - M} = \frac{\alpha(M)}{1 - M}, \quad (1.118)$$

where

$$\alpha(M) = \alpha + \xi M, \quad k = k_0 + \alpha. \tag{1.119}$$

$\alpha(M)$ being the same for waves in both the directions, can be construed to be the 'real' aeroacoustic attenuation constant for a moving medium. The factors $1 \pm M$ in $\alpha^\pm$ as well as $\beta^\pm$ represent only the Doppler effect due to mean flow convection.

The acoustic mass velocity v can now be written as

$$v(z,t) = \frac{1}{Y}\left[C_1 \exp\left(-\frac{\alpha + \xi M + jk}{1+M}z\right) - C_2 \exp\left(+\frac{\alpha + \xi M + jk}{1-M}z\right)\right]\exp(j\omega t). \tag{1.120}$$

The characteristic impedance Y can be constructed heuristically from Equation 1.72, making use of the foregoing remarks, α being replaced by $\alpha(M)$, and the fact observed in Section 1.4 that mean-flow convection does not alter the characteristic impedance. Thus, for the case on hand,

$$Y = Y_0\left\{1 - \frac{\alpha + \xi M}{k_0} + j\frac{\alpha + \xi M}{k_0}\right\}, \tag{1.121}$$

where, as before, Y_0 is the characteristic impedance for plane waves in an inviscid stationary medium given by Equation 1.18.

Equation 1.121 neglects second-order terms like $M\alpha/k$ and $M^2\xi/k$. These terms would further complicate the algebra inasmuch as Y for the forward direction would not be the same as for the backward direction.

It is worth repeating here that the above analysis is oversimplified for the specific purpose of evaluating the aeroacoustic attenuation constant. In particular, Equations 1.108 and 1.111 are not exact because they are one-dimensional.

Thus, Equations 1.116, 1.120 and 1.121 are approximate. Nevertheless, these equations are very useful from an engineering point of view because of their formal similarity with the corresponding equations for the case of the inviscid moving medium and the viscous stationary medium derived in the foregoing section.

1.7 Waves in Ducts with Compliant Walls (Dissipative Ducts)

In all the foregoing sections, the walls of the duct were assumed to be rigid. However, walls of a finite thickness (typical of the sheet metal from which the exhaust mufflers are fabricated) are in general compliant inasmuch as the transverse impedance is finite. Alternatively, the walls of the duct may be lined with an acoustically absorptive material that would, of course, have a finite normal impedance. This latter application is much more important than the former and is discussed at length in Chapter 6.

The normal impedance of a wall lined with an acoustically absorptive layer can be assumed to be independent of z. In other words, the acoustic layer can be assumed to be homogeneous and 'locally reacting'. The same, however, does not apply to unlined metallic walls of the pipes as are used in exhaust mufflers for internal combustion engines, where the wall impedance would vary with z (increasing near the end plates). Nevertheless, in such mufflers the wall thickness is generally substantial so that the impedance is very large or compliance very small.

Neglecting the viscous friction of the medium as relatively insignificant and assuming the mean flow velocity to be constant all over the cross-section, the propagation of waves in compliant ducts would be governed by a 3D wave equation [Equation 1.22 for a stationary medium and Equation 1.91 for a moving medium]. Here we restrict ourselves to the lowest-mode (corresponding to plane wave) analysis of acoustically lined ducts with stationary medium. The relatively minor effect of moving medium is discussed briefly in Chapter 6.

1.7.1 Rectangular Duct with Locally Reacting Lining

For a stationary medium, the general solution to wave equation (1.22) with the Laplacian ∇^2 in terms of Cartesian coordinates [Equation 1.23] is given by Equation 1.25 with wave numbers k_x, k_y, and k_z being related to k_0 as per Equation 1.26.

Let Z_w be the normal impedance of the walls at their exposed boundary and let b and h be the breadth and height of the free section as shown in Figure 1.6. According to the equation of dynamical equilibrium in the x direction, the x component of acoustic particle velocity u_x is related to acoustic pressure p as

$$\rho_0 \frac{\partial u_x}{\partial t} + \frac{\partial p}{\partial x} = 0 \quad (1.122)$$

or

$$u_x = -\frac{\partial p / \partial x}{j \omega \rho_0}. \quad (1.123)$$

Similarly,

$$u_y = -\frac{\partial p / \partial y}{j \omega \rho_0}. \quad (1.124)$$

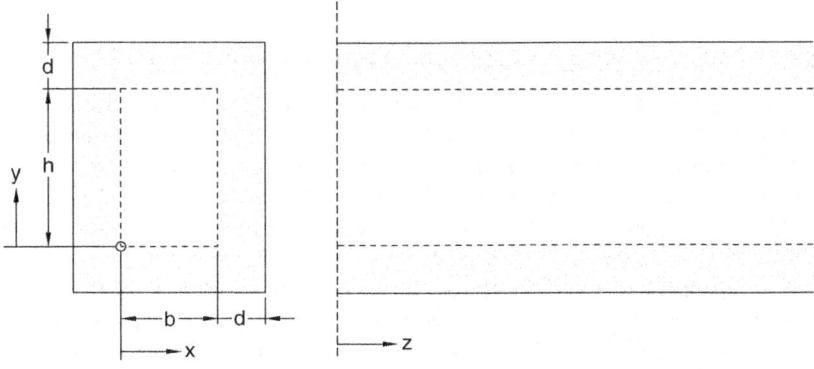

Figure 1.6 Schematic views of an acoustically lined rectangular duct with clear dimensions b and h (cf. Figure 1.1)

Thus, the boundary conditions for a duct with uniform normal wall impedance Z_w would be

$$\frac{p(0, y, z, t)}{-u_x(0, y, z, t)} = \frac{p(b, y, z, t)}{u_x(b, y, z, t)} = Z_{wx}, \quad (1.125)$$

$$\frac{p(x, 0, z, t)}{-u_y(x, 0, z, t)} = \frac{p(x, h, z, t)}{u_y(x, h, z, t)} = Z_{wy}, \quad (1.126)$$

Substituting solution (1.25) and Equations 1.123 and 1.124 in the four boundary conditions (1.125) and (1.126) yields

$$\frac{\omega \rho_0 (1 + C_3)}{-k_x (1 - C_3)} = Z_{wx}, \quad (1.127)$$

$$\frac{\omega \rho_0}{k_x} \frac{e^{-jk_x b} + C_3 e^{+jk_x b}}{e^{-jk_x b} - C_3 e^{+jk_x b}} = Z_{wx}, \quad (1.128)$$

$$\frac{\omega \rho_0 (1 + C_4)}{-k_y (1 - C_4)} = Z_{wy}, \quad (1.129)$$

$$\frac{\omega \rho_0}{k_y} \frac{e^{-jk_y h} + C_4 e^{+jk_y h}}{e^{-jk_y h} - C_4 e^{+jk_y h}} = Z_{wy}. \quad (1.130)$$

Equation 1.127 yields

$$C_3 = \left(\frac{Z_{wx} k_x}{\omega \rho_0} + 1\right) \Big/ \left(\frac{Z_{wx} k_x}{\omega \rho_0} - 1\right), \quad (1.131)$$

Substituting this in Equation 1.128 and rearranging leads to a quadratic in $Z_{wx} k_x / \omega \rho_0$, which in turn yields

$$\frac{Z_{wx} k_x}{\omega \rho_0} = \frac{-\cos k_x b \pm 1}{j \sin k_x b} \quad (1.132)$$

$$= -j \tan \frac{k_x b}{2}, \quad j \cot \frac{k_x b}{2}. \quad (1.133)$$

These two eigen equations can be rewritten in the conventional form

$$\frac{\cot(k_x b/2)}{k_x b/2} = -j \frac{Z_{wx}}{\rho_0 c_0} \frac{1}{k_0 b/2} \quad (1.134a)$$

and

$$\frac{\tan(k_x b/2)}{k_x b/2} = j \frac{Z_{wx}}{\rho_0 c_0} \frac{1}{k_0 b/2}. \quad (1.134b)$$

For the limiting case of rigid unlined walls, $Z_{wx} \to \infty$, $C_3 = 1$, and the two equations yield, respectively,

$$k_x = 0, \quad \frac{2\pi}{b}, \quad \frac{4\pi}{b}, \quad \ldots \tag{1.135a}$$

and

$$k_x = \pi/b, \quad 3\pi/b, \quad 5\pi/b, \quad \ldots \tag{1.135b}$$

Thus, the two equations supply alternate values of the series (1.28a), that is,

$$k_x = m\pi/b, \quad m = 0, 1, 2, 3. \tag{1.136}$$

By analogy, the roots or (eigen values) of Equations 1.134a and 1.134b must be alternating with each other. It can readily be checked that, like the series of roots (1.135a), the roots of the transcendental equation (1.134a) belong to symmetric modes, whereas, like the series of roots (1.135b), the roots of Equation 1.134b represent antisymmetric modes, the symmetry here relating to the axis $x = b/2$.

An identical analysis of Equations 1.129 and 1.130 would show that k_y is given by the transcendental eigen equations

$$\frac{\cot(k_y h/2)}{k_y h/2} = -j \frac{Z_{wy}}{\rho_0 c_0} \frac{1}{k_0 h/2} \tag{1.137a}$$

and

$$\frac{\tan(k_y h/2)}{k_y h/2} = j \frac{Z_{wy}}{\rho_0 c_0} \frac{1}{k_0 h/2}, \tag{1.137b}$$

the roots of which alternate with each other, representing symmetric and antisymmetric modes, respectively, the symmetry being reckoned with respect to the axis $y = h/2$.

Let the infinite roots of Equations 1.134 and 1.137 be

$$k_{x,m}, \quad m = 0, 1, 2, 3, \ldots$$

and

$$k_{y,m}, \quad m = 0, 1, 2, 3, \ldots, \tag{1.138}$$

respectively.

Thus, the general acoustic pressure field equation (1.25) becomes

$$p(x,y,z,t) = \sum_{m=0}^{\infty} \sum_{n=0}^{\infty} \left[e^{-jk_{x,m}x} + \left\{ \frac{Z_{w,x} k_{x,m}/\rho_0 c_0 k_0 + 1}{Z_{w,x} k_{x,m}/\rho_0 c_0 k_0 - 1} \right\} e^{+jk_{x,m}x} \right]$$
$$\times \left[e^{-jk_{y,n}y} + \left\{ \frac{Z_{w,y} k_{y,n}/\rho_0 c_0 k_0 + 1}{Z_{w,y} k_{y,n}/\rho_0 c_0 k_0 - 1} \right\} e^{+jk_{y,n}y} \right] \tag{1.139}$$
$$\left[C_{1,m,n} e^{-jk_{z,m,n}z} + C_{2,m,n} e^{+jk_{z,m,n}z} \right] e^{j\omega t},$$

where $k_{z,m,n}$ is given by the equation

$$k_{z,m,n} = \left\{ k_0^2 - k_{x,m}^2 - k_{y,n}^2 \right\}^{1/2}. \qquad (1.140)$$

On substituting the (m, n) component of Equation 1.139 for acoustic pressure in the momentum equation for the axial direction, evaluating $u_{z,m,n}$, and then summing over m and n, one gets

$$u_{z,m,n}(x, y, z, t) = \sum_{m=0}^{\infty}\sum_{n=0}^{\infty} \left[e^{-jk_{x,m}x} + \left\{ \frac{Z_{w,x}k_{x,m}/\rho_0 c_0 k_0 + 1}{Z_{w,x}k_{x,m}/\rho_0 c_0 k_0 - 1} \right\} e^{+jk_{x,m}x} \right]$$

$$\times \left[e^{-jk_{y,n}y} + \left\{ \frac{Z_{w,y}k_{y,n}/\rho_0 c_0 k_0 + 1}{Z_{w,y}k_{y,n}/\rho_0 c_0 k_0 - 1} \right\} e^{+jk_{y,n}y} \right] \qquad (1.141)$$

$$\times \frac{k_{z,m,n}}{k_0} \frac{1}{\rho_0 c_0} \left[C_{1,m,n} e^{-jk_{z,m,n}z} - C_{2,m,n} e^{+jk_{z,m,n}z} \right] e^{j\omega t}.$$

If all the walls of the duct are not lined with an absorptive material, then the wall impedances $Z_{w,x}$ and $Z_{w,y}$ would be more or less reactive (controlled by mass and elasticity). Then, according to Equations 1.134 and 1.137, k_x and k_y would be real and, as per Equation 1.140, $k_{z,m,n}$ would be real or imaginary, not complex. Thus, the modes that may propagate along an unlined duct with yielding walls would do so without attenuation. In other words, the unlined yielding walls do not introduce axial attenuation.

By the same reasoning it can be seen that ducts lined with acoustically absorptive material (that is, with complex wall impedance) would result in complex values of k_x, k_y, and hence k_z. The imaginary component of k_z would introduce attenuation in the axial direction, and that is the basic principle of dissipative ducts and parallel baffle mufflers discussed at length in Chapter 6.

1.7.2 Circular Duct with Locally Reacting Lining

Waves in a circular duct with stationary medium (see Figure 1.7) are governed by Equation 1.22, with the Laplacian defined in terms of cylindrical polar coordinates according to

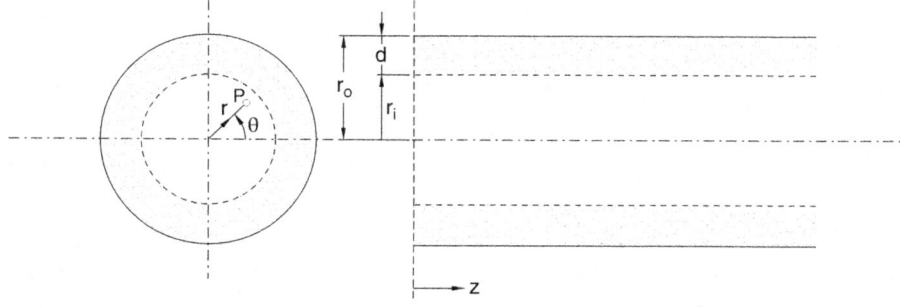

Figure 1.7 Schematic views of an acoustically lined circular duct with clear radius r_i (cf. Figure 1.2)

Equation 1.24; that is,

$$\left[\frac{\partial^2}{\partial t^2} - c_0^2 \left(\frac{\partial^2}{\partial r^2} + \frac{1}{r}\frac{\partial}{\partial r} + \frac{1}{r^2}\frac{\partial^2}{\partial \theta^2} + \frac{\partial^2}{\partial z^2}\right)\right] p = 0. \tag{1.142}$$

Following Section 1.2, the general solution to Equation 1.142 is given by Equation 1.46:

$$p(r,\theta,z,t) = \sum_{m=0}^{\infty}\sum_{n=0}^{\infty} J_m(k_{r,m,n}r) e^{jm\theta} e^{j\omega t} \times \left\{C_{1,m,n} e^{-jk_{z,m,n}z} + C_{2,m,n} e^{+jk_{z,m,n}z}\right\}, \tag{1.143}$$

with $k_{z,mn}$ being determined from Equation 1.47. The notable difference is that $k_{r,m,n}$ is now determined from the boundary condition that the wall ($r = r_i$) has a finite impedance Z_w (the rigid walls have infinite impedance).

The momentum equation in the radial direction

$$\rho_0 \frac{\partial u_r}{\partial t} + \frac{\partial p}{\partial r} = 0 \tag{1.144}$$

yields

$$u_r = -\frac{\partial p/\partial r}{j\omega \rho_0}. \tag{1.145}$$

Therefore,

$$Z_w \equiv \left(\frac{p}{u_r}\right)_{r=r_i} = \frac{-j\omega \rho_0 p}{\partial p/\partial r} \tag{1.146}$$

$$= \frac{-j\omega \rho_0 J_m(k_{r,m,n}r_i)}{k_{r,m,n} J'_m(k_{r,m,n}r_i)}, \tag{1.147}$$

where

$$J'_m(k_{r,m,n}r_i) = \left[\frac{dJ_m(k_{r,m,n}r)}{d(k_{r,m,n}r)}\right]_{r=r_i} \tag{1.148}$$

Thus, $k_{r,m,n}$, $n = 0, 1, 2 \ldots$ are the infinite roots of the transcendental eigen equation

$$\frac{J_m(k_r r_i)}{(k_r r_i) J'_m(k_r r_i)} = j \frac{Z_w}{\rho_0 c_0} \frac{1}{k_0 r_i}. \tag{1.149}$$

It is instructive to compare this equation with Equations 1.134 and 1.137. $J'_m(k_{r,m,n}r_i)$ of Equation 1.149 corresponds to $\cos(k_x b/2)$ in Equation 1.134a, $\sin(k_x b/2)$ in Equation 1.134b, $\cos(k_y h/2)$ in Equation 1.137a, and $\sin(k_y h/2)$ in Equation 1.137b. The correspondence between k_x and k_y, and between r_i, $b/2$ and $h/2$ is of course obvious.

Propagation of Waves in Ducts

Upon substituting the (m, n) component of Equation 1.143 for acoustic pressure in the momentum equation for the axial direction, evaluating $u_{z,m,n}$, and then summing over m and n, one gets the following equation for acoustic particle velocity:

$$u_z(r,\theta,z,t) = \sum_{m=0}^{\infty}\sum_{n=0}^{\infty} J_m(k_{r,m,n}r)e^{jm\theta}e^{j\omega t} \times \frac{k_{z,m,n}}{k_0}\frac{1}{\rho_0 c_0}\{C_1 e^{-jk_{z,m,n}z} - C_2 e^{+jk_{z,m,n}z}\}, \quad (1.150)$$

The remarks following Equation 1.141 on the attenuation of waves along a rectangular duct with compliant walls apply as well to a circular duct.

It is worth noting that, unlike in the z and r directions, for which the solution in general consists of two terms, we have included only $e^{jm\theta}$ for the azimuthal direction; the $e^{-jm\theta}$ term has been omitted. This is because there are no restrictions or discontinuities in the azimuthal direction that would generate waves going in the opposite direction. The spiralling modes represented by $e^{jm\theta}$ can be excited by nonsymmetries in the system such as area discontinuities. In the exhaust systems of reciprocating machinery, therefore, radial as well as azimuthal modes are excited.

Incidentally, for the hypothetical case of axisymmetry, $m=0$ and Equations 1.143 and 1.150 have only a single summation (over n); that is, Equation 1.143 reduces to

$$p(r,\theta,z,t) = \sum_{n=0}^{\infty} J_0(k_{r,n}r)e^{j\omega t} \times \{C_{1,m,n}e^{-jk_{z,m,n}z} + C_{2,m,n}e^{+jk_{z,m,n}z}\}, \quad (1.151)$$

where $k_{r,n}$ is the $(n+1)$th of the root of the eigen equation

$$-\frac{J_0(k_{r,n}r_i)}{(k_{r,n}r_i)J_1(k_{r,n}r_i)} = j\frac{Z_w}{\rho_0 c_0}\frac{1}{k_0 r_i}. \quad (1.152)$$

Impedance of the lining Z_w at the interface $(r = r_i)$ is evaluated later in Section 1.7.4 as a limiting case of the bulk reacting lining.

1.7.3 Rectangular Duct with Bulk Reacting Lining

A bulk reacting lining allows wave propagation inside the lining along the axis of the duct. Wave number of this wave is equal to the axial wave number inside the duct. In fact, all linings are basically bulk reacting in nature. Local reaction is a limiting or special case of the bulk reaction.

In the bulk reacting model, the lining is assumed to be a homogeneous, highly porous, fibrous or foam type material with open pores, (thermal insulation lining material is characterized by closed pores). Its characteristic impedance $Y_w(f)$ and wave number $k_w(f)$ are often given by complex empirical expressions in terms of flow resistivity, E, as shown in Chapter 6 of this monograph. Subscript w connotes wall lining.

The bulk reaction model consists in writing expressions for acoustic pressure and the axial and transverse particle velocity for a forward progressive wave in the air medium inside as well as the lining materials, and equating pressure and transverse particle velocity across the interface. (The effect of thin protective layer [17] and other practical aspects are discussed later in Chapter 6).

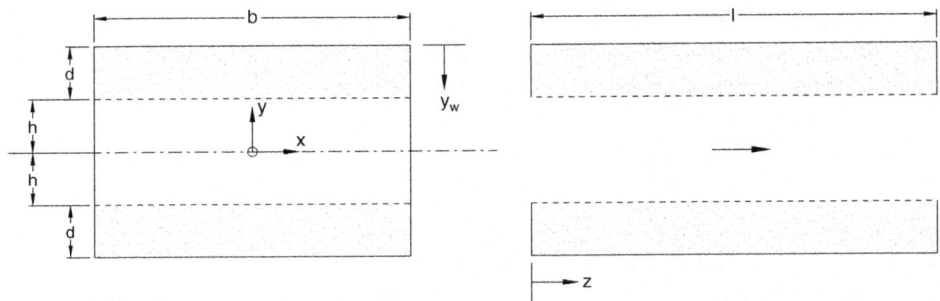

Figure 1.8 Schematic of a bulk reacting rectangular duct lined on two sides

Thus, for a rectangular duct lined on two opposite sides (shown in Figure 1.8), which incidentally represents one unit of the two-unit parallel baffle muffler shown in Figure 1.9, considering the lowest-order mode (corresponding to the plane wave), the field equations are as follows.

Air passage (subscript 0):

$$p(z,y,t) = C_1\left(e^{-jk_{y,0}y} + C_2 e^{jk_{y,0}y}\right)e^{-jk_z z}e^{j\omega t} \qquad (1.153)$$

$$u_{z,0}(z,y,t) = \frac{k_z}{k_0 Y_0} C_1\left(e^{-jk_{y,0}y} + C_2 e^{jk_{y,0}y}\right)e^{-jk_z z}e^{j\omega t} \qquad (1.154)$$

$$u_{y,0}(z,y,t) = \frac{k_{y,0}}{k_0 Y_0} C_1\left(e^{-jk_{y,0}y} - C_2 e^{jk_{y,0}y}\right)e^{-jk_z z}e^{j\omega t} \qquad (1.155)$$

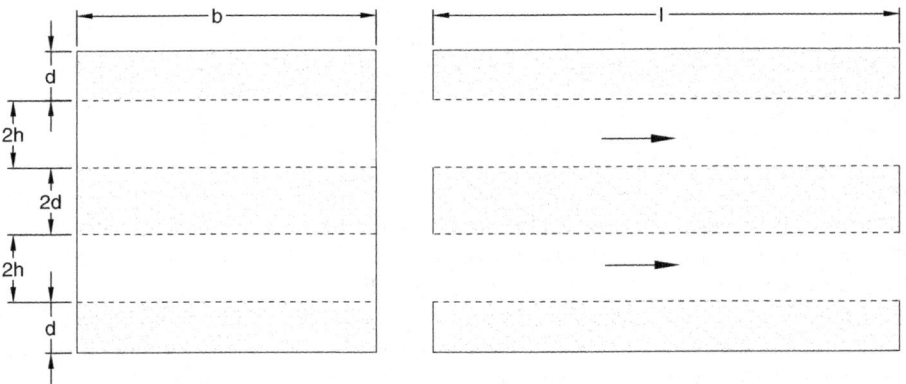

Figure 1.9 Schematic of a parallel baffle muffler consisting of two units of rectangular duct shown in Figure 1.8

Propagation of Waves in Ducts

Inside the wall lining (subscript w):

$$p_w(z, y, t) = C_1 \left(e^{-jk_{y,w}y_w} + C_3 e^{jk_{y,w}y_w}\right) e^{-jk_z z} e^{j\omega t} \tag{1.156}$$

$$u_{z,w}(z, y_w, t) = \frac{k_z}{k_w Y_w} C_1 \left(e^{-jk_{y,w}y_w} + C_3 e^{jk_{y,w}y_w}\right) e^{-jk_z z} e^{j\omega t} \tag{1.157}$$

$$u_{y,w}(z, y_w, t) = \frac{k_{y,w}}{k_w Y_w} C_1 \left(e^{-jk_{y,w}y_w} - C_3 e^{jk_{y,w}y_w}\right) e^{-jk_z z} e^{j\omega t} \tag{1.158}$$

The overriding compatibility conditions:

$$k_{z,0} = k_{z,w} \equiv k_z \tag{1.159}$$

$$k_{y,0} = \left(k_0^2 - k_z^2\right)^{1/2} \tag{1.160}$$

$$k_{y,w} = \left(k_w^2 - k_z^2\right)^{1/2} \tag{1.161}$$

Boundary conditions:
At the center of the duct,

$$u_{y,0}(z, 0, t) = 0 \Rightarrow C_2 = 1 \tag{1.162}$$

At the rigid wall behind the lining,

$$u_{y,w}(z, 0, t) = 0 \Rightarrow C_3 = 1 \tag{1.163}$$

Across the interface,

$$p(z, h, t) = p_w(z, d, t) \tag{1.164}$$

$$u_{y,0}(z, h, t) = -u_{y,w}(z, d, t) \tag{1.165}$$

Making use of Equations 1.153, 1.156, 1.162 and 1.163, Equation 1.164 yields

$$\cos(k_{y,0} h) = \cos(k_{y,w} d) \tag{1.166}$$

Similarly, use of Equations 1.155, 1.158, 1.162 and 1.163 in Equation 1.165 gives

$$\frac{k_{y,0}}{k_0 Y_0} \sin(k_{y,0} h) = -\frac{k_{y,w}}{k_w Y_w} \sin(k_{y,w} d) \tag{1.167}$$

Dividing the two sides of Equation 1.66 with the corresponding sides of Equation 1.167 yields

$$\frac{k_0 Y_0}{k_{y,0}} \cot(k_{y,0} h) = -\frac{k_w Y_w}{k_{y,w}} \cot(k_{y,w} d) \tag{1.168}$$

Here, $k_{y,0}$ and $k_{y,w}$ are given by Equations 1.160 and 1.161, k_w and Y_w are complex functions of frequency, $k_0 = \omega/c_0$ and $Y_0 = \rho_0 c_0$.

Incidentally, for a locally reacting lining, $k_{z,w} = 0$ and therefore Equation 1.161 yields $k_{y,w} = k_w$. Then, Equation 1.168 would reduce to Equation 1.137a, provided

$$Z_w = -jY_w \cot(k_w d) \tag{1.169}$$

which is impedance of the rigid wall transferred to the interface (distance d), as will be shown in Chapter 3.

Like Equation 1.137a, the transcendental eigen equation (1.168) is solved for the common axial wave number, k_z, by means of the Newton–Raphson iteration scheme as discussed at some length in Chapter 6.

1.7.4 Circular Duct with Bulk Reacting Lining

For the circular duct with bulk reacting lining shown in Figure 1.7, the field equations for the lowest-order mode progressive wave corresponding to Equations 1.153–1.158 for rectangular duct would be as follows [17].

Air passage (subscript 0):

$$p(z,r,t) = C_1 J_0(k_{r,0} r) e^{-jk_z z} e^{j\omega t} \tag{1.170}$$

$$u_{z,0}(z,r,t) = \frac{k_z}{\omega \rho_0} C_1 J_0(k_{r,0} r) e^{-jk_z z} e^{j\omega t} \tag{1.171}$$

$$u_{r,0}(z,r,t) = -j \frac{k_{r,0}}{\omega \rho_0} C_1 J_1(k_{r,0} r) e^{-jk_z z} e^{j\omega t} \tag{1.172}$$

Inside the wall lining (subscript w):

$$p_w(z,r,t) = C_2 \{J_0(k_{r,w} r) + C_3 N_0(k_{r,w} r)\} e^{-jk_z z} e^{j\omega t} \tag{1.173}$$

$$u_{z,w}(z,r,t) = \frac{k_z}{\omega \rho_0} C_2 \{J_0(k_{r,w} r) + C_3 N_0(k_{r,w} r)\} e^{-jk_z z} e^{j\omega t} \tag{1.174}$$

$$u_{r,w}(z,r,t) = \frac{-jk_{r,w}}{\omega \rho_w} C_2 \{J_1(k_{r,w} r) + C_3 N_1(k_{r,w} r)\} e^{-jk_z z} e^{j\omega t} \tag{1.175}$$

where J and N denote the Bessel function and Neumann function, respectively. These are often called Bessel functions of the first kind and second kind, respectively. The Neumann function is often denoted by Y. However, in this monograph, Y has been used for characteristic impedance.

The overriding compatibility conditions:

$$k_{z,0} = k_{z,w} = k_z \, (say) \tag{1.176}$$

$$k_z^2 + k_{r,0}^2 = k_0^2 \Rightarrow k_{r,0} = \{k_0^2 - k_z^2\}^{1/2} \tag{1.177}$$

$$k_z^2 + k_{r,w}^2 = k_w^2 \Rightarrow k_{r,w} = \{k_w^2 - k_z^2\}^{1/2} \tag{1.178}$$

Propagation of Waves in Ducts

Boundary conditions:

At the rigid wall behind the lining, (that is, at $r = r_0$),

$$u_{r,w}(z, r_0, t) = 0 \Rightarrow C_3 = -\frac{J_1(k_{r,w} r_0)}{N_1(k_{r,w} r_0)} \tag{1.179}$$

At the interface (that is, at $r = r_i$), in the absence of a thin protective layer or perforated plate,

$$p(z, r_i, t) = p_w(z, r_i, t) \tag{1.180}$$

$$u_{r,0}(z, r_i, t) = u_{r,w}(z, r_i, t) \tag{1.181}$$

which yield the impedance relationship

$$Z_{r,0}(r_i, \omega) = Z_{r,w}(r_i, \omega), \quad Z_r \equiv \frac{p}{u_r} \tag{1.182}$$

On making use of Equations 1.170, 1.172, 1.173, 1.175, 1.180 and 1.181, Equation 1.182 yields

$$j\frac{\omega \rho_0}{k_{r,0}} \frac{J_0(k_{r,0} r_i)}{J_1(k_{r,0} r_i)} = j\frac{\omega \rho_w}{k_{r,w}} \frac{J_0(k_{r,w} r_i) + C_3 N_0(k_{r,w} r_i)}{J_1(k_{r,w} r_i) + C_3 N_1(k_{r,w} r_i)} \tag{1.183}$$

where C_3 is given by Equation 1.179 above.

In Equation 1.183, $j\omega$ has not been cancelled out between the LHS and RHS so as to retain correspondence to Equation 1.182 in terms of the respective impedances $Z_{r,0}$ and $Z_{r,w}$ on the inner and outer side of the interface at $r = r_i$.

It may be noted that

$$\omega \rho_0 = \frac{\omega}{c_0} \cdot \rho_0 c_0 = k_0 Y_0 \tag{1.184}$$

and

$$\omega \rho_w = \frac{\omega}{c_w} \cdot \rho_w c_w = k_w Y_w \tag{1.185}$$

For the limiting case of the locally reacting lining, $k_{z,w} = 0$, and therefore, as per Equation 1.178, $k_{r,w} = k_w$. Then, $Z_{r,w}$ which is given by the right-hand side of Equation 1.183 reduces to the following expression for the locally reacting lining:

$$Z_w = jY_w \frac{J_0(k_w r_i) + C_3' N_0(k_w r_i)}{J_1(k_w r_i) + C_3' N_1(k_w r_i)} \tag{1.186}$$

where C_3' equals C_3 with $(k_{r,w})$ replaced with k_w in Equation 1.179. Thus,

$$C_3' = -\frac{J_1(k_w r_0)}{N_1(k_w r_0)} \tag{1.187}$$

Therefore, Z_w in the eigen equation (1.152) for the locally reacting lining is given by Equation 1.186 with C_3 given by Equation 1.187.

Evaluation of the wave number k_w and characteristic impedance Y_w of the lining material in terms of the flow resistivity, and so on is discussed in some detail later in Chapter 6.

1.8 Three-Dimensional Waves along Elliptical Ducts

Automotive exhaust mufflers are often elliptical in cross-section because of the constraint of clearing (space) under a car with low center of gravity. Three-dimensional analysis of elliptical shells is needed not only to evaluate the cut-off frequency for pure plane wave propagation, but also to analyze short flow-reversal end chambers that are often used in automotive mufflers. Here, we assume the medium to be stationary and inviscid.

For sinusoidal time dependence ($e^{j\omega t}$), that is, working in the frequency domain, the 3D wave equation reduces to the Helmholtz equation

$$(\nabla^2 + k_0^2)p = 0 \qquad (1.188)$$

In terms of the elliptical cylindrical coordinates shown in Figure 1.10, the Laplacian ∇^2 is given by [18]

$$\nabla^2 = \frac{2}{h^2\{\cosh(2\xi) - \cos(2\eta)\}}\left[\frac{\partial^2}{\partial \xi^2} + \frac{\partial^2}{\partial \eta^2}\right] + \frac{\partial^2}{\partial z^2} \qquad (1.189)$$

Here, ξ and η are the radial and angular elliptical coordinates, respectively. Conceptually, these are radial and azimuthal counterparts of the circular polar coordinate system (r, θ) shown in Figure 1.2. Accordingly, the curves of constant ξ define a family of confocal ellipses with semi-major axis. $D_1/2 = h\cosh(\xi)$ and semi-minor axis, $D_2/2 = h\sinh(\xi)$. Each of these ellipses has common foci at $x = \pm h$, where $2h$ is the interfocal distance. Thus, line $F'F$, connecting the two foci, represents the limiting '$\xi = 0$' ellipse. The curves of the constant η denote a family of confocal hyperbolae, as shown in Figure 1.10.

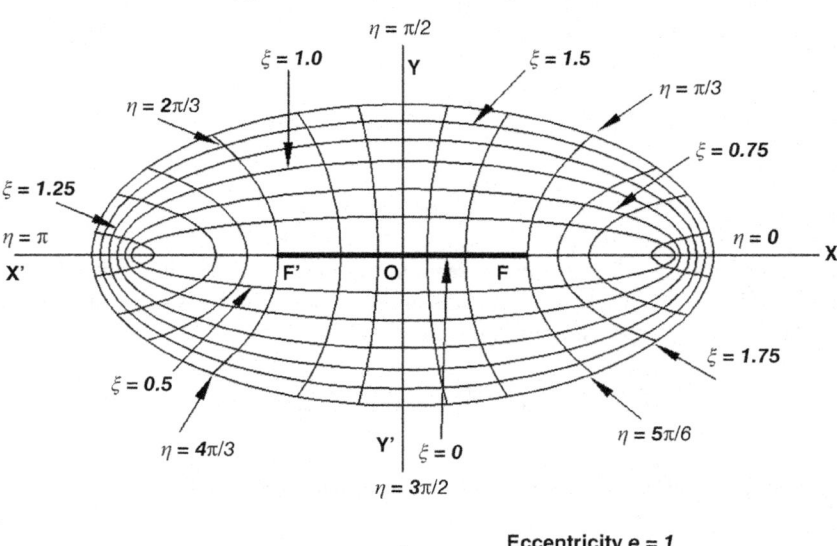

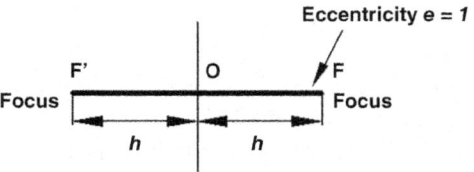

Figure 1.10 An elliptical coordinate system

Incidentally, elliptical cylindrical coordinates are related to the Cartesian coordinates as follows:

$$x = h\cosh(\xi)\cos(\eta,), \quad y = h\sinh(\xi)\sin(\eta), \quad z = z \quad (1.190)$$

where h is semi-interfocal distance. Thus, the $\eta = 0$ and π hyperbolae coincide with X-axis and the $\eta = \pi/2$ and $3\pi/2$ hyperbolae coincide with the Y-axis as shown in Figure 1.10. This behavior is similar to the azimuthal coordinates θ in the cylindrical polar coordinate system (see Figure 1.2).

It may be noted that the tangents at the point of intersection of the family of confocal ellipses and hyperbolae are at right angles, thereby confirming that the elliptical coordinate system is orthogonal and therefore would permit solution by separation of variables. The resulting ordinary differential equations are as follows [18–21]:

$$p(\xi, \eta, z) = p_\xi(\xi) p_\eta(\eta) p_z(z) \quad (1.191)$$

$$\frac{d^2 p_\eta}{d\eta^2} + \{a - 2q\cos(2\eta)\} p_\eta = 0 \quad (1.192)$$

$$\frac{d^2 p_\xi}{d\xi^2} + \{a - 2q\cosh(2\xi)\} p_\eta = 0 \quad (1.193)$$

$$\frac{d^2 p_z}{dz^2} + k_z^2 p_z = 0 \quad (1.194)$$

Equation 1.192 is called Mathieu differential equation, and Equation 1.193 is called the modified Mathieu differential equation. In these two equations,

$$q \equiv \frac{(k_0^2 - k_z^2)h^2}{4} \text{ or } \frac{2\sqrt{q}}{h} = (k_0^2 - k_z^2)^{1/2} \quad (1.195)$$

Thus, q plays the same role as the radial wave number in a circular duct. The constant 'a' is the separation constant which is to be so chosen that solutions are periodic in η, so that $p_\eta(\eta + 2\pi) = p_\eta(\eta)$.

General solutions to Equations 1.192–1.94 are of the type [18]

$$p_\eta(\eta) = C_1 ce_m(\eta, q) + C_2 se_m(\eta, q) \quad (1.196)$$

$$p_\xi(\xi) = C_3 Ce_m(\xi, q) + C_u Se_m(\xi, q) \quad (1.197)$$

$$p_z(z) = C_5 e^{-jk_z z} + C_6 e^{jk_z z} \quad (1.198)$$

where ce_m and se_m are radial and angular Mathieu functions, respectively, and Ce_m and Se_m are the corresponding Modified Mathieu functions. These are given by [18,22] Mathieu functions:

$$\text{even-even:} \quad ce_{2n}(\eta, q) = \sum_{r=0}^{\infty} A_{2r}^{2n} \cos(2r\eta), \quad m = 2n, \quad n = 0, 1, 2, \ldots \quad (1.199)$$

even-odd: $$ce_{2n-1}(\eta, q) = \sum_{r=1}^{\infty} A_{2r-1}^{2n-1} \cos((2r-1)\eta), \quad m = 2n-1, \quad n = 1, 2, 3, \ldots \tag{1.200}$$

odd-even: $$se_{2n}(\eta, q) = \sum_{r=1}^{\infty} B_{2r}^{2n} \sin(2r\eta), \quad m = 2n, \quad n = 1, 2, 3, \ldots \tag{1.201}$$

odd-odd: $$se_{2n-1}(\eta, q) = \sum_{r=1}^{\infty} B_{2r-1}^{2n-1} \sin((2r-1)\eta), \quad m = 2n-1, \quad n = 1, 2, 3, \ldots \tag{1.202}$$

Modified Mathieu functions:

even-even: $$Ce_{2n}(\xi, q) = \sum_{r=0}^{\infty} A_{2r}^{2n} \cosh(2r\eta), \quad m = 2n, \quad n = 0, 1, 2, \ldots \tag{1.203}$$

even-odd: $$Ce_{2n-1}(\xi, q) = \sum_{r=1}^{\infty} A_{2r-1}^{2n-1} \cosh((2r-1)\eta), \quad m = 2n-1, \quad n = 1, 2, 3, \ldots \tag{1.204}$$

odd-even: $$Se_{2n}(\xi, q) = \sum_{r=1}^{\infty} B_{2r}^{2n} \sinh(2r\eta), \quad m = 2n, \quad n = 1, 2, 3, \ldots \tag{1.205}$$

odd-odd: $$Se_{2n-1}(\xi, q) = \sum_{r=1}^{\infty} B_{2r-1}^{2n-1} \sinh((2r-1)\eta), \quad m = 2n-1, \quad n = 1, 2, 3, \ldots \tag{1.206}$$

It may be noted the Modified Mathieu functions are obtained from the corresponding Mathieu functions by replacing circular functions (cos and sine) by the corresponding hyperbolic functions (cosh and sinh). The function names starting with c or C involve cos and cosh functions and therefore are termed 'even', and those starting with s or S involve sine and sinh functions and therefore are termed 'odd'. The second adjective denotes the order (2n: even, 2n-1: odd).

The boundary conditions across the interfocal line in Figure 1.10 are:

Continuity of acoustic pressure: $$p(0, \eta) = p(0, -\eta) \tag{1.207}$$

Continuity of pressure gradient: $$\frac{\partial p(0, \eta)}{\partial \xi} = -\frac{\partial p(0, -\eta)}{\partial \xi} \tag{1.208}$$

Then, it can be shown that acoustic pressure field inside a hollow elliptical chamber is given by [18]

$$p(\xi,\eta,z) = \sum_{m=0}^{\infty} Ce_m(\xi,q) ce_m(\eta,q)\left(C_m^1 e^{-jk_z z} + C_m^2 e^{jk_z z}\right)$$
$$+ \sum_{m=1}^{\infty} Se_m(\xi,q) se_m(\eta,q)\left(S_m^1 e^{-jk_z z} + S_m^2 e^{jk_z z}\right) \quad (1.209)$$

Constants C_m^1, C_m^2, S_m^1, and S_m^2, denote arbitrary constants to be determined from the boundary conditions in the axial direction.

The mode shapes and their propagation are dependent upon the numerical value of the q-parameter given by Equation 1.195. For evaluation of the cut-on frequencies of axial modes corresponding to a certain q valve, let us use the following relations [23]:

$$\frac{2\sqrt{q}}{h} = k_0, \quad \frac{2\sqrt{q}}{e} = k_0 \frac{D_1}{2}, \quad h = \frac{D_1}{2} e, \quad e = \left\{1 - \left(\frac{D_2}{D_1}\right)^2\right\}^{1/2} \quad (1.210)$$

where e is eccentricity of the elliptical section, while D_1 and D_2 are, respectively, the major and minor axes of the elliptical section shown in Figure 1.10.

$q_{m,n}$ and $\bar{q}_{m,n}$, the nth zero of the derivative of the Even and Odd type of the modified Mathieu functions, respectively, of order m, are roots of the following equations characterizing the rigid wall boundary conditions:

$$\left.\frac{dCe_m(\xi,q_{m,n})}{d\xi}\right|_{\xi=\xi_0} = 0 \text{ and } \frac{dSe_m(\xi,\bar{q}_{m,n})}{d\xi} = 0 \quad (1.211)$$

where

$$\xi_0 = \cosh^{-1}(1/e) \quad (1.212)$$

defines the value of the radial elliptical coordinates ξ at the boundary of the elliptical duct or shell shown in Figure 1.10.

By setting $k_z = 0$ in Equation 1.195, the nondimensional cut-on frequencies of the even or odd mode of order m corresponding to the nth parametric zero, that is, $q_{m,n}$ and $\bar{q}_{m,n}$, respectively, are given by

$$k_0(D_1/2)|_{m,n(\text{Even})} = 2\sqrt{q_{m,n}}/e, \quad k_0(D_1/2)|_{m,n(\text{Odd})} = 2\sqrt{\bar{q}_{m,n}}/e. \quad (1.213)$$

Lowson and Bhaskaran [20] tabulated values of the nondimensional frequencies of an elliptical duct in terms of eccentricity e; not in terms of the aspect ratio, D_2/D_1, which would be of greater interest to muffler designers. They did not document nondimensional cut-on frequencies of the purely radial modes and cross modes. Recently, Mimani [23] has produced comprehensive tables of the q-parameters (parametric zeros of the derivative of the modified Mathieu functions). He has also developed interpolating polynomials for any arbitrary value of the aspect ratio, D_2/D_1, and incorporated the convective effect of mean flow.

Table 1.1 The q-parameters and the corresponding nondimensional cut-on frequencies (Extracted from the Mimani Tables [23])

D_2/D_1	Eccentricity e	Even-Odd $m=0, n=1$		Even-Even $m=1, n=1$	
		$q_{m,n}$	$(k_0 D_1/2)_{m,n}$	$q_{m,n}$	$(k_0 D_1/2)_{m,n}$
0.1	0.995	0.8804	1.8861	3.0042	3.4840
0.2	0.98	0.8523	1.8844	2.8999	3.4761
0.3	0.954	0.8056	1.8818	2.7280	3.4628
0.4	0.917	0.7407	1.8781	2.4909	3.4441
0.5	0.866	0.6582	1.8736	2.1918	3.4190
0.6	0.8	0.5585	1.8682	1.8346	3.3862
0.7	0.714	0.4422	1.8622	1.4240	3.3419
0.8	0.6	0.3099	1.8556	0.9680	3.2795
0.9	0.436	0.1623	1.8486	0.4825	3.1871
0.95	0.312	0.0830	1.8449	0.2382	3.1258
0.99	0.141	0.0169	1.8419	0.0469	3.0693

The Mimani tables [23] contain the parametric zeros $q_{m,n}$ and $\bar{q}_{m,n}$, for even and odd modes along with the corresponding nondimensional cut-on frequencies, for different values of the aspect ratio:

$$D_2/D_1 = 0.1, 0.2, \ldots\ldots\ldots 0.8, 0.9, 0.95, 0.99$$

for a hollow elliptical cross-section rigid-wall duct. These are listed below in Table 1.1 for the lowest two modes.

Incidentally, for the $(m=0, n=1)$ mode, $q_{m,n} = (k_0 D_1/2)_{m,n} = 0$. Therefore the $(m=0, n=1)$ mode represents the plane wave, corresponding to the $(0, 0)$ mode of circular duct in Figure 1.4. As D_2/D_1 approaches unity, the nondimensional cut-on frequency of the Even-Odd $(0, 1)$ mode approaches that of the $(1, 0)$ mode of the circular duct $(k_0 D/2 = 1.84)$ and that of the Even-Even $(1, 1)$ mode of the elliptical duct tends to that of the $(2, 0)$ mode of the circular duct $(k_0 D/2 = 3.05)$ in Figure 1.4.

Finally, the cut-off frequency, below which all higher-order modes are cut-off (decay exponentially), of an elliptical cross-section duct of major axis D_1 is given by

$$k_0 D_1/2 \cong 1.86 (\text{within} \pm 1\% \text{accuracy}) \quad (1.214)$$

which compares with $k_0 D/2 = 1.84$ for circular duct (Equation 1.48). Thus, Equation 1.48 can also be used to evaluate the cut-off frequency of an elliptical duct, provided D is replaced by D_1, the major axis of the ellipse, not the geometric mean diameter, $(D_1 D_2)^{1/2}$.

In other words, the cut-off frequency of an elliptical section is lower than the corresponding circular duct with the same equivalent diameter by a factor

$$\frac{(D_1 D_2)^{1/2}}{D_1} = \left(\frac{D_2}{D_1}\right)^{1/2} \quad (1.215)$$

References

1. Morse, P.M. (1948) *Vibration and Sound*, 2nd edn, McGraw-Hill, New York, pp. 305–311.
2. Kinsler, L.E. and Frey, A.R. (1962) *Fundamentals of Acoustics*, John Wiley & Sons, Inc., New York, pp. 441, 97.
3. Morse, P.M. and Ingard, K.U. (1968) *Theoretical Acoustics*, McGraw-Hill, New York, p. 509.
4. Skudrzyk, E. (1968) *The Foundations of Acoustics*, Springer-Verlag, New York, p. 509.
5. Eriksson, L.J. (1980) Higher-order mode effects in circular ducts and expansion chambers. *Journal of the Acoustical Society of America*, **68** (2), 545–550.
6. Kirchhoff, G. (1868) Uber den Einflufs der Warmeleitung in einem gase auf die schallbewegung. *Annalen der Physik und Chemie (Ser. 5)*, **134**, 177–193.
7. Rayleigh, J.W.S. (1945) *The Theory of Sound*, Dover, New York, pp. 317–328.
8. Yuan, S.W. (1969) *Foundations of Fluid Mechanics*, Prentice-Hall of India, pp. 115–123.
9. Kant, S., Munjal, M.L. and Prasanna Rao, D.L. (1974) Waves in branched hydraulic pipes. *Journal of Sound and Vibration*, **37** (4), 507–519.
10. Rschevkin, S.N. (1963) *A Course of Lectures on the Theory of Sound*, Macmillan, New York.
11. Mason, V. (1969) Some experiments on the propagation of sound along a cylindrical duct containing flowing air. *Journal of Sound and Vibration*, **10** (2), 208–226.
12. Fuller, C.R. and Bies, D.A. (1979) The effects of flow on the performance of a reactive acoustic attenuator. *Journal of Sound and Vibration*, **62**, 73–92.
13. Thawani, P.T. (1978) Analytical and Experimental Investigation of the Performance of Exhaust Mufflers with Flow, Ph.D. Thesis, The University of Calgary, Alberta, Canada.
14. Streeter, V.L. (1958) *Fluid Mechanics*, 2nd edn, McGraw-Hill, New York, Chap. 4, Sec. 28.
15. Vennard, J.K. and Street, R.L. (1976) *Elementary Fluid Mechanics (S.I. Version)*, 5th edn, John Wiley & Sons, Inc., New York, Chap. 9.
16. Panicker, V.B. and Munjal, M.L. (1981) Acoustic dissipation in a uniform tube with moving medium. *Journal of the Acoustical Society of India*, **9** (3), 95–101.
17. Munjal, M.L. and Thawani, P.T. (1997) Effect of protective layer on the performance of absorptive ducts. *Noise Control Engineering Journal*, **45** (1), 14–18.
18. McLachlan, N.W. (1947) *Theory and Application of Mathieu Functions*, Oxford University Press, London.
19. Arfken, G.B. and Weber, H.J. (2005) *Mathematical Methods for Physicists*, Academic Press, Elsevier, London.
20. Lowson, M.V. and Bhaskaran, S. (1975) Propagation of sound in elliptic ducts. *Journal of Sound and Vibration*, **38** (2), 185–194.
21. Hong, K. and Kim, J. (1995) Natural mode analysis of hollow and annular elliptical cylindrical cavities. *Journal of Sound and Vibration*, **183** (2), 327–351.
22. Stamnes, J.K. and Spjelkavik, B. (1995) New method for computing eigen functions (Mathieu functions) for scattering by elliptical cylinders. *Pure and Applied Optics*, **4** (3), 251–262.
23. Mimani, A. (March 2012) 1-D and 3-D Analysis of Multi-Port Muffler Configurations with Emphasis on Elliptical Cylindrical Chamber, Ph.D. Thesis, Indian Institute of Science.

2

Theory of Acoustic Filters

An acoustic filter consists of an acoustic element or a set of elements inserted between a source of acoustic signals and the receiver, like atmosphere. Clearly, an acoustic filter is analogous to an electrical filter as well as vibration isolator. An acoustic filter is therefore also called an acoustic transmission line. An exhaust muffler is an acoustic filter except that waves are convected downstream by the moving medium. Mean flow, in fact, affects the waves in a number of ways, and therefore the theory of exhaust muffler (also called aeroacoustic filters) is discussed separately in the following chapters. In the theory of acoustic filters, the medium is assumed to be stationary and the wave propagation one-dimensional (plane wave propagation), governed by the wave Equation 1.5:

$$\frac{\partial^2 p}{\partial t^2} - c_0^2 \frac{\partial^2 p}{\partial z^2} = 0. \tag{2.1}$$

The two variables that characterize the state of acoustic waves are the acoustic pressure, $p(t)$, and the particle velocity, $u(t)$. Here, p is the time-dependent acoustic perturbation over the ambient pressure p_0 and u is the time-dependent velocity of the oscillating particles.

2.1 Units for the Measurement of Sound

The flux of acoustic energy per unit area of a (real or hypothetical) surface is termed acoustic intensity I and equals the time average of the product of p and the normal component of u; that is,

$$I = \overline{p(t)u_n(t)}. \tag{2.2}$$

The total acoustic power radiated by a source, W, can be found by integrating the intensity over a closed hypothetical surface enclosing the source. Thus,

$$W = \oint I dS. \tag{2.3}$$

Acoustics of Ducts and Mufflers, Second Edition. M. L. Munjal.
© 2014 John Wiley & Sons, Ltd. Published 2014 by John Wiley & Sons, Ltd.

The corresponding logarithmic units are:

Sound pressure level

$$L_p \equiv SPL = 20 \log \frac{p_{rms}(N/m^2)}{2 \times 10^{-5}(N/m^2)} \quad dB; \qquad (2.4)$$

Intensity level

$$L_I = 10 \log \frac{I(W/m^2)}{10^{-12}(W/m^2)} \quad dB; \qquad (2.5)$$

Power level

$$L_w = 10 \log \frac{W(Watt)}{10^{-12}(Watt)} \quad dB. \qquad (2.6)$$

Wave front or phase surface is a hypothetical surface where all the particles have the same instantaneous velocity. Thus, the total power flux associated with a plane wave in a duct of cross-sectional area S, where the particle velocity u is axial and hence normal to the wavefront, would be given by

$$W = \overline{Sp \cdot u} = \overline{p \cdot v_v}, \qquad (2.7)$$

where v_v, related to the particle velocity u as

$$v_v \equiv Su, \qquad (2.8)$$

is called the (acoustic) volume velocity.

Defining (acoustic) mass velocity as

$$v = S\rho_0 u, \qquad (2.9)$$

one can write acoustic power flux W as

$$W = \frac{1}{\rho_0} \overline{p \cdot v} \qquad (2.10)$$

$$= \frac{1}{\rho_0} p_{rms} v_{rms} \cos \theta \qquad (2.11)$$

where θ is the phase angle between p and v.

Use of the mass velocity variable v is preferred in hot exhaust systems. This will become more clear in the next chapter. Thus, the two variables adopted in this monograph are acoustic pressure p and acoustic mass velocity v.

For sinusoidal p and v,

$$p = Pe^{j\omega t}, \quad v = Ve^{j\omega t}, \qquad (2.12)$$

where P and V are complex phasors.
$p_{rms} = |P|/\sqrt{2}, \quad v_{rms} = |V|/\sqrt{2}$, and therefore,

$$W = \frac{1}{2\rho_0}Re(PV) = \frac{|V|^2}{2\rho_0}R, \qquad (2.13)$$

where

$$R = Re(Z), Z \equiv p/v = P/V. \qquad (2.14)$$

In the case of reciprocating and rotating machinery, all sound is periodic in nature. As every periodic signal can be looked upon as a Fourier series of sinusoidal signals, the theory of filters (and mufflers) is developed for sinusoidal signals. Therefore, in this chapter as also in the later chapters, unless otherwise stated, all formulation is in the frequency domain. In fact, the exponential time factor $e^{j\omega t}$ is mostly omitted for convenience of writing, and the symbols p and v stand for the (complex) amplitudes of the two state variables.

2.2 Uniform Tube

On defining the characteristic impedance as*

$$Y = \frac{\text{acoustic pressure associated with a progressive wave}}{\text{acoustic mass velocity associated with the progressive wave}}, \qquad (2.15)$$

one gets, from Equations 1.13, 1.17, and 1.18,

$$p = Ae^{-jk_0 z} + Be^{+jk_0 z} \qquad (2.16)$$

$$v = \frac{1}{Y_0}\left(Ae^{-jk_0 z} - Be^{+jk_0 z}\right), \qquad (2.17)$$

where

$$\text{characteristic impedance}, Y_0 = c_0/S, \qquad (2.18)$$

$$\text{wave number}, k_0 = \omega/c_0, \qquad (2.19)$$

---

*Conventionally, the characteristic impedance is denoted by the symbol Z_0. However, in the theory of filters and exhaust mufflers, one needs subscripts for numbering the elements and Z_0 then denotes the radiation impedance of the atmosphere.

In Equations 2.16 and 2.17, the exponential time factor has been absorbed in A and B. Acoustic impedance at any point in the standing wave field is defined as

$$\zeta(z) = \frac{p(z)}{v(z)} = Y_0 \frac{Ae^{-jk_0 z} + Be^{+jk_0 z}}{Ae^{-jk_0 z} - Be^{+jk_0 z}} \quad (2.20)$$

and represents the equivalent impedance of the complete passive subsystem downstream of this point.

Value of the acoustic impedance at $z = 0$ (the beginning of a tube) can be related to that at $z = l$ (the end of the tube) by making use of Equation 2.20 and eliminating A and B as follows (see Figure 2.1a):

$$\zeta(0) = Y_0 \frac{A+B}{A-B}; \quad (2.21)$$

$$\begin{aligned}\zeta(l) &= Y_0 \frac{Ae^{-jk_0 l} + Be^{+jk_0 l}}{Ae^{-jk_0 l} - Be^{+jk_0 l}} \\ &= Y_0 \frac{(A+B)\cos k_0 l - j(A-B)\sin k_0 l}{(A-B)\cos k_0 l - j(A+B)\sin k_0 l}, \\ &= Y_0 \frac{(A+B)/(A-B) - j\tan k_0 l}{1 - (A+B)/(A-B)j\tan k_0 l},\end{aligned} \quad (2.22)$$

or

$$\frac{\zeta(l)}{Y_0} = \frac{\zeta(0)/Y_0 - j\tan k_0 l}{1 - j\zeta(0)/Y_0 \tan k_0 l}. \quad (2.23)$$

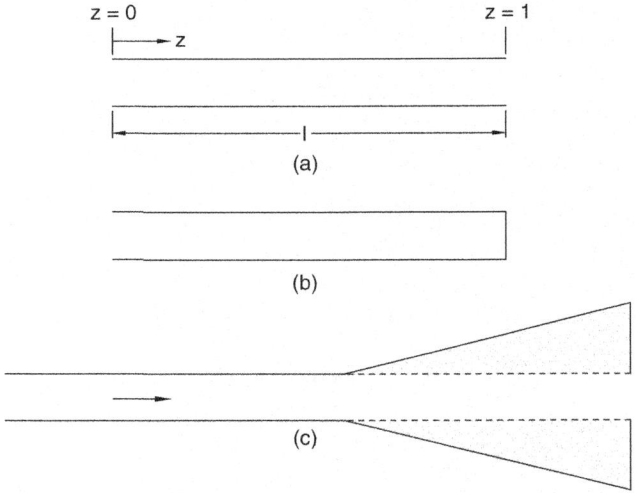

Figure 2.1 A uniform tube with various terminations. (a) A uniform tube of length l. (b) A tube closed at the downstream end. (c) A tube with anechoic termination

Rearranging,

$$\frac{\zeta(0)}{Y_0} = \frac{\zeta(l)/Y_0 + j\tan k_0 l}{1 + j\zeta(l)/Y_0 \tan k_0 l}$$

or

$$\zeta(0) = \frac{\zeta(l)\cos k_0 l + jY_0 \sin k_0 l}{j\zeta(l)/Y_0 \sin k_0 l + \cos k_0 l}. \tag{2.24}$$

Equation 2.23 can also be written in a similar form:

$$\zeta(l) = \frac{\zeta(0)\cos k_0 l - jY_0 \sin k_0 l}{-j\zeta(0)/Y_0 \sin k_0 l + \cos k_0 l}. \tag{2.25}$$

If the end $z = l$ is rigidly closed as shown in Figure 2.1b, then mass velocity $v(l) \to 0$, the acoustic impedance $\zeta(l) \to \infty$, and Equation 2.24 yields

$$\zeta(0)_{rigid\,end} = -jY_0 \cot k_0 l. \tag{2.26}$$

In reality, thin plates used for closing the tubes cannot ensure infinite impedance $\zeta(l)$; in that case the exact relation (2.24) must be used instead of (2.26).

The acoustic behavior at a termination is often described in terms of reflection coefficient R, defined as the ratio of the reflected wave pressure to that of the incident wave:

$$R \equiv |R|e^{j\theta},$$

where $|R|$ and θ are, respectively, the amplitude and phase of the reflection coefficient. Now,

$$R(0) = B/A, \tag{2.27}$$

$$R(l) = \frac{Be^{jk_0 l}}{Ae^{-jk_0 l}}, \tag{2.28}$$

and

$$\zeta = Y_0 \frac{1+R}{1-R}. \tag{2.29}$$

Alternatively,

$$R = \frac{\zeta - Y_0}{\zeta + Y_0}. \tag{2.30}$$

At a rigid termination $v \to 0$, $\zeta \to \infty$, and therefore.

$$R_{rigid\,term.} \to 1. \tag{2.31}$$

Thus, a perfectly rigid termination reflects the incident wave totally (with the same amplitude $|R|$ and phase θ).

Another termination that is often used in research is anechoic termination (Figure 2.1c), characterized by the zero reflection coefficient

$$R_{anech.\,term.} \to 0, \tag{2.32}$$

so that, on making use of relation (2.29),

$$\zeta_{anech.\,term.} \to Y_0. \tag{2.33}$$

Thus, the equivalent impedance of a tube with anechoic termination equals the characteristic impedance of the tube at all points. This is not surprising in that the sound field in such a tube consists of only a forward-moving progressive wave; there is no reflected wave.

2.3 Radiation Impedance

The radiation impedance Z_0 represents the impedance imposed by the atmosphere on acoustic radiation from the end of a tube. This can be evaluated from the three-dimensional acoustic field created by a hypothetical piston located at the radiating end of the tube and vibration with the same particle velocity u_0. The radiation impedance, then is calculated as

$$Z_0 = \frac{\text{average acoustic pressure } p_0 \text{ on the piston}}{\text{acoustic mass velocity } v_0 \text{ of the piston}}, \tag{2.34}$$

where $v_0 = \rho_0 S u_0$.

Like all impedances, Z_0 presumes sinusoidal pressure and mass velocity and is a function of the exciting radiation frequency ω.

For a tube terminating in an 'infinite' flange (hemispherical space) for plane waves [1],

$$Z_0 = R_0 + jX_0; \tag{2.35}$$

$$R_0 = Y_0 \left\{ 1 - \frac{2 J_1(2 k_0 r_0)}{2 k_0 r_0} \right\} = \left\{ \frac{(2 k_0 r_0)^2}{2 \times 4} - \frac{(2 k_0 r_0)^4}{2 \times 4^2 \times 6} + \frac{(2 k_0 r_0)^6}{2 \times 4^2 \times 6^2 \times 8} - \cdots \right\} Y_0; \tag{2.36a}$$

$$X_0 = \left\{ \frac{4}{\pi} \frac{2 k_0 r_0}{3} - \frac{(2 k_0 r_0)^3}{3^2 \times 5} + \frac{(2 k_0 r_0)^5}{3^2 \times 5^2 \times 7} - \cdots \right\} Y_0, \tag{2.36b}$$

where
$Y_0 = c_0/(\pi r_0^2)$, and
$r_0 =$ radius of the tube.

At sufficiently low frequencies such that $k_0 r_0 < 0.5$, these expressions reduce to

$$R_0 = Y_0 \left(\frac{k_0^2 r_0^2}{2} \right); \tag{2.37a}$$

$$X_0 = Y_0 (0.85 k_0 r_0). \tag{2.37b}$$

Fortunately, however, the frequencies of interest are generally low enough so that for typical radii of the tail pipe, $k_0 r_0$ is indeed less than 0.5, and the simplified Equations 2.37a and 2.37b indeed hold.

Radiation pattern from a tube with open end (without any flange) is much more complex to analyze. This has, nevertheless, been accomplished by Levine and Schwinger [2].

The radiation impedance so derived is very complex. However, writing Z_0 in terms of the reflection coefficient R as

$$Z_0 = R_0 + j X_0 = Y_0 \frac{1+R}{1-R}, \tag{2.38}$$

with

$$R = |R| e^{j(\pi - 2k_0 \delta)}, \tag{2.39}$$

a close empirical fit to the values of end correction δ and reflection coefficient modulus $|R|$ are given by the following expressions [3]:

$$\delta / r_0 = 0.6133 - 0.1168 (k_0 r_0)^2, \quad k_0 r_0 < 0.5; \tag{2.40a}$$

$$\delta / r_0 = 0.6393 - 0.1104 \, k_0 r_0, \quad 0.5 < k_0 r_0 < 2; \tag{2.40b}$$

$$|R| = 1 + 0.01336 \, k_0 r_0 - 0.59079 (k_0 r_0)^2 + 0.33576 (k_0 r_0)^3 - 0.06432 (k_0 r_0)^4, 0 < k_0 r_0 < 3.5, \tag{2.41}$$

If δ were zero, the reflection coefficient R would be $-|R|$. Physically, it would indicate a pressure node (or velocity antinode) at the tail pipe end. Equation 2.39 indicates that this mode occurs outside of the tailpipe at a distance of δ. In other words, the tail pipe is extended by δ, as it were. Therefore, δ is termed the end correction.

This is illustrated hereunder with reference to Figure 2.2 where z is reckoned from the free end of an open tube in the opposite direction:

$$p(z) = A e^{jk_0 z} + B e^{-jk_0 z}$$

$$= A e^{jk_0 z} \left\{ 1 + \frac{B}{A} e^{-2jk_0 z} \right\}$$

$$= A e^{jk_0 z} \left\{ 1 + |R| e^{j(\pi - 2k_0 \delta - 2k_0 z)} \right\}$$

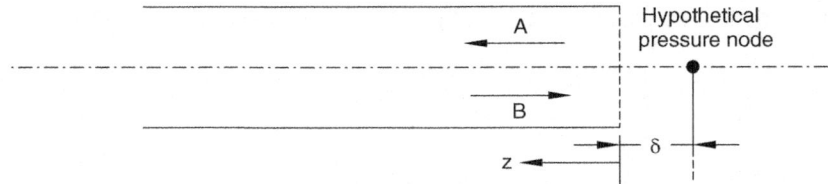

Figure 2.2 The concept of end correction

Assuming $|R| \simeq 1, p(z)$ would tend to zero, if the argument

$$\pi - 2k_0\delta - 2k_0z = \pi$$

or if

$$z = -\delta$$

Thus, the pressure node (or, minimum) would occur outside the duct at about 0.6 times the radius, as shown in Figure 2.2.

At sufficiently low frequencies, such that $k_0r_0 < 0.5$, the radiation impedance can be approximated as

$$Z_0|_{\text{open end}} = Y_0\left(\frac{k_0^2 r_0^2}{4} + j0.6\, k_0 r_0\right), \quad k_0 r_0 < 0.5, \tag{2.42}$$

so that

$$R_0|_{\text{open end}} = Y_0 \frac{k_0^2 r_0^2}{4} = 0.5 R_0|_{\text{infinite flange}} \tag{2.43}$$

$$X_0|_{\text{open end}} = Y_0(0.6\, k_0 r_0) < X_0|_{\text{infinite flange}}. \tag{2.44}$$

R_0, the radiation resistance, is analogous to load resistance in the electrical network theory and is responsible for acoustic radiation from the tail-pipe end. X_0, the radiation reactance, results in a phase difference between p_0 and v_0.

2.4 Reflection Coefficient at an Open End

Reflection coefficient R at a point is related to the acoustic impedance ζ at the point according to Equation 2.30. At the radiation end, ζ equals the radiation impedance Z_0. Thus,

$$R = \frac{Z_0 - Y_0}{Z_0 + Y_0} \tag{2.45}$$

or

$$|R|e^{j\theta} = \frac{R_0 + jX_0 - Y_0}{R_0 + jX_0 + Y_0} = \frac{(0.25k_0^2 r_0^2 - 1) + j(0.6k_0 r_0)}{(0.25k_0^2 r_0^2 + 1) + j(0.6k_0 r_0)}. \tag{2.46}$$

Rationalization of Equation 2.46 yields

$$|R| = \frac{\left\{(0.0625k_0^4r_0^4 + 0.36k_0^2r_0^2 - 1)^2 + (1.2k_0r_0)^2\right\}^{1/2}}{0.0625k_0^4r_0^4 + 0.86k_0^2r_0^2 + 1} \cong 1 - 0.5\,k_0^2r_0^2; \quad (2.47)$$

$$\theta = \tan^{-1}\frac{1.2k_0r_0}{0.0625k_0^4r_0^4 + 0.36k_0^2r_0^2 - 1}$$

$$\cong \tan^{-1}(-1.2\,k_0r_0)$$

$$= \pi - \tan^{-1}(1.2\,k_0r_0). \quad (2.48)$$

Thus, at the open end, the amplitude of the reflection coefficient is nearly unity (only a little less) and the phase angle is slightly less than π radians. Hence it is that an open end reflects the incoming wave almost fully (the remainder, which is very little, is radiated out) but with opposite phase.

2.5 A Lumped Inertance

For wave propagation through a tube of very small length (like a hole through an end plate), as shown in Figure 2.3, there would be little time lag between the two ends ($k_0l \ll 1$). All the medium particles would move together with, say, velocity u. Thus, by Newton's second law of motion,

$$S(p_1 - p_2) = (\rho_0 Sl)du/dt$$

or

$$\Delta p \equiv p_1 - p_2 = \rho_0 lj\omega u$$

or

$$Z \equiv \Delta p/v = j\rho_0 l\omega u/(\rho_0 Su)$$

or

$$Z_{lumped\ inertance} = j\omega\frac{l}{S}. \quad (2.49)$$

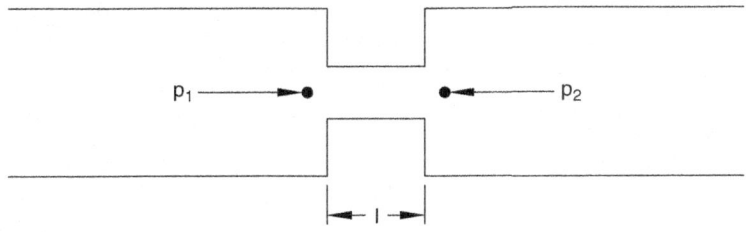

Figure 2.3 A small tube or lumped inertance

By electroacoustic analogies, l/S, which is analogous to a lumped inductance, is called lumped inertance M.

2.6 A Lumped Compliance

A cavity of volume V can allow acoustic motion at its neck (and hence store energy) by contraction. Let an acoustic pressure p be applied at the neck of cross-sectional area S as shown in Figure 2.4, and let the volume decrease by ΔV adiabatically. Then,

$$\frac{p}{p_0} + \gamma \frac{\Delta V}{V} = 0 \qquad (2.50)$$

or

$$p = \frac{-\gamma p_0 \Delta V}{V}. \qquad (2.51)$$

If ζ is the corresponding inward displacement at the neck, then

$$\Delta V = -S\zeta = -\frac{Su}{j\omega}. \qquad (2.52)$$

Therefore,

$$p = \frac{\gamma p_0}{V} \frac{Su}{j\omega} \qquad (2.53)$$

and

$$Z \equiv \frac{p}{v} = \frac{p}{\rho_0 S u} = \frac{\gamma p_0}{\rho_0 V j \omega} \qquad (2.54)$$

or

$$Z_{capacitance} = \frac{1}{j\omega C}, \qquad (2.55)$$

and therefore V/c_0^2 is called the compliance C of the cavity of volume V.

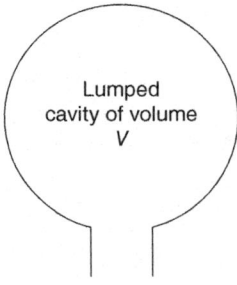

Figure 2.4 A cavity or lumped compliance

2.7 End Correction

If the small tube were exposed to the atmosphere or a large volume, the radiation reactance can be added to the lumped impedance of the tube and the combined reactance can be looked upon as impedance of an extended tube; the hypothetical additional length is then termed the end correction. Thus, Equations 2.37b and 2.49 yield.

$$j\omega(correction\ \Delta l\ due\ to\ a\ flanged\ end)/S = j(0.85k_0r_0)c_0/S$$

or

$$\Delta l_{\text{flanged end}} = 0.85r_0. \tag{2.56}$$

Similarly, Equations 2.44 and 2.49 give

$$\Delta l_{\text{open end}} = 0.6r_0. \tag{2.57}$$

Thus, the effective length of a hole across a plate in Figure 2.3 (or a hole in the periphery of a large-diameter tube) would be

$$l_{ec_{\text{hole}}} = t + 2(0.85r_h) = t + 0.85d_h, \tag{2.58}$$

where t and d_h are the thickness of the plate and the hole diameter, respectively.

In the case of an acoustically long tube, the radiation reactance can be construed as end correction. The tube and radiation impedance can also be considered as two separate acoustical elements in the system (see Section 2.9). It may be noted that the end correction accounts for radiation reactance, the imaginary part of the radiation impedance. Radiation resistance also needs to be considered duly in the analysis (as load resistance).

2.8 Electroacoustic Analogies

From the previous pages, it is obvious that there is more than an incidental correspondence between the impedance approach in acoustical systems and the frequency-domain analysis of electrical transmission networks. The analogies are summarized in the Table 2.1. These analogies are called direct analogies inasmuch as the acoustical force corresponds to

Table 2.1 Electroacoustic analogies

Variable	Acoustic terms		Corresponding electrical terms		
	Symbol	Units	Variable	Symbol	Units
Pressure	p	N/m^2	Electromotive force	e	Volts
Mass velocity	v	kg/s	Current	i	Amperes
Acoustical impedance	Z	$1/ms$	Electrical impedance	Z	Ohms
Resistance	R	$1/ms$	Resistance	R	Ohms
Inertance	M	$1/m$	Inductance	L	Henries
Compliance	C	ms^2	Capacitance	C	Farads

electromotive force and a motion variable (mass velocity) to a motion variable (current). It may also be noted that the terms associated with an acoustically long tube can be expressed in terms of the total inertance M and compliance C of the tube:

Characteristic impedance,

$$Y_0 = \frac{c_0}{S} = \left(\frac{l}{S}\frac{c_0^2}{lS}\right) = \left(\frac{M}{C}\right)^{1/2} \tag{2.59}$$

Wave propagation speed

$$c_0 = 1\left\{\frac{S}{l}\frac{c_0^2}{lS}\right\}^{1/2} = \frac{l}{(MC)^{1/2}} \tag{2.60}$$

These relations are also analogous to those in the electrical network theory.

2.9 Electrical Circuit Representation of an Acoustic System

The analogy between electrical system variables and acoustical ones being so complete, one could make use of the well-established electrical circuit representation for an acoustical filter as shown in Figure 2.5a. Here acoustical pressure p replaces voltage and acoustical mass velocity v replaces current. Source is represented by pressure p_{n+1} (analogous to the open circuit voltage of an electrical source) and an internal impedance Z_{n+1}. Z_0 is the load or radiation impedance. An acoustical filter with n elements separates the source from the load. ζ_n is the acoustical impedance of the passive subsystem (consisting of the filter and load) downstream of point n, and equals p_n/v_n.

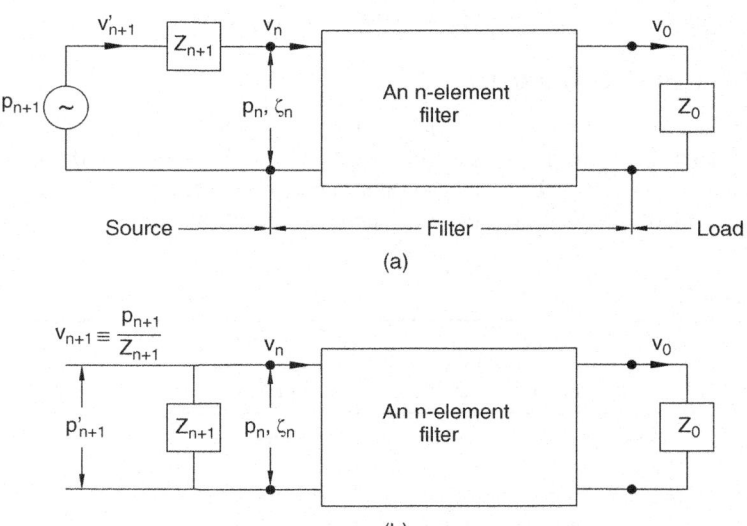

Figure 2.5 A filter with its terminations. (a) Pressure representation for the source. (b) Velocity representation for the source

In general, according to the notation followed here, point r is just upstream of the rth element, all the $n + 2$ elements being always numbered from the load (zeroth element) to the source (numbered $n + 1$) [4].

It may be noted that for a given filter and load (and hence the equivalent impedance ζ_n).

$$v_n = \frac{p_{n+1}}{Z_{n+1} + \zeta_n} \tag{2.61}$$

and that the same value of v_n can be got from the alternative source representation of Figure 2.5b where the source impedance Z_{n+1} is moved over to the shunt position, provided

$$v_{n+1} = p_{n+1}/Z_{n+1}. \tag{2.62}$$

Thus, the two source representations are equivalent insofar as the filter is concerned. The source representation of Figure 2.5a is called the pressure representation and that of Figure 2.5b, the velocity representation. These correspond respectively to the Thevenin theorem and Norton theorem in the electrical network theory.

2.10 Acoustical Filter Performance Parameters

Figure 2.6 shows a typical exhaust muffler or a low-pass acoustic filter, with its terminations. Invariably, the muffler has a small-diameter pipe on either end. The one upstream is called the exhaust pipe and that downstream is called the tail pipe. The middle, larger-diameter portion may be called the muffler proper. In general, for an n-element muffler, the tail pipe would be the first element and the exhaust pipe, the nth.

The performance of an acoustic filter (or muffler) is measured in terms of one of the following parameters:

a. Insertion loss, IL
b. Transmission loss, TL,
c. Level difference, LD, or noise reduction, NR.

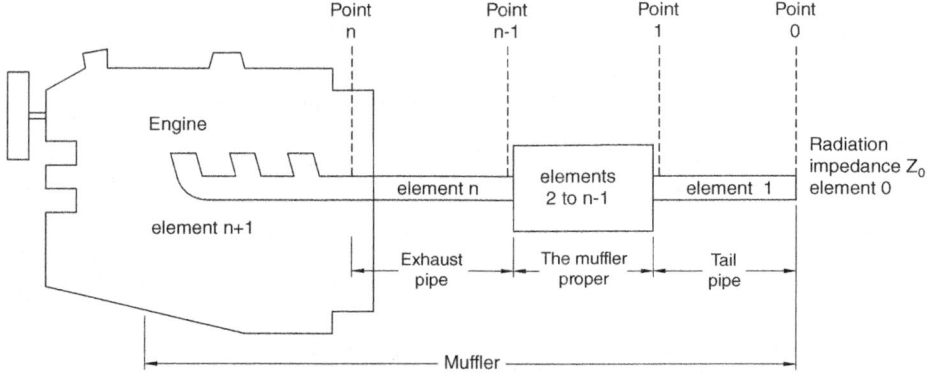

Figure 2.6 Typical engine exhaust system

2.10.1 Insertion Loss, IL

Insertion loss is defined as the difference between the acoustic power radiated without any filter and that with the filter. Symbolically,

$$IL = L_{w1} - L_{w2} \quad (dB) \tag{2.63}$$

$$= 10 \log(W_1/W_2) \quad (dB) \tag{2.64}$$

where subscripts 1 and 2 denote systems without filter and with filter, respectively. Let $Z_{0,1}$ and $Z_{0,2}$ be the corresponding radiation impedances.

Making use of Equation 2.13 and referring to Figure 2.7,

$$W_1 = \frac{1}{2\rho_{0,1}} \left| \frac{p_{n+1}}{Z_{n+1} + Z_{0,1}} \right|^2 R_{0,1}, \tag{2.65}$$

where $R_{0,1}$ is the real component (resistance) of the radiation impedance $Z_{0,1}$. Now, referring to Figure 2.5b for a system with filter and again making use of Equation 2.13 for power flux,

$$W_2 = \frac{1}{2\rho_{0,2}} |v_0|^2 R_{0,2}. \tag{2.66}$$

On substituting these expressions for W_1 and W_2 in Equation 2.64, one gets

$$\begin{aligned} IL &= 10 \log \left[\frac{\rho_{0,2}}{\rho_{0,1}} \frac{R_{0,1}}{R_{0,2}} \left| \frac{p_{n+1}}{Z_{n+1} + Z_{0,1}} \right|^2 \right] \\ &= 20 \log \left[\left(\frac{\rho_{0,2}}{\rho_{0,1}} \frac{R_{0,1}}{R_{0,2}} \right)^{1/2} \left| \frac{Z_{n+1}}{Z_{n+1} + Z_{0,1}} \right| \left| \frac{v_{n+1}}{v_0} \right| \right], \end{aligned} \tag{2.67}$$

Incidentally, for the hypothetical case of zero temperature gradient, $S_n = S_1$, and constant pressure source ($Z_{n+1} \to 0$), it can be readily shown that

$$IL = 20 \log |p_n/p_0|. \tag{2.68}$$

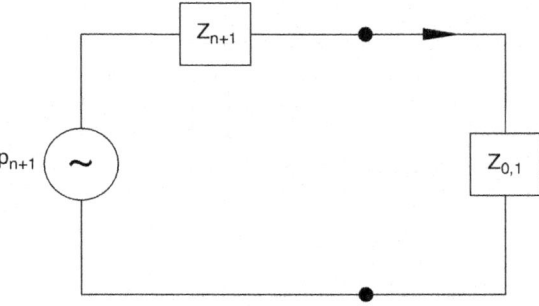

Figure 2.7 A system without any filter

Defining the velocity ratio for a passive subsystem with r elements as [5]

$$VR_r = v_r/v_0, \text{ for } v_0 \neq 0 \text{ and } p_0 = 0. \tag{2.69}$$

one gets [6]

$$IL = 20 \log \left[\left(\frac{\rho_{0,2}}{\rho_{0,1}} \frac{R_{0,1}}{R_{0,2}} \right)^{1/2} \left| \frac{Z_{n+1}}{Z_{n+1} + Z_{0,1}} \right| |VR_{n+1}| \right]. \tag{2.70}$$

An order analysis of Equations 2.69 and 2.70 indicates two very important points in the theory of acoustic filters [7]; namely:

1. Velocity ratio VR_n represents the square root of the inverse ratio of acoustic resistance, $(R_{0,2}/R_n)^{1/2}$,
2. The real action of nondissipative acoustic filters lies in reducing the acoustic resistance seen by the source; without any muffler it sees $R_{0,1}$ and with muffler it sees R_n. The latter must be much smaller for good insertion loss, at desired frequencies.

2.10.2 Transmission Loss, TL

Transmission loss is independent of the source and presumes (or requires) an anechoic termination at the downstream end. It describes the performance of what has been called 'the muffler proper' in Figure 2.6. It is defined as the difference between the power level incident on the muffler proper and that transmitted downstream into an anechoic termination (see Figure 2.8). Symbolically,

$$TL = L_{wi} - L_{wt}. \tag{2.71}$$

In terms of the progressive wave components shown in Figure 2.8,

$$\begin{aligned} TL &= 10 \log \left| \frac{S_n A_n^2}{2} \frac{2}{S_1 A_1^2} \right|, & B_1 = 0 \\ &= 20 \log \left| \frac{A_n}{A_1} \right|, & B_1 = 0. \end{aligned} \tag{2.72}$$

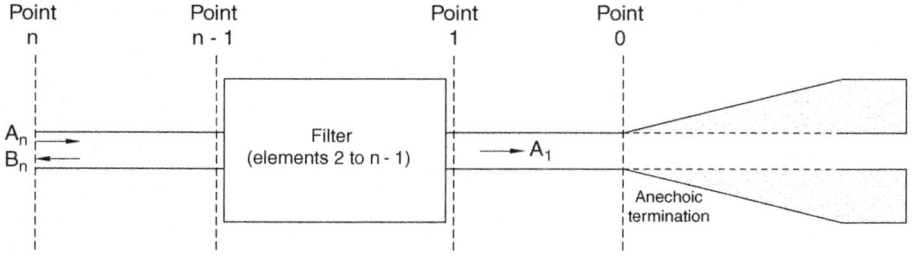

Figure 2.8 Definition of transmission loss: $TL = 20 \log |A_n/A_1|$

because S_n and S_1, the areas of the exhaust pipe and tail pipe, are generally made equal in the experiments for transmission loss.

Thus, TL equals 20 times the logarithm (to the base 10) of the ratio of the acoustic pressure associated with the incident wave (in the exhaust pipe) and that of the transmitted wave (in the tail pipe), with the two pipes having the same cross-sectional area and the tail pipe terminating anechoically [8]. Whereas $A_1 = p_1$ (in view of the anechoic termination, which ensures $B_1 = 0$). A_n cannot be measured directly in isolation from the reflected wave pressure B_n. One has to resort to impedance tube technology (discussed in Chapter 5). Nevertheless, calculations wise there is no difficulty. In terms of the standing wave variables,

$$A_1 = p_1 = Y_1 v_1, \quad Z_0 = Y_1, \qquad (2.73)$$

and

$$A_n = \frac{p_n + Y_n v_n}{2}. \qquad (2.74)$$

With these substitutions, Equation 2.72 can be rewritten as

$$TL = 20 \log \left| \frac{p_n + Y_n v_n}{2 Y_1 v_1} \right|, \quad Z_0 = Y_1. \qquad (2.75)$$

As a progressive wave does not undergo a change in amplitude while moving along a uniform tube, A_n and A_1 can be measured anywhere in the respective pipes. Therefore, for mathematical convenience, lengths l_1 and l_n can be taken as zero. Then,

$$TL = 20 \log \left| \frac{p_n + Y_n v_{n-1}}{2 Y_1 v_1} \right|, \quad Z_0 = Y_1. \qquad (2.76)$$

2.10.3 Level Difference, LD

Level difference (or noise reduction, NR) is the difference in sound pressure levels at two arbitrarily selected points in the exhaust pipe and tail pipe (see Figure 2.9), Symbolically [9],

$$LD = 20 \log |p_n / p_1| \quad (dB). \qquad (2.77)$$

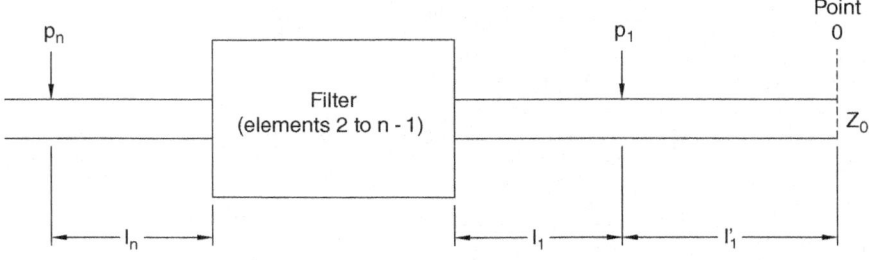

Figure 2.9 Definition of level difference: $LD = 20 \log |p_n / p_1|$

Unlike the transmission loss, the definition of level difference makes use of standing wave pressures and does not require an anechoic termination.

For the purpose of calculations, as p_1 is not known directly, one can write

$$LD = 20 \log \left| \frac{p_n}{p_0} \frac{p_0}{p_1} \right|, \qquad (2.78)$$

Applying the wave relationships for the pipes section of length l'_1 one can evaluate p_0/p_1 as follows.

$$p_1 = A_1 + B_1. \qquad (2.79)$$

Then

$$p_0 = A_1 e^{-jk_0 l'_1} + B_1 e^{+jk_0 l'_1}, \qquad (2.80)$$

$$v_0 = \frac{1}{Y_1} \left(A_1 e^{-jk_0 l'_1} - B_1 e^{+jk_0 l'_1} \right), \qquad (2.81)$$

whence

$$A_1 = \frac{p_0 + v_0 Y_1}{2} e^{jk_0 l'_1}, \qquad (2.82)$$

$$B_1 = \frac{p_0 - v_0 Y_1}{2} e^{-jk_0 l'_1}, \qquad (2.83)$$

Substituting the expressions for A_1 and B_1 in Equation 2.79 yields

$$p_1 = p_0 \cos k_0 l'_1 + j Y_1 v_0 \sin k_0 l'_1, \qquad (2.84)$$

and therefore,

$$\frac{p_1}{p_0} = \cos k_0 l'_1 + \frac{j Y_1}{Z_0} \sin k_0 l'_1. \qquad (2.85)$$

Equations 2.78 and 2.85 yield

$$LD = 20 \log \left| \frac{p_n/p_0}{\cos k_0 l'_1 + j Y_1/Z_0 \sin k_0 l'_1} \right|. \qquad (2.86)$$

Incidentally, if l'_1 were selected to be zero, that is, if the acoustic pressure p_1 were picked up from the end of the tail pipe, then

$$LD(l'_1 \to 0) = 20 \log |p_n/p_0|. \qquad (2.87)$$

Comparing it with Equation 2.68, it can be noticed that level difference is a limiting value of insertion loss for a constant pressure source and l'_1 in Figure 2.9 tending to zero.

2.10.4 Comparison of the Three Performance Parameters

Of the three performance parameters just discussed, insertion loss is clearly the only one that represents the performance of the filter truly, inasmuch as it represents the loss in the radiated power level consequent to insertion of the filter between the source and the receiver (the load). But it requires prior knowledge or measurement of Z_{n+1}, the internal impedance of the source.

Transmission loss does not involve the source impedance and the radiation impedance inasmuch as it represents the difference between incident acoustic energy and that transmitted into an anechoic environment. Being made independent of the terminations, TL finds favor with researchers who are sometimes interested in finding the acoustic transmission behavior of an element or a set of elements in isolation of the terminations. But measurement of the incident wave in a standing wave acoustic field requires use of impedance tube technology, which, as discussed in a later chapter, may be quite laborious, unless one makes use of the two-microphone method with modern instrumentation.

Level difference, or noise reduction, is the difference of SPLs at two points one upstream and one downstream. Like TL, it does not require knowledge of source impedance and, like IL, it does not need anechoic termination. It is therefore, the easiest to measure and calculate and has come to be used widely for experimental corroboration of the calculated transmission behavior of a given set of elements (called the muffler proper in Figure 2.6).

Thus, the various performance parameters have relative advantages and disadvantages. However, in the final analysis, for the user, insertion loss is the only criterion for the performance of a given filter.

2.11 Lumped-Element Representations of a Tube

Electrical circuit representation of one-dimensional acoustical filters is very useful in understanding the behavior of combinations of elements, lumped elements, in particular. This is made possible by studying combinations of their impedances.

Referring to Figure 2.1a, $\zeta(0)$ the equivalent impedance at $z=0$, is related to $\zeta(l)$, the equivalent impedance at $z=l$ by relation (2.24), that is,

$$\zeta(0) = \frac{\zeta(l)\cos k_0 l + jY \sin k_0 l}{j\{\zeta(l)/Y\}\sin k_0 l + \cos k_0 l}, \tag{2.88}$$

At sufficiently low frequencies or for very short tubes, $k_0^2 l^2 \ll 1$. Then, $\cos k_0 l \simeq 1$, $\sin k_0 l \simeq k_0 l$, and Equation 2.88 reduces to

$$\begin{aligned} Lt\, \zeta(0)_{k_0^2 l^2 \ll 1} &= \frac{\zeta(l) + jY k_0 l}{j\zeta(l)k_0 l/Y + 1} \\ &= \frac{\zeta(l) + j\omega l/S}{\zeta(l)j\omega V/c_0^2 + 1} \\ &= \frac{\zeta(l) + Z_M}{\zeta(l)/Z_c + 1}, \end{aligned} \tag{2.89}$$

where use has been made of Equations 2.49 and 2.55 to denote the total inertive impedance of the tube $j\omega l/S$ by Z_M and the total compliance admittance of the tube $j\omega V/c_0^2$ by $1/Z_C$. In the limiting cases, one gets

$$\zeta(0) \xrightarrow[k_0^2 l^2 \ll 1]{\zeta(l) \ll Z_c} \zeta(l) + Z_M; \tag{2.90}$$

$$\zeta(0) \xrightarrow[k_0^2 l^2 \ll 1]{Z_M \ll \zeta(l)} \frac{\zeta(l) Z_c}{\zeta(l) + Z_c}. \tag{2.91}$$

Thus, in the first case, represented by Equation 2.90, the tube acts as a lumped in-line inertance, and Z_M is in series with $\zeta(l)$. In the second case, it acts as a lumped shunt compliance, and Z_c is in parallel with $\zeta(l)$. These two cases are shown in Figure 2.10.

It is important to note here that what matters really is how Z_M and Z_c compare with $\zeta(l)$. In other words, the lumped-element behavior of a tube is a function of what lies downstream of it. In particular, if the tube in question is acoustically small $k_0^2 l^2 \ll 1$ and thin and is sandwiched between two tubes of much larger cross-section (like an orifice plate in a tube), it acts as a lumped inertance with impedance $j\omega l/S$ where l includes the end corrections on both the sides. On the other hand, if an acoustically small tube is sandwiched between tubes of much smaller cross-section (like a small expansion chamber between two tubes), it acts as a lumped cavity or compliance with impedance $c_0^2/(j\omega V)$ where V is the volume of the tube (cavity). In these arguments, it is implied that the equivalent impedance $\zeta(l)$ is of the order of the characteristic impedance of the downstream tube, which in turn is inversely proportional to the area of cross-section ($Y = c_0/S$).

For the typical acoustical filters (mufflers), however, lumped-element approximations do not hold except at very low frequencies. In fact, these are not necessary either, except to understand the basic behavior of low-pass filters.

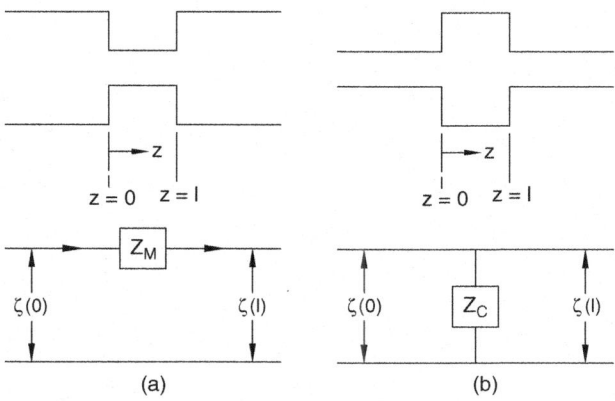

Figure 2.10 Lumped-element representations of a tube. (a) The tube as a lumped in-line inertance. (b) The tube as a lumped shunt compliance

2.12 Simple Area Discontinuities

Two simple types of discontinuities are shown in Figure 2.11. If the diameters of both the tubes are within the limit imposed by pure plane wave propagation [see inequality (1.48)], then acoustic pressure p and acoustic mass velocity $v(=\rho_0 Su)$ remain the same across either of the two discontinuities. Thus,

$$p_2 = p_1, \tag{2.92}$$

$$v_2 = v_1, \tag{2.93}$$

and therefore

$$\zeta_2 \equiv p_2/v_2 = p_1/v_1 = \zeta_1 \tag{2.94}$$

As p, v and ζ all remain unchanged across a simple area discontinuity, it is not represented at all in an equivalent circuit. However, acoustically these area discontinuities are the very foundations of low-pass filters, as will be clear from the following energy analysis.

In terms of the complex amplitudes of the two progressive wave components of the standing waves in the two tubes across the junction, Equations 2.92 and 2.93 can be written as

$$A_2 + B_2 = A_1 + B_1 \tag{2.95}$$

and

$$(A_2 - B_2)/Y_2 = (A_1 - B_1)/Y_1. \tag{2.96}$$

For the special case of anechoic termination, $B_1 = 0$, and then

$$\zeta_2 = \zeta_1 = Y_1. \tag{2.97}$$

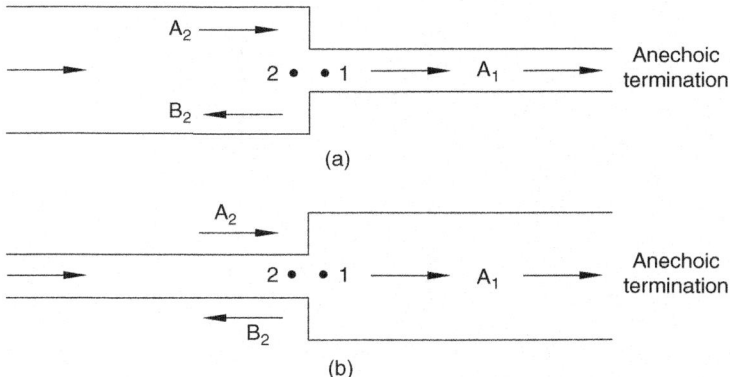

Figure 2.11 Simple area discontinuities. (a) Sudden contraction. (b) Sudden expansion

And the reflection coefficient is given by

$$R_2 \equiv \frac{B_2}{A_2} = \frac{\zeta_2 - Y_2}{\zeta_2 + Y} = \frac{Y_1 - Y_2}{Y_1 + Y_2} = \frac{S_2 - S_1}{S_2 + S_1}. \tag{2.98}$$

Thus, for a sudden contraction $(S_2 > S_1)$, R_2 lies between 0 and 1, and for a sudden expansion $(S_2 < S_1)$, R_2, lies between -1 and 0. Equation 2.13 for power flux yields

$$\text{Incident power:} \quad W_i = |A_2|^2/(2\rho_0 Y_2); \tag{2.99}$$

$$\text{Transmitted power:} \quad W_t = |A_1^2|/(2\rho_0 Y_1); \tag{2.100}$$

$$\text{Reflected power:} \quad W_r = |B_2^2|/(2\rho_0 Y_2); \tag{2.101}$$

The net energy flux in the upstream tube is

$$W_2 = W_i - W_r = \frac{|A_2^2| - |B_2^2|}{2\rho_0 Y_2}, \tag{2.102}$$

and that in the downstream tube (B_1 is assumed to be zero due to anechoic termination) is

$$W_1 = W_t = \frac{|A_1^2|}{2\rho_0 Y_1}. \tag{2.103}$$

Making use of Equations 2.95 and 2.96, it can readily be seen that across either area discontinuity,

$$W_2 = W_1. \tag{2.104}$$

This indicates that in the steady-state a simple area change does not result in any loss of power in the course of transmission, but it does reflect a substantial amount of the incident power back to the source by creating a mismatch of characteristic impedances [refer to Equation 2.98]. Thus, sudden area discontinuities are reflective elements and constitute the very basis of reflective mufflers. As these mufflers do not cause any dissipation of energy, these are also called nondissipative or reactive mufflers.

Incidentally, transmission loss for elements shown in Figure 2.11 can be calculated as follows, making use of the preceding relations:

$$\begin{aligned} TL &= 10 \log \frac{W_i}{W_t} \quad \text{(with anechoic termination)} \\ &= 10 \log \frac{W_i}{W_i - W_r} \\ &= 10 \log \frac{1}{1 - |R_2^2|} \end{aligned} \tag{2.105}$$

or

$$\text{TL} = 10 \log \frac{(S_2 + S_1)^2}{4 S_2 S_1}. \tag{2.106}$$

Thus, TL for a sudden contraction is the same as for a sudden expansion for a pair of tubes of different diameters.

2.13 Gradual Area Changes

Plane wave propagation along a tube of gradually varying area of cross-section is governed by the following equations:

$$\text{Mass continuity:} \quad \frac{\partial \rho}{\partial t} + \rho_0 \frac{\partial u}{\partial z} + \frac{u \rho_0}{S} \frac{dS}{dz} = 0; \quad (2.107)$$

$$\text{Momentum:} \quad \rho_0 \frac{\partial u}{\partial t} + \frac{\partial p}{\partial z} = 0; \quad (2.108)$$

$$\text{Isentropicity:} \quad dp = c_0^2 \, d\rho. \quad (2.109)$$

It is worth noting that only the equation of mass continuity is different from that of uniform area tube. Eliminating u and ρ from the foregoing equations, Equations 2.107–2.109 yield the wave equation

$$\frac{\partial^2 p}{\partial t^2} - c_0^2 \frac{\partial^2 p}{\partial z^2} - \frac{c_0^2}{S(z)} \frac{dS(z)}{dz} \cdot \frac{\partial p}{\partial z} = 0. \quad (2.110)$$

On writing $p(z,t) = p(z)e^{j\omega t}$ as usual, Equation 2.110 reduces to the Webster equation,

$$\frac{d^2 p(z)}{dz^2} + \frac{1}{S(z)} \frac{dS(z)}{dz} \frac{dp(z)}{dz} + k_0^2 p(z) = 0. \quad (2.111)$$

This equation has not yet been solved for a general solution. However, for certain specific shapes [or $S(z)$], this equation can be solved as follows.

2.13.1 Conical Tube

Figure 2.12a shows a conical tube with diameter (or effective diameter) being proportional to z, the distance from the hypothetical apex. Thus,

$$S(z) \propto z^2, \quad (2.112a)$$

so that

$$\frac{1}{S} \frac{dS}{dz} = \frac{2}{z}. \quad (2.112b)$$

With this substitution, Equation 2.111 becomes

$$\frac{d^2 p(z)}{dz^2} + \frac{2}{z} \frac{dp(z)}{dz} + k_0^2 p(z) = 0. \quad (2.113)$$

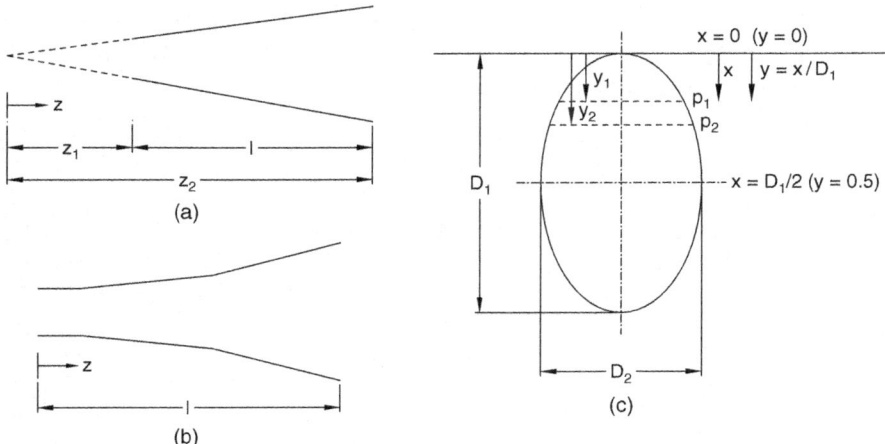

Figure 2.12 Three types of variable-area tubes. (a) A conical tube. (b) An exponential tube. (c) An elliptical cross-section having major diameter as D_1 and minor diameter as D_2

This is the same equation as characterizes spherical waves, with the area of the phase front increasing with the square of the radial distance. Thus, with the transformation $p = q/z$, Equation 2.113 reduces to the one-dimensional Helmholtz equation

$$\frac{d^2 q(z)}{dz^2} + k_0^2 q(z) = 0.$$

Therefore,

$$q(z) = C_1 e^{-jk_o z} + C_2 e^{+jk_o z},$$

and hence

$$p(z,t) = 1/z \{ C_1 e^{-jk_o z} + C_2 e^{+jk_o z} \} e^{j\omega t}. \tag{2.114}$$

Substituting this in the momentum Equation 2.108 yields the following expression for particle velocity:

$$u(z,t) = \frac{j}{\omega \rho_0 z} \left\{ \left(-jk_0 - \frac{1}{z}\right) C_1 e^{-jk_o z} + \left(jk_0 - \frac{1}{z}\right) C_2 e^{+jk_o z} \right\} e^{j\omega t}. \tag{2.115}$$

Now, acoustic mass velocity v can be calculated as follows:

$$v(z,t) = \rho_0 S(z) u(z,t) = \rho_0 S_2 \left(\frac{z}{z_2}\right)^2 u(z,t).$$

Substituting for $u(z,t)$ from Equation 2.115 gives

$$v(z,t) = \frac{z}{Y_2 z_2^2} \left\{ \left(1 - \frac{j}{k_0 z}\right) C_1 e^{-jk_o z} - \left(1 + \frac{j}{k_0 z}\right) C_2 e^{+jk_o z} \right\} e^{j\omega t}, \tag{2.116}$$

where Y_2 is the characteristic impedance at $z = z_2$; that is, $Y_2 = c_0/S_2$.

Alternatively, z may be measured from the left end of the conical duct, as shown in Figure 2.12b for an exponential tube. Then z in Equations 2.115 and 2.116 may be replaced with $z_1 + z$, where, referring to Figure 2.12a,

$$z_1 = \frac{d_0 l}{d_l - d_0}, \quad (2.117)$$

and d_0 and d_l are diameters (or equivalent diameters) of the conical duct at $z = 0$ and $z = l$. Then,

$$z_2 = z_1 + l = \frac{d_l l}{d_l - d_0}. \quad (2.118)$$

2.13.2 Exponential Tube

This tube is characterized by the flare relations

$$r(z) = r(0)e^{mz}, \; S(z) = S(0)e^{2mz} = S(l)e^{-2m(1-z)}. \quad (2.119)$$

Thus,

$$\frac{1}{S(z)} \frac{dS(z)}{dz} = 2m, \quad (2.120)$$

and Equation 2.111 becomes

$$\frac{d^2 p(z)}{dz^2} + 2m \frac{dp(z)}{dz} + k_0^2 p(z) = 0. \quad (2.121)$$

This ordinary differential equation with constant coefficients can be solved easily to obtain

$$p(z, t) = p(z)e^{j\omega t} = e^{-mz}\{C_1 e^{-jk'z} + C_2 e^{+jk'z}\}e^{j\omega t}. \quad (2.122)$$

where,

$$k' = (k_0^2 - m^2)^{1/2}. \quad (2.123)$$

Substituting Equation 2.122 in the momentum Equation 2.108, gives

$$u(z, t) = \frac{e^{-mz}}{\rho_0 c_0} \left\{ \frac{k' - jm}{k_0} C_1 e^{-jk'z} - \frac{k' + jm}{k_0} C_2 e^{+jk'z} \right\} e^{j\omega t}, \quad (2.125)$$

It is obvious that no wave propagation is possible (the acoustic signal would die down exponentially with z, without oscillation) if

$$k_0 \leq m \quad (2.126)$$

or if

$$f_0 \leq \frac{mc_0}{2\pi} \equiv f_c, \qquad (2.127)$$

where f_c is called the cut-off frequency because for frequencies lower than this no power will be transmitted down the horn (the impedance p/v at all positions along the horn being purely reactive for a progressive forward-moving wave) if a driver unit were placed at the throat, as is the practice with loudspeakers. But there are no such implications if a horn-type flare tube were used as one of the intermediate elements in an acoustic filter (to replace the sudden expansion and contraction, for example), because then the constants A and B would also be complex. Thus, flare tubes cannot be used in general to 'block off' wave transmission at any frequencies.

2.13.3 Elliptical Tube

The cross-sectional area $S(x)$ for propagation along the transverse direction in a short elliptical chamber is given as [11].

$$S(x) = (2D_2b/D_1)\left(\sqrt{D_1 x - x^2}\right) \Rightarrow S(y) = (2D_2 b)\left(\sqrt{y - y^2}\right), \quad y = x/D_1. \qquad (2.128)$$

Here D_1 and D_2 are the major axis and minor axis of the elliptical section and b is the breadth of the chamber (perpendicular to the plane of the paper). The coordinate x is being measured from the top of the elliptical chamber as depicted in Figure 2.12c, while y is the nondimensional counterpart of x. The region of interest where one looks for a solution of Equation 2.128 is $0 \leq x \leq D_1$, and in a nondimensional form, one gets $0 \leq y \leq 1$. By making use of Equation 2.128, one gets a nondimensional form of Equation 2.111 as:

$$\frac{d^2 p}{dy^2} + ((1/2 - y)/(y - y^2))\frac{dp}{dy} + \beta^2 p = 0, \beta = k_0 D_1. \qquad (2.129)$$

It is interesting to note that by replacing y with $1 - z$, such that $(0 \leq z \leq 1)$ in Equation 2.129, the differential equation so obtained has the same form as Equation 2.129.

This reveals the symmetry of the solution about $y = 0.5$ or equivalently about $x = D_1/2$. Thus, one need solve Equation 2.129 for the region $0 \leq y \leq 1/2$ only, as the solution for the other half is the same as that obtained by solving Equation 2.129. In fact, we exploit this symmetry property of the differential equation governing the transverse wave propagation in elliptic ducts to find the Frobenius solution of Equation 2.129, computing only as many terms as would give us a satisfactory convergence up to $y = 0.5$.

It is easily discernible from Equation 2.129 that $y = 0$ and 1 are both regular singular points. Thus one may seek a Frobenius solution about either of the points. We choose to find a Frobenius solution [11] about $y = 0$. Towards this end, we assume a series solution of the following form:

$$p(y) = a_0 y^m + a_1 y^{m+1} + a_2 y^{m+2} \ldots \Rightarrow p(y) = \sum_{i=0}^{\infty} a_i y^{m+i}. \qquad (2.130)$$

The index m in Equation 2.130 has to be determined before one finds the co-efficients a_1, $a_2 \ldots$ in terms of a_0. Thus, we substitute Equation 2.130 in Equation 2.129 to get the following:

$$(-(1/2)m + m^2)a_0 y^{m-1} + ((m^2 + (1/2) + (3/2)m) a_1 - m^2 a_0) y^m \\ + (((7/2)m + 3 + m^2)a_2 - (m^2 + 2m + 1)a_1 + \beta^2 a_0) y^{m+1} + \ldots = 0. \quad (2.131)$$

First we equate the co-efficient of y^{m-1} to zero to obtain:

$$m = 0, \quad 1/2. \quad (2.132)$$

From the theory of linear differential equations with variable co-efficients, we know that if the values of the roots m_1 and m_2 of the indicial equation are distinct and do not differ by an integer, we would get two linearly independent solutions corresponding to each value of m.

Thus, in general, one can express the acoustic pressure field $p(y)$ as a linear combination of two solutions so obtained as:

$$p(y) = A F_1(y) + B F_2(y), \quad (2.133)$$

$$F_1(y, \beta) = P_0(\beta) + P_1(\beta) y + P_2(\beta) y^2 + P_3(\beta) y^3 + \ldots, \quad (2.134)$$

$$F_2(y, \beta) = Q_0(\beta) y^{1/2} + Q_1(\beta) y^{3/2} + Q_2(\beta) y^{5/2} + Q_3(\beta) y^{7/2} + \ldots. \quad (2.135)$$

In Equation 2.133, A and B are the two arbitrary constants which can be determined by the application of suitable boundary conditions. The functions $F_1(y)$ and $F_2(y)$ are the linearly independent solutions corresponding to each value of m; that is $m = 0$ and $1/2$, respectively. They comprise of polynomials $P_0(\beta)$, $P_1(\beta)$, $\ldots$ and $Q_0(\beta)$, $Q_1(\beta)$, $\ldots$ in β with the order of polynomials increasing as we go to the higher powers of y. Here we present the first 11 terms of the functions $F_1(y)$ and $F_2(y)$ in Equation 2.133 F or $m = 0$, one gets:

$$P_0(\beta) = 1,$$
$$P_1(\beta) = 0,$$
$$P_2(\beta) = -\beta^2/3,$$
$$P_3(\beta) = -2\beta^2/45,$$
$$P_4(\beta) = -\beta^2/35 + \beta^4/42,$$
$$P_5(\beta) = -32\beta^2/1575 + 5\beta^4/1222, \quad (2.136)$$
$$P_6(\beta) = -32\beta^2/2079 + 7\beta^4/2673 - \beta^6/1386,$$
$$P_7(\beta) = -53\beta^2/4352 + 15\beta^4/7934 - 4\beta^6/29091,$$
$$P_8(\beta) = -64\beta^2/6435 + 23\beta^4/15733 - \beta^6/11397 + \beta^8/83160,$$
$$P_9(\beta) = -27\beta^2/3245 + 11\beta^4/9314 - 13\beta^6/203486 + \beta^8/412178,$$
$$P_{10}(\beta) = -24\beta^2/3383 + 5\beta^4/5084 - 2\beta^6/40033 + \beta^8/647316 - \beta^{10}/7900200.$$

Similarly, for $m = 1/2$, one gets the following polynomial expressions:

$Q_0(\beta) = 1,$
$Q_1(\beta) = 1/6,$
$Q_2(\beta) = 3/40 - \beta^2/5,$
$Q_3(\beta) = 5/112 - 5\beta^2/126,$
$Q_4(\beta) = 35/1152 - 71\beta^2/3240 + \beta^4/90,$
$Q_5(\beta) = 63/2816 - 44\beta^2/2927 + 13\beta^4/5527,$
$Q_6(\beta) = 94/5417 - 24\beta^2/2125 + 12\beta^4/8767 - \beta^6/3510,$
$Q_7(\beta) = 23/1647 - 10\beta^2/1119 + 9\beta^4/9274 - 4\beta^6/64063,$
$Q_8(\beta) = 30/2597 - 25\beta^2/3416 + 19\beta^4/25408 - \beta^6/26890 + \beta^8/238680,$
$Q_9(\beta) = 52/5327 - 27\beta^2/4394 + 7\beta^4/11583 - \beta^6/37359 + \beta^8/1065491,$
$Q_{10}(\beta) = 92/10965 - 6\beta^2/1141 + \beta^4/1984 - \beta^6/47878 + \beta^8/1766099 - \beta^{10}/25061400.$
(2.137)

2.14 Extended-Tube Resonators

Four types of extended-tube resonators are shown in Figure 2.13. Assuming that the plane wave condition is fulfilled by all the tubes, it can be observed that acoustic pressures at the junction would be equal, that is,

$$p_3 = p_1 = p_2, \quad (2.138)$$

and for the indicated positive directions, the continuity of acoustic mass flux would yield

$$v_3 = v_1 + v_2, \quad (2.139)$$

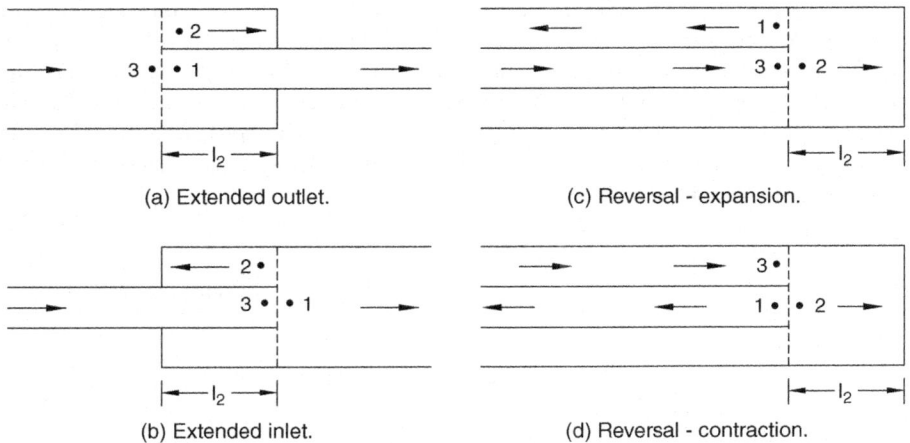

(a) Extended outlet. (c) Reversal - expansion.

(b) Extended inlet. (d) Reversal - contraction.

Figure 2.13 Extended-tube resonators

Further, the resonator cavity can be represented at the junction by an equivalent impedance

$$Z_2 = p_2/v_2, \quad (2.140)$$

where Z_2 is given by Equations 2.24 or 2.26, that is,

$$Z_2 = Y_2 \frac{\zeta_{end} \cos k_0 l_2 + jY_2 \sin k_0 l_2}{j\zeta_{end} \sin k_0 l_2 + Y_2 \cos k_0 l_2},$$

$$\underset{\zeta_{end} \to \infty}{Lt} Z_2 = -jY_2 \cot k_0 l_2. \quad (2.141)$$

Alternatively,

$$Z_2 = Y_2 \frac{1 + R_{end} e^{-2jk_0 l_2}}{1 - R_{end} e^{-2jk_0 l_2}} \quad (2.142)$$

$$\underset{R_{end} \to 1}{Lt} Z_2 = -jY_2 \cot k_0 l_2. \quad (2.143)$$

where ζ_{end} is the normal impedance and R_{end} the reflection coefficient of the end plate Equations 2.138–2.140 may be rearranged as

$$v_3/p_3 = v_1/p_1 + v_2/p_2$$

or as

$$1/\zeta_3 = 1/\zeta_1 + 1/Z_2. \quad (2.144)$$

Equations 2.138, 2.139 and 2.144 indicate that an extended-tube resonator would be represented as a branch (shunt) element, as shown in Figure 2.14, in the equivalent circuit [4].

At certain frequencies, Z_2 would tend to zero, the branch element would act as a short circuit, and no acoustic power would be transmitted to the downstream side (tube 1). All the incoming power flux would thus seem to be used to resonate the closed-end cavity, giving it the name branch resonator. However, in fact, an unlined (nonabsorptive) rigid-wall resonator acts primarily as an impedance mismatch element, ensuring that hardly any power leaves the source at the resonance frequencies.

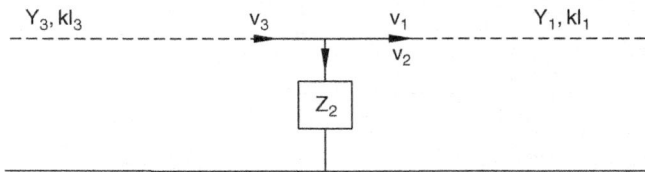

Figure 2.14 A typical branch resonator

If the end walls of the resonator cavity are not rigid enough, these would be set into strong vibration at resonance frequencies and would radiate sound either downstream into the filter or directly to the atmosphere, as the case may be. In the first case, the end plates should be made rigid enough, and in the second (when they are exposed to the atmosphere), these should be lined with acoustically absorptive material.

For rigid end plate, resonance would occur when $\cot k_0 l_2 \to 0$ or when

$$k_0 l_2 = (2n+1)\pi/2, \qquad n = 0, 1, 2, \ldots \tag{2.145}$$

or when

$$l_2 = (2n+1)\lambda/4, \qquad n = 0, 1, 2, \ldots \tag{2.146}$$

Therefore, extended-tube resonators are also called quarter-wave resonators.

In actual practice, however, the end wall of the tubular cavity is not rigid enough to have a reflection coefficient of unity. The actual value of R_{end} is generally around 0.95 [12].

2.15 Helmholtz Resonator

A branch resonator of the type shown in Figure 2.15 is known as Helmholtz resonator. It consists of a small branch tube (called a neck) and a cavity: the former represents a lumped inertance and the latter, a lumped compliance. The equivalent circuit representation is the same as in Figure 2.14. The branch impedance Z_2 consists of the neck inertance and cavity compliance in series.

Adding the radiation impedance on either side of the neck tube gives

$$Z_2 = j\omega \frac{l_{eq}}{S_n} + \frac{1}{j\omega V_c/c_0^2} + 2Y_n \frac{k_0^2 r_n^2}{2}, \tag{2.147}$$

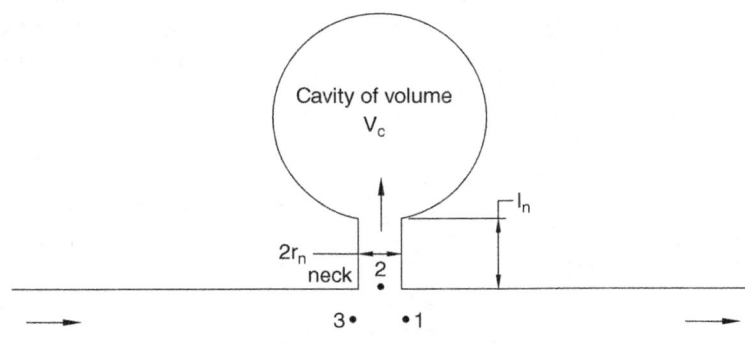

Figure 2.15 Helmholtz resonator

where as per Equation 2.58,

$$l_{eq} = l_n + t_w + 1.7r_n \tag{2.148}$$

t_w = thickness of the wall of the propagation tube,
r_n = radius of the neck,
$Y_n = c_0/S_n$, and
$S_n = \pi r_n^2$.

Equation 2.147 can be rewritten in the form

$$Z_2 = j\left\{\omega \frac{l_{eq}}{S_n} - \frac{c_0^2}{\omega V_c}\right\} + \frac{\omega^2}{\pi c_0}. \tag{2.149}$$

On neglecting the radiation resistance term, the branch impedance Z_2 would tend to zero when

$$\omega = c_0 \left(\frac{S_n}{l_{eq} V_c}\right)^{1/2}. \tag{2.150}$$

This is the resonance frequency (in radians per second) of the resonator at which it would yield very high transmission loss, the amplitude of it being limited only by the radiation resistance term in Z_2. For this resonator also, p_3 and v_3 are related to p_1 and v_1 as per Equations 2.138 and 2.139.

2.16 Concentric Hole-Cavity Resonator

Such a resonator, as shown in Figure 2.16, consists of an annular tubular cavity communicating with the center tube (propagating tube) through a number of holes on its periphery. Like extended-tube resonators and the Helmholtz resonator, the concentric hole-cavity resonator is also represented in the equivalent circuit by a shunt impedance as shown in Figure 2.14, and the upstream state variables, p_3 and v_3 are related to the downstream variables to p_1 and v_1 by the same equations, namely, (2.138) and (2.139), which in fact characterize any branch resonator.

The difference, of course, lies in the expression for branch impedance Z_2. This would be similar to expression (2.147) barring the fact that the annular cavity is acoustically long ($k_0 l_a$ and $k_0 l_b$ are not $\ll 1$); it is made of two quarter-wave resonators in parallel and the neck

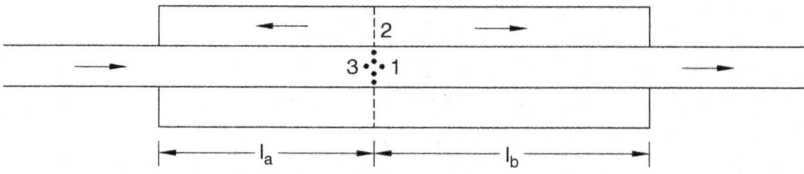

Figure 2.16 Concentric hole-cavity resonator

Theory of Acoustic Filters

length is equal to the thickness of the communicating hole. Thus, for rigid end plates [4,8],

$$Z_2 = \frac{1}{n_h}\left\{j\omega\frac{l_{eq}}{S_h} + \frac{\omega^2}{\pi c_0}\right\} + \frac{(-jY_c\cot k_0 l_a)(-jY_c\cot k_0 l_b)}{(-jY_c\cot k_0 l_a)+(-jY_c\cot k_0 l_b)}$$

$$= \frac{1}{n_h}\left\{j\omega\frac{l_{eq}}{S_h} + \frac{\omega^2}{\pi c_0}\right\} - jY_c\frac{1}{\tan k_0 l_a + \tan k_0 l_b},$$

(2.151)

where subscripts h and c stand for hole and cavity, respectively, and

n_h = number of holes in a row,
$S_h = \pi d_h^2/4$,
$Y_c = c_0/S_c$.
$l_{eq} = t_w + 0.85 d_h$, and
t_w = wall thickness.

Of course, expression (2.151) reduces to (2.149) in the lower frequency range when $\tan k_0 l_a \simeq k_0 l_a$, $\tan k_0 l_b \simeq k_0 l_b$. Thus the Helmholtz resonator of Figure 2.15 is only the lumped-element approximation of the concentric hole-cavity resonator of Figure 2.16. The resonance frequency of the latter, like that of the former, corresponds to the state at which the impedance due to the inertance of the holes is equal and opposite to the impedance due to the compliance of the annular cavity. From Equation 2.151, the frequency (called the tuned frequency) can be seen to be given by the transcendental equation

$$\frac{1}{n_h}\frac{\omega l_{eq}}{S_h} = \frac{Y_c}{\tan k_0 l_a + \tan k_0 l_b}.$$

(2.152)

Incidentally, n_h the number of holes in a circumferential row, appears in the denominator, as impedance of n_h holes would be $1/n_h$ times that of a single hole. This feature lends considerable flexibility to the design of a concentric hole-cavity resonator.

2.17 An Illustration of the Classical Method of Filter Evaluation

The classical method of filter evaluation consists in writing down the relations connecting the two ends of each of the tubular elements and the various element junctions, and solving the same simultaneously to evaluate the insertion loss or transmission loss or level difference across the filter.

Figure 2.17 shows a line diagram of a typical straight-through low-pass acoustic filter consisting of nine elements ($n = 9$). Insertion loss for this particular filter is given by Equation 2.70, with $n = 9$, Z_{10} being the internal impedance of the source (assumed to be known a priori).

On assuming that the area of cross-section of the exhaust pipe (element 9) is equal to that of the tail pipe (element 1), as is generally the case, and noting that medium density $\rho_{0,2}$ would be equal to $\rho_{0,1}$ (in the absence of any hot exhaust gases flowing through the filter), one gets from Equation 2.42

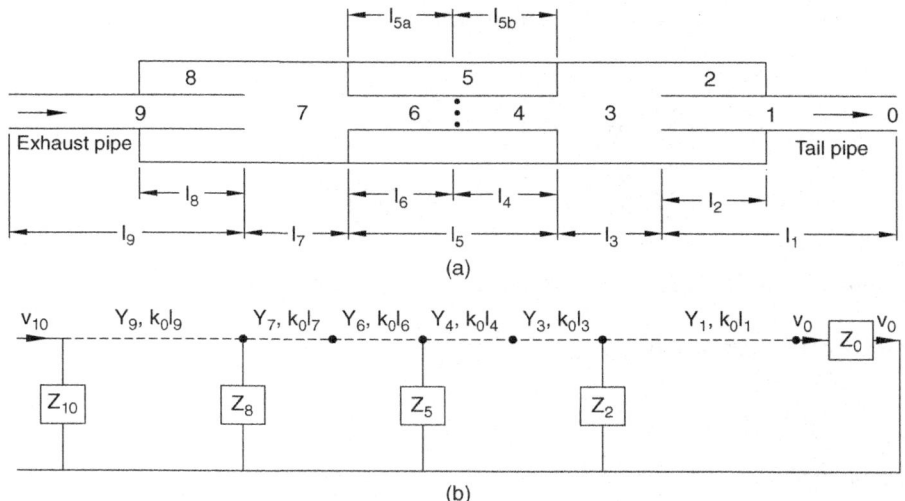

Figure 2.17 A typical acoustic filter and its equivalent circuit. (a) Line diagram of a filter. (b) The equivalent circuit of the filter

$$Z_{0,1} = Z_{0,2} = Z_0 = R_0 + jX_0 = Y_1 \left(\frac{k_0^2 r_1^2}{4} + j0.6k_0 r_1 \right)$$

and

$$\frac{\rho_{0,2} R_{0,1}}{\rho_{0,1} R_{0,2}} = 1,$$

where
$Y_1 = c_0/S_1$,
$S_1 = \pi r_1^2$ (for a round tube), and
$k_0 = \omega/c_0$.

Equation 2.70 for insertion loss becomes

$$IL(\omega) = 20 \log \left| \frac{Z_{10}}{Z_{10} + Z_0} \frac{v_{10}}{v_0} \right|,$$

where v_{10} and v_0 are as shown in Figure 2.17b.

The velocity ratio v_{10}/v_0 can be calculated by making use of the governing equations for the various constituent elements, as follows [4]. The direction of wave propagation (the positive direction of z) is from the source to the radiation end (from element 10 to element 0). Let A_i and B_i denote complex amplitudes of the forward-moving and backward-moving progressive components of the standing waves in the ith tubular element. On making use of

Theory of Acoustic Filters

relations (2.16), (2.17), and (2.20) for a tubular element; (2.92) and (2.93) for sudden expansion and sudden contractions; (2.138), (2.139) and (2.144) for branch elements, namely, extended inlet, extended outlet, hole-cavity resonator, and the source impedance; and (2.141) and (2.151) for resonator impedances, one gets

$$v_{10} = v_9 + \frac{p_9}{Z_{10}}$$
$$= \frac{A_9 - B_9}{Y_9} + \frac{A_9 + B_9}{Z_{10}}, \quad \text{(i)}$$

$$A_9 e^{-jk_0 l_9} + B_9 e^{+jk_0 l_9} = A_7 + B_7, \quad \text{(ii)}$$

$$\frac{1}{Y_9} \frac{A_9 e^{-jk_0 l_9} - B_9 e^{+jk_0 l_9}}{A_9 e^{-jk_0 l_9} + B_9 e^{+jk_0 l_9}} = \frac{1}{Y_7} \frac{A_7 - B_7}{A_7 + B_7} + \frac{1}{Z_8}, \quad \text{(iii)}$$

$$A_7 e^{-jk_0 l_7} + B_7 e^{+jk_0 l_7} = A_6 + B_6, \quad \text{(iv)}$$

$$\frac{A_7 e^{-jk_0 l_7} - B_7 e^{+jk_0 l_7}}{Y_7} = \frac{A_6 - B_6}{Y_6}, \quad \text{(v)}$$

$$A_6 e^{-jk_0 l_6} + B_6 e^{+jk_0 l_6} = A_4 + B_4, \quad \text{(vi)}$$

$$\frac{1}{Y_6} \frac{A_6 e^{-jk_0 l_6} - B_6 e^{+jk_0 l_6}}{A_6 e^{-jk_0 l_6} + B_6 e^{+jk_0 l_6}} = \frac{1}{Y_4} \frac{A_4 - B_4}{A_4 + B_4} + \frac{1}{Z_5}, \quad \text{(vii)}$$

$$A_4 e^{-jk_0 l_4} + B_4 e^{+jk_0 l_4} = A_3 + B_3, \quad \text{(viii)}$$

$$\frac{A_4 e^{-jk_0 l_4} - B_4 e^{+jk_0 l_4}}{Y_4} = \frac{A_3 - B_3}{Y_3}, \quad \text{(ix)}$$

$$A_3 e^{-jk_0 l_3} + B_3 e^{+jk_0 l_3} = A_1 + B_1, \quad \text{(x)}$$

$$\frac{1}{Y_3} \frac{A_3 e^{-jk_0 l_3} - B_3 e^{+jk_0 l_3}}{A_3 e^{-jk_0 l_3} + B_3 e^{+jk_0 l_3}} = \frac{1}{Y_1} \frac{A_1 - B_1}{A_1 + B_1} + \frac{1}{Z_2}, \quad \text{(xi)}$$

$$Y_1 \frac{A_1 e^{-jk_0 l_1} + B_1 e^{+jk_0 l_1}}{A_3 e^{-jk_0 l_1} - B_1 e^{+jk_0 l_1}} = Z_0, \quad \text{(xii)}$$

$$v_0 = \frac{A_1 e^{-jk_0 l_1} - B_1 e^{jk_0 l_1}}{Y_1} \quad \text{(xiii)}$$

where, for rigid end plates,

$$Z_2 = -jY_2 \cot k_o l_2,$$

$$Z_5 = \frac{1}{n_h}\left\{j\omega\frac{l_{eq}}{S_h} + \frac{\omega^2}{\pi c_0}\right\} - jY_5\frac{1}{\tan k_o l_{5a} + \tan k_o l_{5b}},$$

$$Z_8 = -jY_8 \cot k_o l_8.$$

There are 11 equations [Equations ii to xii] for 12 variables $(A_1, B_1, A_3, B_3, A_4, B_4, A_6, B_6, A_7, B_7, A_9$ and $B_9)$. These equations can be rearranged and solved in terms of any one of these variables, say, A_1. Then Equations i and xiii yield, respectively, v_{10} and v_0 in terms of A_1. Finally, in evaluating the velocity ratio v_{10} and v_0, the unknown A_1 cancels out. For numerical analysis, A_1, can be set to arbitrary number like unity. Then one gets 11 linear algebraic inhomogeneous equations with complex coefficients for simultaneous solution by one of the standard subroutines on a computer. In fact, the coefficient matrix turns out to be tridiagonal because of sequential relationships.

It is worth noting that in the preceding way of writing equations, the classical method of filter evaluation would in general involve simultaneous solution of $2n_t - 1$ equations, where n_t is the number of tubular elements. Of course, if there is more than one lumped element between any two consecutive tubular elements, these have to be replaced by an equivalent lumped impedance in the equivalent circuit [4].

2.18 The Transfer Matrix Method

The velocity ratio VR_{n+1} in Equation 2.70, the only term that depends upon the filter composition, can also be evaluated – and much more readily – through the use of transfer matrices for the different constituents elements [5]. The transfer matrix (also called transmission matrix or four-pole parameter representation) has been known for a long time [13,14]. It can be observed from Figure 2.17 that all the elements of a straight-through low-pass filter can be represented by one of the three types of elements shown in Figure 2.18.

2.18.1 Definition of Transfer Matrix

Adopting acoustic pressure p and mass velocity v as the two state variables, the following matrix relation can be written so as to relate state variables on the two sides of the element subscripted r in the equivalent circuit (see Figure 2.19):

$$\begin{bmatrix} p_r \\ v_r \end{bmatrix} = \begin{bmatrix} \text{A } 2 \times 2 \text{ transfer} \\ \text{matrix for the } r^{\text{th}} \text{ element} \end{bmatrix} \begin{bmatrix} p_{r-1} \\ v_{r-1} \end{bmatrix} \qquad (2.153)$$

$[p_r\ v_r]$ is called the state vector at the upstream point r, and $[p_{r-1}\ v_{r-1}]$ is called the state vector at the downstream point $r - 1$. The transfer matrix for the rth element can be denoted by $[T_r]$.

Writing this matrix as

$$\begin{bmatrix} A_{11} & A_{12} \\ A_{21} & A_{22} \end{bmatrix}, \qquad (2.154)$$

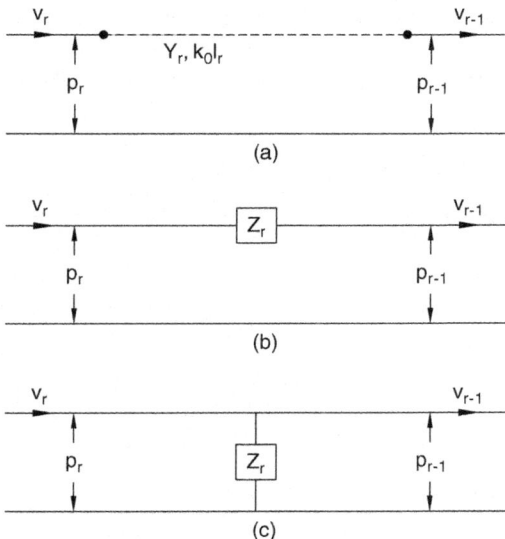

Figure 2.18 The three basic types of elements in an equivalent circuit. (a) A distributed element. (b) An in-line lumped element. (c) A shunt lumped element

It may be seen readily from the transfer matrix relation (2.153) that

$$A_{11} = \frac{p_r}{p_{r-1}}\bigg|_{v_{r-1}=0}, \quad A_{12} = \frac{p_r}{v_{r-1}}\bigg|_{p_{r-1}=0}, \quad A_{21} = \frac{v_r}{p_{r-1}}\bigg|_{v_{r-1}=0}, \quad A_{22} = \frac{v_r}{v_{r-1}}\bigg|_{p_{r-1}=0}. \quad (2.155)$$

These relations for the four-pole parameters $A_{11}, A_{12}, A_{21},$ and A_2 lend to each of them individual physical significance. For example A_{11} is the ratio of the upstream pressure and downstream pressure for the hypothetical case of the downstream end being rigidly fixed ($\zeta_{r-1} \to \infty$), and A_{12} is the ratio of the upstream pressure to the velocity at the downstream end for the hypothetical case of the downstream end being totally free or unconstrained ($\zeta_{r-1} \to 0$). Here ζ_{r-1} denotes the equivalent acoustic impedance at point $r - 1$, looking downstream, it being assumed that there is no source between points 0 and r. Symbolically, $\zeta_r = p_r/v_r$.

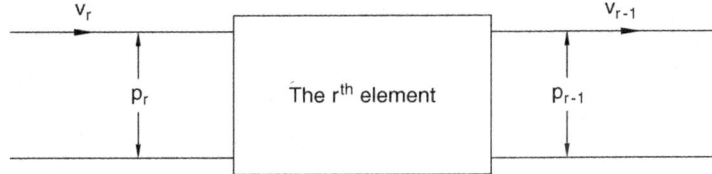

Figure 2.19 General representation of an element for transfer matrix

The preceding expressions for the four-pole parameters enable one to measure each of these parameters for an element or subsystem separately in the laboratory (see Chapter 5), or calculate numerically (by finite element method; see Chapter 7) where exact solutions are not available.

2.18.2 Transfer Matrix of a Uniform Tube

Upon making use of the standing wave relations (2.16) and (2.17), one gets

$$\begin{aligned}
p_r &= A_r + B_r; \\
v_r &= (A_r - B_r)/Y_r; \\
p_{r-1} &= A_r e^{-jk_0 l_r} + B_r e^{+jk_0 l_r} \\
&= (A_r + B_r)\cos k_0 l_r - j(A_r - B_r)\sin k_0 l_r \\
&= p_r \cos k_0 l_r - jY_r v_r \sin k_0 l_r;
\end{aligned} \qquad (2.156)$$

$$\begin{aligned}
v_{r-1} &= \{A_r e^{-jk_0 l_r} - B_r e^{+jk_0 l_r}\}/Y_r \\
&= \frac{A_r - B_r}{Y_r}\cos k_0 l_r - j\left(\frac{A_r + B_r}{Y_r}\right)\sin k_0 l_r \\
&= v_r \cos k_0 l_r - j\frac{p_r}{Y_r}\sin k_0 l_r.
\end{aligned} \qquad (2.157)$$

Equations 2.156 and 2.157 can be written in the matrix form

$$\begin{bmatrix} p_{r-1} \\ v_{r-1} \end{bmatrix} = \begin{bmatrix} \cos k_0 l_r & -jY_r \sin k_0 l_r \\ -j/Y_r \sin k_0 l_r & \cos k_0 l_r \end{bmatrix} \begin{bmatrix} p_r \\ v_r \end{bmatrix}, \qquad (2.158)$$

which can be inverted to obtain the desired transfer matrix relation

$$\begin{bmatrix} p_r \\ v_r \end{bmatrix} = \begin{bmatrix} \cos k_0 l_r & jY_r \sin k_0 l_r \\ j/Y_r \sin k_0 l_r & \cos k_0 l_r \end{bmatrix} \begin{bmatrix} p_{r-1} \\ v_{r-1} \end{bmatrix} \qquad (2.159)$$

for the distributed element of Figure 2.18a.

2.18.3 A General Method for Derivation of Transfer Matrix

A general method for deriving the transfer matrix from the basic equations is as follows. In general, pressure $p(z)$ and velocity $v(z)$ may be arranged in the matrix form

$$\begin{bmatrix} p(z) \\ v(z) \end{bmatrix} = \begin{bmatrix} G_{11}(z) & G_{12}(z) \\ G_{21}(z) & G_{22}(z) \end{bmatrix} \begin{bmatrix} C_1 \\ C_2 \end{bmatrix} \qquad (2.160)$$

or

$$\{S(z)\} = [G(z)]\{C\}, \qquad (2.161)$$

where C_1 and C_2 are arbitrary constants (see, for example Equations 2.114 and 2.116.

The state vectors $S(0)$ and $S(l)$ may now be written as follows:

$$\{S(0)\} = [G(0)]\{C\}; \tag{2.162}$$

$$\{S(l)\} = [G(l)]\{C\}. \tag{2.163}$$

Eliminating the constant vector $\{C\}$ from Equations 2.162 and 2.163 yields

$$\{S(0)\} = [G(0)][G(l)]^{-1}\{S(l)\}. \tag{2.164}$$

Hence, the transfer matrix is given by the general relation

$$[T] = [G(0)][G(l)]^{-1} \tag{2.165}$$

Formula (2.165) proves to be handy where matrix [G] is not very simple, as would happen for variable area ducts, perforated elements, and so on, discussed later in this chapter and the next chapter.

2.18.4 Transfer Matrices of Lumped Elements

For an in-line lumped element shown in Figure 2.18b, one can write directly (by observation) the relations

$$p_r = p_{r-1} + Z_r \cdot v_{r-1}, \tag{2.166}$$

$$v_r = v_{r-1}, \tag{2.167}$$

which yield the desired transfer matrix relation

$$\begin{bmatrix} p_r \\ v_r \end{bmatrix} = \begin{bmatrix} 1 & Z_r \\ 0 & 1 \end{bmatrix} \begin{bmatrix} p_{r-1} \\ v_{r-1} \end{bmatrix}. \tag{2.168}$$

For a branch lumped element shown in Figure 2.18c, one can observe that

$$p_r = p_{r-1}, \tag{2.169}$$

$$v_r = p_{r-1}/Z_r + v_{r-1}. \tag{2.170}$$

These equations yield the desired transfer matrix relation

$$\begin{bmatrix} p_r \\ v_r \end{bmatrix} = \begin{bmatrix} 1 & 0 \\ 1/Z_r & 1 \end{bmatrix} \begin{bmatrix} p_{r-1} \\ v_{r-1} \end{bmatrix}. \tag{2.171}$$

Equations 2.159, 2.168, and 2.171 yield the following transfer matrices:

Distributed element (uniform tube) (Figure 2.18a)

$$\begin{bmatrix} \cos k_0 l_r & (jY_r)\sin k_0 l_r \\ j/jY_r \sin k_0 l_r & \cos k_0 l_r \end{bmatrix}; \tag{2.172}$$

Lumped in-line element (Figure 2.18b)

$$\begin{bmatrix} 1 & Z_r \\ 0 & 1 \end{bmatrix}; \qquad (2.173)$$

Lumped shunt element (Figure 2.18c)

$$\begin{bmatrix} 1 & 0 \\ 1/Z_r & 1 \end{bmatrix}. \qquad (2.174)$$

As both p and v remain unchanged across sudden expansion and sudden contraction, the transfer matrices for these elements are unity matrices.

2.18.5 Transfer Matrices of Variable Area Tubes

One element that does not fall in any of the preceding three basic elements is the flare tube. It is used rarely in exhaust mufflers or air ducts, but is invariably there in the intakes and nacelles of turbofan engines. The transfer matrix for such a tube can be derived in the same manner as for a uniform tube. Thus, for a conical tube (Figure 2.12a), Equations 2.114 and 2.116 yield the transfer matrix relation.

$$\begin{bmatrix} p(z_1) \\ v(z_1) \end{bmatrix} = \begin{bmatrix} \dfrac{z_2}{z_1}\cos k_0 l - \dfrac{1}{k_0 z_1}\sin k_0 l & jY_2\dfrac{z_2}{z_2}\sin k_0 l \\ \left\{\dfrac{j}{Y_2}\dfrac{z_1}{z_2}\left(1+\dfrac{1}{k_0^2 z_1 z_2}\sin k_0 l\right)\right. & \left\{\dfrac{1}{k_0 z_2}\sin k_0 l + \dfrac{z_1}{z_2}\cos k_0 l\right\} \\ \left. -\dfrac{j}{k_0 z_2 Y_2}\left(1-\dfrac{z_1}{z_2}\right)\cos k_0 l\right\} & \end{bmatrix} \begin{bmatrix} p(z_2) \\ v(z_2) \end{bmatrix} \qquad (2.175)$$

This transfer matrix may also be obtained by means of Equation 2.165. However, application of the formula (2.165) requires z to be reckoned from the beginning of the element (Figure 2.12b), not from the hypothetical apex (Figure 2.12a). Therefore, Equation 2.165 may be rewritten as

$$[T] = [G(z_1)][G(z_2)]^{-1} \qquad (2.176)$$

where $G(z)$, defined by Equation 2.160, may be obtained from Equations 2.114 and 2.116. Thus,

$$G_{11}(z) = \frac{1}{z}e^{-jk_0 z}, \quad G_{12}(z) = \frac{1}{z}e^{jk_0 z},$$

$$G_{21}(z) = \frac{z}{Y_2 z_2^2}\left\{1 - \frac{j}{k_0 z}\right\}e^{-jk_0 z}, \quad G_{22}(z) = -\frac{z}{Y_2 z_2^2}\left\{1 + \frac{j}{k_0 z}\right\}e^{jk_0 z},$$

Substituting matrix $[G(z)]$ from Equation 2.176 in the formula (2.165), and noting that

$$Y_2 = Y(l) = Y(0)\frac{S(0)}{S(l)} = Y(0)\left(\frac{z_1}{z_2}\right)^2 = Y(0)\left(\frac{z_1}{z_1+l}\right)^2 \qquad (2.177)$$

yields (after considerable algebra) Equation 2.175, where $z_2 = z_1 + l$ and z_1 is given by Equation 2.117. Incidentally, Equation 2.176 is particularly suitable for computation.

The transfer matrix of the conical tube of Figure 2.12a may also be written in terms of diameters d_0, d_l and slope m, defined as

$$m = \frac{d_l - d_0}{l}, \qquad (2.178)$$

where d_0 and d_l are diameters of the conical tube at $z = 0$ and $z = l$.

Noting that

$$z_1 = \frac{d_0}{m}, z_2 = \frac{d_l}{m}, Y_2 = Y(l) = Y(0)\left(\frac{d_0}{d_l}\right)^2, \qquad (2.179)$$

Equation 2.175 becomes

$$\begin{bmatrix} p(0) \\ v(0) \end{bmatrix} = \begin{bmatrix} \dfrac{d_l}{d_0}\cos(k_0 l) - \dfrac{m}{k_0 d_0}\sin(k_0 l) & jY(0)\dfrac{d_0}{d_l}\sin(k_0 l) \\ \left\{\dfrac{j}{Y(0)}\dfrac{d_l}{d_0}\left(1 + \dfrac{m^2}{k_0^2 d_0 d_l}\right)\sin(k_0 l) \right. & \left\{\dfrac{m}{k_0 d_l}\sin(k_0 l) + \dfrac{d_0}{d_l}\cos(k_0 l)\right\} \\ \left. -\dfrac{jm}{k_0 d_0 Y(0)}\dfrac{d_l}{d_0}\left(1 - \dfrac{d_0}{d_l}\right)\cos(k_0 l)\right\} & \end{bmatrix} \begin{bmatrix} p(l) \\ v(l) \end{bmatrix}. \qquad (2.180)$$

Similarly, for exponential tube (Figure 2.12b), Equations 2.122 and 2.125 lead to the transfer matrix relation

$$\begin{bmatrix} p(0) \\ v(0) \end{bmatrix} = \begin{bmatrix} e^{ml}\left(\cos k'l - \dfrac{m}{k'}\sin k'l\right) & je^{-ml}\dfrac{k_0}{k'}Y(0)\sin k'l \\ \dfrac{j}{Y(0)}e^{ml}\dfrac{k_0}{k'}\sin k'l & e^{-ml}\left(\cos k'l + \dfrac{m}{k'}\sin k'l\right) \end{bmatrix} \begin{bmatrix} p(l) \\ v(l) \end{bmatrix} \qquad (2.181)$$

This can also be written in terms of $Y(l)$ by making use of the relations

$$S(l) = S(0)e^{2ml} \text{ or } Y(0) = Y(l)e^{2ml}. \qquad (2.182)$$

$Y(l)$ corresponds to Y_2 for a conical duct. In this form, the transfer matrix relations (2.180) and (2.181) are identical to those derived by Lung and Doige [17].

Third type of flare tube is an elliptical duct shown in Figure 2.12c. Based upon the Frobenius solution of Equation 2.129, we derive the transfer matrix relation between the upstream point (P_1) and the downstream point (P_2). The coordinates of these sections are y_1 and y_2 respectively, measured from the top of the elliptical section, as shown in Figure 2.12c.

We first derive the acoustic particle velocity $u(y)$ from Equations 2.133 making use of the momentum equation. Thus,

$$\rho_0 c_0 u(y) = -(1/j\beta)\bigl(AF'_1(y) + BF'_2(y)\bigr), \qquad (2.183)$$

where

$$F'_1(y) = \frac{dF_1(y)}{dy}, \; F'_2(y) = \frac{dF_2(y)}{dy}. \qquad (2.184)$$

Making use of Equations 2.133, 2.183 and 2.165, one gets the transfer matrix between section 1 and section 2 (as shown in Figure 2.12c of the elliptical section) as [11].

$$\left\{ \begin{array}{c} p(y_1) \\ \rho_0 c_0 u(y_1) \end{array} \right\} = [R_{y_1 y_2}]_{2 \times 2} \left\{ \begin{array}{c} p(y_2) \\ \rho_0 c_0 u(y_2) \end{array} \right\}, \qquad (2.185)$$

where

$$[R_{y_1 y_2}] = \frac{1}{\kappa} \begin{bmatrix} F_1(y_1)F'_2(y_2) - F_2(y_1)F'_1(y_2) & \left(\dfrac{\beta}{j}\right)\bigl(F_1(y_2)F_2(y_1) - F_1(y_1)F_2(y_2)\bigr) \\ -\left(\dfrac{j}{\beta}\right)\bigl(F'_2(y_1)F'_1(y_2) - F'_2(y_2)F'_1(y_1)\bigr) & F_1(y_2)F'_2(y_1) - F_2(y_2)F'_1(y_1) \end{bmatrix}$$

$$(2.186)$$

$$\kappa = F_1(y_2)F'_2(y_2) - F'_1(y_2)F_2(y_2). \qquad (2.187)$$

The transfer matrix $[R_{y_1 y_2}]$ relates the acoustic pressure and particle velocity scaled by the characteristic impedance of the medium, that is $\rho_0 c_0 u(y)$. In terms of the acoustic pressure and mass velocity $v(y)$, one gets the following relation:

$$\left\{ \begin{array}{c} p(y_1) \\ v(y_1) \end{array} \right\} = [T_{y_1 y_2}] \left\{ \begin{array}{c} p(y_2) \\ v(y_2) \end{array} \right\}, \qquad (2.188)$$

where

$$[T_{y_1 y_2}] = \begin{bmatrix} R_{y_1 y_2}(1,1) & Y_2 R_{y_1 y_2}(1,2) \\ R_{y_1 y_2}(2,1)/Y_1 & (Y_2/Y_1) R_{y_1 y_2}(2,2) \end{bmatrix}, \quad Y_1 = c_0/S(y_1), \quad Y_2 = c_0/S(y_2).$$

$$(2.189)$$

Thus, $[T_{y_1 y_2}]$ would signify the transfer matrix between the upstream section P_1 and the downstream section P_2 as shown in Figure 2.12c, while the symbol Y_i stands for the characteristic impedance at the ith cross-section, $i = 1$ and 2. It is also worthwhile to mention

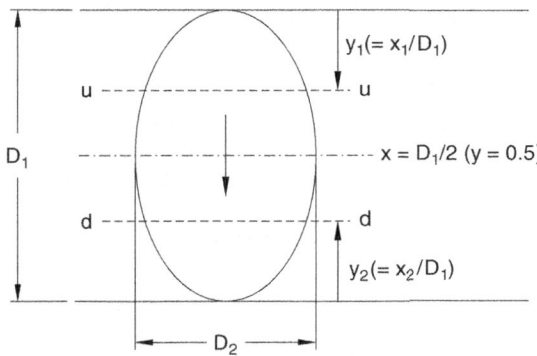

Figure 2.20 Upstream section u-u and downstream section d-d for evaluation of transfer matrix across an elliptical duct (cf. Figure 2.12c)

here that the transfer matrix relations derived in Equations 2.185–2.189 are expressed in terms of the function fields $F_1(y)$ and $F_2(y)$, that converge for $y \leq 0.5$. Hence, the transfer matrices derived above are valid for $y_1 \leq 0.5$ and $y_2 \leq 0.5$, with $y_2 \geq y_1$, thereby indicating that both y_1 and y_2 should be in the same half of the elliptical section.

When the upstream point and downstream point are located on the opposite sides of the mean (center) line as distances y_1 and y_2 from the respective ends as shown in Figure 2.20, then one may proceed as follows [11].

$$\begin{Bmatrix} p_1 \\ v_1 \end{Bmatrix} = [T_{y_1 m}] \begin{Bmatrix} p_m \\ v_m \end{Bmatrix}; \tag{2.190a}$$

$$\begin{Bmatrix} p_2 \\ v_2 \end{Bmatrix} = [T_{y_2 m}] \begin{Bmatrix} p_m \\ v_m \end{Bmatrix}. \tag{2.190b}$$

Combining, Equations 2.190a and 2.190b and noting that the directions of v_m and v_2 in Equation (2.190) are now reversed, we get:

$$\begin{Bmatrix} p_1 \\ v_1 \end{Bmatrix} = \frac{1}{det} [T_{y_1 m}] \begin{bmatrix} T_{y_2 m}(2,2) & T_{y_2 m}(1,2) \\ T_{y_2 m}(2,1) & T_{y_2 m}(1,1) \end{bmatrix} \begin{Bmatrix} p_2 \\ v_2 \end{Bmatrix}, \tag{2.191}$$

where $det = T_{y2m}(1,1)\, T_{y2m}(2,2) - T_{y2m}(2,1)\, T_{y2m}(1,2)$

It can be noted that all types of elements can be represented by the general four-pole form shown in Figure 2.19. For this reason, the four parameters of a transfer matrix are also sometimes called four-pole parameters.

2.18.6 Overall Transfer Matrix of the System

In general, for a dynamical filter consisting of n elements, one gets a general block diagram as shown in Figure 2.5b. The transfer matrix relation for this can be written by successive application of definition (2.153):

$$\{S_{n+1}\} = [T_{n+1}][T_n] \cdots [T_r] \cdots [T_1]\{S_0\}, \qquad (2.192)$$

where the state vector

$$\{S_0\} = \begin{bmatrix} p_0 \\ v_0 \end{bmatrix}$$

can be rewritten as

$$\{S_0\} = \begin{bmatrix} p_0 \\ v_0 \end{bmatrix} = \begin{bmatrix} 1 & Z_0 \\ 0 & 1 \end{bmatrix} \begin{bmatrix} 0 \\ v_0 \end{bmatrix}, \qquad (2.193)$$

because

$$p_0/v_0 = Z_0.$$

Thus, the radiation impedance Z_0 can be placed in the in-line position, and then the state vector $\{0\ v_0\}$ refers to a point downstream of Z_0 as shown in Figure 2.21 [5].

As v_{n+1} is related to the source pressure p_{n+1} independently, the pressure across Z_{n+1} is indicated in Figure 2.21 by p'_{n+1}. Thus,

$$\{S_{n+1}\} = \begin{bmatrix} p'_{n+1} \\ v_{n+1} \end{bmatrix} \qquad (2.194)$$

and Z_{n+1} being a branch impedance, is represented by the transfer matrix

$$[T_{n+1}] = \begin{bmatrix} 1 & 0 \\ 1/Z_{n+1} & 1 \end{bmatrix} \qquad (2.195)$$

Of course, as one needs only the velocity ratio v_{n+1}/v_0 in evaluation of the insertion loss of the filter, p'_{n+1} would not be involved as such. Incidentally, as can be observed from Figure 2.21, $p'_{n+1} = p_n$.

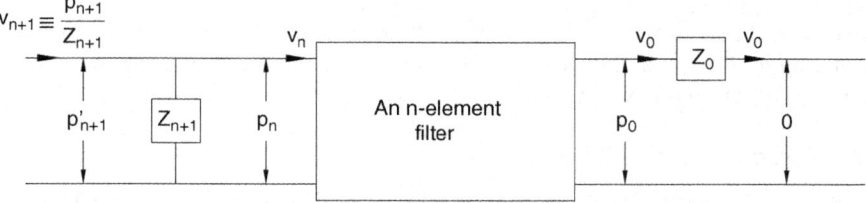

Figure 2.21 A general block diagram of a one-dimensional filter

Theory of Acoustic Filters

As an illustration of the transfer matrix method, for the filter of Figure 2.17, one gets the following transfer matrix relation:

$$\begin{bmatrix} p'_{10} \\ v_{10} \end{bmatrix} = \begin{bmatrix} 1 & 0 \\ 1/Z_{10} & 1 \end{bmatrix} \begin{bmatrix} \cos k_0 l_9 & jY_9 \sin k_0 l_9 \\ j/Y_9 \sin k_0 l_9 & \cos k_0 l_9 \end{bmatrix} \begin{bmatrix} 1 & 0 \\ 1/Z_8 & 1 \end{bmatrix} \begin{bmatrix} \cos k_0 l_7 & jY_7 \sin k_0 l_7 \\ j/Y_7 \sin k_0 l_7 & \cos k_0 l_7 \end{bmatrix}$$

$$\times \begin{bmatrix} \cos k_0 l_6 & jY_6 \sin k_0 l_6 \\ j/Y_6 \sin k_0 l_6 & \cos k_0 l_6 \end{bmatrix} \begin{bmatrix} 1 & 0 \\ 1/Z_5 & 1 \end{bmatrix} \begin{bmatrix} \cos k_0 l_4 & jY_4 \sin k_0 l_4 \\ j/Y_4 \sin k_0 l_4 & \cos k_0 l_4 \end{bmatrix}$$

$$\times \begin{bmatrix} \cos k_0 l_3 & jY_3 \sin k_0 l_3 \\ j/Y_3 \sin k_0 l_3 & \cos k_0 l_3 \end{bmatrix} \begin{bmatrix} 1 & 0 \\ 1/Z_2 & 1 \end{bmatrix} \begin{bmatrix} \cos k_0 l_1 & jY_1 \sin k_0 l_1 \\ j/Y_1 \sin k_0 l_1 & \cos k_0 l_1 \end{bmatrix} \begin{bmatrix} 1 & Z_0 \\ 0 & 1 \end{bmatrix} \begin{bmatrix} 0 \\ v_0 \end{bmatrix}. \quad (2.196)$$

If transfer matrices for simple area discontinuities were unity matrices, this would not alter the product matrix; these therefore have been left out in Equation 2.196.

By performing the multiplication of all the matrices in Equation 2.196, one would obtain the resultant transfer matrix relation

$$\begin{bmatrix} p'_{10} \\ v_{10} \end{bmatrix} = \begin{bmatrix} E_{11} & E_{12} \\ E_{21} & E_{22} \end{bmatrix} \begin{bmatrix} 0 \\ v_0 \end{bmatrix}, \quad (2.197)$$

where $[E_{ij}]$ is the resultant transfer matrix. Finally, the required velocity ratio v_{10}/v_0 is given by the second-row-second-column element of the resultant transfer matrix,

$$v_{10}/v_0 = E_{22}. \quad (2.198)$$

Thus, in general, for an n-element filter, the velocity ratio VR_{n+1} required in Equation 2.70 for insertion loss is equal to second-row-second-column element (E_{22}) of the resultant transfer matrix. Evaluation of VR_{n+1} therefore, consists in

i. making an equivalent circuit for the system,
ii. writing down transfer matrices for all the elements, starting from the source impedance and ending with the radiation impedance, and
iii. multiplying these matrices sequentially, keeping track of only the second-row elements E_{21} and E_{22} of the continued product.

2.18.7 Evaluation of TL in Terms of the Four-Pole Parameters

Transmission loss of a filter can also be found from elements of the overall transfer matrix of elements 2 to $n - 1$ constituting the filter, as shown in Figure 2.8. Let

$$\begin{bmatrix} p_{n-1} \\ v_{n-1} \end{bmatrix} = \begin{bmatrix} T_{11} & T_{12} \\ T_{21} & T_{22} \end{bmatrix} \begin{bmatrix} p_1 \\ v_1 \end{bmatrix}, \quad (2.199)$$

where [T] is the product transfer matrix,

$$p_{n-1} = A_n + B_n,$$
$$v_{n-1} = (A_n - B_n)/Y_n,$$
$$p_1 = A_1 + B_1 = A_1 \text{(as } B_1 = 0\text{), and}$$
$$v_1 = (A_1 - B_1)/Y_1 = A_1/Y_1.$$

Thus,

$$A_n = (p_{n-1} + Y_n v_{n-1})/2 \tag{2.200}$$

$$= \left[\left(T_{11}A_1 + T_{12}\frac{A_1}{Y_1}\right) + Y_n\left(T_{21}A_1 + T_{22}\frac{A_1}{Y_1}\right)\right]/2 \tag{2.201}$$

$$\frac{A_n}{A_1} = \frac{1}{2}\left[T_{11} + \frac{T_{12}}{Y_1} + Y_n T_{21} + \frac{Y_n}{Y_1}T_{22}\right]. \tag{2.202}$$

Therefore,

$$TL = 20 \log\left[\left(\frac{Y_1}{Y_n}\right)^{1/2}\left|\frac{T_{11} + T_{12}/Y_1 + Y_n T_{21} + (Y_n/Y_1)T_{22}}{2}\right|\right]. \tag{2.203}$$

Incidentally, insertion loss reduces to transmission loss in the limit, as shown hereunder. If the source impedance Z_{n+1} were to be equal to the characteristic impedance Y_n and the radiation impedance Z_0 were to be equal to the characteristic impedance Y_1, then Equation 2.70 for IL would yield

$$IL = 20 \log\left|\frac{Y_n}{Y_n + Y_n}VR_{n+1}\right|. \tag{2.204}$$

Velocity ratio $VR_{n+1} = v_{n+1}/v_0$ may be expressed in terms of the product matrix as follows

$$\begin{bmatrix} p'_{n+1} \\ v_{n+1} \end{bmatrix} = \begin{bmatrix} 1 & 0 \\ 1/Z_{n+1} & 1 \end{bmatrix} \begin{bmatrix} C_n & jY_n S_n \\ j/Y_n S_n & C_n \end{bmatrix} \begin{bmatrix} T_{11} & T_{12} \\ T_{21} & T_{22} \end{bmatrix} \begin{bmatrix} C_1 & jY_1 S_1 \\ j/Y_1 S_1 & C_1 \end{bmatrix} \begin{bmatrix} p_0 \\ v_0 \end{bmatrix}, \tag{2.205}$$

where $C \equiv \cos kl$ and $S \equiv \sin kl$.
Substituting

$$p_0/v_0 = Z_0 = Y_1$$

and

$$Z_{n+1} = Y_n$$

into the preceding matrix equation and multiplying out yields

$$\begin{bmatrix} p'_{n+1} \\ v_{n+1} \end{bmatrix} = e^{jkl_1} \begin{bmatrix} C_n & jY_n S_n \\ e^{jkl_n}/Y_n & e^{jkl_n} \end{bmatrix} \begin{bmatrix} T_{11} & T_{12} \\ T_{21} & T_{22} \end{bmatrix} \begin{bmatrix} Y_1 v_0 \\ v_0 \end{bmatrix}, \quad (2.206)$$

whence

$$|VR_{n+1}| = \left| \frac{v_{n+1}}{v_0} \right| = \left| \frac{Y_1}{Y_n} T_{11} + \frac{T_{12}}{Y_n} + T_{21} Y_1 + T_{22} \right|. \quad (2.207)$$

$$IL = 20 \log \left[\left(\frac{Y_1}{Y_n} \right) \left| \frac{T_{11} + T_{12}/Y_1 + T_{21} Y_n + (Y_n/Y_1) T_{22}}{2} \right| \right]$$

$$= TL \quad \text{if} \quad Y_n = Y_1$$

Hence,

$$IL = TL \text{ in the limit } Z_{n+1} = Y_n; Z_0 = Y_1. \quad (2.208)$$

The condition $Y_n = Y_1$ simply requires that the area of the upstream pipe equals that of the downstream pipe, which is generally true.

Here, it may be relevant to recall that LD (or NR) is the limiting value of IL for a constant pressure source and zero temperature gradient.

2.19 TL of a Simple Expansion Chamber Muffler

A simple expansion chamber muffler (see Figure 2.22) is the most basic muffler configuration. It is used as a benchmark for more complex (and efficient) muffler configurations. It is, therefore, analyzed here explicitly in order to derive its transmission loss expression, draw its TL curve against the nondimensional frequency, $(k_0 l)$, and discuss its characteristic domes and troughs.

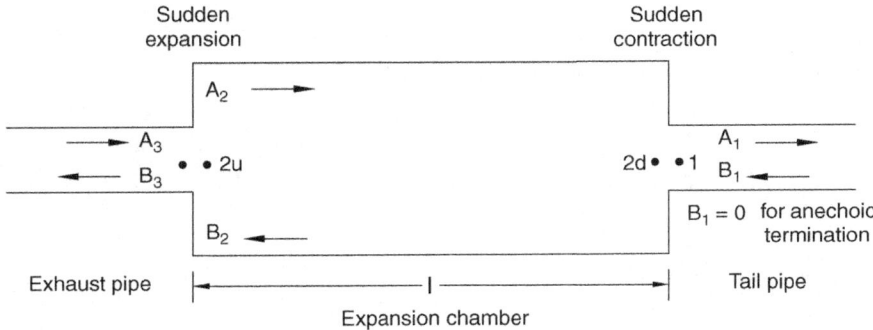

Figure 2.22 Schematic of a simple expansion chamber muffler

The classical method of analysis proceeds as follows.
Equating pressure and mass velocity at the sudden expansion,

$$A_3 + B_3 = A_2 + B_2 \qquad \text{(i)}$$

$$\frac{A_3 - B_3}{Y_3} = \frac{A_2 - B_2}{Y_2} \qquad \text{(ii)}$$

Equating pressure and mass velocity at the sudden contraction,

$$A_2 e^{-jk_0 l} + B_2 e^{jk_0 l} = A_1 \qquad \text{(iii)}$$

$$\frac{A_2 e^{-jk_0 l} - B_2 e^{jk_0 l}}{Y_2} = \frac{A_1}{Y_1} \qquad \text{(iv)}$$

Assuming the exhaust pipe diameter to be equal to that of the tail pipe,

$$S_3 = S_1 \Rightarrow Y_3 = Y_1 \qquad \text{(v)}$$

Equations iii and iv can be solved simultaneously in order to get A_2 and B_2 in terms of A_1, and then Equations i and ii may be solved in order to obtain A_3 in terms of A_1.

Finally,

$$TL = 20 \log \left| \frac{A_3}{A_1} \right|. \qquad \text{(vi)}$$

The transfer matrix method consists in multiplying the transfer matrices of sudden expansion, expansion chamber (a uniform tube) and sudden contraction, successively, in order to obtain the overall transfer matrix. Thus,

$$\begin{bmatrix} p_3 \\ v_3 \end{bmatrix} = \begin{bmatrix} 1 & 0 \\ 0 & 1 \end{bmatrix} \begin{bmatrix} \cos k_0 l & jY \sin k_0 l \\ j/Y_2 \sin k_0 l & \cos k_0 l \end{bmatrix} \begin{bmatrix} 1 & 0 \\ 0 & 1 \end{bmatrix} \begin{bmatrix} p_1 \\ v_1 \end{bmatrix}.$$

$$= \begin{bmatrix} \cos k_0 l & jY_2 \sin k_0 l \\ j/Y_2 \sin k_0 l & \cos k_0 l \end{bmatrix} \begin{bmatrix} p_1 \\ v_1 \end{bmatrix}. \qquad \text{(vii)}$$

Making use of Equation 2.203, we get

$$TL = 20 \log \left| \frac{\cos k_0 l + j\frac{Y_2}{Y_1} \sin k_0 l + j\frac{Y_3}{Y_2} \sin k_0 l + \frac{Y_3}{Y_1} \cos k_0 l}{2} \right|. \qquad \text{(viii)}$$

Let the area expansion ratio

$$\frac{S_2}{S_3} = \frac{S_2}{S_1} = m. \quad (m \geq 1) \qquad \text{(ix)}$$

Then,

$$\frac{Y_2}{Y_1} = \frac{S_1}{S_2} = \frac{1}{m}, \quad \frac{Y_3}{Y_2} = \frac{S_2}{S_3} = m, \quad \frac{Y_3}{Y_1} = \frac{S_1}{S_3} = 1, \qquad \text{(x)}$$

and Equation viii yields

$$TL = 10 \log \left\{ \cos^2 k_0 l + \frac{1}{4}\left(m + \frac{1}{m}\right)^2 \sin^2 k_0 l \right\} \qquad \text{(xi)}$$

This simplifies to

$$TL = 10 \log \left\{ 1 + \frac{1}{4}\left(m - \frac{1}{m}\right)^2 \sin^2 k_0 l \right\} \quad (dB) \qquad \text{(xii)}$$

Incidentally, it may readily be checked that Equation vi also yields the same expression (xii) for TL of a simple expansion chamber.

Equation xii is plotted in Figure 2.23 against the nondimensional frequency parameter (or Helmholtz number) $k_0 l$ with area ratio m as a parameter.

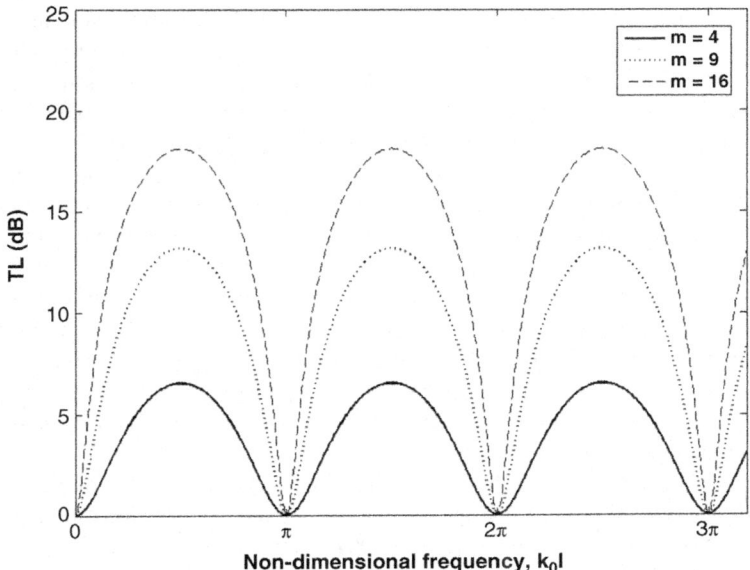

Figure 2.23 TL of the simple expansion chamber muffler of Figure 2.22

2.20 An Algebraic Algorithm for Tubular Mufflers

This section deals with an algebraic algorithm for prediction of Velocity Ratio, VR_{n+1}, without having to solve a number of algebraic equations or multiply a number of transfer matrices successively. This is applied to mufflers that consist of tubes and lumped elements shown in Figure 2.18, with transfer matrices given by Equations 2.172–2.174.

2.20.1 Development of the Algorithm

As observed earlier, the evaluation of the velocity ratio VR_{n+1} consists in determining the composition of E_{22}, the second-row-second-column element of the resultant transfer matrix. VR_{n+1} would be the algebraic sum of a number of terms, the form of each term being

$$\{j^{some\ power}\} \left\{\begin{array}{c} a\ rational \\ fraction\ of\ Ys \\ and\ Zs \end{array}\right\} \left\{\begin{array}{c} a\ product\ of\ circular\ functions \\ corresponding\ to\ the\ distributed\ elements \end{array}\right\} \quad (2.209)$$

The first factor determines the sign of the term. The second is referred to here as 'the Y-Z combination' and the third, as 'the circular function product'.

VR_{n+1}, being dimensionless, would be the sum of a number of nonzero terms, each of which must be individually dimensionless. Therefore, the number of Ys and/or Zs in the numerator of a combination would be equal to the number of Ys and/or Zs in the denominator.

From Expressions (2.172, 2.173 and (2.174)), it may be observed that Y_r or Z_r appears only in the coupling positions. Where it appears in the first-row-second-column position, it is in the numerator, where it appears in the second-row-first-column position, it is in the denominator. From the algebra of matrix multiplication, it can be argued that the highest subscripted Y or Z in the Y-Z combination of a term shall appear only in the denominator and the lowest subscripted Y or Z shall appear only in the numerator. In addition, the intermediate Ys and Zs would alternate between the numerator and the denominator of a combination.

There will always be one term without the Y-Z combination. This term will be unity, or the continued product $\Pi \cos kl_r$, where r stands for the subscripts of all the distributed elements.

As may be observed from expressions (2.172, 2.173 and (2.174)), the Z of an in-line lumped element will not appear in the denominator of a combination and that of a shunt element will not appear in the numerator. No such restriction applies to Ys of the distributed elements.

Each constituent term must take one element from the transfer matrix of each of the distributed elements. It would have $\cos kl_r$, or $Y_r \sin kl_r$, or $1/Y_r \sin kl_r$ corresponding to the rth distributed element. Hence, each term would be the product of the Y-Z combination and a product of $\cos kl_r$, where r stands for all the distributed elements not represented in the Y-Z combination, and $\sin kl_s$, where s stands for all the distributed elements present in the Y-Z combination of the term.

It can be observed directly from the transfer matrices of the three types of elements that the Ys are always accompanied by a j in the numerator. Hence, the sign of a term with q Ys is given by $(j)^q$, where q is the number of Ys in the Y-Z combination of the term.

Thus, we see that the product of circular functions and the sign of a term are fixed by the composition and the number of Ys in the Y-Z combination. The problem therefore reduces to one of finding a way of writing the probable Y-Z combinations. With Y_r and Z_r referred to by its subscript r, the probable Y-Z combinations can be written out in terms of the probable combinations of the positional subscripts 0 to $n+1$.

So, we have to look only for even-membership (0, 2, 4, 6 ...) combinations out of integers (positional subscripts) 0 to $n+1$, arranged in two rows in such a way that the subscripts of Zs that can appear in the numerator are in the first row, those that can appear in the denominator are in the second row, and the subscripts of the distributed elements are in both the rows. The problem then consists in writing out the 'permissible' even-membership combinations out of such an array of integers, drawing alternately from the top row and the bottom row.

The sufficiency of the preceding restraints on 'permissibility' of a combination has been established [15] by evaluating the total number of permissible combinations of positional subscripts 0 to $n+1$, and comparing it with the actual number of terms for the general system. For a general n-element filter with nd distributed elements, nl lumped elements grouped into s_1 groups of one element each, s_2 groups of two elements each, ..., s_r groups of r elements each, such that $nl = \sum(rs_r)$, the total number of terms ST comprising VR_{n+1} is given by the following expression [15],

$$ST = 2^{nd+1-ns}(3)^{s_1}(5)^{s_2}(8)^{s_3}\ldots(S_r)^{s_r}, \qquad (2.210)$$

where ns, the total number of discrete sets of lumped elements, is equal to $\sum s_r$ and s_r is given by the iterative relation (Fibonocci series)

$$S_r = S_{r-1} + S_{r-2}, \qquad (2.211)$$

with the first two terms of the series being $S_1 = 3$ and $S_2 = 5$.

2.20.2 Formal Enunciation and Illustration of the Algorithm

The algorithm is enunciated hereunder as a sequence of discrete steps and illustrated for a five-element filter of Figure 2.24.

i. From the analogous circuit of the given system, prepare a two-row array of integers 0 to $n+1$ in the ascending order such that the subscripts of the in-line elements are in the first row, branch elements are in the second row, and distributed elements are in both the rows.

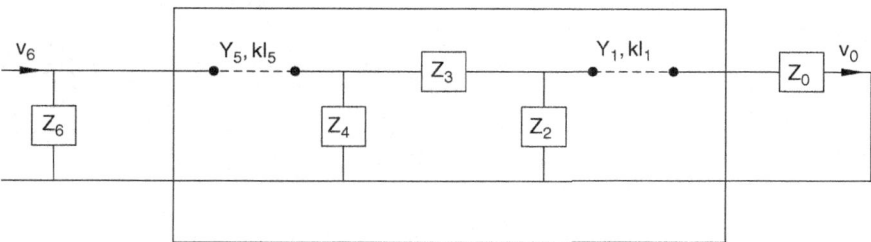

Figure 2.24 An illustration of the algorithm

For the system of Figure 2.23, the array would be

$$\begin{bmatrix} 0 & 1 & & 3 & & 5 \\ & 1 & 2 & & 4 & 5 & 6 \end{bmatrix} \qquad (2.212)$$

ii. Write all possible even-membership combinations of the array such that in a combination the first integer is from the first row, the second (higher than the first) is from the second row, the third (higher than the second) is from the first, and so on. As each combination is of even membership, the highest integer of the combination would be always from the second row. The membership $2q$ of a combination is given by

$$0 \leq 2q \leq n+2 \qquad (2.213)$$

where q is an integer.

For the array (2.212), the combinations would consist of 0, 2, 4 and 6 membership (as $n+2=7$). The permissible combinations are

a. the null combination;
b. 0_1, 0_2, 0_3, 0_4, 0_5, 0_6, 1_2, 1_4, 1_5, 1_6, 3_4, 3_5, 3_6, 5_6;
c. $0_1 3_4$, $0_1 3_5$, $0_1 3_6$, $0_1 5_6$, $0_2 3_4$, $0_2 3_5$, $0_2 3_6$, $0_2 5_6$, $0_4 5_6$, $1_2 3_4$, $1_2 3_5$, $1_2 3_6$, $1_2 5_6$, $1_4 5_6$, $3_4 5_6$;
d. $0_1 3_4 5_6$, $0_2 3_4 5_6$, $1_2 3_4 5_6$ (2.214)

The total number of combinations is 32, as could be checked against formula (2.210) with $nd = 2$, $s_1 = 0$, $s_2 = 0$, $s_3 = 1$ and $ns = 1$.

iii.
a. Replace the subscripts by their parent Ys and Zs such that the Ys corresponding to the first-flow subscripts are in the numerator and those corresponding to the second row are in the denominator.
b. Affix a sign given by $(j)^q$ where q is the number of Ys in the Y-Z combination.
c. Multiply the Y-Z combination with the repeated products $\sin kl_r$ and $\cos kl_s$, where r and s take on all values corresponding to all the Ys included or not included in the Y-Z combination under consideration.

On applying this to the combinations of (2.214), one gets

$$VR_{n+1} = VR_6 = \frac{v_6}{v_0} = C_1 C_5 + j\frac{Z_0}{Y_1} S_1 C_5 + \frac{Z_0}{Z_2} C_1 C_5 + \frac{Z_0}{Z_4} C_1 C_5 + j\frac{Z_0}{Y_5} C_1 S_5$$

$$+ \frac{Z_0}{Z_6} C_1 C_5 + \ldots + j\frac{Y_5}{Z_6} C_1 S_5 + j\frac{Z_0 Z_3}{Y_1 Z_4} S_1 C_5 + \ldots + j\frac{Z_3 Y_5}{Z_4 Z_6} C_1 S_5$$

$$- \frac{Z_0 Z_3 Y_5}{Y_1 Z_4 Z_6} S_1 S_5 + j\frac{Z_0 Z_3 Y_5}{Z_2 Z_4 Z_6} C_1 S_5 - \frac{Y_1 Z_3 Y_5}{Z_2 Z_4 Z_6} S_1 S_5,$$

(2.215)

where $C_1 \equiv \cos kl_1$, $C_5 \equiv \cos kl_5$, $S_1 \equiv \sin kl_1$, $S_5 \equiv \sin kl_5$

It may be noted that the preceding algorithm does not impose any restriction on the composition of the filter except that all elements fall under one of the three types shown in Figure 2.18.

2.21 Synthesis Criteria for Low-Pass Acoustic Filters

An algebraic algorithm developed from a heuristic study of the transfer matrix multiplication permits the set of most significant terms constituting the velocity ratio VR_{n+1} to be identified from a knowledge of the relative magnitudes of the impedances of various elements comprising a particular filter configuration. This feature makes the algorithm a potential tool in a first approach to a rational synthesis of one-dimensional dynamical filters [16]. The foregoing algorithm has been used to study the comparative performances of the reactive elements of an acoustic filter and thereby to synthesize a low-pass, one dimensional, straight-through acoustic filter so as to satisfy the following requirements:

(maximum) length of the central, larger-diameter part $= 75$ cm,
(maximum) internal diameter of the central part $= 12$ cm,
(minimum) internal diameter of any tube $= 3$ cm.
(These requirements are typical of an automotive muffler).

A realistic compression of the filter performance of two different acoustic elements would involve comparison of the insertion loss characteristics of two filters identical in all aspects except that one particular element of the first has been replaced by another in the second. A reactive straight-through acoustic filter would comprise one or more of the following elements: (a) expansion chamber, (b) extended-tube expansion chamber, (c) extended-tube resonator, (d) hole-cavity resonator.

A muffler with three expansion chambers of lengths l_2, l_4 and l_6 is shown in Figure 2.25. $VR_{n+1}(n=7)$ for the filter would be an algebraic sum of a large number of terms corresponding to the even-membership Y-Z combinations of the array

$$\begin{bmatrix} 0 & 1 & 2 & 3 & 4 & 5 & 6 & 7 \\ 1 & 2 & 3 & 4 & 5 & 6 & 7 & 8 \end{bmatrix}, \qquad (2.216)$$

according to the foregoing algorithm. For an expansion ratio of $(12/3)^2 = 16$,

$$Y_1 = Y_3 = Y_5 = Y_7 \qquad (2.217)$$

and each is equal to 16 times

$$Y_2 = Y_4 = Y_6. \qquad (2.218)$$

Thus, the terms having Y_2, Y_4 and Y_6 in the denominator (Y_8 would be in the denominator only) and Y_1, Y_3, Y_5 and Y_7 in the numerator of their Y-combinations will be of a higher order than others.

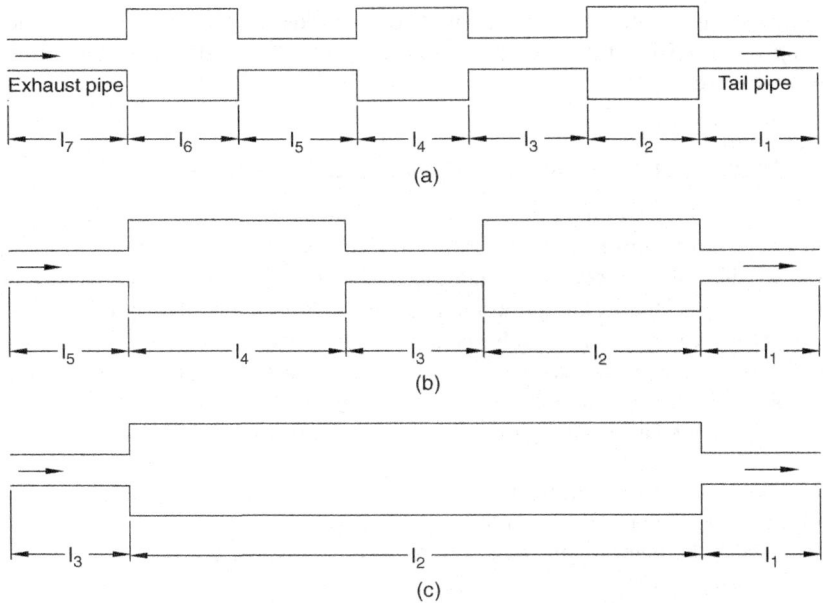

Figure 2.25 Expansion chamber filters. (a) Three expansion chambers. (b) Two expansion chambers. (c) Single expansion chamber

Again, Z_0 which can appear in the numerator of a Y-Z combination, is much smaller than Y_1, Y_3, Y_5 and Y_7 at lower frequencies of interest. The significant terms will therefore correspond to the Y-Z combinations of the array

$$\begin{bmatrix} 1 & 3 & 5 & 7 \\ 2 & 4 & 6 & 8 \end{bmatrix} \tag{2.219}$$

The same line of argument would suggest that each of the highest-order Y-Z combinations would have all of the three lower-order Ys (Y_2, Y_4 and Y_6) in the denominator accompanied by suitable Ys in the numerator. With this additional restraint, valid even-membership combinations of the array (2.219) may easily be seen to be

$$\frac{Y_1 Y_3 Y_5}{Y_2 Y_4 Y_6} \quad \text{and} \quad \frac{Y_1 Y_3 Y_5 Y_7}{Y_2 Y_4 Y_6 Z_8} \tag{2.220}$$

and, thus,

$$\begin{aligned} VR_{n+1} \simeq & -\frac{Y_1 Y_3 Y_5}{Y_2 Y_4 Y_6} \sin kl_1 \sin kl_2 \sin kl_3 \sin kl_4 \sin kl_5 \sin kl_6 \cos kl_7 \\ & -j\frac{Y_1 Y_3 Y_5 Y_7}{Y_2 Y_4 Y_6 Z_8} \sin kl_1 \sin kl_2 \sin kl_3 \sin kl_4 \sin kl_5 \sin kl_6 \sin kl_7. \end{aligned} \tag{2.221}$$

Alternatively, relation (2.221) may be written as

$$VR_{n+1} \equiv W \cdot X \qquad (2.222)$$

where

$$W = -\frac{Y_1 Y_3 Y_5}{Y_2 Y_4 Y_6} \sin kl_2 \sin kl_3 \sin kl_4 \sin kl_5 \sin kl_6 \qquad (2.223)$$

and

$$X = j\frac{Y_7}{Z_8} \sin kl_1 \sin kl_7 + \sin kl_1 \cos kl_7 \qquad (2.224)$$

or

$$X = \sin kl_1 \left(j\frac{Y_n}{Z_{n+1}} \sin kl_n + \cos kl_n\right). \qquad (2.225)$$

It may be noted that, in a way, W represents the effect of the central portion, and X that of the source impedance, tail pipe, and exhaust pipe. When the source impedance Z_{n+1} is fixed,

$$X = X(l_1, l_2) \qquad (2.226)$$

with l_1 plus l_n fixed, the optimum combination of l_1 and l_n may be found to maximize X at a given frequency or over a small range of frequencies.

In what follows, the effect of expansion chambers on W or VR_{n+1}/X is considered. Thus, for the three-expansion-chamber filter of Figure 2.25a,

$$\left.\frac{VR_{n+1}}{X}\right]_3 \simeq -\frac{Y_1 Y_3 Y_5}{Y_2 Y_4 Y_6} \sin kl_2 \sin kl_3 \sin kl_4 \sin kl_5 \sin kl_6. \qquad (2.227)$$

Similarly, for the filters with two and one expansion chambers (Figure 2.25b and c),

$$\left.\frac{VR_{n+1}}{X}\right]_2 \simeq \frac{Y_1 Y_3}{Y_2 Y_4} \sin kl_2 \sin kl_3 \sin kl_4 \qquad (2.228)$$

and

$$\left.\frac{VR_{n+1}}{X}\right]_1 \simeq -\frac{Y_1}{Y_2} \sin kl_2, \qquad (2.229)$$

respectively. The order of magnitude of the Y-combinations of the significant terms in VR_{n+1}/X for the three cases may be observed from Equations 2.227, 2.228 and 2.229 to be, respectively, $(16)^3$, $(16)^2$ and 16. The product of the accompanying circular functions depends on the number of expansion chambers and the frequency. In fact, the number of sine functions is equal to twice the number of expansion chambers minus one. At low frequencies (of the order of 100 Hz) all the sine functions would be positive and small. As the number of expansion chambers increases, with the total length of the central part fixed (at 75 cm), length of the expansion chambers and connecting tubes are correspondingly reduced. This leads to a considerable decrease in the order of magnitude of the product $\sin kl_2 \sin kl_3 \ldots \sin kl_n$ at low frequencies. At higher frequencies, the order of magnitude of Y-Z combinations is the deciding factor, and therefore insertion loss would be improved as the number of expansion chambers increases. Thus, for a given overall length, insertion loss increases at high frequencies and decreases at low frequencies. At a given frequency, however, the best insertion loss would be obtained by a definite number of expansion chambers.

The observations made from an examination of the most significant terms are by and large borne out by the exact values obtained on a computer.

It may also be observed from Equations 2.227, 2.228, and 2.229 that a higher expansion ratio would result in an increase in the magnitude of insertion loss at all frequencies. This agrees with the experimental findings [8] and is true in fact for most muffler configurations.

A similar exercise with other types of elements yields the following additional synthesis criteria [16].

i. An extended-tube expansion chamber gives better overall velocity ratio, VR_{n+1}, and hence insertion loss, IL, as compared to a simple expansion chamber of equivalent length.
ii. Over a small frequency band, increased insertion loss may be obtained by means of a row of holes opening into an annular cavity, tuned preferably at the higher end of the band. The effective width of such a hole-cavity combination may be increased by means of a larger cavity.

A hole-cavity combination is more flexible than, and therefore may be preferred to, an isolated extended-tube resonator.

References

1. Rayleigh, L. (1894) *The Theory of Sound*, 2nd edn, Macmillan, London, Art. 307-314.
2. Levine, M. and Schwinger, J. (1948) On the radiation of sound from an unflanged circular pipe. *Physical Review*, **73** (4), 383–406.
3. Davies, P.O.A.L., Bento Coelho, J.L. and Bhattacharya, M. (1980) Reflection coefficients for an unflanged pipe with flow. *Journal of Sound and Vibration*, **72** (4), 543–546.
4. Sreenath, A.V. and Munjal, M.L. (1970) Evaluation of noise attenuation due to exhaust mufflers. *Journal of Sound and Vibration*, **12** (1), 1–19.
5. Munjal, M.L., Sreenath, A.V. and Narasimhan, M.V. (1973) Velocity ratio in the analysis of linear dynamical systems. *Journal of Sound and Vibration*, **26** (2), 173–191.
6. Munjal, M.L. (1977) Exhaust noise and its control. *Shock and Vibration Digest*, **9** (8), 21–32.
7. Munjal, M.L. (1980) A new look at the performance of reflective exhaust mufflers, DAGA 80, München.
8. Davis Jr., D.D., Stokes, M., Moore, D. and Stevens, L. (1954) Theoretical and experimental investigation of mufflers with comments on engine exhaust muffler design, NACA Rept. 1192.
9. Crocker, M.J. (1977) Internal combustion engine exhaust muffling. *Noise-Con'*, **77**, 331–358.
10. Morse, P.M. (1948) *Vibration and Sound*, 2nd edn, McGraw-Hill, New York, pp. 265–285.
11. Mimani, A. and Munjal, M.L. (2011) Transverse plane wave analysis of short elliptical chamber mufflers: an analytical approach. *Journal of Sound and Vibration*, **330**, 1472–1489.
12. Alfredson, R.J. and Davies, P.O.A.L. (1971) Performance of exhaust silencer components. *Journal of Sound and Vibration*, **15** (2), 175–196.

13. Igarashi, J. *et al.*, Fundamentals of acoustic silencers, *Aero. Res. Inst. Univ. Tokyo*, **I. R**ep. No. 339, 223-241 (Dec. 1958); **II R**ep. No. 344, 67-85 (May 1959); and **III. R**ep. No. 351, 17-31, (Feb. 1960).
14. Fukuda, M. (1963) A study on the exhaust mufflers of internal combustion engines. *Bulletin of JSME*, **6**, 22.
15. Munjal, M.L., Sreenath, A.V. and Narasimhan, M.V. (1973) An algebraic algorithm for the design and analysis of linear dynamical systems. *Journal of Sound and Vibration*, **26** (2), 193–208.
16. Munjal, M.L., Narasimhan, M.V. and Sreenath, A.V. (1973) A rational approach to the synthesis of one-dimensional acoustic filters. *Journal of Sound and Vibration*, **29** (3), 263–280.
17. Lung, T.Y. and Doige, A.G. (1983) A time-averaging transient testing method for acoustic properties of piping system and mufflers with flow. *Journal of the Acoustical Society of America*, **73** (3), 867–876.

3

Flow-Acoustic Analysis of Cascaded-Element Mufflers

A muffler in the intake or exhaust system of a reciprocating internal combustion engine differs from a classical one-dimensional acoustic filter in a number of ways. An appreciation of it would require some understanding of the basic thermodynamic cycle of an internal combustion engine and, in particular, the exhaust process.

3.1 The Exhaust Process

Typical processes constituting the thermodynamic cycle of a four-stroke internal combustion engine (Figure 3.1) are sketched in Figure 3.2. There are, of course, some variations between a spark-ignition engine cycle and a compression-ignition engine cycle. However, there is no qualitative difference between the two. The lower part of the figure shows the exhaust process, with the piston moving in and the exhaust valve open. The area enclosed by the clockwise-traversed part of the cycle in the pressure-volume diagram represents the positive work done by the gases on the piston (and hence on the crankshaft); this has been marked by a + sign in the figure. A small area enclosed by the counter clock-wise-traversed part of the cycle is the negative work that is done by the piston (and hence by the crankshaft) on the gases in trying to push them out. The average pressure in the exhaust pipe during the exhaust stroke is called the mean exhaust pressure; the term 'back pressure' is used to denote the difference between this and the ambient pressure.

Back pressure is responsible for pumping losses (negative work) and reduction in volumetric efficiency due to delayed suction of the fresh charge. Therefore, it is obvious that the higher the back pressure, the less net power is available on the crankshaft and the more is the specific fuel consumption.

The exhaust valve opens a few degrees before the piston reaches the bottom dead center (BDC) during the expansion stroke of a four-stroke cycle engine and closes a few degrees after the piston reaches the top dead center (TDC). Thus, out of $720°$ of crankshaft motion (during which four strokes of a thermodynamic cycle are completed), the exhaust valve remains open for about $240°$ only. Therefore burnt-out gases are exhausted only during about

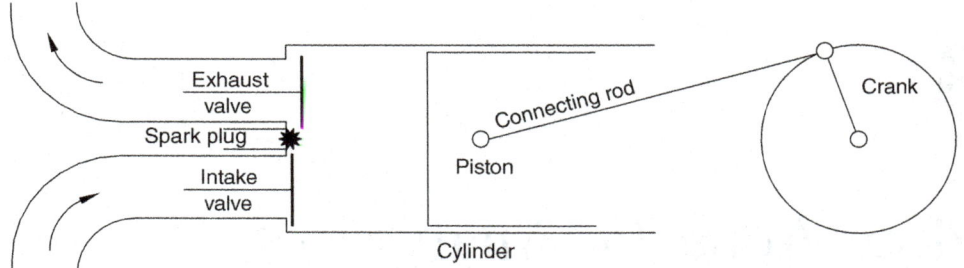

Figure 3.1 A four-stroke spark ignition (gasoline) engine

one-third of the total cycle time. The rest of the time the exhaust pipe has a closed-end termination on the engine side for pressure pulses continuing to move up and down, with the other termination being the atmosphere (or an exhaust muffler, if there is one). Thus, at the atmosphere termination of the exhaust pipe, there appear exhaust pulses with a frequency equal to the number of cycles (equal to half the number of revolutions of the crankshaft in a four-stroke-cycle engine) per second. This frequency is called the firing frequency per cylinder, F. The exhaust noise appears at F and its integral multiples. Thus F is also called the fundamental frequency.

Figures 3.3 and 3.4 show the speed dependence of the overall exhaust noise level and typical spectra thereof for a six-cylinder, two-stroke-cycle diesel engine which developed approximately 200 b.h.p. (brake horsepower) or 146 kW at a crankshaft speed of 2200 RPM (revolutions per minute), measured at 3 ft or 0.91 m from the outlet of the exhaust pipe fitted

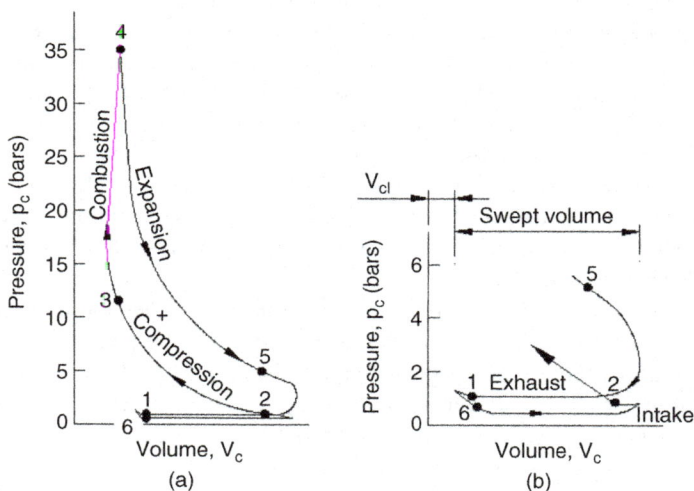

Figure 3.2 Thermodynamic processes of a four-stroke gasoline engine. (a) Engine indicator diagram. (b) Gas exchange diagram

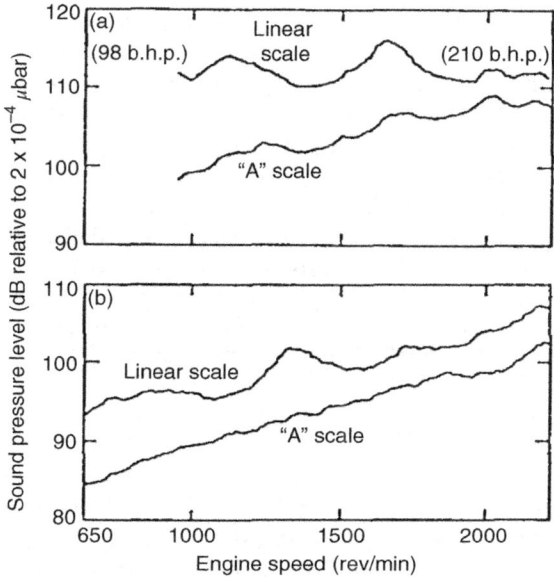

Figure 3.3 Variation of exhaust noise with speed. (a) With full load. (b) With no load (Reproduced with permission from [1])

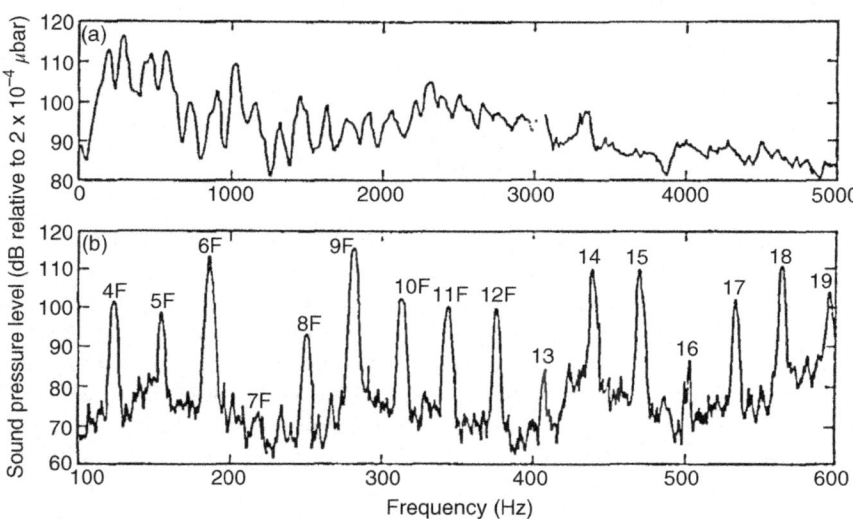

Figure 3.4 Typical spectrum of exhaust noise 3 ft from outlet, with crankshaft speed of 1880 RPM and fundamental frequency of 31.4 Hz. (a) 50-Hz filter. (b) 10-Hz filter (Reproduced with permission from [1])

with a simple expansion chamber silencer [1]. The observations shown in Figures 3.3 and 3.4 are typical of such engines.

It may be noted that the exhaust noise consists of a large number of discrete frequency components – all integral multiples of the firing frequency F. These discrete frequency components are found up into the kilohertz region of the spectrum and show that the contribution from the broad-band jet noise of the gas flow (the peak of which would lie between 200 and 500 Hz for typical tail pipes) is insignificant.

All exhaust (or for that matter, intake) noise appears at the cylinder firing frequency and its multiples, which is related to the crankshaft rotational speed as follows:

$$\text{Individual cylinder firing frequency, } F_{cyl} = \frac{RPM}{60} \times \frac{2}{n_{st}}; \tag{3.1}$$

$$\text{Multi-cylinder engine firing frequency } F_{eng} = F_{cyl} \times n_{cyl}, \tag{3.2}$$

where $n_{st} = 2$ for a two-stroke-cycle engine, and 4 for a four-stroke-cycle engine; and n_{cyl} = number of cylinders in a multi-cylinder engine.

For example, F in Figure 3.4 denotes the individual cylinder firing frequency, $F_{cyl} = RPM/120$ Hz.

It is convenient and customary to describe these discrete frequencies in terms of Speed Order (SO) related to frequency f as follows:

$$\text{speed order of frequency } f = \frac{f}{RPM/60}. \tag{3.3}$$

Substituting this relationship in Equations 3.1 and 3.2 yields

$$\text{speed order of frequency } F_{cyl} = \frac{2}{n_{st}}. \tag{3.4}$$

$$\text{speed order of } F_{eng} = \frac{2}{n_{st}} \times n_{cyl} \tag{3.5}$$

Thus, all exhaust (or intake) noise of a multi-cylinder engine appears at speed order of $2n_{cyl}/n_{st}$ and its integral multiples, provided acoustic lengths of all runners to the exhaust (or intake) manifold are equal. This is, in fact, a very important design feature and will be discussed at some length in the last chapter of this monograph. Here, it would suffice to indicate that the abscissa in Figure 3.4 should be speed order, rather than the mid-frequency of a constant bandwidth or constant percentage bandwidth filter. These days, real-time dual-channel (or multi-channel) analyzers are available wherein one channel gets its signal from the crankshaft rotation, and making use of the built-in tracking frequency multiplier (filter) provides the abscissa with speed order proportional to F_{cyl} or F_{eng}, as required. One such spectrum is shown in Figure 3.5 for the unmuffled exhaust noise at 0.5 m from the tail pipe end of a four cylinder turbocharged engine and the corresponding naturally aspirated engine of 2.5 litre capacity running at 4000 RPM [2].

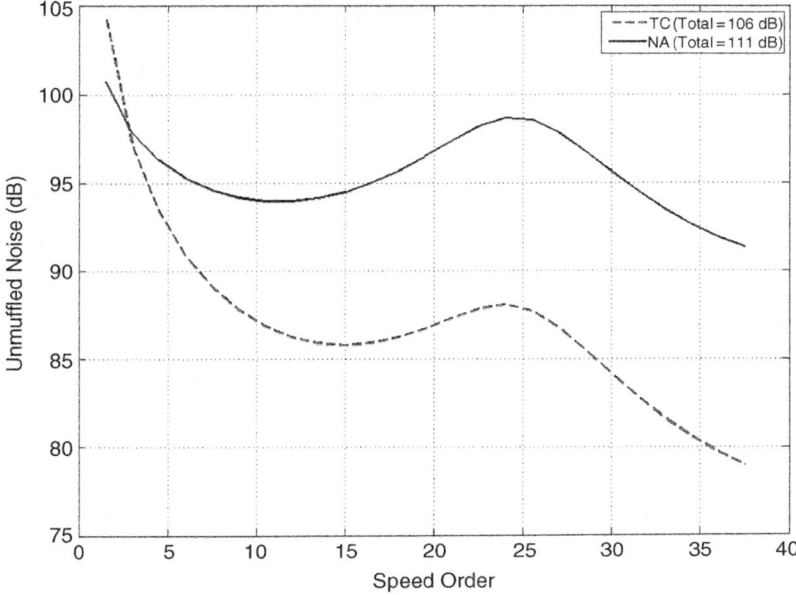

Figure 3.5 The unmuffled sound pressure level (SPL) spectra for naturally aspirated (NA) and turbocharged (TC) diesel engine

3.2 Finite Amplitude Wave Effects

Typical values of the maximum levels of sound pressure in exhaust pipes have been found to be of the order of 155 dB at 300 Hz and about 135 dB at 1000 Hz. An appreciation of the finite amplitude effects in the exhaust pipe can be obtained from the time required for a sinusoidal wave with amplitude Δp and frequency f to steepen-up into a shock wave [1]:

$$T = \frac{p_0}{5f\Delta p} \qquad (3.6)$$

or 0.065 s for 155 dB at 300 Hz, and 0.2 s for 135 dB at 1000 Hz.

The corresponding distances traveled by a sinusoidal wave before shock formation at an average sound speed of 500 m/s would be as much as 32.5 m and 100 m, respectively.

Though the preceding relation neglects dissipation and is a very indirect index of the significance of finite amplitude effects, it is obvious that for typical exhaust systems (which are rarely longer than 5 m), the linearized theory would suffice for analysis of the conditions in the tail pipe. Nevertheless, the finite wave analysis, which, unlike the acoustic theory, is a time-domain analysis, has its uses. In fact, it is an alternative to the acoustic theory and offers information that cannot be provided at all by the acoustic analysis, which works only in the frequency domain. More about it later. In this chapter, the frequency domain 1D acoustic theory developed in Chapter 2 is extended to exhaust mufflers where the convective as well as dissipative effect of mean flow is duly accounted for.

3.3 Mean Flow and Acoustic Energy Flux

The exhaust gas mass flux through the exhaust pipe equals the rate at which gases are being forced out by the piston(s):

$$m = \rho_0 SU = \rho_0 \int_{r=0}^{r_0} u(r) 2\pi r \, dr, \qquad (3.7)$$

where
$S = r_0^2$,
U = velocity of mean flow averaged over the cross-section of the round pipe.

Defining the mean-flow Mach number M as

$$M(r) = \bar{u}(r)/c_0, \qquad (3.8)$$

where c_0 is the sound speed of the gas medium; and if M_0 is the mean-flow Mach number in the center of the pipe, one gets the relation [3]

$$M(r) = M_0 \left(\frac{r_0 - r}{r_0}\right)^{1/7} \qquad (3.9)$$

for fully developed turbulent flows at Reynold's numbers of the order of 10^5, which are typical of exhaust pipes of internal combustion engines. The average value of the Mach number M has been observed to be $0.85 \times M_0$ (the midstream Mach number), which corresponds to a value of $1/8.5$ for the exponent in Equation 3.9. This discrepancy is not serious, as the solution of the flow-acoustic equations is found to be dependent primarily on the value of the space-average Mach number, M.

An expression for the acoustic intensity for waves in moving media follows from the conservation of energy. For a closed surface embedded in the fluid, the total outward energy flow is

$$E = \int_s N_i dS_i, \qquad (3.10)$$

where N_i is the instantaneous energy flux normal to the surface element dS_i per unit area in the ith direction, or parallel to the coordinate x_i. Thus [4]

$$N_i = Jm_i \qquad (3.11)$$

and

$$E = \int_s Jm_i \, dS_i, \qquad (3.12)$$

where J is the stagnation enthalpy given by

$$J = h + \frac{1}{2}V_1^2. \tag{3.13}$$

and m_i is the ith component of the mass flux per unit area:

$$m_i = \rho V_i, \tag{3.14}$$

If the flow is irrotational and of uniform entropy over S, it follows that the time-averaged value $\langle J \rangle$ (averaged over a wave period $2\pi/\omega$) is also uniform over S. If, in addition, the mass of fluid contained in S remains constant on average so that

$$\int_s \langle m_i' \rangle \, dS_i = 0, \tag{3.15}$$

then Equation 3.12 gives the time-average energy flow out of S as

$$\langle E \rangle = \int_s \langle J' m_i' \rangle dS_i, \tag{3.16}$$

where J' and m_i' are acoustic perturbations on J and m_i.

Thus, the net acoustic energy flux per unit area, I_i, in the ith direction is given by the following expression [5]:

$$I_i = \langle J' m_i' \rangle. \tag{3.17}$$

Now, according to Equation 3.13, for one-dimensional flow and wave propagation only along the axis of the pipe, writing V_i as a sum of mean component U and perturbation component u,

$$V_i = (U + u)\delta_{i1}, \tag{3.18}$$

one gets

$$J' = h' + Uu$$

or, according to Appendix B,

$$J' = p'/\rho_0 + Ts' + Uu. \tag{3.19}$$

Here the entropy fluctuation s' can be neglected (provided the flow over S is free from entropy gradients). Thus

$$J' = p'/\rho_0 + Uu \tag{3.20}$$

and

$$m'_i = \rho_0 u + \rho' U. \tag{3.21}$$

With these substitutions in Equation 3.17, the acoustic intensity in a pipe with mean flow can be written as

$$I = \langle p'u \rangle + \frac{U}{\rho_0}\langle p'\rho' \rangle + U\rho_0\langle u^2 \rangle + U^2\langle u\rho' \rangle. \tag{3.22}$$

Here p' and u would in general be functions of the radius. The total power flow through a tube is obtained by integrating Equation 3.22 over the cross-section of the tube.

In an exhaust pipe, flow shear, viscosity, and yielding of the walls of the pipe all occur simultaneously. These make the particle velocity (both mean and fluctuating) and pressure functions of radius even for the so-called (0, 0) mode. Consequently, a knowledge of this dependence is a prerequisite for the integration of I over the cross-section. It has been shown [1], however, that for rigid-walled tubes, the effect of shear and viscosity on net energy flux is small for all frequencies and can, for practical purposes, be ignored in exhaust systems. When the walls yield, however, the influence is not necessarily small. Fortunately, for typical pipe wall thicknesses used in commercial systems, the reduction in net flux due to shear and yielding compared with the rigid-wall, zero-shear case is only about 1 dB. Under these circumstances, it is simplest to assume that the mean-flow velocity profile is invariant. Acoustic intensity can therefore be calculated from Equation 3.22, with U being understood as mean velocity averaged over the cross-section, p being taken as constant, and u also being averaged over the cross-section.

Thus the total acoustic energy flux through a cross-section can be written [dropping the primes in Equation 3.22]

$$W = \int_S I dS = \frac{1}{\rho_0}\left[\langle pv \rangle + \frac{M}{Y_0}\langle p^2 \rangle + MY_0\langle v^2 \rangle + M^2\langle pv \rangle\right], \tag{3.23}$$

where

$v =$ acoustic mass velocity $= \int_S dS \rho_0 u$,

ρ' has been replaced by p'/c_0^2 (or p/c_0^2, dropping the primes),

$M = U/c_0$ (averaged over the cross-section S), and

$Y_0 = c_0/S$, the characteristic impedance defined as the ratio of acoustic pressure and mass velocity of a progressive wave.

If A is the amplitude of the forward wave and RA that of the reflected wave, where R is the reflection coefficient at the exhaust-tail-pipe termination, then

$$p = A(1 + R), \qquad v = \frac{A}{Y_0}(1 - R), \tag{3.24}$$

and Equation 3.23 yields

$$W = W(M) = \frac{1}{2\rho_0} \frac{|A|^2}{Y_0} \left\{ (1+M)^2 - |R|^2 (1-M)^2 \right\}. \tag{3.25a}$$

If convective effect of the mean flow were neglected, one would get

$$W(0) = \frac{1}{2\rho_0} \frac{|A|^2}{Y_0} \left(1 - |R|^2\right). \tag{3.25b}$$

As $W(M) > W(0)$ for all M, the error that would be caused by neglecting mean flow for a given (or measured) reflection coefficient would be positive. It would be all the more significant around $|R| \simeq 1$.

$$|R|_{max} = \frac{1+M}{1-M}. \tag{3.26}$$

So, in the presence of mean flow, the reflection coefficient can be more than unity, as was earlier observed by Mechel *et al.* [6].

Incidentally, it may also be noted that in view of Equation 3.25, the TL expression for a moving medium, corresponding to Equation 2.203 for a stationary medium, would be given by

$$TL = 20 \log \left[\frac{1+M_n}{1+M_1} \left(\frac{Y_1}{Y_n}\right)^{1/2} \left| \frac{T_{11} + T_{12}/Y_1 + Y_n T_{21} + (Y_n/Y_1) T_{22}}{2} \right| \right] \tag{3.27}$$

Alfredson and Davies [1] measured, *A*, *R* and *M* just within a tail pipe (using it as an impedance tube), found *W* transmitted to the atmosphere from Equation 3.25 and, by assuming spherical wave propagation, calculated *SPL* at a distance r in the far field.

Significantly, they found that their predictions were within 2 dB of their experimental observations spanning frequencies of 228 to 1325 Hz ($ka = 0.218$ to 1.267 for $M = 0.078$ and 0.171). However, all the acoustic power that leaves the tail pipe does not appear in the far field as such. It is considerably attenuated while getting out of the jet. This would also have a bearing on the reflection coefficient. This is discussed in Section 3.5.

3.4 Aeroacoustic State Variables

As has been discussed in the preceding section, the acoustic energy flux through a tube (defined as acoustic perturbation of the thermokinetic energy flux) can be written as the time average of the product of mass flux perturbation and stagnation enthalpy perturbation.

$$W = \overline{m' J'}, \tag{3.28}$$

where

$$m' = (\rho_0 + \rho)S(U + u) - \rho_0 SU$$
$$= \rho_0 Su + \rho SU \text{ (neglecting second-order terms)}$$
$$= v + \frac{p}{c_0^2} SU$$
$$= v + \frac{U}{c_0}\frac{S}{c_0}p$$
$$= v + M\frac{p}{Y_0}$$

(3.29)

and

$$J' = \frac{p_0 + p}{\rho_0} + \frac{1}{2}(U + u)^2 - \left(\frac{p_0}{\rho_0} + \frac{1}{2}U^2\right)$$
$$= \frac{p}{\rho_0} + Uu \text{ (neglecting second-order terms)}$$
$$= \frac{1}{\rho_0}\left(p + \frac{U \cdot c_0}{c_0 S}(\rho_0 Su)\right)$$
$$= \frac{1}{\rho_0}(p + MY_0 v).$$

(3.30)

As stagnation enthalpy J is equal to stagnation pressure divided by ambient density, J' can be looked upon as perturbation stagnation pressure divided by ambient density. Let this be called aeroacoustic pressure, p_c, Similarly, m', being perturbation mass flux, can be called aeroacoustic mass velocity, v_c. Thus, using subscript c for convection [7],

$$W_c = \frac{1}{\rho_0} \overline{v_c \cdot p_c},$$

(3.31)

where aeroacoustic state variables p_c and v_c are linearly related to classical acoustic variables (for stationary medium) as

$$p_c = p + MY_0 v$$

(3.32)

$$v_c = v + \frac{Mp}{Y_0},$$

(3.33)

or as

$$\begin{bmatrix} p_c \\ v_c \end{bmatrix} = \begin{bmatrix} 1 & MY_0 \\ M/Y_0 & 1 \end{bmatrix} \begin{bmatrix} p \\ v \end{bmatrix}.$$

(3.34)

If A is the amplitude of the forward wave and B that of the reflected wave at a particular point in a tube (say, at $z = 0$), then

$$p = A + B, \qquad v = \frac{A - B}{Y_0},$$

$$p_c = p + MY_0 v = A(1 + M) + B(1 - M) = A_c + B_c, \qquad (3.35)$$

$$v_c = v + \frac{Mp}{Y_0} = \frac{A(1 + M) + B(1 - M)}{Y_0} = \frac{A_c - B_c}{Y_0}, \qquad (3.36)$$

and

$$W_c = \frac{1}{\rho_0} \overline{p_c v_c} = \frac{1}{2\rho_0 Y_0} \left(|A^2|(1 + M)^2 - |B^2|(1 - M)^2 \right)$$

$$= \frac{1}{2\rho_0 Y_0} \left(|A_c^2| - |B_c^2| \right). \qquad (3.37)$$

Defining reflection coefficients

$$R \equiv B/A, \qquad R_c \equiv B_c/A_c, \qquad (3.38)$$

one gets

$$W_c = \frac{|A_c^2|}{2\rho_0 Y_0} \left(1 - |R_c^2| \right) \qquad (3.39)$$

and

$$R_c = \frac{B(1 - M)}{A(1 + M)} = R \frac{1 - M}{1 + M}, \qquad (3.40)$$

In the case of convected waves, the reflection coefficient

$$|R_c| \leq 1 \qquad (3.41)$$

or

$$\frac{|R|(1 - M)}{1 + M} \leq 1$$

or

$$|R| \leq \frac{1 + M}{1 - M}. \qquad (3.42)$$

Thus, for moving medium, the reflection coefficient measured with static pressure probes can exceed unity. Aeroacoustic pressure can be picked up only through a total or stagnation-pressure probe facing the flow.

It is obvious from the preceding expressions that the forward wave gets strengthened with mean flow by a factor $1 + M$ for pressure and $(1 + M)^2$ for power associated with it, while the reflected wave against flow gets weakened, as it were, by a factor $1 - M$ and $(1 - M)^2$ for the power associated with it.

3.5 Aeroacoustic Radiation

The aeroacoustic power flux through a tail pipe (Figure 3.6) at the exit point T (just inside the opening) is given by Equation 3.37

$$W_T = \frac{1}{2\rho_0 Y_0} \left(|A^2|(1+M)^2 - |B^2|(1-M)^2 \right)$$

$$= \frac{S}{2\rho_0 c_0} \left(|A_c^2| - |B_c^2| \right) \quad (3.43)$$

$$\frac{S}{2\rho_0 c_0} |A_c^2| \left(1 - |R_c^2| \right) = W_0 \left(1 - |R_c^2| \right),$$

where $S|A_c^2|/2\rho_0 c_0$ is the incident power, W_0 (say).

The acoustic energy W_T that escapes from the pipe is partitioned between two distinct disturbances in the exterior fluid. The first of these is the free-space radiation, whose directivity is equivalent to that produced by monopole (and dipole) sources. Second, essentially incompressible vortex waves are excited by the shedding of vorticity from the pipe tip, and may be associated with the large-scale instabilities of the jet. The interaction of the outcoming wave and the jet absorbs a substantial part of the power of the wave and exerts

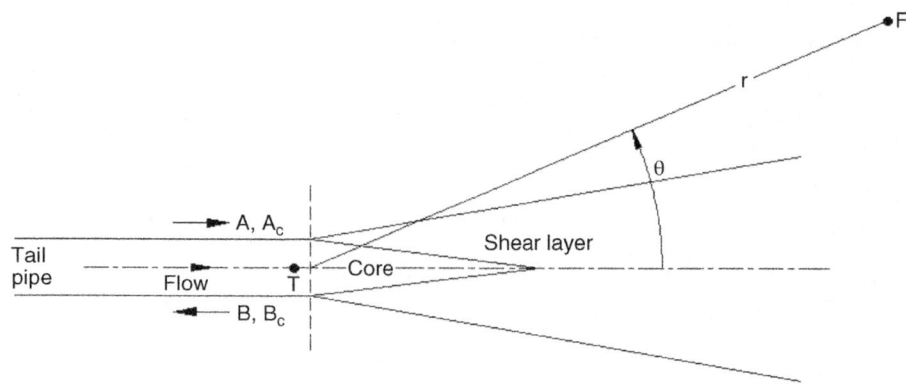

Figure 3.6 Interaction of acoustic field with mean-flow shear

a back reaction on the wave [8–11]. The overall effect of it is reflected in the following expression [10] for a cold jet:

$$W_F = (4\pi r^2) I_F \simeq W_T \cdot \frac{(kr_0)^2}{2M + (kr_0)^2};\qquad(3.44)$$

Subscript F stands for free field (see Figure 3.6), and it has been assumed that the exit Mach number is small enough, that is

$$M^2 \ll 1, \qquad k^4 r_0^4 \ll 1,$$

and the jet is cold (at ambient temperature).

It is obvious from Equation 3.44 that W_F, the total power radiated to the far field, is always less than W_T, the power leaving the tail pipe. The difference, which is plotted in Figure 3.7, can be seen to be significant only at very low values of the Helmholtz number. The experimentally observed values of attenuation [9] are still lower (nearer to zero) than those plotted in the figure. This interaction of sound wave and jet, which attenuates the wave as it comes out, amplifies the broad-band jet noise. However, this is inconsequential inasmuch as the jet noise (even after amplification) remains insignificantly low as compared to the (attenuated) sound wave.

The preceding formulae hold only for a cold jet characterized by the medium within the pipe having the same temperature, density, and velocity of wave propagation as the ambient medium outside.

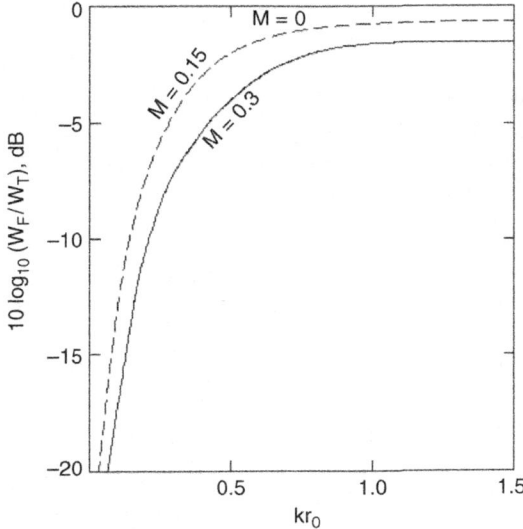

Figure 3.7 The attenuation of the wave in the jet (Reproduced with permission from [9])

Some empirical expressions for radiation resistance in the presence of mean flow are as follows (for $M^2 \ll 1$):

Mechel et al. [6]

$$\frac{R_0(M)}{Y} = \frac{R_0}{Y} - 1.1M = \frac{k^2 r_0^2}{4} - 1.1M; \tag{3.45}$$

Ingard and Singhal [12]

$$\frac{R_0(M)}{Y} = \frac{R_0}{Y} - M; \tag{3.46}$$

Panicker and Munjal [13]

$$\frac{R_0(M)}{Y} = \frac{R_0}{Y} - 2M^2; \tag{3.47}$$

where R_0 is the radiation resistance for a stationary medium.

As Equations 3.45–3.47 hold only for $M < 0.25$, the effect of M (i.e. difference between $R_0(M)$ and R_0) as indicated by Equation 3.47 is less than that implied in Equations 3.45 and 3.46.

As can be observed from Equations 3.45–3.47, the resistive part of the radiation impedance decreases with mean flow and can even be negative at sufficiently low values of kr_0 and hence frequency. However, as remarked earlier, in the presence of mean flow, what matters or counts for noise radiation is

$$Z_{c,0} = Y \frac{1 + R_c}{1 - R_c}, \tag{3.48}$$

the real part of the aeroacoustic radiation impedance remains positive at all frequencies and in fact is more than its counterpart for a stationary medium.

The corresponding values of the reflection coefficient $R(M)$ are given by the following experimental (empirical) formulae:

Mechel et al. [6]

$$|R(M)| = |R|(1 + 2M); \tag{3.49}$$

Panicker and Munjal [13]

$$|R(M)| = |R|\left(1 + 2.5M^2\right) \tag{3.50}$$

The comments made earlier on radiation impedance $R_0(M)$ also apply to the amplitude of the reflection coefficient $|R(M)|$.

3.6 Insertion Loss

With the aeroacoustic state variables replacing static ones in the expressions for acoustic power flux, the insertion loss expression (2.70) can be written as

$$IL_c = 20 \log \left[\left(\frac{\rho_{0,2}}{\rho_{0,1}} \right)^{1/2} \left(\frac{R_{c,0,1}}{R_{c,0,2}} \right)^{1/2} \left| \frac{Z_{c,n+1}}{Z_{c,0,1} + Z_{c,n+1}} \right| |VR_{c,n+1}| \right], \qquad (3.51)$$

where all notations correspond to the static medium case, that is,

$$Z_{c,0} = R_{c,0} + jX_{c,0}; \qquad (3.52)$$

subscripts 1 and 2 stand for without muffler and with muffler, respectively;

$$VR_{c,n+1} = \frac{v_{c,n+1}}{v_{c,0}}, \qquad (3.53)$$

defined with respect to the modified circuit of Figure 3.8; and $Z_{c,n+1}$ is the impedance of the exhaust source at the entrance of the exhaust muffler, defined with respect to convective or aeroacoustic state variables, p_c and v_c.

The velocity ratio $VR_{c,n+1}$ defined by Equation 3.53 can be evaluated in terms of transfer matrices for different constituent elements from the relation

$$\begin{bmatrix} p'_{c,n+1} \\ v_{c,n+1} \end{bmatrix} = [T_{c,n+1}] [T_{c,n}] \cdots [T_{c,r}] \cdots [T_{c,1}] [T_{c,0}] \begin{bmatrix} 0 \\ v_{c,0} \end{bmatrix}, \qquad (3.54)$$

where

$$v_{c,n+1} = p_{c,n+1}/Z_{c,n+1}, \qquad p'_{c,n+1} = p_{c,n}; \qquad (3.55)$$

$$[T_{c,n+1}] = \begin{bmatrix} 1 & 0 \\ 1/Z_{c,n+1} & 1 \end{bmatrix}; \qquad (3.56)$$

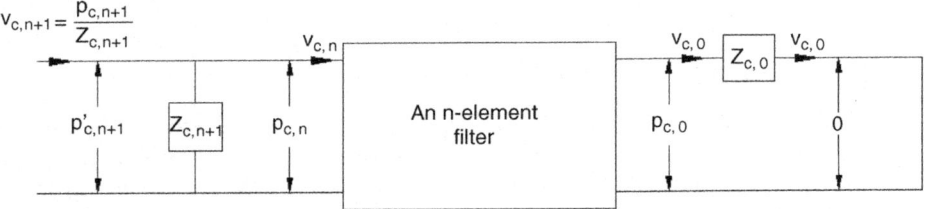

Figure 3.8 Analogous representation of a muffler and its terminations

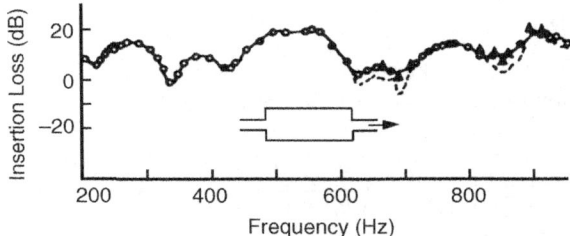

Figure 3.9 Insertion loss of expansion chamber with flow (M = 0.15). ----, Theoretical, using measured impedance; ▲, theoretical, using measured impedance and damping — O — O — O —, experimental (Reproduced with permission from [14])

$$[T_{c,0}] = \begin{bmatrix} 1 & Z_{c,0} \\ 0 & 1 \end{bmatrix}; \tag{3.57}$$

$$Z_{c,0} = \frac{p_{c,0}}{v_{c,0}} = \frac{p_0 + M_1 Y_1 v_0}{v_0 + M_1 p_0/Y_1} = \frac{Z_0(M_1) + M_1 Y_1}{1 + M_1 Z_0(M_1)/Y_1}. \tag{3.58}$$

M_1 is the Mach number of mean flow in the tail pipe U_1/c_1. On writing the product matrix as

$$\begin{bmatrix} A_{11} & A_{12} \\ A_{21} & A_{22} \end{bmatrix}$$

one gets

$$VR_{c,n+1} = \frac{v_{c,n+1}}{v_{c,0}} = A_{22}. \tag{3.59}$$

Thus the required velocity ratio is equal to the second-row-second-column element of the product of all the $n + 2$ transfer matrices from source to radiation load.

Doige and Thawani [14] have demonstrated the validity of the concept of convective source impedance $Z_{c,n+1}$ experimentally. They measured the no-flow impedance of the source Z_{n+1} with a standing wave impedance tube (see Chapter 5), calculated $Z_{c,n+1}$, and made use of relation (3.51) to obtain insertion loss of a typical large expansion chamber. Good agreement was obtained for the flow rate used ($M = 0.15$), as shown in Figure 3.9.

3.7 Transfer Matrices for Tubular Elements

Transfer matrices are derived in the following subsections for various aeroacoustic elements, taking into account the convective as well as dissipative effects of mean flow. These are derived with respect to the aeroacoustic variables p_c and v_c. The corresponding transfer matrices with respect to acoustic variables p and v can be readily derived from these matrices, making use of relation (3.34).

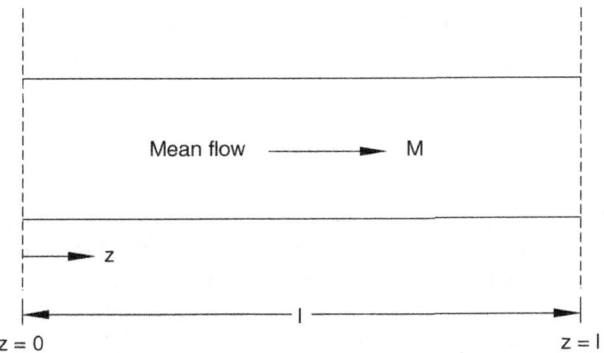

Figure 3.10 A uniform tube with mean flow

3.7.1 Uniform Tube

Acoustic pressure p, mass velocity v, and the corresponding characteristic impedance Y for a uniform tube (Figure 3.10) with viscous medium and turbulent incompressible mean flow are given by relations (1.116), (1.120), and (1.121).

Absorbing the time dependence in the constant, and neglecting $(\alpha(M)/k)^2$ w.r.t. 1, these relations become

$$p(z) = Ae^{-jk^+z} + Be^{+jk^-z}; \tag{3.60}$$

$$v(z) = \frac{1}{Y}\left\{Ae^{-jk^+z} - Be^{+jk^-z}\right\}; \tag{3.61}$$

$$Y = Y_0\left\{1 - \frac{\alpha(M)}{k_0} + j\frac{\alpha(M)}{k_0}\right\}; \tag{3.62}$$

where

$$k^+ = \frac{k_0 - j\alpha(M)}{1+M} = \frac{k_0 - j\alpha(M)}{1-M^2}(1-M) = k_c(1-M); \tag{3.63}$$

$$k^- = \frac{k_0 - j\alpha(M)}{1-M} = \frac{k_0 - j\alpha(M)}{1-M^2}(1+M) = k_c(1+M); \tag{3.64}$$

$$\alpha(M) = \alpha + \xi M; \tag{3.65}$$

$$\xi = F/2d; \tag{3.66}$$

$F =$ Froude's friction factor (see e.g. [15,16], d = diameter of the tube, and

$$k_c = \frac{k_0 - j\alpha(M)}{1-M^2}. \tag{3.67}$$

Equations 3.60 and 3.61, when rewritten in the form

$$p(z) = e^{jMk_c z}\left(Ae^{-jk_c z} + Be^{+jk_c z}\right), \tag{3.68}$$

$$v(z) = \frac{e^{jMk_c z}}{Y}\left(Ae^{-jk_c z} - Be^{+jk_c z}\right), \tag{3.69}$$

may be observed to be very similar to, and in fact reduce to, Equations 2.16 and 2.17 for stationary nonviscous medium. Thus, on making use of the approach outlined in Equations 2.156–2.159, Equations 3.68 and 3.69 yield the transfer matrix relation [7]

$$\begin{bmatrix} p \\ v \end{bmatrix}_{z=0} = e^{-jMk_c l}\begin{bmatrix} \cos k_c l & jY \sin k_c l \\ (j/Y)\sin k_c l & \cos k_c l \end{bmatrix}\begin{bmatrix} p \\ v \end{bmatrix}_{z=l}. \tag{3.70}$$

The corresponding relation in terms of the convective state variables can be written in terms of the transformation matrix (3.34):

$$\begin{bmatrix} p_c \\ v_c \end{bmatrix}_{z=0} = \begin{bmatrix} 1 & MY_0 \\ M/Y_0 & 1 \end{bmatrix}\begin{bmatrix} p \\ v \end{bmatrix}_{z=0}$$

$$= \frac{e^{-jMk_c l}}{1-M^2}\begin{bmatrix} 1 & MY_0 \\ M/Y_0 & 1 \end{bmatrix}\begin{bmatrix} \cos k_c l & jY \sin k_c l \\ (j/Y)\sin k_c l & \cos k_c l \end{bmatrix}\begin{bmatrix} 1 & -MY_0 \\ -M/Y_0 & 1 \end{bmatrix}\begin{bmatrix} p_c \\ v_c \end{bmatrix}_{z=l} \tag{3.71}$$

For nonviscous medium,

$$Y = Y_0,$$

$$k_c = \frac{k_0}{1-M^2} \equiv k_{c0},$$

and the transfer matrix relation (3.71) reduces to [7]

$$\begin{bmatrix} p_c \\ v_c \end{bmatrix}_{z=0} = e^{-jMk_c l}\begin{bmatrix} \cos k_{c0} l & jY_0 \sin k_{c0} l \\ (j/Y_0)\sin k_{c0} l & \cos k_{c0} l \end{bmatrix}\begin{bmatrix} p_c \\ v_c \end{bmatrix}_{z=l}. \tag{3.72}$$

Formal similarity of Equations 3.70 and 3.72 indicates that for a tube with nonviscous (or ideal) moving medium, the transfer matrix defined with respect to the convective variables p_c and v_c is the same as that defined with respect to the acoustic variables p and v, that is,

$$e^{-jMk_{c0} l}\begin{bmatrix} \cos k_{c0} l & jY_0 \sin k_{c0} l \\ (j/Y_0)\sin k_{c0} l & \cos k_{c0} l \end{bmatrix}. \tag{3.73}$$

3.7.2 Extended-Tube Elements

There are four types of such elements, as shown in Figure 2.13. Unlike in the case of a static medium, where static pressure (and hence perturbation pressure) is constant across an area discontinuity, the stagnation pressure (and hence the perturbation p_c) decreases across an area

Table 3.1 Stagnation pressure loss coefficient

Element	K
Sudden contraction and extended outlet [15,16]	$(1 - S_d/S_u)/2$
Sudden expansion and extended inlet [15,16]	$[(S_d/S_u) - 1]^2$
Reversal cum expansion [18]	$(S_d/S_u)^2$
Reversal cum contraction [18]	0.5

discontinuity. As the flow passes a sudden area change, a part of the flow-acoustic energy is converted into heat, which manifests itself as an increase in entropy. This increase in entropy can be accounted for through a parameter that can be evaluated by means of the measured coefficients of the loss in stagnation pressure for incompressible flows $(M^2 \ll 1)$. Thus [17,18]

$$p_{s,3} = p_{s,1} + K\left(\frac{1}{2}\rho_0 U_1^2\right). \tag{3.74}$$

or

$$p_{0,3} + \frac{1}{2}\rho_0 U_3^2 = p_{0,1} + \frac{1}{2}\rho_0 U_1^2 + K\left(\frac{1}{2}\rho_0 U_1^2\right). \tag{3.75}$$

The loss coefficient K has been measured for steady flow for various types of area discontinuities. It is given in Table 3.1, where area change is assumed to be substantial. Here S_u and S_d are areas of cross-section of the upstream tube and downstream tube, respectively.

Equation 3.75, when perturbed, becomes

$$p_{0,3} + p_3 + \frac{1}{2}\rho_0(U_3 + u_3)^2 = p_{0,1} + p_1 + \frac{1}{2}\rho_0(U_1 + u_1)^2 + \frac{1}{2}K\rho_0(U_1 + u_1)^2. \tag{3.76}$$

Subtraction of Equation 3.75 from (3.76) and neglecting of the second-order terms yields

$$p_3 + \rho_0 U_3 u_3 = p_1 + \rho_0 U_1 u_1 + K\rho_0 U_1 u_1$$

or

$$p_3 + M_3 Y_3 v_3 = p_1 + M_1 Y_1 v_1 + K M_1 Y_1 v_1$$

or, in terms of aeroacoustic state variables,

$$p_{c,3} = p_{c,1} + KM_1 Y_1 \frac{v_{c,1} - M_1 p_{c,1}/Y_1}{1 - M_1^2} \tag{3.77}$$

or

$$p_{c,3} = \left(1 - \frac{KM_1^2}{1 - M_1^2}\right) p_{c,1} + \left(\frac{KM_1 Y_1}{1 - M_1^2}\right) v_{c,1}. \tag{3.78}$$

For all practical purposes $M_1 \sim 0.15$ and therefore M_1^2 can be neglected with respect to unity; that is, $M_1^2 \ll 1$.

Then Equation 3.78 reduces to

$$p_{c,3} = p_{c,1} + KM_1 Y_1 v_{c,1} \tag{3.79}$$

As shown in Appendix B, the upstream and downstream aeroacoustic pressure variables $p_{c,3}$ and $p_{c,1}$ are related to the downstream entropy fluctuations s_1 by relations (B.6) [19]:

$$p_{c,3} = p_{c,1} + \frac{p_0 s_1}{R}. \tag{3.80}$$

Now, from Equation B.17 of Appendix B [19],

$$\rho_3 = \frac{p_3}{c_0^2} \quad \text{and} \quad \rho_1 = \frac{p_1 - s_1 p_0 / C_v}{c_0^2}. \tag{3.81}$$

Fortunately, however, entropy fluctuations introduce additional terms that are of the order of M_1^2 (16, 17), and therefore, for practical purposes, Equation 3.81 may be replaced by the isentropocity relations

$$\rho_3 = \frac{p_3}{c_0^2} \quad \text{and} \quad \rho_1 = \frac{p_1}{c_0^2} \tag{3.82}$$

Now equations of mass continuity for steady flow and for the case of aeroacoustic perturbations are

$$\rho_0 S_3 U_3 = \rho_0 S_1 U_1 \tag{3.83}$$

and

$$(\rho_0 + \rho_3) S_3 (U_3 + u_3) = (\rho_0 + \rho_1) S_1 (U_1 + u_1) + (\rho_0 + \rho_2) S_2 u_2, \tag{3.84}$$

respectively.

Subtracting Equation 3.83 from (3.84) and neglecting second-order terms yields

$$\rho_0 S_3 u_3 + \rho_3 S_3 U_3 = \rho_0 S_1 u_1 + \rho_1 S_1 U_1 + \rho_0 S_2 u_2. \tag{3.85}$$

Substituting Equation 3.82 in Equation 3.85, proceeding as above, neglecting terms involving M_1^2 yields

$$v_{c,3} = v_{c,1} + v_2 \tag{3.86}$$

Derivation of a transfer matrix relation between the upstream point 3 and the downstream point 1 requires another relation for elimination of v_2 from Equation 3.86. The equation of momentum balance comes in handy here.

$$p_0 S_3 + \rho_0 S_3 U_3^2 + C_1 \left(p_0 S_1 + \rho_0 S_1 U_1^2 \right) + C_2 p_0 S_2 = 0, \tag{3.87}$$

Table 3.2 Constants C_1 and C_2

Element	C_1	C_2
Extended outlet (Figure 2.13a)	-1	-1
Extended inlet (Figure 2.13b)	-1	$+1$
Reversal-expansion (Figure 2.13c)	$+1$	-1
Reversal-contraction (Figure 2.13d)	$+1$	-1

where the constants C_1 and C_2 are listed below in Table 3.2. In fact, these constants are related by the area relationship

$$S_3 + C_1 S_1 + C_2 S_2 = 0 \tag{3.88}$$

With acoustic perturbation, the corresponding momentum equation would be

$$(p_0 + p_3)S_3 + (\rho_0 + \rho_3)S_3(U_3 + u_3)^2 + C_1\left\{(p_0 + p_1)S_1 + (\rho_0 + \rho_1)S_1(U_1 + u_1)^2\right\} \\ + C_2(p_0 + p_2)S_2 = 0. \tag{3.89}$$

Upon subtracting Equation 3.87 from (3.89) and neglecting the second and higher-order terms, one obtains

$$S_3 p_3 + 2\rho_0 S_3 U_3 u_3 + \rho_3 S_3 U_3^2 + C_1\left\{S_1 p_1 + 2\rho_0 S_1 U_1 u_1 + \rho_1 S_1 U_1^2\right\} + C_2 S_2 p_2 = 0 \tag{3.90}$$

Substituting Equation 3.82 in Equation 3.90, proceeding as above, neglected terms of the order of M_1^2 we get

$$S_3\left(p_{c,3} + M_3 Y_3 v_{c,3}\right) + C_1 S_1\left\{p_{c,1} + M_1 Y_1 v_{c,1}\right\} + C_2 S_2 p_2 = 0. \tag{3.91}$$

Now,

$$p_2/v_2 = Z_2, \tag{3.92}$$

where, for a rigid-end cavity,

$$Z_2 = -jY_2 \cot k_0 l_2. \tag{3.93}$$

Upon substituting in Equation 3.91 for p_2 from Equation 3.92, v_2 from Equation 3.86, and $p_{c,3}$ from Equation 3.79, making use of Equation 3.88, and rearranging, one obtains

$$v_{C,3} = \frac{1}{C_2 S_2 Z_2 + S_3 M_3 Y_3}\left[\{C_2 S_2\}p_{c,1} + \{C_2 S_2 Z_2 - M_1 Y_1(C_1 S_1 + K S_3 v_{c,1})\}\right] \tag{3.94}$$

Finally, Equations 3.79 and 3.94 may be rearranged in the matrix form

$$\begin{bmatrix} p_{c,3} \\ v_{c,3} \end{bmatrix} \simeq \begin{bmatrix} 1 & KM_1Y_1 \\ \dfrac{C_2S_2}{C_2S_2Z_2 + S_3M_3Y_3} & \dfrac{C_2S_2Z_2 - M_1Y_1(C_1S_1 + KS_3)}{C_2S_2Z_2 + S_3M_3Y_3} \end{bmatrix} \begin{bmatrix} p_{c,1} \\ v_{c,1} \end{bmatrix} \quad (3.95)$$

This transfer matrix applies to each of the four extended-tube resonators of Figure 2.14, provided appropriate loss factor K is taken from Table 3.1 and the relevant C_1 and C_2 are read from Table 3.2.

It is worth noting here that for each of these resonators, the transfer matrix of Equation 3.95 reduces to

$$\begin{bmatrix} 1 & 0 \\ 1/Z_2 & 1 \end{bmatrix}$$

for the case of stationary medium. This tallies with the results of Chapter 2 and provides a necessary check.

3.7.3 Simple Area Discontinuities

The elements of sudden contraction and sudden expansion shown in Figure 2.11 fall into this class. Upon comparing Figure 2.11a and 2.11b with Figure 2.13a and 2.13b, respectively, it can be observed that sudden contraction is the limiting case of extended outlet and sudden expansion is the limiting case of extended inlet when length l_2 tends to zero or, as per Equation 3.93, when Z_2 tends to infinity. Therefore, the transfer matrix relation for these two elements can be derived from Equation 3.95 by letting $1/Z_2$ tends to zero and numbering the upstream point as 2 (instead of 3), Thus, one gets

$$\begin{bmatrix} p_{c,2} \\ v_{c,2} \end{bmatrix} \simeq \begin{bmatrix} 1 & KM_1Y_1 \\ 0 & 1 \end{bmatrix} \begin{bmatrix} p_{c,1} \\ v_{c,1} \end{bmatrix} \quad (3.96)$$

For the case of a stationary medium, the foregoing transfer matrix reduces to a unity matrix, which again checks with the analysis of the preceding chapter.

3.7.4 Physical Behavior of Area Discontinuities

Derivation of the transfer matrices above is based on the basic equations formulated by Alfredson and Davies [20] who hypothesized that across a sudden area discontinuity, the stagnation pressure would drop resulting in entropy fluctuations. This results in a lumped inline aeroacoustic resistance KM_1Y_1 in the transfer matrices (3.95) and (3.96) in the first-row second-column position.

Another way of looking at the effect of area discontinuities is that they would excite higher order acoustic modes [21–24]. Lung and Doige [24] wrote the transfer matrix for a sudden area change as

$$\begin{bmatrix} 1 & j\omega L \\ 0 & 1 \end{bmatrix}, \quad (3.97)$$

where the quantity L is the inertance [21] accounting for the inductance effect of the discontinuity:

$$L = l_{ec}/(\pi r_0^2), \tag{3.98}$$

where

$$l_{ec} = \frac{8r_0}{3\pi} H(\alpha), \tag{3.99}$$

$$H(\alpha) \simeq 1 - \alpha, \tag{3.100}$$

and

$$\alpha = \frac{\text{radius of the smaller tube, } r_0}{\text{radius of the larger tube}}. \tag{3.101}$$

They compared their predictions with experiments and also with Munjal's predictions making use of relation (3.96), and concluded that their higher-order-mode hypothesis gives more accurate results than the entropy fluctuation hypothesis [24]. There is, however, a basic difference in the two transfer matrices: The entropy fluctuation introduces an aeroacoustic resistance KM_1Y_1, whereas the higher-order-mode excitation introduces inertive impedance $j\omega L$. Finally multiplication of the transfer matrices (3.96) and (3.97) yields

$$\begin{bmatrix} 1 & KM_1Y_1 + j\omega L \\ 0 & 1 \end{bmatrix} \tag{3.102}$$

3.8 Perforated Elements with Two Interacting Ducts

The analysis of perforated-element mufflers started in 1978, when Sullivan and Crocker [25] suggested an analytical approach to predict the transmission loss (*TL*) of concentric-tube resonators. They solved the coupled equations, writing the acoustic field in the annular cylindrical cavity as an infinite summation of natural modes satisfying the rigid-wall boundary conditions at the two ends. This method was intrinsically weak inasmuch as

a. one needs to select an appropriate number of modes of every case,
b. the modes cannot be described in terms of simple circular functions for nonrigid boundary conditions – in fact, for yielding end walls one may not be able to find an infinite set of orthogonal modal functions – and
c. the method cannot be applied to cross-flow elements.

Notwithstanding these limitations of the analytical technique, Sullivan and Crocker's predictions were amply corroborated by experimental observations for the case of stationary medium.

Sullivan [26,27] followed it up by presenting a segmentation procedure for modeling all types of perforated element mufflers. In this method, each segment is described by a separate

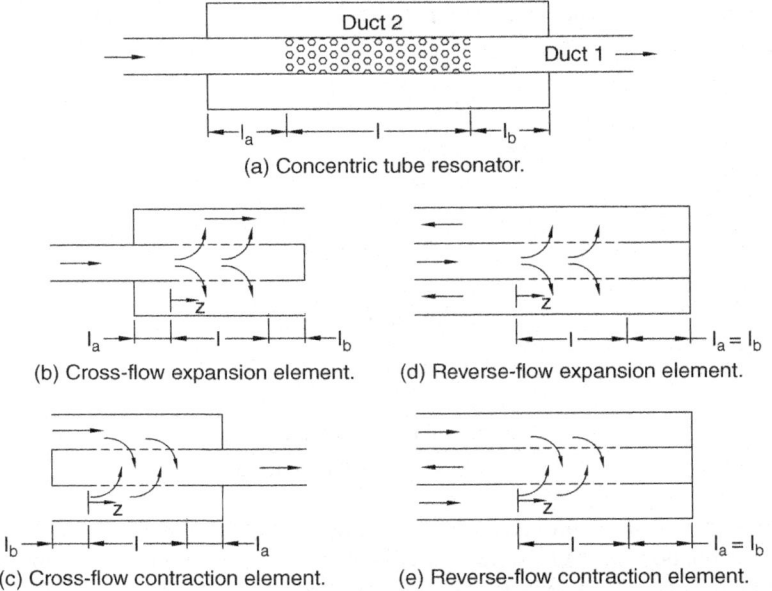

(a) Concentric tube resonator.

(b) Cross-flow expansion element.

(c) Cross-flow contraction element.

(d) Reverse-flow expansion element.

(e) Reverse-flow contraction element.

Figure 3.11 Two-duct perforated elements

transfer matrix. This method, though wider in application, presumes that a perforated element would behave as if it were physically divided in several segments. This arbitrary discretization of a uniformly perforated element is intrinsically unsound, and for better prediction one has to go on increasing the number of segments. Nevertheless, Sullivan's predictions tallied well with experimental findings for different perforated elements.

The following paragraphs describe a generalized decoupling approach, with actual (unequal) mean-flow Mach numbers in the adjoining tubes [22].

For concentric-tube resonators as well as cross-flow elements (Figure 3.11), represented by the common section of Figure 3.12, the mass continuity and momentum balance equations

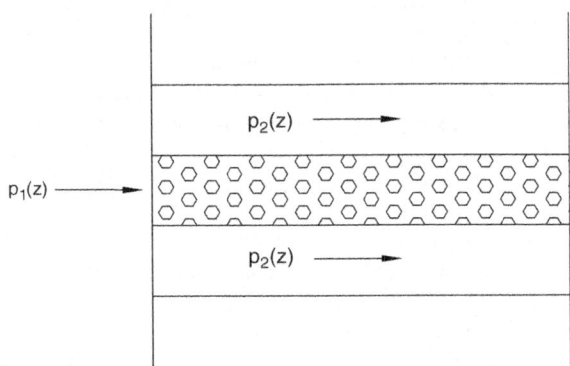

Figure 3.12 The common two-duct perforated section

may be written as [25]

$$U_1 \frac{\partial \rho_1}{\partial z} + \rho_0 \frac{\partial u_1}{\partial z} + \frac{4\rho_0}{d_1} u = -\frac{\partial \rho_1}{\partial t}, \qquad (3.103)$$

$$\rho_0 \left(\frac{Du_1}{Dt}\right) = \frac{-\partial p_1}{\partial z}, \qquad (3.104)$$

for the inner tube of diameter d_1 and

$$U_2 \frac{\partial \rho_2}{\partial z} + \rho_0 \frac{\partial u_2}{\partial z} - \frac{4d_1 \rho_0}{d_2^2 - d_1^2} u = -\frac{\partial \rho_2}{\partial t}, \qquad (3.105)$$

$$\rho_0 \left(\frac{Du_2}{Dt}\right) = -\frac{\partial p_2}{\partial z}, \qquad (3.106)$$

for the outer tube of diameter d_2, where ρ_0, U_1 and U_2 denote the time-average density and axial flow velocities in the inner tube and outer tube (cavity), respectively, while $u_1, u_2, \rho_1, \rho_2, p_1$ and p_2 denote the fluctuations in axial particle velocity, density and pressure in the inner tube and outer tube, respectively, and u is the fluctuating radial particle velocity in the perforations.

Assuming uniform perforate impedance ρ_0, c_0, ζ, the radial particle velocity at the perforations is related to the pressure difference across the perforations as

$$u(z) = [p_1(z) - p_2(z)]/(\rho_0 c_0 \zeta). \qquad (3.107)$$

Assuming that the process is isentropic and that the time dependence of all variables in Equations 3.103–3.106 is harmonic, and eliminating $\rho_1, \rho_2, u, u_1,$ and u_2, yields the following coupled differential equations:

$$\begin{bmatrix} D^2 + \alpha_1 D + \alpha_2 & \alpha_3 D + \alpha_4 \\ \alpha_5 D + \alpha_6 & D^2 + \alpha_7 D + \alpha_8 \end{bmatrix} \begin{bmatrix} p_1(z) \\ p_2(z) \end{bmatrix} = \begin{bmatrix} 0 \\ 0 \end{bmatrix} \qquad (3.108a)$$

or

$$[A(D)]\{p\} = \{0\}, \qquad (3.108b)$$

where

$$\alpha_1 = -\frac{jM_1}{1 - M_1^2}\left(\frac{k_a^2 + k_0^2}{k_0}\right), \quad \alpha_2 = \frac{k_a^2}{1 - M_1^2},$$

$$\alpha_3 = \frac{jM_1}{1 - M_1^2}\left(\frac{k_a^2 - k_0^2}{k_0}\right), \quad \alpha_4 = -\left(\frac{k_a^2 - k_0^2}{1 - M_1^2}\right),$$

$$\alpha_5 = \frac{jM_2}{1 - M_2^2}\left(\frac{k_b^2 - k_0^2}{k_0}\right), \quad \alpha_6 = -\left(\frac{k_b^2 - k_0^2}{1 - M_2^2}\right),$$

$$\alpha_7 = -\frac{jM_2}{1 - M_2^2}\left(\frac{k_b^2 + k_0^2}{k_0}\right), \quad \alpha_8 = \frac{k_b^2}{1 - M_2^2},$$

$$k_0 = \frac{\omega}{c_0}, \quad M_1 = \frac{U_1}{c_0}, \quad M_2 = \frac{U_2}{c_0}, \quad k_a^2 = k_0^2 - \frac{j4k_0}{d_1 \zeta},$$

and

$$k_b^2 = k_0^2 - \frac{j4k_0 d_1}{(d_2^2 - d_1^2)\zeta} \quad \text{and} \quad D = \frac{d}{dz}.$$

The second-order Equation 3.108 can be rearranged as a set of four simultaneous first-order equations as

$$\begin{bmatrix} 0 & 0 & 1 & 0 \\ 0 & 0 & 0 & 1 \\ 1 & 0 & \alpha_1 & \alpha_3 \\ 0 & 1 & \alpha_5 & \alpha_7 \end{bmatrix} \begin{bmatrix} p_1'' \\ p_2'' \\ p_1' \\ p_2' \end{bmatrix} + \begin{bmatrix} -1 & 0 & 0 & 0 \\ 0 & -1 & 0 & 0 \\ 0 & 0 & \alpha_2 & \alpha_4 \\ 0 & 0 & \alpha_6 & \alpha_8 \end{bmatrix} \begin{bmatrix} p_1' \\ p_2' \\ p_1 \\ p_2 \end{bmatrix} = \begin{bmatrix} 0 \\ 0 \\ 0 \\ 0 \end{bmatrix}. \quad (3.109)$$

By defining

$$p_1' = y_1, \quad p_2' = y_2, \quad p_1 = y_3, \text{ and } p_2 = y_4, \quad (3.110)$$

Equation 3.109 reduces to a more convenient form:

$$\begin{bmatrix} -1 & 0 & D & 0 \\ 0 & -1 & 0 & D \\ D & 0 & \alpha_1 D + \alpha_2 & \alpha_3 D + \alpha_4 \\ 0 & D & \alpha_5 D + \alpha_6 & \alpha_7 D + \alpha_8 \end{bmatrix} \begin{bmatrix} y_1 \\ y_2 \\ y_3 \\ y_4 \end{bmatrix} = \begin{bmatrix} 0 \\ 0 \\ 0 \\ 0 \end{bmatrix} \quad (3.111a)$$

or

$$[\Delta]\{y\} = \{0\}. \quad (3.111b)$$

Equation 3.111 is transformed to the principal variables $\Gamma_1, \Gamma_2, \Gamma_3$, and Γ_4, as

$$\begin{bmatrix} D - \beta_1 & 0 & D & 0 \\ 0 & D - \beta_2 & 0 & 0 \\ 0 & 0 & D - \beta_3 & 0 \\ 0 & 0 & 0 & D - \beta_4 \end{bmatrix} \begin{bmatrix} \Gamma_1 \\ \Gamma_2 \\ \Gamma_3 \\ \Gamma_4 \end{bmatrix} = \begin{bmatrix} 0 \\ 0 \\ 0 \\ 0 \end{bmatrix}, \quad (3.112)$$

where the $\beta's$ are the zeros of the characteristic polynomial $|\Delta|$.

Equation 3.112 is the desired decoupled equation. The principal state variables $\Gamma_1, \Gamma_2, \Gamma_3$, and Γ_4 are related to the generalized variables y_1, y_2, y_3, and y_4 through the eigenmatrix $[\psi]$ as

$$\{y\} = [\psi]\{\Gamma\}, \quad (3.113)$$

where

$$\psi_{1,i} = 1, \psi_{2,i} = -\frac{\beta_i^2 + \alpha_1 \beta_i + \alpha_2}{\alpha_3 \beta_i + \alpha_4},$$
$$\psi_{3,i} = 1/\beta_i,$$
$$\psi_{4,i} = \psi_{2,i}/\beta_i = \psi_{2,i}\psi_{3,i},$$

and $i = 1, 2, 3, 4$.

The general solutions to Equation 3.112 can be written as

$$\Gamma_1(z) = C_1 e^{\beta_1 z}, \quad \Gamma_2(z) = C_2 e^{\beta_2 z}.$$
$$\Gamma_3(z) = C_3 e^{\beta_3 z} \quad \text{and} \quad \Gamma_4(z) = C_4 e^{\beta_4 z}.$$
(3.114)

Now, Equations 3.104 and 3.106 may be used to obtain expressions for $u_1(z)$ and $u_2(z)$. Then one can write

$$\begin{bmatrix} p_1(z) \\ p_2(z) \\ \rho_0 c_0 u_1(z) \\ \rho_0 c_0 u_2(z) \end{bmatrix} = [A(z)] \begin{bmatrix} C_1 \\ C_2 \\ C_3 \\ C_4 \end{bmatrix},$$
(3.115)

where

$$A_{1,i} = \psi_{3,i} e^{\beta_i z},$$
$$A_{2,i} = \psi_{4,i} e^{\beta_i z},$$
$$A_{3,i} = -\frac{\psi_{1,i} e^{\beta_i z}}{jk_0 + M_2 \beta_i},$$
$$A_{4,i} = -\frac{\psi_{2,i} e^{\beta_i z}}{jk_0 + M_2 \beta_i},$$

and $i = 1, 2, 3, 4$ for the respective columns of $[A(z)]$.

Finally, the pressures and velocities at $z = 0$ can be related to those at $z = l$ through the transfer matrix relation

$$\begin{bmatrix} p_1(0) \\ p_2(0) \\ \rho_0 c_0 u_1(0) \\ \rho_0 c_0 u_2(0) \end{bmatrix} = [T] \begin{bmatrix} p_1(l) \\ p_2(l) \\ \rho_0 c_0 u_1(l) \\ \rho_0 c_0 u_2(l) \end{bmatrix},$$
(3.116)

where 4×4 transfer matrix $[T]$ is given by

$$[T] = [A(0)][A(l)]^{-1}.$$
(3.117)

The desired 2×2 transfer matrix for a particular two-duct element may be obtained from $[T]$ by making use of the appropriate upstream and downstream variables and two boundary conditions characteristic of the element. Leaving out the details of the elimination/simplification process, the final results are given hereunder for the various two-duct elements [28].

3.8.1 Concentric-Tube Resonator

Boundary conditions (Figure 3.11a)

$$Z_2(0) = \frac{p_2(0)}{-u_2(0)} = -j\rho_0 c_0 \cot(k_0 l_a), \quad (3.118a)$$

$$Z_2(l) = \frac{p_2(l)}{u_2(l)} = -j\rho_0 c_0 \cot(k_0 l_b). \quad (3.118b)$$

The transfer matrix relation

$$\begin{bmatrix} p_1(0) \\ \rho_0 c_0 u_1(0) \end{bmatrix} = \begin{bmatrix} T_a & T_b \\ T_c & T_d \end{bmatrix} \begin{bmatrix} p_1(l) \\ \rho_0 c_0 u_1(l) \end{bmatrix} \quad (3.119)$$

where [28]

$$T_a = T_{11} + A_1 A_2, \qquad T_b = T_{13} + B_1 A_2,$$
$$T_c = T_{31} + A_1 B_2, \qquad T_d = T_{33} + B_1 B_2,$$
$$A_1 = (X_1 T_{21} - T_{41})/F_1, \quad B_1 = (X_1 T_{23} - T_{43})/F_1,$$
$$A_2 = T_{12} + X_2 T_{14}, \qquad B_2 = T_{32} + X_2 T_{34},$$
$$F_1 = T_{42} + X_2 T_{44}, -X_1(T_{22} + X_2 T_{24}),$$
$$X_1 = -j\tan(k_0 l_a) \text{ and } X_2 = j\tan(k_0 l_b).$$

3.8.2 Cross-Flow Expansion Element

Boundary conditions (Figure 3.11b)

$$Z_2(0) = \frac{p_2(0)}{-u_2(0)} = -j\rho_0 c_0 \cot(k_0 l_a), \quad (3.120a)$$

$$Z_1(l) = \frac{p_1(l)}{u_1(l)} = -j\rho_0 c_0 \cot(k_0 l_b). \quad (3.120b)$$

The transfer matrix relation

$$\begin{bmatrix} p_1(0) \\ \rho_0 c_0 u_1(0) \end{bmatrix} = \begin{bmatrix} T_a & T_b \\ T_c & T_d \end{bmatrix} \begin{bmatrix} p_2(l) \\ \rho_0 c_0 u_2(l) \end{bmatrix}, \quad (3.121)$$

where [28]

$$T_a = T_{12} + A_1 A_2, \quad T_a = T_{14} + B_1 A_2,$$
$$T_c = T_{32} + A_1 B_2, \quad T_a = T_{34} + B_1 B_2,$$
$$A_1 = (X_1 T_{22} - T_{42})/F_1, \quad B_1 = (X_1 T_{24} - T_{44})/F_1,$$
$$A_2 = T_{11} + X_2 T_{13}, \quad B_2 = T_{31} + X_2 T_{33},$$
$$F_1 = T_{41} + X_2 T_{43}, -X_1(T_{21} + X_2 T_{23}),$$
$$X_1 = -j\tan(k_0 l_a) \text{ and } X_2 = j\tan(k_0 l_b).$$

3.8.3 Cross-Flow Contraction Element

Boundary conditions (Figure 3.11c)

$$Z_1(0) = \frac{p_1(0)}{-u_1(0)} = -j\rho_0 c_0 \cot(k_0 l_b), \qquad (3.122a)$$

$$Z_2(l) = \frac{p_2(l)}{u_2(l)} = -j\rho_0 c_0 \cot(k_0 l_a). \qquad (3.122b)$$

The transfer matrix relation

$$\begin{bmatrix} p_2(0) \\ \rho_0 c_0 u_2(0) \end{bmatrix} = \begin{bmatrix} T_a & T_b \\ T_c & T_d \end{bmatrix} \begin{bmatrix} p_1(l) \\ \rho_0 c_0 u_1(l) \end{bmatrix}, \qquad (3.123)$$

where

$$T_a = T_{21} + A_1 A_2, \quad T_a = T_{23} + B_1 A_2,$$
$$T_c = T_{41} + A_1 B_2, \quad T_a = T_{43} + B_1 B_2,$$
$$A_1 = (X_1 T_{11} - T_{31})/F_1, \quad B_1 = (X_1 T_{13} - T_{33})/F_1,$$
$$A_2 = T_{22} + X_2 T_{24}, \quad B_2 = T_{42} + X_2 T_{44},$$
$$F_1 = T_{32} + X_2 T_{34}, -X_1(T_{12} + X_2 T_{14}),$$
$$X_1 = -j\tan(k_0 l_a) \text{ and } X_2 = j\tan(k_0 l_b).$$

3.8.4 Some Remarks

The preceding transfer matrices for perforated elements have been derived in the form

$$\begin{bmatrix} p_1 \\ \rho_0 c_0 u_1 \end{bmatrix} = \begin{bmatrix} T_a & T_b \\ T_c & T_d \end{bmatrix} \begin{bmatrix} p_2 \\ \rho_0 c_0 u_2 \end{bmatrix}, \qquad (3.124)$$

This can be rewritten as

$$\begin{bmatrix} p_1 \\ Y_1 v_1 \end{bmatrix} = \begin{bmatrix} T_a & T_b \\ T_c & T_d \end{bmatrix} \begin{bmatrix} p_2 \\ Y_2 v_2 \end{bmatrix}, \qquad (3.125)$$

and, finally, in the usual form

$$\begin{bmatrix} p_1 \\ v_1 \end{bmatrix} = \begin{bmatrix} T_a & Y_2 T_b \\ \dfrac{T_c}{Y_1} & \dfrac{Y_2 T_4}{Y_1} \end{bmatrix} \begin{bmatrix} p_2 \\ v_2 \end{bmatrix}, \qquad (3.126)$$

where Y_1 and Y_2 are characteristic impedances of the upstream and downstream tubes. Finally, the required transfer matrix defined with respect to convective-state variables is found by combining Equation 3.126 with Equation 3.34 as

$$\begin{bmatrix} 1 & M_1 Y_1 \\ M_1/Y_1 & 1 \end{bmatrix} \begin{bmatrix} T_a & Y_2 T_b \\ T_c/T_1 & T_d Y_2/Y_1 \end{bmatrix} \begin{bmatrix} 1 & M_2 Y_2 \\ M_2/Y_2 & 1 \end{bmatrix}^{-1}. \qquad (3.127)$$

At this stage, a comparison of Munjal *et al.*'s distributed parameter approach [28] and Sullivan's segmentation approach [26,27] would be in order. The former is more elegant inasmuch as it yields a closed-form solution and treats a uniformly perforated element as a distributed element. However, the mean-flow velocity, which decreases as more and more of the flow crosses over to the annular duct through the perforate, is assumed to be constant (at its average value, say) in the distributed parameter approach. Sullivan's segmentation approach has an edge here because the decrease in the mean-flow velocity from segment to segment is easily accounted for. The limitation of the distributed parameter approach introduces a small error in the convective effect of mean flow, which in any case is relatively small [28]. Much more important is the effect of perforate impedance, the expressions for which are given in the following section. This depends on mean-flow velocity through the holes, and this velocity is assumed to be uniform in both the approaches.

Finally, it may be noted that instead of decoupling the two (or three) second order differential equations, one could combine them into one fourth-(or sixth-) order equation and solve it for four (or six) complex roots directly by computer using readily available subroutines. However, the foregoing decoupling analysis is aimed at derivation of closed-form expressions for the four-pole parameters of various two-duct and three-duct perforated elements. The resulting transfer matrices may then be combined with those of other elements upstream and downstream of the perforated elements) to evaluate the overall performance of the muffler.

3.9 Acoustic Impedance of Perforates

As may be noted from the preceding section, the acoustic impedance of perforates is the most important parameter in the aeroacoustic analysis of perforated-element mufflers. It is a complex function of several physical variables, namely porosity (assumed to be uniform), mean-flow velocity through the holes or grazing the holes, diameter, and thickness of the tube and diameter of the holes. As it is very difficult to model analytically the interaction of flow and waves thorough holes located near one another, direct measurement of the aeroacoustic impedance of perforates has been resorted to.

As a result of his experiments at different frequencies and varying Reynolds number of the flow (through the orifices), Ingard and Labate [29] experimentally showed that nonlinear behavior of orifice impedance is due to the interaction between the sound field and circulatory effects. Later, Ingard and Ising [30] conducted experiments (on an orifice) in which acoustic particle velocity was measured directly by means of a hot-wire anemometer. They made a significant observation: the influence of the steady flow on the orifice impedance in the linear range is quite similar to the influence of the acoustic particle velocity amplitude u_0 in the nonlinear range in the absence of steady flow [30]. In order to conceptualize the behavior of the acoustic impedance of an orifice, Ronneberger [31] attempted measurement of the impedance of an orifice in the wall of flow duct (grazing flow impedance), and observed that the resistance of the orifice at large values of the parameter U_0/r_0 is independent of frequency and increases linearly with flow velocity U_0 along the duct of r_0. He attributed this behavior to the building up of a thin flow shear layer above the orifice.

A detailed theoretical analysis of the acoustic impedance of perforated plates (with a reasonably large number of orifices in close proximity to one another), with experimental corroboration, was provided by Melling [32] for medium and high sound pressure levels (corresponding to the linear and nonlinear ranges). His studies however were limited to stationary media. Dean [33] made use of his two-microphone method for *in situ* measurement of impedance of orifice in ducts carrying mean (grazing) flows for a large number of perforated samples. But the presence of a honeycomb in the backing cavity makes his results unsuitable for automotive applications.

Sullivan and Crocker attempted to measure the perforated impedance for grazing flow using the two-microphone method, but they reported results for stationary media only [25]. Later, Sullivan [34] did a good job of reviewing the existing literature and identifying the best available impedance formulae for the case of zero mean flow as well as through-flow. These are as follows:

Perforates with cross-flow [27]

$$\zeta = \frac{p}{\rho_0 c_0 u} = \left[0.514 \frac{d_1 M}{l\sigma} + j0.95 k_0(t + 0.75 d_h)\right]/\sigma, \qquad (3.128)$$

where

d_1 is diameter of the perforated tube,
M is the mean-flow Mach number in the tube,
l is the length of perforate,
σ is porosity,
f is frequency,
t is the thickness of the perforated tube, and
d_h is the hole diameter.

Perforates in stationary medium [25,26] (linear case)

$$\zeta = \left[6 \times 10^{-3} + jk_0(t + 0.75 d_h)\right]/\sigma, \qquad (3.129)$$

For the case of grazing flow, however, the stationary medium impedance was suggested with the implicit assumption that the mean flow does not enter the cavity. Later, Sullivan, by the use of rather simple experiments involving measurement of mean flow in the cavity and sound pressure variations in tube and cavity (radial as well as circumferential), determined that the net mass flow in the cavity of a concentric resonator was small, but the resulting mean

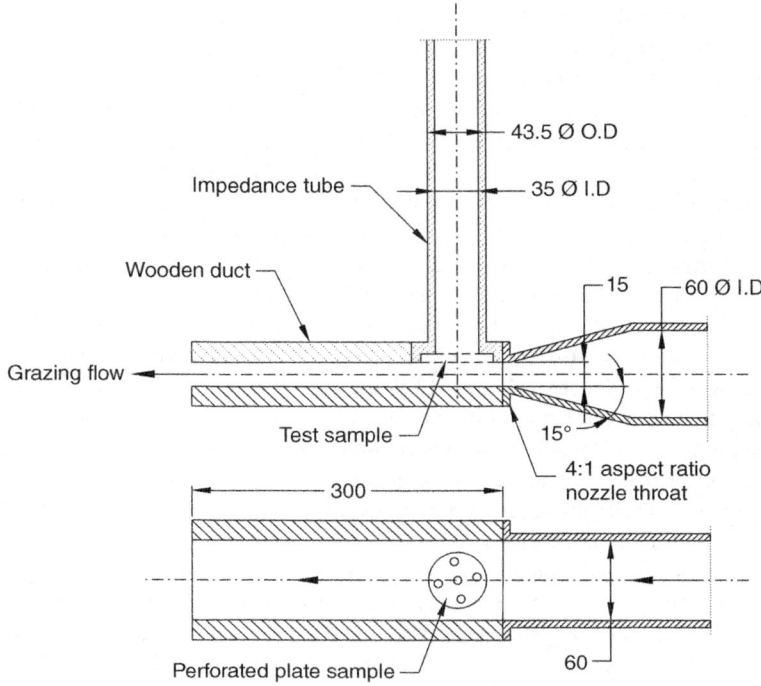

Figure 3.13 Measurement of the impedance of a perforated plate in the presence of grazing flow

flow through the perforations was not small relative to the particle velocity in the orifice [34]. This may have a significant influence on orifice impedance through the mechanism of the discharge coefficient. Sullivan also observed that spatial variations of acoustic pressure in the transverse direction of either tube or cavity are small provided the tube is perforated axisymmetrically and uniformly.

Obviously, there was a need for a comprehensive empirical formula for grazing-flow impedance of perforates for use in the analysis of a concentric-tube resonator. Making use of an experimental setup as shown in Figure 3.15, Rao and Munjal [35] evaluated impedance of a perforated plate by measuring the impedance without the plate and subtracting this value from all the measured values with the perforated plate in position (see Figure 3.13). The acoustic impedance of a perforate is a function of many geometrical and operating parameters, namely, grazing-flow Mach number M, porosity σ, plate thickness t, and hole (or orifice) diameter d_h.

These parameters were varied one at a time, in steps, as follows:

$M = 0.05, \underline{0.1}, 0.15, 0.2$
$\sigma = 0.0309, \underline{0.0412}, 0.072, 0103$
$t = 1/24, \underline{1/16}, 1/8$ in.
$d_h = 1.75, 2.5, \underline{3.5}, 5.0, 7.0$ mm.

The underlined values indicate the default option values when some other parameter is being varied. Least squares fits were obtained for the dependence of normalized impedance of the perforate on each of the four parameters. The resulting expressions were combined to obtain the following empirical formula:

Perforates with grazing flow [35]

$$\zeta = \left[7.337 \times 10^{-3}(1 + 72.23M) + j2.2245 \times 10^{-5}(1 + 51t)(1 + 204d_h)f\right]/\sigma. \quad (3.130)$$

where t and d_h are in meters.

The perforate impedance formula (3.130) was substantiated for use in the analysis of mufflers with concentric-tube resonators by calculating the noise reduction of such a muffler (with uniformly perforated tube), making use of this formula and comparing it with measured values of noise reduction. The two tallied quite well at all frequencies (for which only plane-wave propagation would be there), indicating thereby the applicability of the empirical formula (3.130) of the grazing-flow impedance to concentric-tube resonators.

It is obvious from expression (3.128)–(3.130) that mean flow affects mainly the resistive part of the perforate impedance and that its contribution dominates over the basic stationary-medium acoustic resistance.

3.10 Matrizant Approach

This is a very elegant and powerful alternative for derivation of transfer matrices [36]. Based on the matrix calculus theory [37], it can be explained simply as follows:

A first-order linear ordinary differential equation

$$\frac{dy}{dz} = A(z)y \quad (3.131)$$

can be rewritten as

$$\frac{dy}{y} = A(z)dz,$$

which on integration would yield

$$\ln y \big|_{y(0)}^{y(l)} = \int_0^l A(z)dz$$

or

$$y(0) = y(l)e^{-\int_0^l A(z)dz}. \quad (3.132)$$

If y were replaced by a state vector $\{y\}$ and $A(z)$ by a coefficient matrix $[A(z)]$, then Equations 3.131 and 3.132 would be replaced by

$$\{y\}' = [A(z)]\{y\} \quad (3.133)$$

and

$$\{y(0)\} = \left[e^{-\int_0^l [A(z)dz]}\right]\{y(l)\}, \tag{3.134}$$

respectively. Equation 3.134 is obviously a transfer matrix relation between the state vectors at $z = 0$ and $z = l$. If the coefficient matrix $A(z)$ were constant (i.e. independent of the axial coordinate z), then the transfer matrix in Equation 3.134 would reduce to

$$\left[e^{-[A]l}\right],$$

which is given by [37]

$$\left[e^{-[A]l}\right] = [\psi]^{-1}\left\lceil e^{-\beta l}\right\rceil[\psi], \tag{3.135}$$

where $[\psi]$ is the modal matrix, $\beta's$ are eigenvalues of the coefficient matrix [A], and $\lceil e^{-\beta l}\rceil$ is a diagonal matrix. Thus, the transfer matrix may be obtained directly from the coefficient matrix [A] of Equation 3.133 if the basic equations of mass continuity and momentum balance are arranged in the canonical form with the state variables p and u being normalized as p and $\rho_0 c_0 u$, so that that product $[A]l$ is non-dimensional. Of course, ρ, the density variable in the mass continuity equations, must first be replaced with p/c_0^2, and the local time derivative $\partial/\partial t$ with $j\omega$.

For example, for plane waves along a uniform tube with stationary inviscid medium, for harmonic time dependence $\exp(j\omega t)$, Equations 1.1 and 1.2 may be written as

$$\rho_0 \frac{du}{dz} + \frac{j\omega}{c_0^2}p = 0 \quad \text{and} \quad \rho_0 j\omega u + \frac{dp}{dz} = 0 \tag{3.136a,b}$$

or

$$\rho_0 c_0 \frac{du}{dz} + jk_0 p = 0 \quad \text{and} \quad jk_0 \rho_0 c_0 u + \frac{dp}{dz} = 0. \tag{3.137a,b}$$

These two equations may now be rearranged in the canonical form

$$\begin{bmatrix} p \\ \rho_0 c_0 u \end{bmatrix}' = \begin{bmatrix} 0 & -jk_0 \\ -jk_0 & 0 \end{bmatrix}\begin{bmatrix} p \\ \rho_0 c_0 u \end{bmatrix}, \tag{3.138}$$

where (') denotes differential operator d/dz, and as usual, wave number $k_0 = \omega/c_0$.

Comparing Equation 3.138 with (3.133), it may be noted that

$$[A] = \begin{bmatrix} 0 & -jk_0 \\ -jk_0 & 0 \end{bmatrix} \text{ and } [\beta I - [A]] = \begin{bmatrix} \beta & jk_0 \\ jk_0 & \beta \end{bmatrix}. \tag{3.139a,b}$$

Its eigenvalues may be seen to be

$$\beta_1 = -jk_0 \text{ and } \beta_2 = jk_0, \tag{3.140a}$$

and the eigenmatrix $[\psi]$ is given by

$$[\psi] = \begin{bmatrix} 1 & 1 \\ 1 & -1 \end{bmatrix}. \tag{3.140b}$$

Now, use of the matrix Equation 1.135 yields the following transfer matrix for the uniform tube:

$$[T] = \begin{bmatrix} 1 & 1 \\ 1 & -1 \end{bmatrix}^{-1} \begin{bmatrix} e^{jk_0 l} & 0 \\ 0 & e^{-jk_0 l} \end{bmatrix} \begin{bmatrix} 1 & 1 \\ 1 & -1 \end{bmatrix} = \begin{bmatrix} \cos k_0 l & j \sin k_0 l \\ j \sin k_0 l & \cos k_0 l \end{bmatrix}. \tag{3.141}$$

Thus,

$$\begin{bmatrix} p(0) \\ \rho_0 c_0 u(0) \end{bmatrix} = \begin{bmatrix} \cos k_0 l & j \sin k_0 l \\ j \sin k_0 l & \cos k_0 l \end{bmatrix} \begin{bmatrix} p(l) \\ \rho_0 c_0 u(l) \end{bmatrix}. \tag{3.142}$$

As

$$\rho_0 c_0 u = \frac{c_0}{S} \cdot \rho_0 S u = Y_0 v,$$

Equation 3.142 may be rewritten as

$$\begin{bmatrix} p(0) \\ v(0) \end{bmatrix} = \begin{bmatrix} \cos k_0 l & jY_0 \sin k_0 l \\ j/Y_0 \sin k_0 l & \cos k_0 l \end{bmatrix} \begin{bmatrix} p(l) \\ v(l) \end{bmatrix}, \tag{3.143}$$

which tallies with the transfer relation (2.159).

It may be noted that this Matrizant approach makes use of basic equations and therefore does not involve derivation and solution of the wave equation. This is a very distinct advantage for derivation of transfer matrices of complex elements like perforated elements involving 2, 3, 4 or several interacting ducts. This is illustrated in the following section for three interacting ducts.

3.11 Perforated Elements with Three Interacting Ducts

For the three-duct section shown in Figure 3.14, which is common to the three-duct muffler elements shown in Figure 3.15, the mass continuity and momentum equations may be written as [25]

$$\rho_0 \frac{\partial u_1}{\partial z} + U_1 \frac{\partial \rho_1}{\partial z} + \frac{4}{d_1} \rho_0 u_{1,2} = -\frac{\partial \rho_1}{\partial t} \tag{3.144}$$

and

$$\rho_0 \frac{Du_1}{Dt} = -\frac{\partial p_1}{\partial z} \quad (3.145)$$

for the inner duct of diameter d_1;

$$\rho_0 \frac{\partial u_2}{\partial z} + U_2 \frac{\partial \rho_2}{\partial z} - \frac{4d_1}{d_2^2 - d_1^2 - d_3^2} \rho_0 u_{1,2} + \frac{4d_3}{d_2^2 - d_1^2 - d_3^2} \rho_0 u_{2,3} = -\frac{\partial \rho_2}{\partial t} \quad (3.146)$$

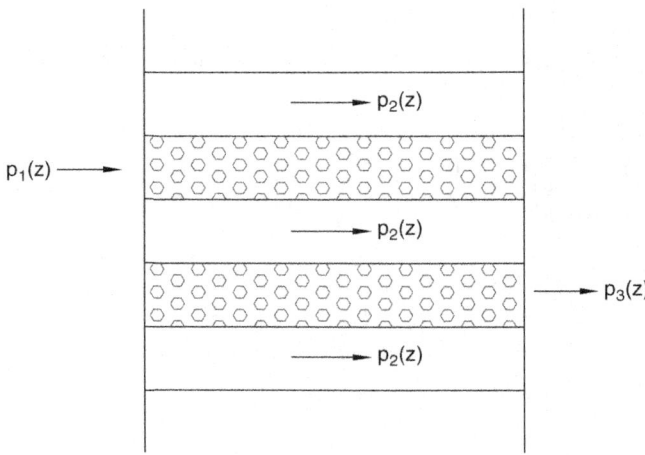

Figure 3.14 The common three-duct perforated section

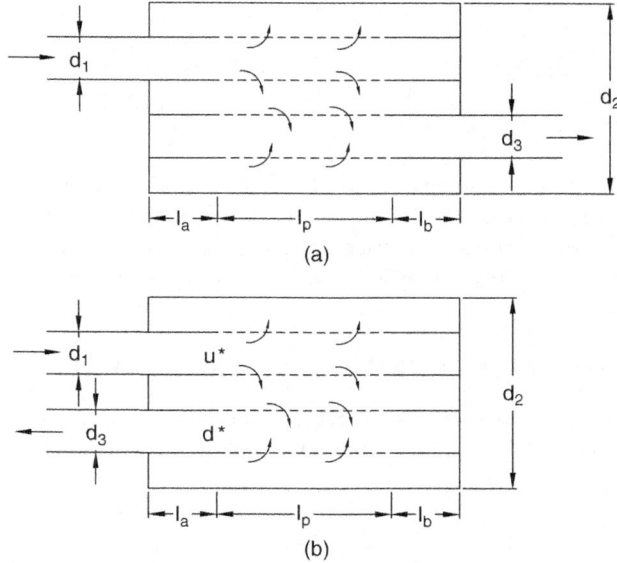

Figure 3.15 Three-duct muffler components. (a) Cross-flow expansion chamber. (b) Reverse-flow expansion chamber

and

$$\rho_0 \frac{Du_2}{Dt} = -\frac{\partial p_2}{\partial z} \qquad (3.147)$$

for the outer duct or diameter d_2; and

$$\rho_0 \frac{\partial u_3}{\partial z} + U_3 \frac{\partial \rho_3}{\partial z} - \frac{4}{d_3} \rho_0 u_{2,3} = -\frac{\partial \rho_3}{\partial t} \qquad (3.148)$$

and

$$\rho_0 \frac{Du_3}{Dt} = -\frac{\partial p_3}{\partial z} \qquad (3.149)$$

for the outer duct or diameter d_3;
The radial momentum equations at the interfaces of duct 1 and duct 3 are

$$u_{1,2} = \frac{p_1 - p_2}{\rho_0 c_0 \zeta_1} \qquad (3.150)$$

and

$$u_{2,3} = \frac{p_2 - p_3}{\rho_0 c_0 \zeta_2}. \qquad (3.151)$$

These equations may be observed to be formally similar to those for the two-duct section of Figure 3.12. Thus the complete analysis proceeds on the same lines as for the two-duct elements.

Replacing $\partial/\partial t$ with $j\omega$, and $\rho_1, \rho_2,$ and ρ_3 with $p/c_0^2, p/c_0^2$ and p_3/c_0^2, and making use of Equations 3.150 and 3.151, Equations 3.144–3.149 may be arranged in the canonical form

$$\begin{bmatrix} p_1 \\ p_2 \\ p_3 \\ \rho_0 c_0 u_1 \\ \rho_0 c_0 u_2 \\ \rho_0 c_0 u_3 \end{bmatrix}' = \begin{bmatrix} A_{11} & A_{12} & A_{13} & A_{14} & A_{15} & A_{16} \\ A_{21} & A_{22} & A_{23} & A_{24} & A_{25} & A_{26} \\ A_{31} & A_{32} & A_{33} & A_{34} & A_{35} & A_{36} \\ A_{41} & A_{42} & A_{43} & A_{44} & A_{45} & A_{46} \\ A_{51} & A_{52} & A_{53} & A_{54} & A_{55} & A_{56} \\ A_{61} & A_{62} & A_{63} & A_{64} & A_{65} & A_{66} \end{bmatrix} \begin{bmatrix} p_1 \\ p_2 \\ p_3 \\ \rho_0 c_0 u_1 \\ \rho_0 c_0 u_2 \\ \rho_0 c_0 u_3 \end{bmatrix} \qquad (3.152)$$

This process is considerably simplified by neglecting the convective effect of mean flow while retaining its effect on the perforate impedance expressions for cross flow (Equation 3.128 as well as grazing flow (Equation 3.130). The neglect of convective effects of mean flow appears to overestimate TL upto 5% (in decibels) [28]. Thus, we let $U_1 = U_2 = U_3$ tend to zero, and accordingly, $D/Dt \rightarrow \partial/\partial t$ in Equations 3.144–3.149. With these simplifications or approximations, Equations 3.144–3.149 may be written as follows:

$$\rho_0 c_0 \frac{\partial u_1}{\partial z} = -\left(jk_0 + \frac{4}{d_1 \zeta_1}\right) p_1 + \frac{4}{d_1 \zeta_1} p_2, \qquad (3.153)$$

$$\frac{dp_1}{dz} = -jk_0(\rho_0 c_0 u_1), \tag{3.154}$$

$$\rho_0 c_0 \frac{\partial u_2}{\partial z} = \frac{4d_1}{(d_2^2 - d_1^2 - d_3^2)\zeta_1} p_1 - \frac{4\left(\frac{d_3}{\zeta_2} + \frac{d_1}{\zeta_1}\right) + jk_0}{d_2^2 - d_1^2 - d_3^2} p_2 + \frac{4d_3}{(d_2^2 - d_1^2 - d_3^2)\zeta_2} p_3, \tag{3.155}$$

$$\frac{dp_2}{dz} = -jk_0 \cdot (\rho_0 c_0 u_2), \tag{3.156}$$

$$\rho_0 c_0 \frac{\partial u_3}{\partial z} = \frac{4}{d_1 \zeta_2} p_2 - \left(jk_0 + \frac{4}{d_3 \zeta_2}\right) p_3, \tag{3.157}$$

$$\frac{dp_3}{dz} = -jk_0 \cdot (\rho_0 c_0 u_3). \tag{3.158}$$

Thus, elements of the coefficient matrix [A] in Equation 3.152 above are given by the following expressions:

$$A_{14} = -jk_0, \quad A_{41} = -\left(jk_0 + \frac{4}{d_1 \zeta_1}\right), \quad A_{42} = \frac{4}{d_1 \zeta_1}, \quad A_{25} = -jk_0,$$

$$A_{51} = \frac{4d_1}{(d_2^2 - d_1^2 - d_3^2)\zeta_1}, \quad A_{52} = -\frac{jk_0 + 4\left(\frac{d_3}{\zeta_2} + \frac{d_1}{\zeta_1}\right)}{d_2^2 - d_1^2 - d_3^2}, \quad A_{53} = \frac{4d_3}{(d_2^2 - d_1^2 - d_3^2)\zeta_2},$$

$$A_{36} = -jk_0, \quad A_{62} = \frac{4}{d_3 \zeta_2}, \quad A_{63} = -\left(jk_0 + \frac{4}{d_3 \zeta_2}\right).$$

$$\tag{3.159}$$

The remaining 26 elements of the coefficient matrix are all zero. Substituting the resultant coefficient matrix [A] of Equation 3.152 in Equation 3.134 yields

$$\begin{bmatrix} p_1(0) \\ p_2(0) \\ p_3(0) \\ \rho_0 c_0 u_1(0) \\ \rho_0 c_0 u_2(0) \\ \rho_0 c_0 u_3(0) \end{bmatrix} = [T] \begin{bmatrix} p_1(0) \\ p_2(0) \\ p_3(0) \\ \rho_0 c_0 u_1(l) \\ \rho_0 c_0 u_2(l) \\ \rho_0 c_0 u_3(l) \end{bmatrix} \tag{3.160}$$

where

$$[T] = \left[e^{-[A]l}\right]. \tag{3.161}$$

The transfer matrix $[T]$ across the perforate, between $z = 0$ and $z = l$, in Figure 3.15 may be evaluated in FORTRAN as an eigenvalue problem by means of Equation 3.135, or directly in MATLAB® making use of Equation 3.161 by means of the library function EXPM.

The desired 2×2 transfer matrix for a particular three-duct element may be obtained from $[T]$, making use of the appropriate upstream and downstream variables and four boundary conditions characteristics of the element. Leaving out the details of the elimination/simplification process, the final results are given hereunder for the two three-duct elements of Figure 3.15 [38].

3.11.1 Three-Duct Cross-Flow Expansion Chamber Element

Referring to the configuration of Figure 3.15a, the boundary conditions are:

$$Z_2(0) = \frac{p_2(0)}{-u_2(0)} = -j\rho_0 c_0 \cot(k_0 l_a) \quad (3.162a)$$

$$Z_3(0) = \frac{p_3(0)}{-u_3(0)} = -j\rho_0 c_0 \cot(k_0 l_a) \quad (3.162b)$$

$$Z_1(l) = \frac{p_1(l)}{u_1(l)} = -j\rho_0 c_0 \cot(k_0 l_b) \quad (3.162c)$$

$$Z_2(l) = \frac{p_2(l)}{u_2(l)} = -j\rho_0 c_0 \cot(k_0 l_b). \quad (3.162d)$$

Use of these boundary conditions on Equation 3.160 yields

$$\begin{bmatrix} p_1(0) \\ \rho_0 c_0 u_1(0) \end{bmatrix} = \begin{bmatrix} T_a & T_b \\ T_c & T_d \end{bmatrix} \begin{bmatrix} p_3(l) \\ \rho_0 c_0 u_3(l) \end{bmatrix} \quad (3.163)$$

where [38]

$$\begin{aligned}
T_a &= TT_{1,2} + A_3 C_3, & T_b &= TT_{1,4} + B_3 C_3, \\
T_c &= TT_{3,2} + A_3 D_3, & T_4 &= TT_{3,4} + B_3 D_3, \\
A_3 &= (TT_{2,2} X_2 - TT_{4,2})/F_2, \\
B_3 &= (TT_{2,4} X_2 - TT_{4,4})/F_2, \\
C_3 &= TT_{1,1} + X_1 TT_{1,3}, \quad D_3 = TT_{3,1} + TT_{3,3} X_1, \\
F_2 &= TT_{4,1} + X_1 TT_{4,3}, - X_2 (TT_{2,1} + X_1 TT_{2,3}),
\end{aligned} \quad (3.164)$$

and $[TT]$ is an intermediate 4×4 matrix defined as

$$\begin{bmatrix} p_1(0) \\ p_2(0) \\ \rho_0 c_0 u_1(0) \\ \rho_0 c_0 u_2(0) \end{bmatrix} = [TT] \begin{bmatrix} p_2(l) \\ p_3(l) \\ \rho_0 c_0 u_2(l) \\ \rho_0 c_0 u_3(l) \end{bmatrix} \quad (3.165)$$

with

$$TT_{1,1} = A_1 A_2 + T_{1,2}, \quad TT_{1,2} = B_1 A_2 + T_{1,3},$$
$$TT_{1,3} = C_1 A_2 + T_{1,5}, \quad TT_{1,4} = D_1 A_2 + T_{1,6},$$
$$TT_{2,1} = A_1 B_2 + T_{2,2}, \quad TT_{2,2} = B_1 B_2 + T_{2,3},$$
$$TT_{2,3} = C_1 B_2 + T_{2,5}, \quad TT_{2,4} = D_1 B_2 + T_{2,6},$$
$$TT_{3,1} = A_1 C_2 + T_{4,2}, \quad TT_{3,2} = B_1 C_2 + T_{4,3},$$
$$TT_{3,3} = C_1 C_2 + T_{4,5}, \quad TT_{3,4} = D_1 C_2 + T_{4,6},$$
$$TT_{4,1} = A_1 D_2 + T_{5,2}, \quad TT_{4,2} = B_1 D_2 + T_{5,3},$$
$$TT_{4,3} = C_1 D_2 + T_{5,5}, \quad TT_{4,4} = D_1 D_2 + T_{5,6}, \qquad (3.166)$$
$$A_1 = (T_{3,2} X_2 - T_{6,2})/F_1, \quad B_1 = (T_{3,3} X_2 - T_{6,3})/F_1,$$
$$C_1 = (T_{3,5} X_2 - T_{6,5})/F_1, \quad D_1 = (T_{3,6} X_2 - T_{6,6})/F_1,$$
$$A_2 = T_{1,1} + T_{1,4} X_1, \quad B_2 = T_{2,1} + T_{2,4} X_1,$$
$$C_2 = T_{4,1} + T_{4,4} X_1, \quad D_2 = T_{5,1} + T_{5,4} X_1,$$
$$F_1 = T_{6,1} + X_1 T_{6,4} - X_2 (T_{3,1} + X_1 T_{3,4})$$
$$X_1 = j \tan(k_0 l_b), \text{ and}$$
$$X_2 = -j \tan(k_0 l_a).$$

3.11.2 Three-Duct Reverse Flow Expansion Chamber Element

Boundary conditions relevant to the configuration of Figure 3.15b are:

$$Z_2(0) = \frac{p_2(0)}{-u_2(0)} = -j\rho_0 c_0 \cot(k_0 l_a) \qquad (3.167a)$$

$$Z_1(l) = \frac{p_1(l)}{u_1(l)} = -j\rho_0 c_0 \cot(k_0 l_b) \qquad (3.167b)$$

$$Z_2(l) = \frac{p_2(l)}{u_2(l)} = -j\rho_0 c_0 \cot(k_0 l_b) \qquad (3.167c)$$

$$Z_3(l) = \frac{p_3(l)}{u_3(l)} = -j\rho_0 c_0 \cot(k_0 l_b). \qquad (3.167d)$$

The transfer matrix relation

$$\begin{bmatrix} p_1(0) \\ \rho_0 c_0 u_1(0) \end{bmatrix} = \begin{bmatrix} T_a & -T_b \\ T_c & -T_d \end{bmatrix} \begin{bmatrix} p_3(0) \\ \rho_0 c_0 u_3(0) \end{bmatrix}, \qquad (3.168)$$

where

$$T_a = B_{1,1}D_{1,1} + B_{1,2}D_{2,1} + B_{1,3}D_{3,1},$$
$$T_b = B_{1,1}D_{1,2} + B_{1,2}D_{2,2} + B_{1,3}D_{3,2},$$
$$T_c = B_{4,1}D_{1,1} + B_{4,2}D_{2,1} + B_{4,3}D_{3,1},$$
$$T_d = B_{4,1}D_{1,2} + B_{4,2}D_{2,2} + B_{4,3}D_{3,2},$$
$$B_{i1,i2} = T_{i1,i2} + X_1 T_{i1,i2+3},$$
$$i1 = 1, 2, \ldots, 6,$$
$$i2 = 1, 2, 3,$$
$$X_1 = j \tan(k_0 l_b),$$
$$D_{1,1} = C_{1,1}D_{2,1} + C_{1,2}D_{3,1},$$
$$D_{1,2} = C_{1,1}D_{2,2} + C_{1,2}D_{3,2},$$
$$D_{2,1} = C_{3,2}/F_4, \quad D_{2,2} = -C_{2,2}/F_4,$$
$$D_{3,1} = -C_{3,1}/F_4, \quad D_{3,2} = C_{2,1}/F_4,$$
$$F_4 = C_{2,1}C_{3,2} - C_{2,2}C_{3,1},$$
$$C_{1,1} = (B_{5,2} - X_2 B_{2,2})/F_3, \quad C_{1,2} = (B_{5,3} - X_2 B_{2,3})/F_3,$$
$$C_{2,1} = B_{3,2} + C_{1,1}B_{3,1}, \quad C_{2,2} = B_{3,3} + C_{1,2}B_{3,1},$$
$$C_{3,1} = B_{6,2} + C_{1,1}B_{6,1}, \quad C_{3,2} = B_{6,3} + C_{1,2}B_{6,1},$$
$$F_3 = X_2 B_{2,1} - B_{5,1}, \text{ and}$$
$$X_2 = -j \tan(k_0 l_a).$$

(3.169)

Finally, Equations 3.124–3.127 may be applied (replacing subscript 2 with 3 everywhere) in order to obtain transfer matrices in terms of the dimensional classical state variables p and v and the corresponding convective state variables p_c and v_c.

3.12 Other Elements Constituting Cascaded-Element Mufflers

Some of the other elements that constitute cascaded-element mufflers are shown in Figures 3.16–3.35.

The numbers in square brackets at the end of the captions of Figures 3.16–3.34 refer to the original sources (papers) where the transfer matrices have been derived between the upstream

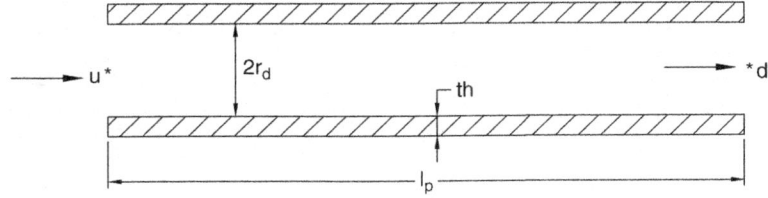

Figure 3.16 Hose (or a uniform area tube with compliant wall), analyzed in [39,40]

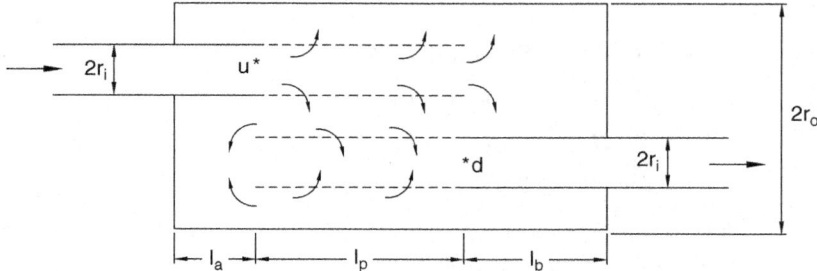

Figure 3.17 Three-duct, cross-flow, open-end perforated element, analyzed in [41]

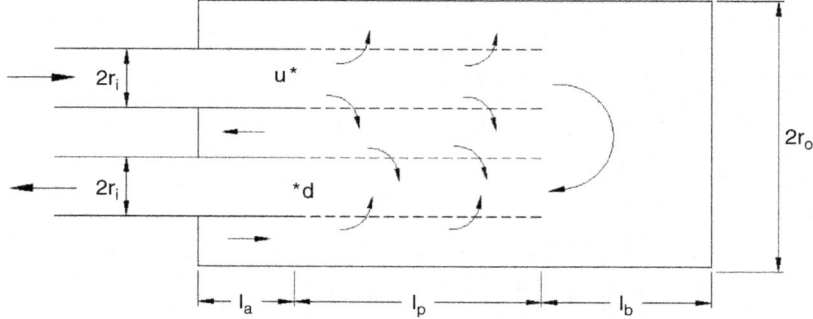

Figure 3.18 Reverse-flow, three-duct, open-end perforated element, analyzed in [41]

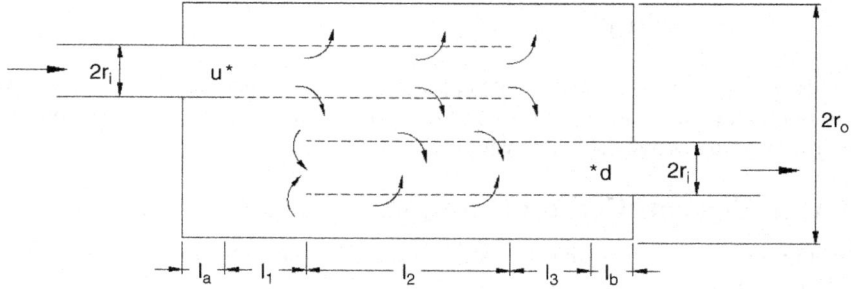

Figure 3.19 Extended (non-overlapping) perforation, cross-flow, open-end element, analyzed in [41]

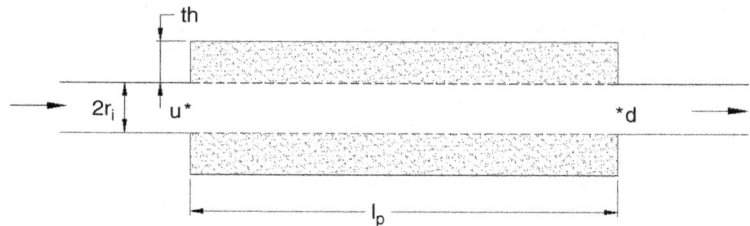

Figure 3.20 Acoustically lined duct, analyzed in [42–44]

Flow-Acoustic Analysis of Cascaded-Element Mufflers

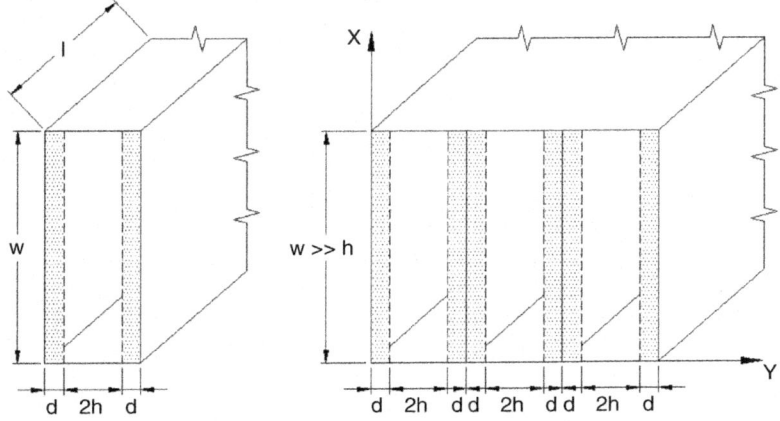

Figure 3.21 Parallel baffle muffler and a constituent rectangular duct lined on two sides, analyzed in [42–44]

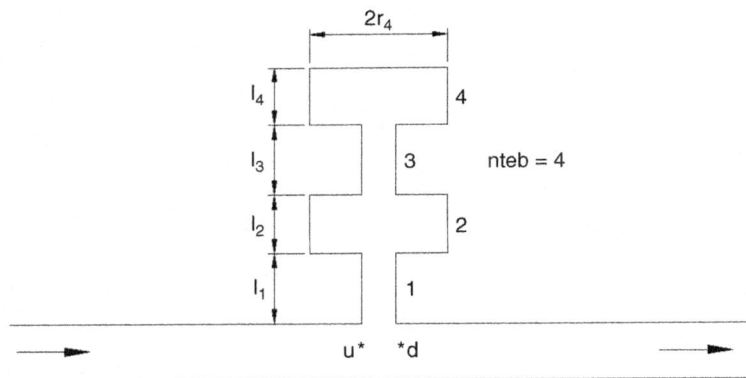

Figure 3.22 Branch sub-system

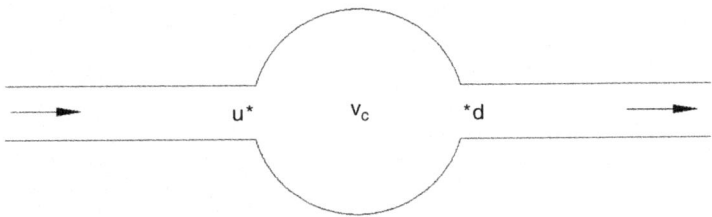

Figure 3.23 Inline cavity of volume V_c

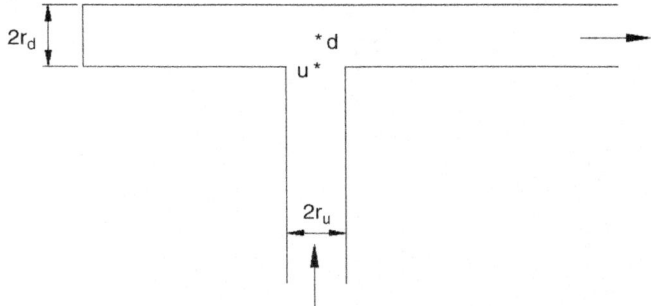

Figure 3.24 Side inlet, analyzed in [45]

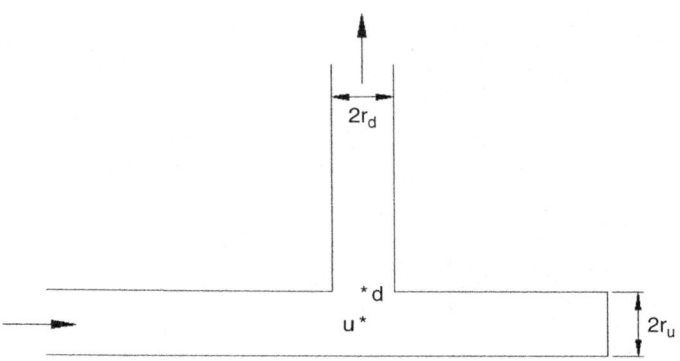

Figure 3.25 Side outlet, analyzed in [45]

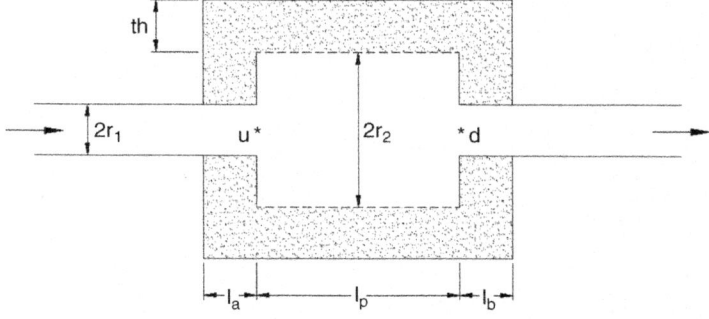

Figure 3.26 Acoustically lined plenum chamber, analyzed in [46]

Flow-Acoustic Analysis of Cascaded-Element Mufflers

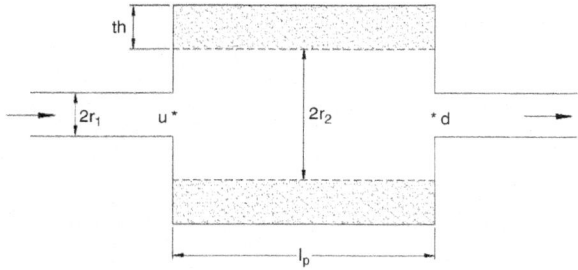

Figure 3.27 Lined wall simple expansion chamber, analyzed in [47,46]

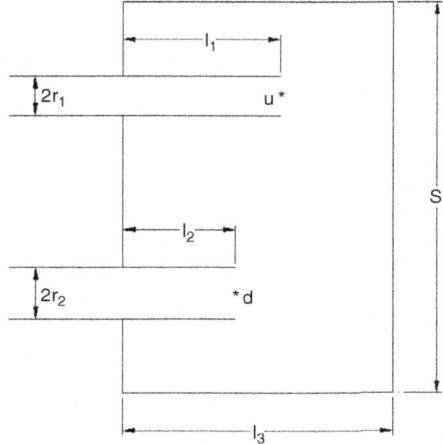

Figure 3.28 Extended tube reversal chamber, analyzed in [48,49]

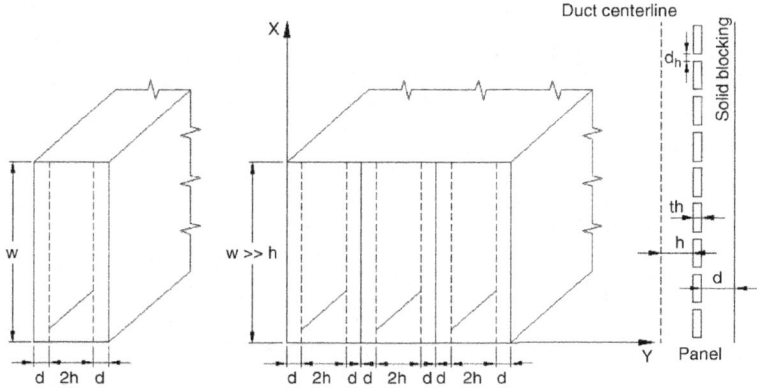

Figure 3.29 Micro-perforated Helmholtz panel parallel baffle muffler and a constituent rectangular duct, analyzed in [50,42]

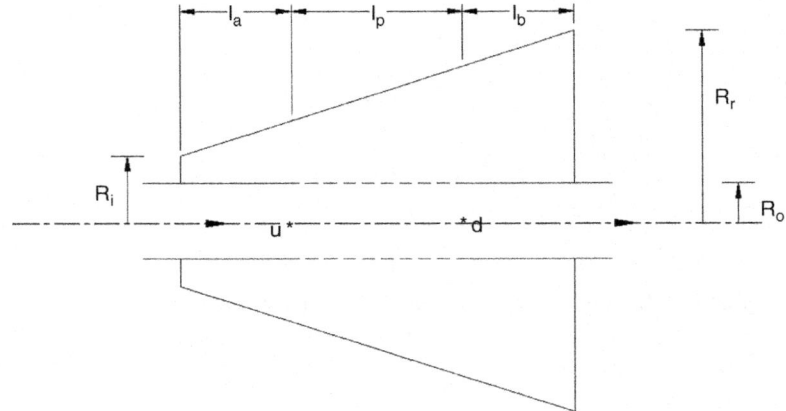

Figure 3.30 Conical concentric tube resonator (CCTR), analyzed in [51]

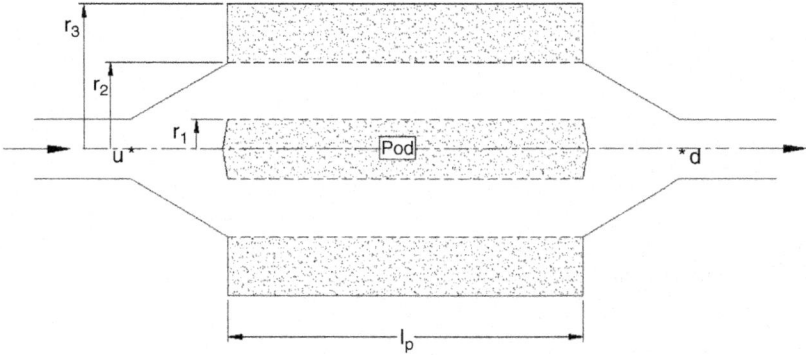

Figure 3.31 Pod silencer, analyzed in [52]

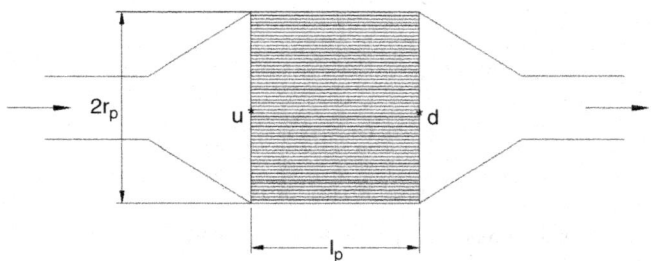

Figure 3.32 Catalytic converter (capillary-tube monolith), analyzed in [53]

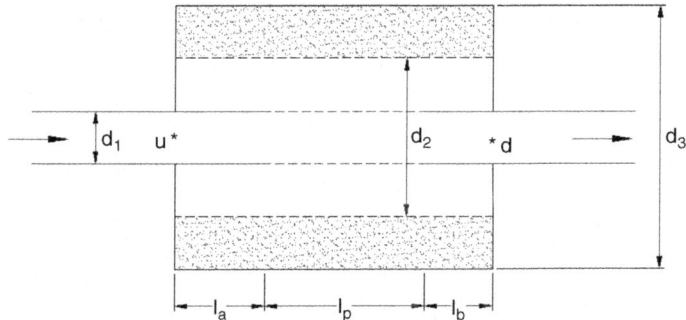

Figure 3.33 Annular air gap lined duct, analyzed in [47]

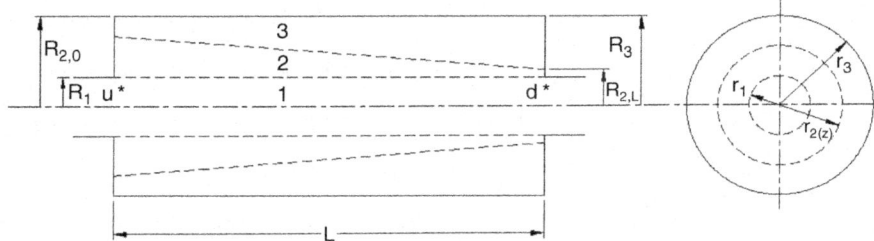

Figure 3.34 Conical concentric tube resonator (CCTR) with three interacting ducts, analyzed in [54]

point u and downstream point d. Many of them have been compiled in the Formulas of Acoustics [42]. These are skipped here for want of space. Nevertheless, these constitute subprograms in a comprehensive transfer matrix based program (TMMP) developed by Munjal et al. [55]. We shall revert to some of them in the final chapter where we try to synthesize efficient muffler configurations.

References

1. Alfredson, R.J. and Davies, P.O.A.L. (1970) The radiation of sound from the engine exhaust. *Journal of Sound and Vibration*, **13**(4), 389–408.
2. Hota, R.N. and Munjal, M.L. (2008) Approximate empirical expressions for aeroacoustic source strength level of the exhaust system of the compression ignition engines. *International Journal of Aeroacoustics*, **7**(3 & 4), 349–371.
3. Schlichting, H. (1953) *Boundary Layer Theory*, McGraw-Hill, New York.
4. Cantrell, R.H. and Hart, R.W. (1964) Interaction between sound and flow in acoustic cavities: mass, moment and energy consideration. *Journal of the Acoustical Society of America*, **36**, 697–706.
5. Morfey, C.L. (1971) Sound transmission and generation in ducts with flow. *Journal of Sound and Vibration*, **14**(1), 37–55.
6. Mechel, F.P., Schilz, W.M. and Dietz, J. (1965) Akustische impedanz eiber luftdurchstromten offnung. *Acustica*, **15**, 199–206.
7. Munjal, M.L. (1975) Velocity ratio cum transfer matrix method for the evaluation of a muffler with mean flow. *Journal of Sound and Vibration*, **39**(1), 105–119.

8. Munt, R.M. (1977) The interaction of sound with a subsonic jet issuing from a semi-infinite cylindrical pipe. *Journal of Fluid Mechanics*, **83**(4), 609–640.
9. Moore, C.J. (1977) The role of shear layer instability waves in jet exhaust noise. *Journal of Fluid Mechanics*, **80**, 321–367.
10. Howe, M.S. (1979) Attenuation of sound in a low mach number nozzle flow. *Journal of Fluid Mechanics*, **91**(2), 209–229.
11. Bechert, D.W., Michel, U. and Pfizenmaier, E. (1977) Experiments on the transmission of sound through jets, AIAA 4th Aeroacoustics Conference, 1277–1278.
12. Ingard, U. and Singhal, V.K. (1974) Sound attenuation in turbulent pipe flow. *Journal of the Acoustical Society America*, **55**(3), 535–538.
13. Panicker, V.B. and Munjal, M.L. (1982) Radiation impedance of an unflanged pipe with mean flow. *Noise Control Engineering Journal*, **18**(2), 48–51.
14. Doige, A.G. and Thawani, P.T. (1979) Muffler transmission from transmission matrices. *NOISECON*, **79**, 245–254.
15. Sreeter, V.L. (1958) Chap. 4, Sec. 28, *Fluid Mechanics*, 2nd edn, McGraw-Hill, New York.
16. Vennard, J.K. and Street, R.L. (1976) Chap. 9, *Elementary Fluid Mechanics (S.I. Version)*, 5th edn, John Wiley & Sons Inc., New York.
17. Panicker, V.B. and Munjal, M.L. (1981) Aeroacoustic analysis of straight-through mufflers with simple and extended tube expansion chambers. *Journal of the Indian Institute of Science*, **63**(A), 1–19.
18. Panicker, V.B. and Munjal, M.L. (1981) Aeroacoustic analysis of mufflers with flow reversals. *Journal of the Indian Institute of Science*, **63**(A), 21–38.
19. Mungur, P. and Gladwell, G.M.L. (1971) Acoustic wave propagation in a sheared fluid contained in a duct. *Journal of Sound and Vibration*, **9**(1), 335–372.
20. Alfredson, R.J. and Davies, P.O.A.L. (1971) Performance of exhaust silencer components. *Journal of Sound and Vibration*, **15**, 175–196.
21. Karal, F.C. (1953) The analogous impedance for discontinuities and constrictions of circular cross section. *Journal of the Acoustical Society of America*, **25**(2), 327–334.
22. Eriksson, L.J. (1980) Higher order mode effects in circular ducts and expansion chambers. *Journal of the Acoustical Society of America*, **68**, 545–550.
23. Eriksson, L.J. (1982) Effect of inlet/outlet locations on higher order modes in silencers. *Journal of the Acoustical Society of America*, **72**(4), 1208–1211.
24. Lung, T.Y. and Doige, A.G. (1983) A time-averaging transient testing method for acoustic properties of piping systems and mufflers with flow. *Journal of the Acoustical Society of America*, **73**(3), 867–876.
25. Sullivan, J.W. and Crocker, M.J. (1978) Analysis of concentric tube resonators having unpartitioned cavities. *Journal of the Acoustical Society of America*, **64**, 207–215.
26. Sullivan, J.W. (1979) A method of modeling perforated tube muffler components I: theory. *Journal of the Acoustical Society of America*, **66**, 772–778.
27. Sullivan, J.W. (1979) A method of modeling perforated tube muffler components II: applications. *Journal of the Acoustical Society of America*, **66**, 779–788.
28. Munjal, M.L., Narayana Rao, K. and Sahasrabudhe, A.D. (1987) Aeroacoustic analysis of perforated muffler components. *Journal of Sound and Vibration*, **114**(2), 173–188.
29. Ingard, U. and Labate, S. (1950) Acoustic circulation effects and the nonlinear impedance of orifices. *Journal of the Acoustical Society of America*, **22**, 211–218.
30. Ingard, U. and Ising, H. (1967) Acoustic nonlinearity of an orifice. *Journal of the Acoustical Society of America*, **42**(1), 6–17.
31. Ronneberger, D. (1972) The acoustical impedance of holes in the wall of flow ducts. *Journal of Sound and Vibration*, **24**(1), 133–150.
32. Melling, T.H. (1973) The acoustic impedance of perforates at medium and high sound pressure levels. *Journal of Sound and Vibration*, **29**(1), 1–65.
33. Dean, P.D. (1973) An in situ method of wall acoustic impedance measurements in flow ducts. *Journal of Sound and Vibration*, **29**( ), 1–65.
34. Sullivan, J.W. (1984) Some gas flow and acoustic pressure measurements inside a concentric-tube resonator. *Journal of the Acoustical Society of America*, **76**(2), 479–484.
35. Rao, K.N. and Munjal, M.L. (1986) Experimental Evaluation of impedance of perforates with grazing flow. *Journal of Sound and Vibration*, **108**(2), 283–295.

36. Dokumaci, E. (1996) Matrizant approach to acoustic analysis of perforated multiple pipe mufflers carrying meanflow. *Journal of Sound and Vibration*, **191**(4), 505–518.
37. Frazer, R.A., Duncan, W.J. and Collar, A.R. (1952) *Elementary Matrices and Some Applications to Dynamics and Differential Equations*, Cambridge University, Cambridge, England.
38. Narayana Rao, K. and Munjal, M.L. (1986) Noise reduction with perforated three-duct muffler components, *Sadhana. Proceedings (in Engg. Sciences) of the Indian Academy of Sciences*, **9**(4), 255–269.
39. Munjal, M.L. and Thawani, P.T. (1996) Acoustic performance of hoses – a Parametric study. *Noise Control Engineering Journal*, **44**(6), 274–280.
40. Munjal, M.L. and Thawani, P.T. (1997) Prediction of the vibro-acoustic transmission loss of planar hose-pipe systems. *Journal of the Acoustical Society of America*, **101**(5), 2524–2535.
41. Gogate, G.R. and Munjal, M.L. (1995) Analytical and experimental aeroacoustic studies of open-ended three-duct perforated elements used in mufflers. *Journal of the Acoustical Society of America*, **97**(5), 2919–2927.
42. Munjal, M.L. (2008) Muffler Acoustics, Chapter K, in *Formulas of Acoustics*, 2nd edn. (ed. F.P. Mechel), Springer-Verlag, Berlin.
43. Munjal, M.L. and Thawani, P.T. (1997) Effect of protective layer on the performance of absorptive ducts. *Noise Control Engineering Journal*, **45**(1), 14–18.
44. Panigrahi, S.N. and Munjal, M.L. (2005) Comparison of various methods for analyzing lined circular ducts. *Journal of Sound and Vibration*, **285**(4–5), 905–923.
45. Munjal, M.L. (1997) Plane wave analysis of side inlet/outlet chamber mufflers with mean flow. *Applied Acoustics*, **52**(2), 165–175.
46. Vijayasree, N.K. and Munjal, M.L. (2012) On an integrated transfer matrix method for multiply connected mufflers. *Journal of Sound and Vibration*, **331**, 1926–1938.
47. Munjal, M.L. and Venkatesham, B. (2002) Analysis and design of an annular airgap lined duct for hot exhaust systems, Proceedings of the IUTAM Symposium on Designing for Quietness, Kluwer Academic Publishers, Dordrecht, Netherlands, 1–19.
48. Munjal, M.L. (1997) Analysis of a flush tube three-pass perforated element muffler by means of transfer matrices. *International Journal of Acoustics and Vibration*, **2**(2), 63–68.
49. Munjal, M.L. (December 1997) Analysis of extended-tube three-pass perforated element muffler by means of transfer matrices, Proceedings of the Fifth International Congress on Sound and Vibration, ICSV'5, Adelaide, Australia, Vol. III, 1707–1714.
50. Wu, M.Q. (1997) Micro-perforated panels for duct silencing. *Noise Control Engineering Journal*, **45**, 69–77.
51. Kar, T. and Munjal, M.L. (2004) Analysis and design of conical concentric tube resonators. *Journal of the Acoustical Society of America*, **116**(1), 74–83.
52. Munjal, M.L. (2003) Analysis and design of pod silencers. *Journal of Sound and Vibration*, **262**(3), 497–507.
53. Selamet, A., Easwaran, V., Novak, J.M. and Kach, R.A. (1998) Wave attenuation in catalytical converters: reactive versus dissipative effects. *Journal of the Acoustical Society of America*, **103**, 935–943.
54. Kar, T., Sharma, P.P.R. and Munjal, M.L. (2006) Analysis of multiple-duct variable area perforated-tube resonators. *International Journal of Acoustics and Vibration*, **11**(1), 19–26.
55. Munjal, M.L., Panigrahi, S.N. and Hota, R.N. (July 9–12 2007) FRITAmuff: A comprehensive platform for prediction of unmuffled and muffled exhaust noise of I.C. Engines. 14th International Congress on Sound and Vibration (ICSV14), Cairns, Australia.

4

Flow-Acoustic Analysis of Multiply-Connected Perforated Element Mufflers

The overall transfer matrix of a cascaded-element muffler may be obtained through successive multiplication of the transfer matrices of the constituent elements. However, there are mufflers that do not permit this convenience; there is more than one path connecting any two given points. One such configuration is the Herschel-Quincke tube [1,2] shown in Figure 4.1.

4.1 Herschel-Quincke Tube Phenomenon

It may be noted from Figure 4.1 that the downstream point 'd' is connected to the upstream point 'u' in two ways; one straight (3–4) of length l_2 and the other curved (2–5) of length l_3. For plane wave propagation along both the paths, the following relationships would hold:

$$p_u = p_2 = p_3 \qquad (4.1, 4.2)$$

$$v_u = v_2 + v_3 \qquad (4.3)$$

$$\begin{bmatrix} p_2 \\ v_2 \end{bmatrix} = \begin{bmatrix} \cos kl_3 & jY_3 \sin kl_3 \\ (j/Y_3)\sin kl_3 & \cos kl_3 \end{bmatrix} \begin{bmatrix} p_5 \\ v_5 \end{bmatrix} \qquad (4.4)$$

$$\begin{bmatrix} p_3 \\ v_3 \end{bmatrix} = \begin{bmatrix} \cos kl_2 & jY_2 \sin kl_2 \\ (j/Y_2)\sin kl_2 & \cos kl_2 \end{bmatrix} \begin{bmatrix} p_4 \\ v_4 \end{bmatrix} \qquad (4.5)$$

$$p_d = p_4 = p_5 \qquad (4.6, 4.7)$$

$$v_d = v_4 + v_5 \qquad (4.8)$$

Acoustics of Ducts and Mufflers, Second Edition. M. L. Munjal.
© 2014 John Wiley & Sons, Ltd. Published 2014 by John Wiley & Sons, Ltd.

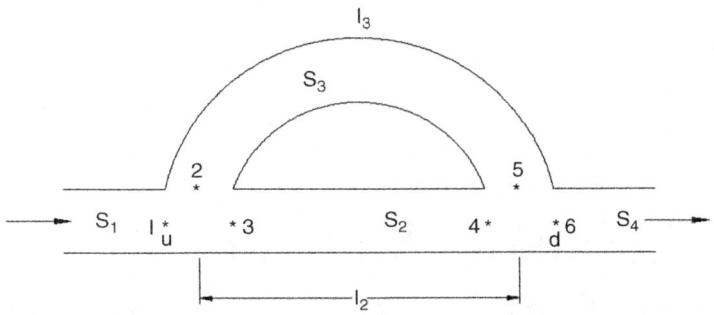

Figure 4.1 Herschel-Quincke tube, analyzed in [1,2]

Equations 4.1–4.8 represent 10 linear algebraic equations in 12 state variables. These have to be solved simultaneously – a feature that is common to all multiply-connected elements – in order to obtain the desired transfer matrix:

$$\begin{bmatrix} p_u \\ v_u \end{bmatrix} = \begin{bmatrix} T_{11} & T_{12} \\ T_{21} & T_{22} \end{bmatrix} \begin{bmatrix} p_d \\ v_d \end{bmatrix}, \tag{4.9}$$

where

$$T_{11} = (C_2 Y_3 S_3 + C_3 Y_2 S_2)/\det, \tag{4.10}$$

$$T_{12} = (j Y_2 S_2 Y_3 S_3)/\det, \tag{4.11}$$

$$T_{21} = j \left[(Y_2 S_2 + Y_3 S_3) \left\{ \frac{S_2}{Y_2} + \frac{S_3}{Y_3} \right\} + (C_2 - C_3)^2 \right] / \det, \tag{4.12}$$

$$T_{22} = (C_2 Y_3 S_3 + C_3 Y_2 S_2)/\det, \tag{4.13}$$

$\det = Y_2 S_2 + Y_3 S_3$, $S_2 \equiv \sin(kl_2)$, $S_3 \equiv \sin(kl_3)$, $C_2 \equiv \cos(kl_2)$, $C_3 \equiv \cos(kl_3)$, and Y_2 and Y_3 are the characteristic impedances of the two branches in Figure 4.1.

For this rather simple configuration, the four-pole parameters T_{11}, T_{12}, T_{21}, and T_{22} have been obtained making use of sequential elimination.

If the two branches are of equal area of cross-section, then $Y_2 = Y_3 = Y_0$ (say), and the overall transfer matrix of Equation 4.9 would become

$$\frac{1}{\sin kl_2 + \sin kl_3} \begin{bmatrix} \sin k(l_2 + l_3) & jY_0 \sin kl_2 \sin kl_3 \\ (j/Y_0)(\sin kl_2 + \sin kl_3)^2 + (\cos kl_2 - \cos kl_3)^2 & \sin k(l_2 + l_3) \end{bmatrix} \tag{4.14}$$

If we invert the transfer matrix of Equation 4.14 to express p_d and v_d in terms of p_u and v_u, then all elements of the resultant matrix will be proportional to '$\sin kl_2 + \sin kl_3$'.

At the frequencies at which $\sin kl_2 + \sin kl_3 = 0$, p_d and v_d would tend to zero; that is, there would be no sound at the downstream point, d. These frequencies are given by

$$\sin kl_3 = -\sin kl_2 \qquad (4.15)$$

Values of the wave number k satisfying Equation 4.15 are given by

$$kl_3 = kl_2 + (2m - 1)\pi, m = 1, 2, 3, \ldots \qquad (4.16a)$$

or

$$kl_3 = 2n\pi - kl_2, \quad n = 1, 2, 3, \ldots \qquad (4.16b)$$

This has the following physical explanation.

A plane progressive harmonic wave in the curved branch of Figure 4.1 will reach the downstream junction in l_3/c_0 seconds whereas that in the straight section will reach there in l_2/c_0 seconds. The phase difference between the two is given by $kl_3 - kl_2$. When this equals π or its odd integral multiples, or the sum of the phases $kl_3 + kl_2$ equals an even multiple of π, then the progressive waves in the two branches would cancel each other. TL at each of these frequencies would tend to infinity. The resultant peaks in the TL curve would therefore occur when Equation 4.16 is satisfied, or at the frequencies given by

$$f_m = \frac{2m-1}{2}\frac{c_0}{l_3 - l_2}, \quad m = 1, 2, 3, \ldots \qquad (4.17)$$

or

$$f_n = n\frac{c_0}{l_3 + l_2}, \quad n = 1, 2, 3, \ldots \qquad (4.18)$$

This is illustrated in the TL curve in Figure 4.2 for $c_0 = 346\,m/s$, $l_2 = 0.5\,m$ and $l_3 = \pi l_2/2$ (semi-circular arc) $= 0.7854\,m$. For these data, as per Equation 4.17, TL peaks would occur at 606.2 Hz, 1818 Hz, 3031 Hz, ... and, as per Equation 4.18, TL peaks would occur at 269.2 Hz, 538.4 Hz, 808 Hz, 1077 Hz, 1346 Hz, 1615 Hz, 1884 Hz, 2153 Hz, 2423 Hz, 2692 Hz, 2961 Hz, 3230 Hz, 3500 Hz, ... These sharp peaks may be observed in the TL curve of Figure 4.2.

Incidentally, if the two branches are of equal length, that is, if $l_2 = l_3 = l$ (say), then it may readily be checked that Equation 4.14 reduces to

$$\begin{bmatrix} p_u \\ v_u \end{bmatrix} = \begin{bmatrix} \cos kl & j(Y_0/2)\sin kl \\ j\sin kl/(Y_0/2) & \cos kl \end{bmatrix} \begin{bmatrix} p_d \\ v_d \end{bmatrix} \qquad (4.19)$$

Thus, two parallel branches of equal length behave together as a single tube of half the characteristic impedance, or double the cross-section (Y_0 is inversely proportional to the cross-section of the tube), which is more or less obvious.

This rather simple principle of wave cancellation has been used knowingly or unknowingly in several configurations of the present-day automotive mufflers that combine the advantage

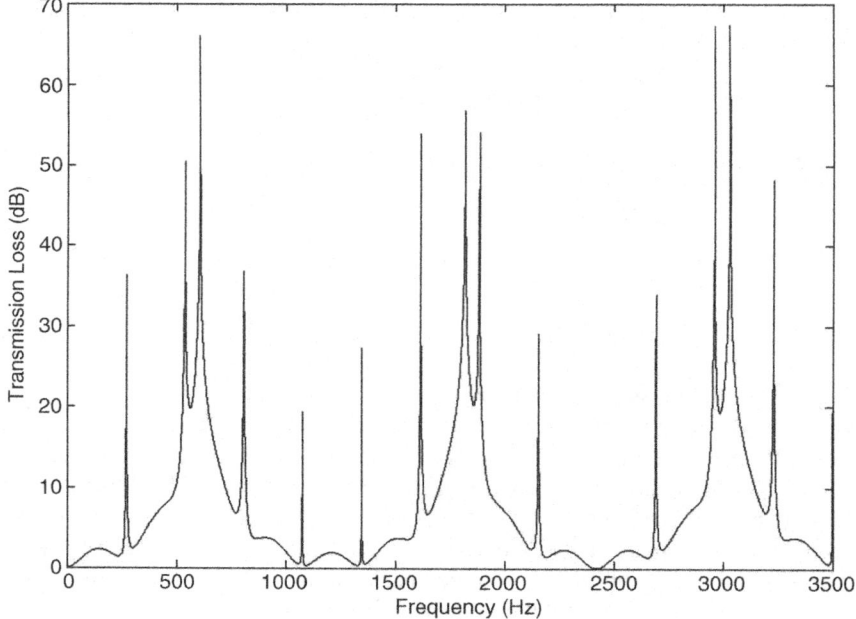

Figure 4.2 Typical TL of the Herschel-Quincke tube

of wideband TL or IL with remarkably low back-pressure. One such configuration is shown in Figure 4.3 where it may be noted that waves travel from point 'u' to 'd' not only across the interacting perforates but also after traversing the two end-chambers. To analyze this configuration we need to first learn to analyze a perforated element with several interacting ducts. This is discussed in the next section.

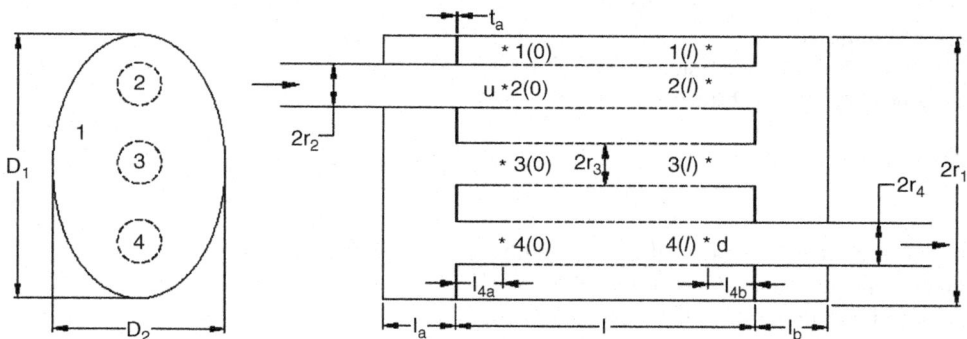

Figure 4.3 Flush-tube three-pass perforated element chamber $\left(r_1 = 1/2(D_1 D_2)^{1/2}\right)$

4.2 Perforated Element with Several Interacting Ducts

In practice, there are automotive mufflers with more than three interacting ducts. For example, Figure 4.3 shows a muffler with four interacting ducts. Incidentally, it is called a three-pass double flow-reversal muffler and was analyzed numerically by Selamet *et al.* [3] and analytically by Munjal [4] for stationary medium. Later, Kar and Munjal [5] presented a generalized analysis of a perforated element with any number of (say, n) interacting ducts (see Figure 4.4). Recently, the matrizant adaptation of it was developed by Bhushan and Munjal [6]. This is presented below and illustrated by means of the three-pass muffler of Figure 4.3. This element is characterized by n-1 perforated ducts of diameters $d_2, d_3, \ldots, d_n$ inside a bigger one of (equivalent) diameter d_1, as shown in Figure 4.4.

This n-duct element may be modeled by n equations of mass continuity and n equations of momentum balance as follows.

Continuity equations:
For the ith duct ($i = 2, 3, \ldots, n$):

$$\rho_0 \frac{\partial u_i}{\partial z} + U_i \frac{\partial \rho_i}{\partial z} - \frac{4}{d_i} \rho_0 \ u_{i,1} = -\frac{\partial \rho_i}{\partial t}. \tag{4.20}$$

For the first duct (outer duct):

$$\rho_0 \frac{\partial u_1}{\partial z} + U_1 \frac{\partial \rho_1}{\partial z} + \sum_{m=2}^{m=n} \frac{4 d_m}{d_1^2 - \sum_{j=2}^{n} d_j^2} \rho_0 \ u_{m,1} = -\frac{\partial \rho_1}{\partial t}. \tag{4.21}$$

Momentum equations:
For the ith duct ($i = 1, \ldots, n$) in the axial direction,

$$\frac{\partial p_i}{\partial z} = -\rho_0 \frac{\partial u_i}{\partial t} - \rho_0 U_i \frac{\partial u_i}{\partial z}. \tag{4.22a}$$

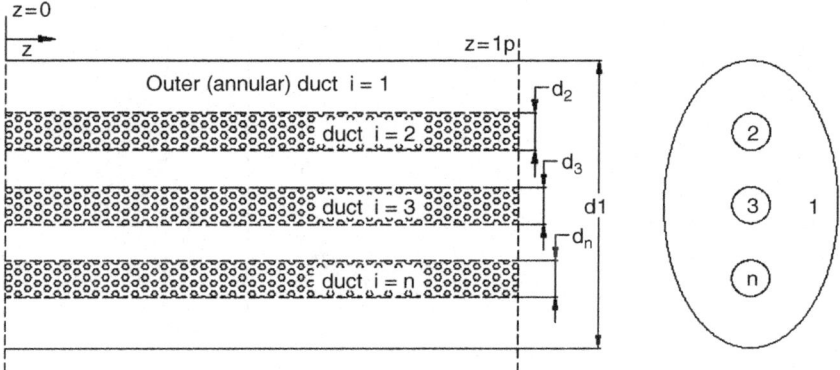

Figure 4.4 A perforated element with n interacting ducts, analyzed in [5]

and in the radial direction,

$$u_{i,1} = \frac{p_1 - p_i}{\rho_0 c_0 \zeta_i}. \tag{4.22b}$$

Assuming harmonic time dependence $(e^{j\omega t})$ and isentropicity $(dp = c_0^2 d\rho)$ we get as follows.
Continuity equations:
For the ith duct ($i = 2, 3, \ldots, n$):

$$M_i \frac{dp_i}{dz} + \rho_0 c_0 \frac{du_i}{dz} = \frac{4}{d_i \zeta_i} p_1 - \left(jk_0 + \frac{4}{d_i \zeta_i}\right) p_i; \tag{4.23}$$

For the first duct:

$$M_1 \frac{dp_1}{dz} + \rho_0 c_0 \frac{du_1}{dz} = -jk_0 p_1 + \sum_{m=2}^{m=n} \frac{4 d_m (p_1 - p_m)}{\zeta_m \left(d_1^2 - \sum_{j=2}^{j=n} d_j^2\right)}; \tag{4.24}$$

Momentum equations:

$$\frac{dp_j}{dz} + M_j \rho_0 c_0 \frac{du_i}{dz} = -jk_0 \rho_0 c_0 u_i, \quad i = 1, 2, \ldots, n. \tag{4.25}$$

Equations 4.23–4.25 may be rearranged in a differential matrix form as follows.

$$\begin{bmatrix} 1 & M_1 & 0 & 0 & \cdots & 0 & 0 & \cdots & 0 & 0 \\ M_1 & 1 & 0 & 0 & \cdots & 0 & 0 & \cdots & 0 & 0 \\ 0 & 0 & 1 & M_2 & \cdots & 0 & 0 & \cdots & 0 & 0 \\ 0 & 0 & M_2 & 1 & \cdots & 0 & 0 & \cdots & 0 & 0 \\ \vdots & \vdots & \vdots & \vdots & \vdots & \vdots & \vdots & & \vdots & \vdots \\ 0 & 0 & 0 & 0 & \cdots & 1 & M_i & \cdots & 0 & 0 \\ 0 & 0 & 0 & 0 & \cdots & M_i & 1 & \cdots & 0 & 0 \\ \vdots & \vdots & \vdots & \vdots & & \vdots & \vdots & & \vdots & \vdots \\ 0 & 0 & 0 & 0 & \cdots & 0 & 0 & \cdots & 1 & M_n \\ 0 & 0 & 0 & 0 & \cdots & 0 & 0 & \cdots & M_n & 1 \end{bmatrix}_{2n \times 2n} \begin{bmatrix} \frac{dp_1}{dz} \\ \rho_0 c_0 \frac{du_1}{dz} \\ \frac{dp_2}{dz} \\ \rho_0 c_0 \frac{du_2}{dz} \\ \vdots \\ \frac{dp_i}{dz} \\ \rho_0 c_0 \frac{du_i}{dz} \\ \vdots \\ \frac{dp_n}{dz} \\ \rho_0 c_0 \frac{du_n}{dz} \end{bmatrix}_{2n \times 1} =$$

$$\begin{bmatrix} -\left(jk_0 + \sum_{m=2}^{m=n}\dfrac{4d_m}{\varsigma_m\left(d_1^2 - \sum_{j=2}^{j=n}d_j^2\right)}\right) & -jk_0 & 0 & 0 & \cdots & 0 & 0 & \cdots & 0 & 0 \\ \dfrac{4}{d_2\varsigma_2} & 0 & \dfrac{4d_2}{\varsigma_1\left(d_1^2 - \sum_{j=2}^{j=n}d_j^2\right)} & 0 & \cdots & \dfrac{4d_i}{\varsigma_j\left(d_1^2 - \sum_{j=2}^{j=n}d_j^2\right)} & 0 & \cdots & \dfrac{4d_n}{\varsigma_n\left(d_1^2 - \sum_{j=2}^{j=n}d_j^2\right)} & 0 \\ & & & -jk_0 & \cdots & 0 & 0 & \cdots & 0 & 0 \\ & & -\left(jk_0 + \dfrac{4}{d_2\varsigma_2}\right) & 0 & \cdots & 0 & 0 & \cdots & 0 & 0 \\ \vdots & \vdots & \vdots & \vdots & \vdots & \vdots & \vdots & \vdots & \vdots & \vdots \\ 0 & 0 & 0 & 0 & \cdots & 0 & -jk_0 & \cdots & 0 & 0 \\ \dfrac{4}{d_i\varsigma_i} & 0 & 0 & 0 & \cdots & -\left(jk_0 + \dfrac{4}{d_i\varsigma_i}\right) & 0 & \cdots & 0 & 0 \\ \vdots & \vdots & \vdots & \vdots & \vdots & \vdots & \vdots & \vdots & \vdots & \vdots \\ 0 & 0 & 0 & 0 & \cdots & 0 & 0 & \cdots & 0 & -jk_0 \\ \dfrac{4}{d_n\varsigma_n} & 0 & 0 & 0 & \cdots & 0 & 0 & \cdots & -\left(jk_0 + \dfrac{4}{d_n\varsigma_n}\right) & 0 \end{bmatrix}_{2n \times 2n}$$

$$\times \begin{bmatrix} p_1 \\ \rho_0 c_0 u_1 \\ p_2 \\ \rho_0 c_0 u_2 \\ \vdots \\ p_i \\ \rho_0 c_0 u_i \\ \vdots \\ p_n \\ \rho_0 c_0 u_n \end{bmatrix}_{2n \times 1} \tag{4.26}$$

Premultiplying the matrix Equation 4.26 with inverse of the mean flow matrix, $[M]$, it can be reduced to the canonical form, $\{\hat{V}'\} = [H]\{\hat{V}\}$. Then, making use of the matrizant relationship Equation 3.135, we obtain a transfer matrix relationship between the normalized state variables p and $\rho_0 c_0 u$. As $\rho_0 c_0 u = (c_0/S)(\rho_0 S u) = Yv$, the desired relationship between the state variables p and v may be obtained by dividing all the even-numbered rows of matrix $[H]$ by Y_i and multiplying the even-numbered columns with Y_i, where $i = 2, 4, \ldots, 2n$ represents the ith duct.

$$\begin{bmatrix} p_1(0) \\ v_1(0) \\ p_2(0) \\ v_2(0) \\ \vdots \\ p_i(0) \\ v_i(0) \\ \vdots \\ p_n(0) \\ v_n(0) \end{bmatrix} = \begin{bmatrix} A_{11} & A_{12} & A_{13} & A_{14} & \cdots & A_{1(2i)} & \cdots & A_{1(2n-1)} & A_{1(2n)} \\ A_{21} & A_{22} & A_{23} & A_{24} & \cdots & A_{2(2i)} & \cdots & A_{2(2n-1)} & A_{2(2n)} \\ A_{31} & A_{32} & A_{33} & A_{34} & \cdots & A_{3(2i)} & \cdots & A_{3(2n-1)} & A_{3(2n)} \\ A_{41} & A_{42} & A_{43} & A_{44} & \cdots & A_{4(2i)} & \cdots & A_{4(2n-1)} & A_{4(2n)} \\ \vdots & \vdots & \vdots & \vdots & \vdots & \vdots & \vdots & \vdots & \vdots \\ A_{(2i-1)1} & A_{(2i-1)2} & A_{(2i-1)3} & A_{(2i-1)4} & \cdots & A_{(2i-1)(2i)} & \cdots & A_{(2i-1)(2n-1)} & A_{(2i-1)(2n)} \\ A_{(2i)1} & A_{(2i)2} & A_{(2i)3} & A_{(2i)4} & \cdots & A_{(2i)(2i)} & \cdots & A_{(2i)(2n-1)} & A_{(2i)(2n)} \\ \vdots & \vdots & \vdots & \vdots & \vdots & \vdots & \vdots & \vdots & \vdots \\ A_{(2n-1)1} & A_{(2n-1)2} & A_{(2n-1)3} & A_{(2n-1)4} & \cdots & A_{(2n-1)(2i)} & \cdots & A_{(2n-1)(2n-1)} & A_{(2n-1)(2n)} \\ A_{(2n)1} & A_{(2n)2} & A_{(2n)3} & A_{(2n)4} & \cdots & A_{(2n)(2i)} & \cdots & A_{(2n)(2n-1)} & A_{(2n)(2n)} \end{bmatrix} \begin{bmatrix} p_1(l) \\ v_1(l) \\ p_2(l) \\ v_2(l) \\ \vdots \\ p_i(l) \\ v_i(l) \\ \vdots \\ p_n(l) \\ v_n(l) \end{bmatrix}$$

$$\tag{4.27}$$

4.3 Three-Pass Double-Reversal Muffler

A typical three-pass muffler with two flush-tube flow-reversal end-chambers is shown in Figure 4.3, where r_1 is the equivalent radius of the elliptical cross-section with major axis equal to D_1 and minor axis equal to D_2. Thus,

$$r_1 = \frac{(D_1 D_2)^{1/2}}{2}. \tag{4.28}$$

We proceed to derive the 2×2 transfer matrix between points u and d in Figure 4.3. We start with the matrix Equation 4.27 with n = 4.

$$\begin{bmatrix} p_1(0) \\ v_1(0) \\ p_2(0) \\ v_2(0) \\ p_3(0) \\ v_3(0) \\ p_4(0) \\ v_4(0) \end{bmatrix}_{8 \times 1} = [A]_{8 \times 8} \begin{bmatrix} p_1(l) \\ v_1(l) \\ p_2(l) \\ v_2(l) \\ p_3(l) \\ v_3(l) \\ p_4(l) \\ v_4(l) \end{bmatrix}_{8 \times 1}, \tag{4.29}$$

and make use of the relevant boundary conditions and the end-chambers transfer matrices to eliminate $p_1(0), v_1(0), p_3(0), v_3(0), p_4(0), v_4(0), p_1(l), v_1(l), p_2(l), v_2(l), p_3(l)$ and $v_3(l)$, and thereby derive the desired matrix between the state variable $p_2(0)$ and $v_2(0)$ at the upstream point u and $p_4(l)$ and $v_4(l)$ at the downstream point d.

Making use of the boundary conditions

$$v_1(0) = \frac{X_{4a}}{Y_1} p_1(0), X_{4a} = -j \tan(k_0 l_{4a}) \tag{4.30}$$

and

$$v_1(l) = \frac{X_{4b}}{Y_1} p_1(l), X_{4b} = j \tan(k_0 l_{4b}) \tag{4.31}$$

$p_1(0), v_1(0), p_1(l)$ and $v_l(l)$ may be eliminated in a systematic manner as follows [6].

Using Equations 4.30 and 4.31, $v_1(0)$ and $v_1(l)$ may be eliminated from Equation 4.29 and we get

$$\begin{bmatrix} p_1(0) \\ (X_{4a}/Y_1)p_1(0) \\ p_2(0) \\ v_2(0) \\ p_3(0) \\ v_3(0) \\ p_4(0) \\ v_4(0) \end{bmatrix}_{8 \times 1} = [A]_{8 \times 8} \begin{bmatrix} p_1(l) \\ (X_{4b}/Y_1)p_1(l) \\ p_2(l) \\ v_2(l) \\ p_3(l) \\ v_3(l) \\ p_4(l) \\ v_4(l) \end{bmatrix}_{8 \times 1}. \tag{4.32}$$

Equation 4.32 may be rewritten in the following form:

$$\begin{bmatrix} 1 & 0 & 0 & 0 & 0 & 0 & 0 \\ X_{4a}/Y_1 & 0 & 0 & 0 & 0 & 0 & 0 \\ 0 & 1 & 0 & 0 & 0 & 0 & 0 \\ 0 & 0 & 1 & 0 & 0 & 0 & 0 \\ 0 & 0 & 0 & 1 & 0 & 0 & 0 \\ 0 & 0 & 0 & 0 & 1 & 0 & 0 \\ 0 & 0 & 0 & 0 & 0 & 1 & 0 \\ 0 & 0 & 0 & 0 & 0 & 0 & 1 \end{bmatrix}_{8 \times 7} \begin{bmatrix} p_1(0) \\ p_2(0) \\ v_2(0) \\ p_3(0) \\ v_3(0) \\ p_4(0) \\ v_4(0) \end{bmatrix}_{7 \times 1}$$

$$= [A]_{8 \times 8} \begin{bmatrix} 1 & 0 & 0 & 0 & 0 & 0 & 0 \\ X_{4b}/Y_1 & 0 & 0 & 0 & 0 & 0 & 0 \\ 0 & 1 & 0 & 0 & 0 & 0 & 0 \\ 0 & 0 & 1 & 0 & 0 & 0 & 0 \\ 0 & 0 & 0 & 1 & 0 & 0 & 0 \\ 0 & 0 & 0 & 0 & 1 & 0 & 0 \\ 0 & 0 & 0 & 0 & 0 & 1 & 0 \\ 0 & 0 & 0 & 0 & 0 & 0 & 1 \end{bmatrix}_{8 \times 7} \begin{bmatrix} p_1(l) \\ p_2(l) \\ v_2(l) \\ p_3(l) \\ v_3(l) \\ p_4(l) \\ v_4(l) \end{bmatrix}_{7 \times 1} . \quad (4.33)$$

Denoting the two 8×7 matrices in Equation 4.33 by [B] and [C], it may be rewritten as

$$[B]_{8 \times 7} \begin{bmatrix} p_1(0) \\ p_2(0) \\ v_2(0) \\ p_3(0) \\ v_3(0) \\ p_4(0) \\ v_4(0) \end{bmatrix}_{7 \times 1} = [C]_{8 \times 7} \begin{bmatrix} p_1(l) \\ p_2(l) \\ v_2(l) \\ p_3(l) \\ v_3(l) \\ p_4(l) \\ v_4(l) \end{bmatrix}_{7 \times 1} \quad (4.34)$$

This can further be rearranged so as to transfer $p_1(l)$ to the left-hand side. Thus,

$$[D]_{8 \times 8} \begin{bmatrix} p_1(0) \\ p_2(0) \\ v_2(0) \\ p_3(0) \\ v_3(0) \\ p_4(0) \\ v_4(0) \\ p_1(l) \end{bmatrix}_{8 \times 1} = [E]_{8 \times 6} \begin{bmatrix} p_2(l) \\ v_2(l) \\ p_3(l) \\ v_3(l) \\ p_4(l) \\ v_4(l) \end{bmatrix}_{6 \times 1} , \quad (4.35)$$

Premultiplying Equation 4.35 with $[D^{-1}]$ we obtain

$$\begin{bmatrix} p_1(0) \\ p_2(0) \\ v_2(0) \\ p_3(0) \\ v_3(0) \\ p_4(0) \\ v_4(0) \\ p_1(l) \end{bmatrix}_{8\times 1} = [F]_{8\times 6} \begin{bmatrix} p_2(l) \\ v_2(l) \\ p_3(l) \\ v_3(l) \\ p_4(l) \\ v_4(l) \end{bmatrix}_{6\times 1}, \qquad (4.36)$$

where, making use of the MATLAB® notation,

$$[D]_{8\times 8} = [B - C(:,1)]_{8\times 8}, \qquad (4.37)$$

$$[E]_{8\times 6} = [C(:,2:7)]_{8\times 6}, \qquad (4.38)$$

and

$$[F] = [D]^{-1}_{8\times 8}[E]_{8\times 6} \qquad (4.39)$$

Retaining only a part of Equation 4.36, (dropping the equations for $p_1(0)$ and $p_1(l)$, we can write

$$\begin{bmatrix} p_2(0) \\ v_2(0) \\ p_3(0) \\ v_3(0) \\ p_4(0) \\ v_4(0) \end{bmatrix}_{6\times 1} = [G]_{6\times 6} \begin{bmatrix} p_2(l) \\ v_2(l) \\ p_3(l) \\ v_3(l) \\ p_4(l) \\ v_4(l) \end{bmatrix}_{6\times 1}, \qquad (4.40)$$

where, making use of the MATLAB® notation,

$$[G]_{6\times 6} = [F(2:7,:)] \qquad (4.41)$$

Thus, the boundary conditions (4.30) and (4.31) have been used to reduce the 8×8 matrix $[A]$ to a 6×6 matrix $[G]$.

In the next section we shall derive transfer matrix $[H]$ for the right-hand side end chamber and $[J]$ for the left-hand side end chamber (see Figure 4.3):

$$\begin{bmatrix} p_2(l) \\ v_2(l) \end{bmatrix} = [H]_{2\times 2} \begin{bmatrix} p_3(l) \\ v_3(l) \end{bmatrix} \qquad (4.42)$$

$$\begin{bmatrix} p_3(0) \\ v_3(0) \end{bmatrix} = [J]_{2\times 2} \begin{bmatrix} p_4(0) \\ v_4(0) \end{bmatrix} \quad (4.43)$$

At this stage, assuming the same to be available, let us make use of Equations 4.42 and 4.43 to eliminate $p_3(0), v_3(0), p_4(0), v_4(0), p_2(l), v_2(l), p_3(l)$, and $v_3(l)$ from Equation (4.40).

The LHS vector of Equation 4.40 may be rewritten as follows to incorporate [J]:

$$\begin{bmatrix} p_2(0) \\ v_2(0) \\ p_3(0) \\ v_3(0) \\ p_4(0) \\ v_4(0) \end{bmatrix}_{6\times 1} = \begin{bmatrix} 1 & 0 & 0 & 0 \\ 0 & 1 & 0 & 0 \\ 0 & 0 & [J]_{2\times 2} & \\ 0 & 0 & & \\ 0 & 0 & 1 & 0 \\ 0 & 0 & 0 & 1 \end{bmatrix}_{6\times 4} \begin{bmatrix} p_2(0) \\ v_2(0) \\ p_4(0) \\ v_4(0) \end{bmatrix}_{4\times 1} \equiv [K]_{6\times 4} \begin{bmatrix} p_2(0) \\ v_2(0) \\ p_4(0) \\ v_4(0) \end{bmatrix} \quad (\text{say}) \quad (4.44)$$

Similarly, in order to incorporate [H], the RHS vector of Equation 4.40 may be rewritten as follows

$$\begin{bmatrix} p_2(l) \\ v_2(l) \\ p_3(l) \\ v_3(l) \\ p_4(l) \\ v_4(l) \end{bmatrix}_{6\times 1} = \begin{bmatrix} [H]_{2\times 2} & 0 & 0 \\ & & 0 & 0 \\ 1 & 0 & 0 & 0 \\ 0 & 1 & 0 & 0 \\ 0 & 0 & 1 & 0 \\ 0 & 0 & 0 & 1 \end{bmatrix} \begin{bmatrix} p_3(l) \\ v_3(l) \\ p_4(l) \\ v_4(l) \end{bmatrix} \equiv [L]_{6\times 4} \begin{bmatrix} p_3(l) \\ v_3(l) \\ p_4(l) \\ v_4(l) \end{bmatrix} \quad (\text{say}) \quad (4.45)$$

Using Equations 4.44 and 4.45, Equation 4.40 can be written as

$$[K]_{6\times 4} \begin{bmatrix} p_2(0) \\ v_2(0) \\ p_4(0) \\ v_4(0) \end{bmatrix}_{4\times 1} = [M]_{6\times 4} \begin{bmatrix} p_3(l) \\ v_3(l) \\ p_4(l) \\ v_4(l) \end{bmatrix}_{4\times 1} \quad (4.46)$$

where

$$[M]_{6\times 4} = [G]_{6\times 6} [L]_{6\times 4} \quad (4.47)$$

To retain only the downstream variables $p_4(l)$ and $v_4(l)$ on the right-hand side we can rewrite Equation 4.46 as

$$[N]_{6\times 6} \begin{bmatrix} p_2(0) \\ v_2(0) \\ p_4(0) \\ v_4(0) \\ p_3(l) \\ v_3(l) \end{bmatrix} = [P]_{6\times 2} \begin{bmatrix} p_4(l) \\ v_4(l) \end{bmatrix} \quad (4.48)$$

where, making use of the MATLAB® notation,

$$[N]_{6\times 6} = [K - M(:, 1:2)] \quad (4.49)$$

$$[P]_{6\times 2} = [M(:, 3:4)]_{6\times 2} \quad (4.50)$$

Premultiplying Equation 4.48 with $[N]^{-1}$ yields

$$\begin{bmatrix} p_2(0) \\ v_2(0) \\ p_4(0) \\ v_4(0) \\ p_3(l) \\ v_3(l) \end{bmatrix} = [Q]_{6\times 2} \begin{bmatrix} p_4(l) \\ v_4(l) \end{bmatrix} \quad (4.51)$$

where

$$[Q]_{6\times 2} = [N]_{6\times 6}^{-1}[P]_{6\times 2} \quad (4.52)$$

Adopting the first two rows of the matrix Equation 4.51 yields the transfer matrix relation between the upstream point 'u' and the downstream point 'd':

$$\begin{bmatrix} p_2(0) \\ v_2(0) \end{bmatrix} = [T]_{2\times 2} \begin{bmatrix} p_4(l) \\ v_4(l) \end{bmatrix} \quad (4.53)$$

where, using the MATLAB® notation, the desired transfer matrix $[T]$ is given by

$$[T]_{2\times 2} = [Q(1:2,:)] \quad (4.54)$$

Incidentally, the elimination process represented by Equations 4.44–4.54 could be replaced by the partitioned matrix approach [3] as an alternative. However, the systematic approach presented above in this section is more generic and is particularly suited for MATLAB® programming.

In the next section we discuss derivations of the end-chamber transfer matrices $[H]$ and $[J]$ of Equations 4.42 and 4.43.

4.4 Flow-Reversal End Chambers

The flow-reversal end chambers are used often in automotive exhaust mufflers due to logistic constraints. The muffler configuration of Figure 4.3 makes use of two such chambers (of axial length l_a and l_b). If the end-cavity lengths l_a and l_b are of sufficient length such that

$$\frac{L}{D_1} \geq 0.5 + 0.5\frac{d}{D_1}, L \equiv l_a, l_b, \quad (4.55)$$

then plane waves in the end cavity will be primarily in the axial direction (see Figure 4.4). In this case, the state variables at $z = 0$ in Figure 4.3 in duct 3 may be related to those in duct 4 across the left-hand cavity by means of the relation (4.43) where $[J]$ is the product of the transfer matrices [4] of

a. uniform duct of radius r_3 and length $\delta_{31} + l_{4a} + t_a$,
b. reversal-expansion element (see Figure 2.13c),
c. sudden contraction element (see Figure 2.11a), and
d. uniform duct of radius r_4 and length $\delta_{41} + t_a + l_{4a}$,

where δ_{31} and δ_{41} are the end corrections given by

$$\delta_{31} = 0.6 r_3 (1 - 1.25 r_3 / r_a), \tag{4.56a}$$

$$\delta_{41} = 0.6 r_4 (1 - 1.25 r_4 / r_a) r_a = \left(r_1^2 - r_2^2\right)^{1/2} \tag{4.56b}$$

and t_a is thickness of the intermediate baffle plate (see Figure 4.5a).

Similarly, for the RHS end chamber, referring again to Figure 4.3, where $[H]$ is the product of the transfer matrices [4] of

a. uniform duct of radius r_2 and length $\delta_{21} + l_{4b} + t_b$,
b. reversal-expansion element (see Figure 2.13c)
c. sudden contraction element, (see Figure 2.11a) and
d. uniform duct of radius r_3 and length $\delta_{31} + t_b + l_{4b}$,

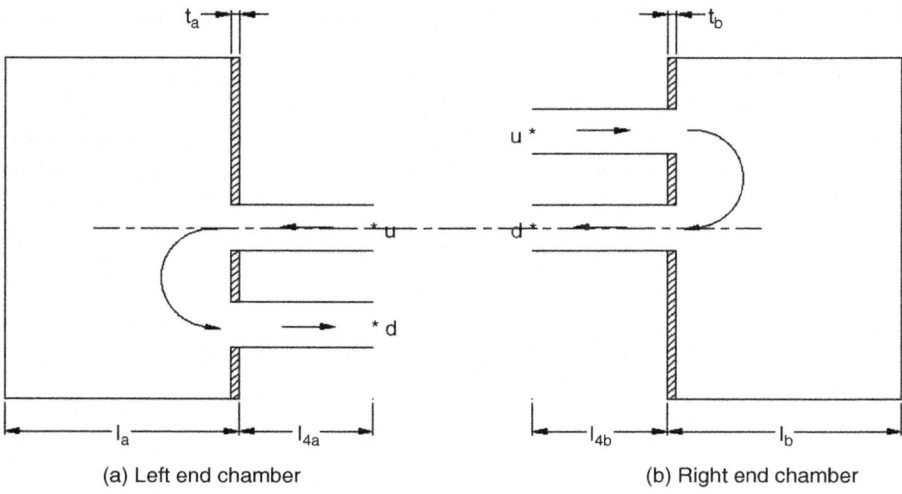

(a) Left end chamber   (b) Right end chamber

Figure 4.5 The two end chambers of Figure 4.3, presumed to be axially long enough to support axial plane waves

where δ_{21} is the end correction given by

$$\delta_{21} = 0.6r_2(1 - 1.25r_2/r_b), \; r_b = \left(r_1^2 - r_4^2\right)^{1/2}. \tag{4.56c}$$

However, if the end-chamber lengths l_a and l_b are small enough (see Figure 4.5) to satisfy the following inequality

$$\frac{L}{D_1} \leq 0.25 + 0.25\frac{d}{D_1}, \quad L \equiv l_a, l_b, \tag{4.57}$$

then the wave propagation will be primarily in the transverse direction. In this case, the transverse wave propagation would encounter a variable-area duct with cross-sectional area given by Equation 2.128, the corresponding Helmholtz equation would be given by Equation 2.129, and the acoustic pressure field would be given by Equation 2.133 along with Equations 2.134–2.137. Thus, in terms of the nondimensional space variable, $y = x/D_1$, we have [7]

$$p(y) = AF_1(y) + BF_2(y). \tag{4.58}$$

and

$$\rho_0 c_0 u(y) = \frac{j}{\beta}\left[AF'_1(y) + BF'_2(y)\right], \tag{4.59}$$

where

$$\beta = k_0 D_1, \quad F'_1(y) = \frac{dF_1(y)}{dy}, \quad F'_2(y) = \frac{dF_2(y)}{dy}. \tag{4.60}$$

We may recall from Section 2.13.3 that functions $F_1(y)$ and $F_2(y)$ are Frobenius polynomial functions given by Equations 2.134–2.137.

Impedance of the variable-area closed-end quarter-wave resonator cavity at $y = y_1 (y_1 = x_1/D_1)$ is evaluated from Equations 4.58 and 4.59 as follows.

At the closed end $(y = x = 0), u = 0$. Expression (2.135) for $F_2(y)$ indicates that $F'_2(0) \to \infty$. Therefore, B in Equations 4.58 and 4.59 must be zero. With this simplification, dividing Equation 4.58 by Equation 4.59 yields

$$\frac{p(y_1)}{\rho_0 c_0 u(y_1)} = -j\beta \frac{F_1(y_1)}{F'_1(y_1)}, \tag{4.61}$$

whence, noting that $\rho_0 c_0 u(y_1) = Y_1 v_1(y_1)$, impedance $Z(y_1)$ looking into the cavity is given by

$$Z(y_1) = \frac{p(y_1)}{-v(y_1)} = j\beta Y_1 \frac{F_1(y_1)}{F''_1(y_1)}, \tag{4.62}$$

where $Y_1 = \frac{c_0}{S_1}$, $S_1 = 2l_e D_2 \left(y_1 - y_1^2\right)^{1/2}$ (as per Equation 2.128).

If $x = D_1/2$ or $y = 0.5$, that is, if the inlet/outlet pipe is located in the middle, then

$$Z \equiv Z(0.5) = j\beta Y_m \frac{F_1(0.5)}{F_1'(0.5)}, \quad Y_m = \frac{c_0}{S_m} = \frac{c_0}{l_e D_2}, \quad (4.63)$$

where l_e is the axial length of the end chamber. Thus,

$l_e = l_a$ for the LHS end chamber (Figure 4.5a), and
$l_e = l_b$ for the RHS end chamber (Figure 4.5b).

Incorporating the branch resonator impedance $Z(y_1)$, transfer matrix of the $x = x_1$ junction in Figure 4.5(b) may be evaluated on the lines of the extended-inlet junction. Thus, adopting the transfer matrix relation (3.95) with $C_1 = -1$ and $C_2 = +1$ from Table 3.2, loss coefficient, $K = [(S_d/S_u) - 1]^2$ from Table 3.1 with $S_u = \pi d_2^2/4$ and $S_d = 2D_2 l_b (y_1 - y_1^2)^{1/2}$ (as per Equation 2.128, making use of Equation 3.34 to convert the convective state variables p_c and v_c into the classical state variables p and v adopted here, and neglecting terms involving M^2 and higher-order terms in keeping with the practice adopted throughout this monograph, we get after considerable algebra [8]

$$\begin{bmatrix} p_1 \\ v_1 \end{bmatrix} = \frac{1}{Z_{eff\,1}} \begin{bmatrix} Z_2(y_1) - K_{1u} M_u Y_u & M_u Y_u Z_2(y) K_{2u} \\ 1 & Z_1(y_1) - M_u Y_u K_{3u} \end{bmatrix} \begin{bmatrix} p_3 \\ v_3 \end{bmatrix}, \quad (4.64)$$

where, referring to points 1, 2 and 3 in Figure 4.6(c),

$$Z_{eff\,1} = Z_2(y_1) + \frac{S_u}{S_2} M_u Y_u;$$

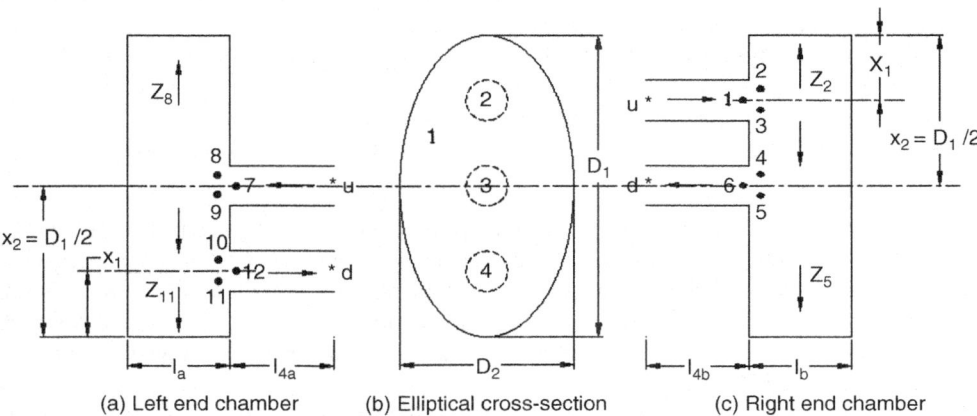

(a) Left end chamber (b) Elliptical cross-section (c) Right end chamber

Figure 4.6 The two end chambers of Figure 4.3, presumed to be axially short to support transverse plane waves

$$Z_2(y_1) = j\beta \frac{F_1(y_1)}{F_1'(y_1)} = Z_m \text{ (as per Equation 4.62 above);}$$

$$K_{1u} = 1 - \frac{S_u}{S_2}; \; K_{2u} = 2\frac{S_u}{S_2}\left(1 - \frac{S_u}{S_2}\right); \; K_{3u} = \frac{S_u}{S_2}\left(1 - 4\frac{S_u}{S_2}\right); \quad (4.65)$$

$$S_2 = 2l_b D_2 \left(y_1 - y_1^2\right)^{1/2}, \quad y_1 = x_1/D_1.$$

Similarly, at the outlet junction of the right end-chamber, referring to points 4, 5 and 6 in Figure 4.6(c), we get [8]

$$\begin{bmatrix} p_4 \\ v_4 \end{bmatrix} = \frac{1}{Z_{\text{eff } 2}} \begin{bmatrix} Z_5(y_2) - K_{1d}M_d Y_d & M_d Y_d Z_5(y_2) K_{2d} \\ 1 & Z_5(y_2) + M_d Y_d K_{3d} \end{bmatrix} \begin{bmatrix} p_d \\ v_d \end{bmatrix}, \quad (4.66)$$

where

$$Z_{\text{eff } 2} = Z_5(y_2) - M_d Y_d \left(\frac{S_d}{S_5}\right)^2;$$

$$Z_5(y_2) = j\beta \frac{F_1(0.5)}{F_1'(0.5)} = Z_m$$

$$K_{1d} = 2\left(\frac{S_d}{S_5}\right)^2; \; K_{2d} = 0.5\left(3 - \frac{S_d}{S_5}\right), \; K_{3d} = 1.5\left(1 - \frac{S_d}{S_5}\right) \quad (4.67)$$

$$S_5 = l_b D_2, \quad y_2 = x_2/D_1 = 0.5$$

Thus, for transverse plane waves in the short right end chamber of Figure 4.6(c), the transfer matrix [H] between the upstream point 'u' and downstream point 'd' is the product of the transfer matrices of

a. uniform duct of radius r_2 and length $\delta_{21} + l_{4b} + t_b$ (as for the axial wave model above),
b. transfer matrix (4.64) for the inlet junction (1–2–3),
c. transfer matrix (2.189) of the elliptical duct between points 3 and 4 with $y_1 = x_1/D_1$ and $y_2 = 0.5$,
d. transfer matrix (4.66) for the outlet junction (4–5–6), and
e. uniform duct of radius r_3 and length $\delta_{31} + t_b + l_{4b}$ (as for the axial wave model above).

Similarly, the transfer matrix (J) between the upstream point 'u' and downstream point 'd' of the left end chamber may be computed by replacing Z_2 with Z_8(or Z_m), Z_5(or Z_m) with Z_{11}(or Z_2), l_b with l_a, t_b with t_a, and l_{4b} with l_{4a}. In fact, for a symmetric chamber (with $r_2 = r_3 = r_4$, $l_a = l_b$, $t_a = t_b$ and $l_{4a} = l_{4b}$, which is common practice), it may be noted that the overall transfer matrix of the left end chamber [J] will be inverse of the transfer matrix of the right end chamber [H]; that is $[J] = [H]^{-1}$.

For such a symmetric three-pass double-reversal chamber, $M_u = M_d = M_g$, where M_g is the grazing component of the mean flow Mach number, as will be shown in the next section, which deals with flow distribution and the stagnation pressure drops.

4.5 Meanflow Lumped Resistance Network Theory

Mean flow affects the perforate impedance as shown in Equations 3.128 and 3.129. In muffler configurations, particularly of the type of Figure 4.3, grazing flow and bias flow occur simultaneously, and then the acoustic impedance of a perforate is given by the expression [9]

$$\zeta = \vartheta + j\chi \tag{4.68}$$

where

$$\vartheta = \mathrm{Re}\left\{j\frac{k}{\sigma C_D}\left[\frac{t}{F(\mu')} + \frac{\delta_{re} f_{\mathrm{int}}}{F(\mu)}\right]\right\} + \frac{1}{\sigma}\left[1 - \frac{2J_1(kd)}{kd}\right] + 0.3\frac{M_g}{\sigma} + 1.15\frac{M_b}{\sigma C_D} \tag{4.69}$$

$$\chi = \mathrm{Im}\left\{j\frac{k}{\sigma C_D}\left[\frac{t}{F(\mu')} + 0.5\frac{d}{F(\mu)}\right]\right\} \tag{4.70}$$

$$K = \left(-\frac{j\omega}{\nu}\right)^{0.5}, \quad K' = \left(-\frac{j\omega}{\nu'}\right)^{0.5}, \quad F(Kd) = 1 - \frac{4J_1(Kd/2)}{Kd.J_0(Kd/2)} \tag{4.71}$$

$$\delta_{re} = 0.2\,d + 200\,d^2 + 16000\,d^3, \quad f_{\mathrm{int}} = 1 - 1.47\,\sigma^{0.5} + 0.47\,\sigma^{1.5} \tag{4.72}$$

In Equations 4.68–4.72, t is the orifice thickness, d is the orifice diameter, σ is the porosity, k is the wave number ω/c, c is the speed of sound, C_D is the orifice discharge coefficient, J is the Bessel function, $\nu = \mu/\rho_0$ is the kinematic viscosity, ρ_0 is the fluid density, μ is the adiabatic dynamic viscosity, $\mu' = 2.179\,\mu$, M_g is the grazing flow Mach number, and M_b is the bias flow Mach number inside the holes of the perforate. Equation 4.72 is a correction factor for the orifice interaction effects.

Evaluation of the bias flow Mach number, M_b and the grazing flow Mach number, M_g in each of the perforated pipes in a multiply-connected perforated-element muffler may be carried out readily by means of a lumped flow resistance network making use of the electrical circuit analogy with Kirchhoff's first law for the nodes or junctions and second law for the closed loops.

Flow resistance R of the lumped element is defined here as

$$R = \frac{\Delta p}{Q|Q|}, \text{ or } R = \frac{\Delta p}{Q^2}, \tag{4.73}$$

if directionality is immaterial or understood. Here, Δp is the stagnation pressure drop across the element and Q is the volume flow rate passing through the element.

This definition is substantially different from the definition of electrical resistance which is defined as voltage drop across the element divided by the current passing through the element. If Δp is written in terms of the dynamic head, $H = \frac{1}{2}\rho_0 U^2$, and loss coefficient ε, as is the practice in fluid mechanics of incompressible flows, then

$$\Delta P = \varepsilon\left(\frac{1}{2}\rho_0 U^2\right), U = \frac{Q}{S}, \tag{4.74}$$

where S is the area of cross-section (m^2) through which the flow passes at the rate of $Q(m^3/s)$.

Combining Equations 4.73 and 4.74 yields the following expression for flow resistance in terms of the loss coefficient ε and the flow passage area, S:

$$R = \varepsilon \frac{\rho_0}{2S^2}. \tag{4.75}$$

Resistances given by Elnady et al. [9] are as follows:

1. Open end flow resistance:

$$R_{open} = \varepsilon \frac{\rho_0}{2S^2}, \tag{4.76}$$

where S is area of pipe, and the loss coefficient $\varepsilon = 1.0$ for outlet (expansion) and 0.5 for inlet (contraction).

2. Perforate flow resistance for cross flow (or through flow or bias flow):

$$R_{CF} \equiv R_{perforate} = \frac{\rho_0}{2(C_D S_p)^2}, \tag{4.77}$$

where C_D is coefficient of discharge of the orifices or holes; $C_D = 0.8$ (assumed), and

$$S_p = \text{perforate area} = \pi d l_p \sigma. \tag{4.78}$$

Here, d, l_p and σ are, respectively, diameter, length and porosity of the perforate.

For the cross-flow expansion element shown in Figure 4.7a, the following empirical expression has been derived for the normalized pressure drop as a function of the Open Area Ratio (OAR) and porosity (σ) [10]:

$$\Delta P/H = 3.252(OAR)^{-1.391} \sigma^{0.018}, 0.3 < OAR < 2.2, 0.055 < \sigma < 0.21. \tag{4.79a}$$

This is shown in Figure 4.8. It may be noted from Equation 4.79a that the normalized stagnation pressure drop has a relatively weak dependence on porosity outside OAR which is proportional to porosity. Thus, for practical purposes, Equation 4.79a may be simplified to

$$\Delta P/H = 3.136(OAR)^{-1.391}, 0.3 < OAR < 2.2. \tag{4.79b}$$

In Equation 4.79, the open area ratio, OAR, is defined as the ratio of the total area of the pores or holes or orifices of the perforate to the cross-sectional area of the perforated elements. Porosity is related to the center-to-center distance between the holes. OAR and porosity are in turn related to each other as follows:

$$OAR = \frac{4L_p \sigma}{d} = n_h \left(\frac{d_h}{d}\right)^2, \tag{4.80}$$

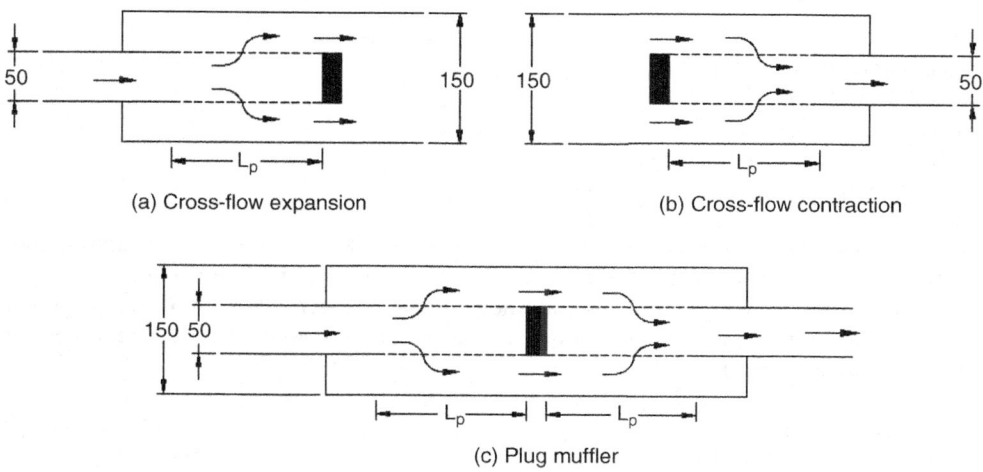

Figure 4.7 Schematics and dimensional details of the plug muffler and its constituent cross-flow elements

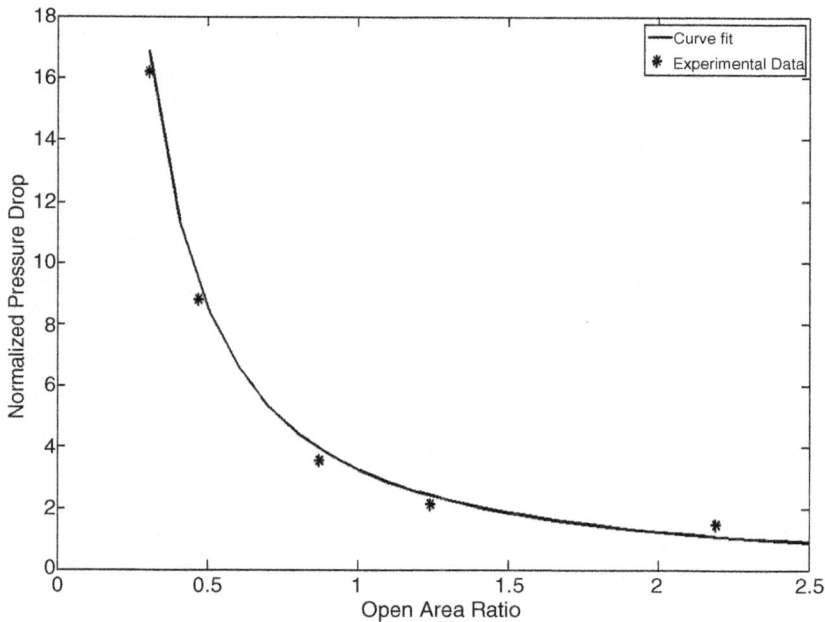

Figure 4.8 Pressure drop versus OAR for cross-flow expansion (Reproduced with permission from [10])

where

L_p is length of the perforate,
σ is porosity of the perforate,
d is diameter of the inner pipe,
d_h is diameter of each hole, and
n_h is the total number of the holes in the perforate.

Knowing the OAR, porosity and diameter of the pipe and holes, the other parameters like length of the perforate and number of holes can be calculated from Equation 4.80.

For the cross-flow contraction element shown in Figure 4.7b, the following empirical expression has been derived for the normalized stagnation pressure drop as a function of the open area ratio and porosity [10]:

$$\Delta P/H = 2.31(OAR)^{-1.5818}\sigma^{0.019}, 0.31 < OAR < 1.66, 0.055 < \sigma < 0.13. \quad (4.81a)$$

This is shown in Figure 4.9. It may be observed from Equation 4.81a that the normalized pressure drop across a cross-flow contraction element has a weak dependence on porosity outside OAR which is proportional to porosity. Thus, for practical purposes, Equation 4.81a may be simplified to

$$\Delta P/H = 2.208(OAR)^{-1.5818}, 0.31 < OAR < 1.66 \quad (4.81b)$$

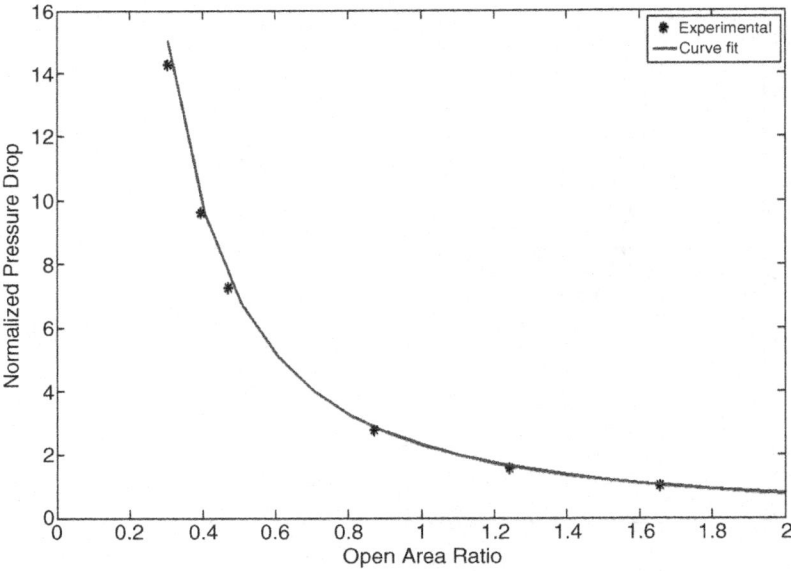

Figure 4.9 Pressure drop versus OAR for cross-flow contraction (Reproduced with permission from [10])

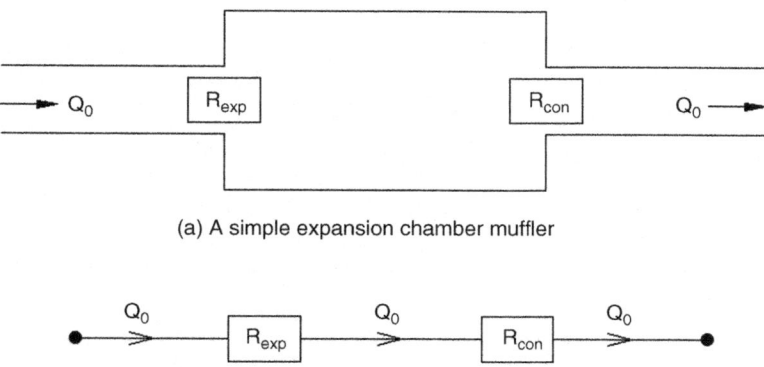

(a) A simple expansion chamber muffler

(b) Equivalent flow resistance network for simple expansion chamber

Figure 4.10 Equivalent impedance of two resistances in series

To develop a methodology for constructing the network for a given muffler configuration and analyzing the network to determine the overall pressure drop across a complex muffler, let us first consider a simple expansion chamber muffler as shown in Figure 4.10(a).

The equivalent flow resistance circuit for the muffler of Figure 4.10(a) is shown in Figure 4.10(b). Here, R_{exp} and R_{con} represent the lumped resistances due to sudden expansion and sudden contraction, respectively. Conservation of mass is valid in the approach and frictional pressure drop across the chamber wall is neglected. For incompressible flow, the outgoing flow rate would be equal to the incoming flow rate.

The resistances due to sudden expansion and contraction are in series, as seen from the network in Figure 4.10(b). The total resistance or the equivalent resistance of the circuit is the sum of the two resistances and hence the total pressure drop can be calculated from the equivalent resistance

$$R_{eq} = R_{exp} + R_{con} \tag{4.82}$$

and the stagnation pressure drop across the chamber of Figure 4.10a is given by

$$\Delta P = R_{eq} Q_0^2 = (R_{exp} + R_{con}) Q_0^2 \tag{4.83}$$

Thus, the resistances in series can simply be added as in an electrical equivalent circuit. When it comes to resistances in parallel, however, there is a complication due to the nonlinearity in Equation 4.73. Let us consider a general circuit with resistances in parallel as shown in Figure 4.11, and derive the relation for the equivalent resistance.

By Kirchhoff's first law, the volume velocity entering a node should be equal to that leaving a node. Hence,

$$Q_0 = Q_1 + Q_2 \tag{4.84}$$

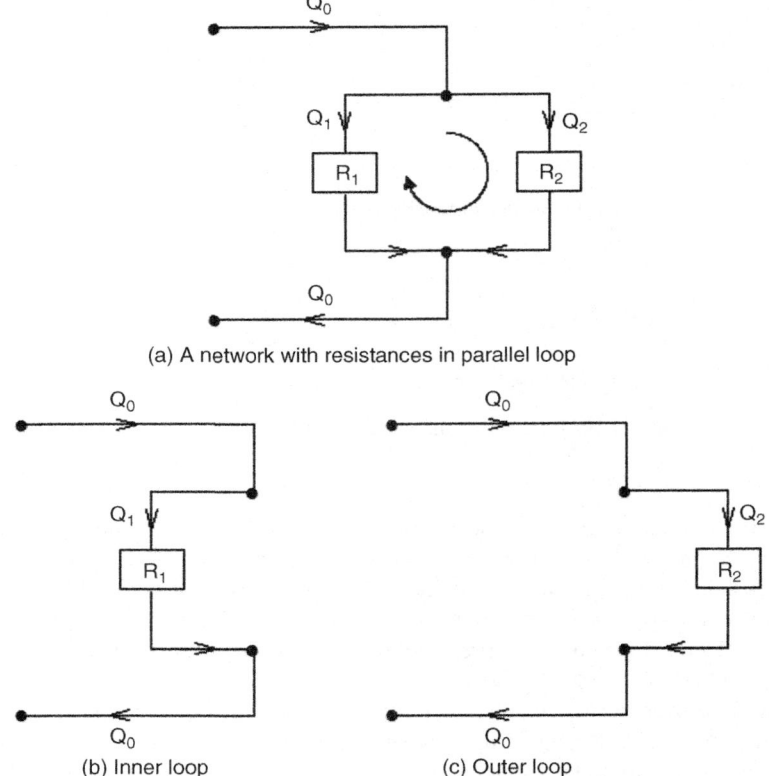

Figure 4.11 Equivalent impedance of two resistances in parallel

From Kirchhoff's second law, for the closed loop,

$$Q_1|Q_1|R_1 - Q_2|Q_2|R_2 = 0 \qquad (4.85)$$

Assuming the flow to be positive, we get

$$Q_1^2 R_1 = Q_2^2 R_2 \qquad (4.86)$$

The total pressure drop across the muffler can be defined in terms of the equivalent resistance (see Figure 4.11). But this should be equal to the pressure drop across the inner loop and also the outer loop. Hence, we get

$$\Delta P = \Delta P_{inner\ loop} = \Delta P_{outer\ loop} \qquad (4.87)$$

or

$$Q_0^2 R_{eq} = Q_1^2 R_1 = Q_2^2 R_2 \qquad (4.88)$$

From Equation 4.88, we get

$$Q_1 = Q_0 \sqrt{\frac{R_{eq}}{R_1}} \quad Q_2 = Q_0 \sqrt{\frac{R_{eq}}{R_2}} \tag{4.89}$$

Substituting Equation 4.89 in Equation 4.84 we get

$$Q_0 = Q_0 \sqrt{\frac{R_{eq}}{R_1}} + Q_0 \sqrt{\frac{R_{eq}}{R_2}} \tag{4.90}$$

Therefore,

$$\sqrt{\frac{1}{R_{eq}}} = \sqrt{\frac{1}{R_1}} + \sqrt{\frac{1}{R_2}} \tag{4.91a}$$

or

$$R_{eq} = \frac{R_1 R_2}{R_1 + R_2 + 2\sqrt{R_1 R_2}} \tag{4.91b}$$

Incidentally, Equation 4.91b indicates that if $R_1 = R_2 = R_0$ (say), then $R_{eq} = R_0/4$, not $R_0/2$, which would be the value for linear element circuits. It has a very important (and useful) implication: back-pressure of a multiply-connected muffler will be very little because of the parallel flow paths. In particular, for n parallel paths of total area of cross-section equal to that of the incoming or outgoing pipe, the stagnation pressure drop due to wall friction will be equal to $1/n^2$ times that of the single pipe of the same length and equivalent area of cross-section.

Now, the resistances of any complex muffler can be converted into one single equivalent resistance using formulae (4.82) and (4.91) for resistances in series and parallel, respectively. Once the equivalent resistance is obtained, the total pressure drop can be easily calculated as shown in the next section.

4.6 Meanflow Distribution and Back Pressure Estimation

The lumped-element network theory described above is illustrated below for three different multiply-connected perforated-element mufflers.

4.6.1 A Chamber with Three Interacting Ducts

An expansion chamber with three interacting ducts, shown in Figure 4.12(a), can be represented in an equivalent resistance network, as shown in Figure 4.12(b). The network consists of resistances due to sudden area changes (expansion and contraction) and resistances due to the flow through perforates. As indicated before, the grazing flow resistance and the wall friction resistance are relatively insignificant and may be neglected.

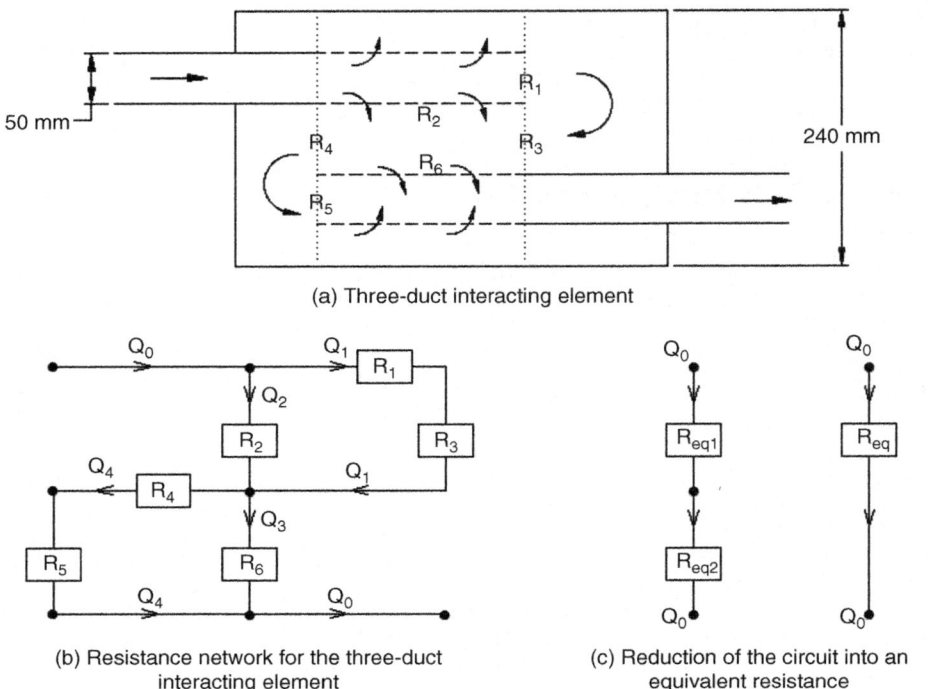

Figure 4.12 Application of the network theory to a three-duct interacting element

Q_0 is the input volume flow which is usually measured upstream of the muffler. It is related to the flow velocity U and Mach number M by

$$Q_0 = UA_0 = McA_0 \quad (4.92)$$

where c is the speed of sound and A_0 is the inlet duct cross-sectional area. Assuming incompressible flow, Kirchhoff's first law can be applied. For the analogous circuit shown in Figure 4.12(b), this gives

$$Q_0 = Q_1 + Q_2 \quad (4.93)$$

$$Q_0 = Q_3 + Q_4 \quad (4.94)$$

Application of the Kirchhoff's second law to the two closed loops in Figure 4.12(b) yields

$$R_1 Q_1 |Q_1| + R_3 Q_1 |Q_1| - R_2 Q_2 |Q_2| = 0 \quad (4.95)$$

$$R_4 Q_4 |Q_4| + R_5 Q_4 |Q_4| - R_6 Q_3 |Q_3| = 0 \quad (4.96)$$

Equations 4.93 to 4.96 represent a set of four nonlinear equations. This system has four equations and four unknowns (Q_1, Q_2, Q_3 and Q_4). Following Elnady et al. [9] these equations would have to be solved simultaneously to get the flow distribution. However, this is unnecessary as shown below.

The network has six resistances. The resistances R_1 and R_4 are due to sudden area expansion, R_3 and R_5 due to sudden contractions, and R_2 and R_6 are the flow resistances across perforates. The resistance due to wall frictional losses has been neglected here. The resistances can be calculated using the formulae given in Section 4.5 above.

Once the resistances are known, the total pressure drop can be calculated by determining the equivalent resistance of the circuit. The given circuit can be reduced to a circuit with equivalent resistance as shown in Figure 4.12(c).

Now, use of Figure 4.12(b) and Equations 4.82 and 4.91b yields

$$R_{eq} = R_{eq1} + R_{eq2} \tag{4.97}$$

where

$$R_{eq1} = \frac{(R_1 + R_3)R_2}{(R_1 + R_3) + R_2 + 2\sqrt{(R_1 + R_3)R_2}} \tag{4.98}$$

and

$$R_{eq2} = \frac{(R_4 + R_5)R_6}{(R_4 + R_5) + R_6 + 2\sqrt{(R_4 + R_5)R_6}} \tag{4.99}$$

The total pressure drop can now be calculated as

$$\Delta P = R_{eq} Q_0^2 \tag{4.100}$$

The pressure drop thus calculated has been compared with the measured values. The given configuration has been validated for different open area ratios, OAR. Figure 4.13 shows the experimental validation for OAR $= 0.871$. It is clear that predictions of the network approach presented here compare well with those observed experimentally.

4.6.2 Three-Pass Double-Reversal Chamber

Figure 4.14 shows the mean flow distribution in a three-pass double reversal chamber. It is obvious that for a symmetrical configuration with the three perforated ducts having the same porosity as well as diameter, the bias flow Q_b getting out of duct 2 into the annulus 1 would enter duct 4, and the grazing flow Q_g in all the three ducts would be the same except for the change in directions, as shown in Figure 4.14.

Thus,

$$Q_5 = -Q_1 = -Q_b \text{ and } Q_3 = 0. \tag{4.101}$$

Now,

$$Q_0 = Q_b + Q_g, \tag{4.102}$$

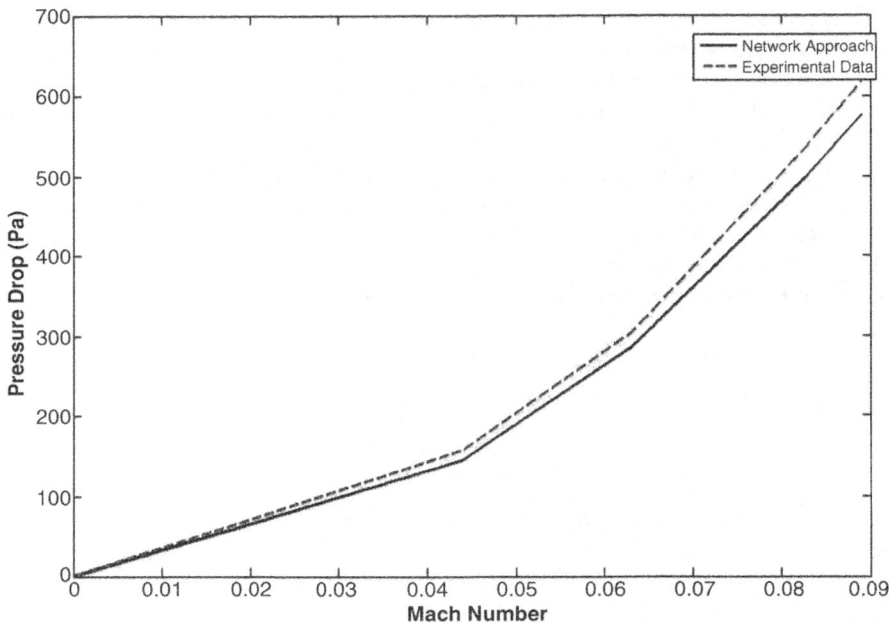

Figure 4.13 Pressure flow curve for the three duct muffler of Figure 4.12(a) with OAR = 0.871 (Reproduced with permission from [10])

and Q_b and Q_g are related to each other as

$$Q_b^2 R_{CF} = Q_g^2 (R_e + R_c), \tag{4.103}$$

where the sudden expansion resistance R_e, sudden contraction resistance R_c, and the cross-flow resistance R_{CF} are given by Equations 4.76 and 4.77, respectively.

Simultaneous solution of Equations 4.102 and 4.103 yields

$$Q_g = \frac{Q_0}{1 + \left(\frac{R_e + R_c}{R_{CF}}\right)^{1/2}} \quad \text{and} \quad Q_b = \frac{Q_0}{1 + \left(\frac{R_{CF}}{R_e + R_c}\right)^{1/2}} \tag{4.104}$$

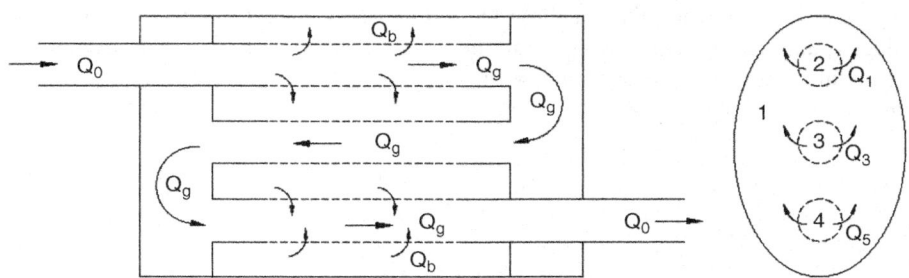

Figure 4.14 Flow distribution in a -pass double-reversal muffler

Thus, the mean flow distribution in the three-pass double-reversal chamber is known without having to solve the five nonlinear equations simultaneously. Finally, the total back-pressure would be the sum of the pressure drops across the ducts 2 and 4. The two are in fact equal to each other, and therefore the total pressure drop would be given by

$$\Delta p = 2 Q_0^2 R_{eq} \qquad (4.105)$$

where the equivalent resistance R_{eq} is given by Equation 4.91b with $R_1 = R_{CF}$ and $R_2 = R_e + R_c$.

The heuristic analysis presented above, however, may not always be feasible and we may have to write and solve a number of nonlinear algebraic equations simultaneously, as shown below.

4.6.3 A Complex Muffler Configuration

The flow resistance network is now applied to a complex muffler configuration shown in Figure 4.15(a), adopted from Elnady et al. [9]. It can be represented in an equivalent flow

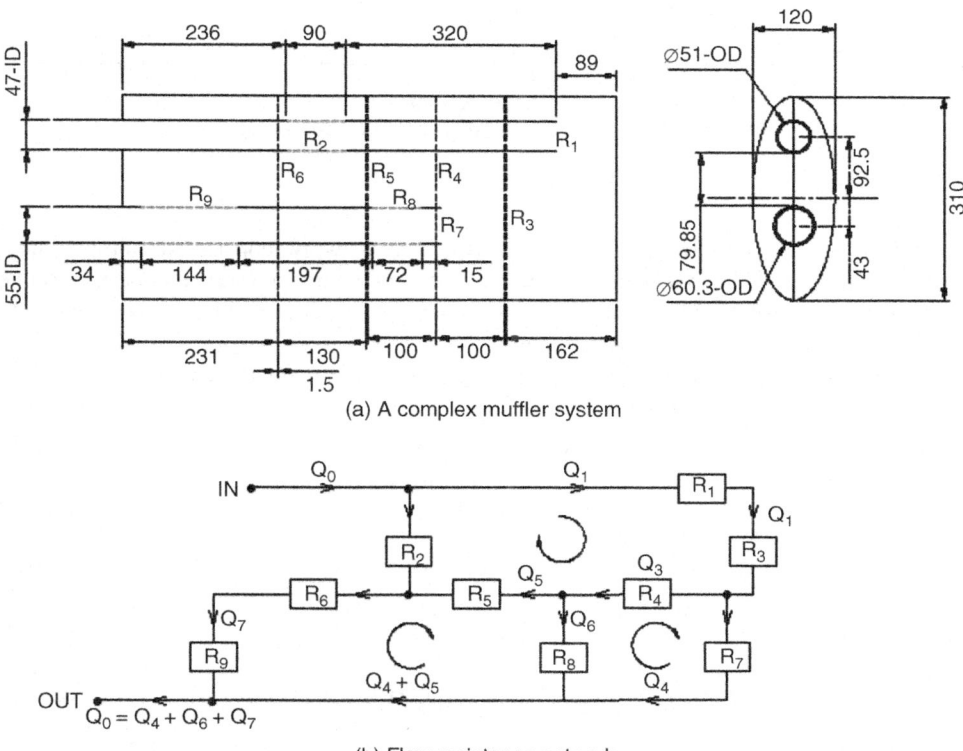

(a) A complex muffler system

(b) Flow resistance network

Figure 4.15 Illustration of the network theory for a complex multiply connected muffler configuration (Reproduced with permission from [10])

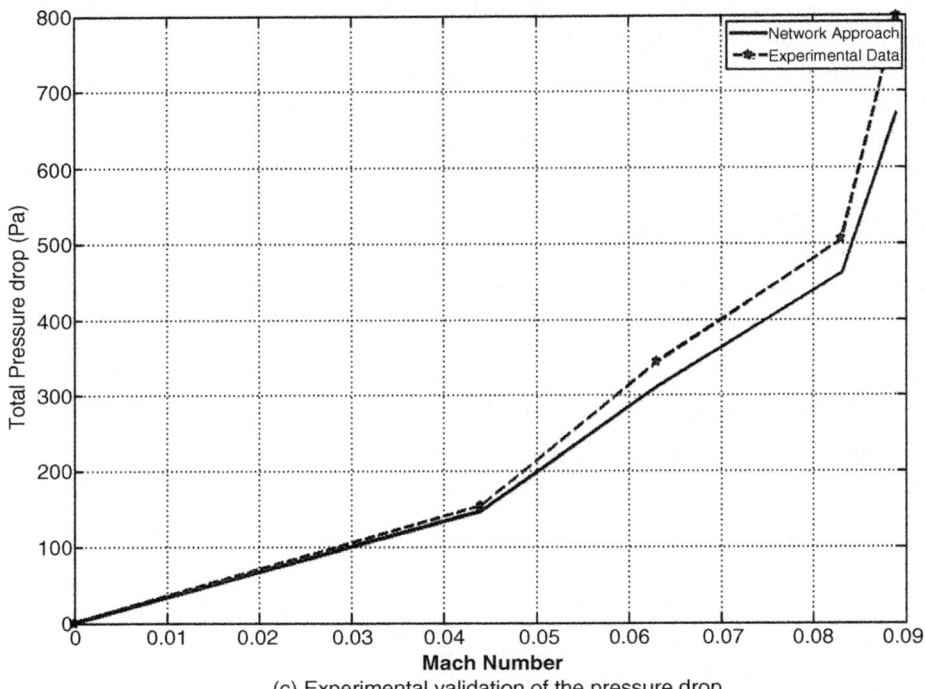

(c) Experimental validation of the pressure drop

Figure 4.15 (*Continued*)

resistance network as shown in Figure 4.15(b). The network consists of lumped flow resistances due to sudden area changes (expansion R_1 and contraction R_4) and resistances due to flow through perforates (R_2, R_8 and R_9) and baffles (R_3, R_5 and R_6). The resistance due to wall frictional losses has been neglected here.

For the circuit given in Figure 4.15(b), applying Kirchhoff's first and second law, equations at the junctions can be written as [9]

$$Q_0 = Q_1 + Q_2 \tag{4.106}$$

$$Q_1 = Q_3 + Q_4 \tag{4.107}$$

$$Q_3 = Q_5 + Q_6 \tag{4.108}$$

$$Q_7 = Q_2 + Q_5 \tag{4.109}$$

$$R_1 Q_1 |Q_1| + R_3 Q_1 |Q_1| + R_4 Q_3 |Q_3| + R_5 Q_5 |Q_5| - R_2 Q_2 |Q_2| = 0 \tag{4.110}$$

$$R_8 Q_6 |Q_6| - R_5 Q_5 |Q_5| - R_6 Q_7 |Q_7| - R_9 Q_7 |Q_7| = 0 \tag{4.111}$$

$$R_7 Q_4 |Q_4| - R_8 Q_6 |Q_6| - R_4 Q_3 |Q_3| = 0 \tag{4.112}$$

The '*fsolve*' built-in function in Matlab can be used to solve this system in the form of $f(x) = 0$, where

$$f(\overline{Q}) = \begin{bmatrix} Q_0 - Q_1 - Q_2 \\ Q_1 - Q_3 - Q_4 \\ Q_3 - Q_5 - Q_6 \\ Q_7 - Q_2 - Q_5 \\ R_1 Q_1 |Q_1| + R_3 Q_1 |Q_1| + R_4 Q_3 |Q_3| + R_5 Q_5 |Q_5| - R_2 Q_2 |Q_2| \\ R_8 Q_6 |Q_6| - R_5 Q_5 |Q_5| - R_6 Q_7 |Q_7| - R_9 Q_7 |Q_7| \\ R_7 Q_4 |Q_4| - R_8 Q_6 |Q_6| - R_4 Q_3 |Q_3| \end{bmatrix} \qquad (4.113)$$

and

$$\overline{Q} = [Q_1, Q_2, Q_3, Q_4, Q_5, Q_6, Q_7]^T \qquad (4.114)$$

Thus, Equations 4.106 to 4.112 can be solved in Matlab for volume velocities (Q_1 to Q_7) in terms of Q_0. This gives the volume velocities (flow distribution) at each element. Once the volume velocities are known, the back-pressure can be calculated readily by considering any loop of the circuit and calculating the pressure drop across that loop. Considering the loop with resistances R_2, R_6 and R_9 of Figure 4.15(b) yields the following expression for back-pressure ΔP:

$$\Delta P = R_2 Q_2 |Q_2| + R_6 Q_7 |Q_7| + R_9 Q_7 |Q_7| \qquad (4.115)$$

The muffler configuration of Figure 4.15(a) (adopted from Ref. [9]) was fabricated and tested for stagnation pressure drop in the laboratory. The calculated pressure drop has been compared with the measurements in Figure 4.15(c) [10]. There is obviously a good agreement between the two. The baffle's resistance depends on the coefficient of discharge (C_D), which in turn depends on Mach number. The C_D value decreases with the increasing velocity. But this has been taken as constant in the analysis, possibly leading to a discrepancy at higher Mach flows.

Thus, the lumped resistance network theory enables us to predict the total back-pressure as well as the mean flow distribution of the complex multiply-connected muffler configurations.

4.7 Integrated Transfer Matrix Approach

The Integrated Transfer Matrix (ITM) method presented here [11] suggests a transfer matrix based approach to analyze a general commercial muffler. Here, the given muffler is divided into a number of sections based on area discontinuities, perforated surfaces, baffles, or absorptive linings. An integrated transfer matrix relates the state variables across the entire cross-section of the muffler shell, as one moves along the axis, and can be partitioned appropriately in order to relate the state variables of different tubes constituting the cross-section. An '$n \times n$' transfer matrix is obtained relating the upstream and downstream state variables. The size of the matrix $n \times n$ would depend on the number of the individual acoustic elements which can be identified in each section, for which the transfer matrix relations are already available in the literature as well as in Chapters 2 and 3 of this monograph. The constituent muffler sections are then combined appropriately into a single final matrix, with the aid of the boundary conditions, to get the relation between the inlet and outlet state variables, and transmission loss can then be calculated.

The present method, thus, gives a more comprehensive way of analyzing complex multiply connected muffler configurations with much less pre-computational effort and faster execution. It also deals effectively with perforated elements and absorptive linings, and incorporates the convective as well as dissipative effects of mean flow.

In the ITM approach, the given muffler is divided into a number of sections. The division of muffler into sections is carried out on the basis of the relative changes in the muffler geometry such as sudden area changes, perforated pipes, cross baffles or boundaries. Once the muffler is dissected into sections, each section is assigned a transfer matrix relating the upstream and downstream state variables of the section. The size of the state vector and hence the transfer matrix depends on the number of smallest individual elements constituting the cross-section.

The ITM method is illustrated hereunder for three different muffler configurations, starting with a muffler with partially overlapping perforated tubes, followed by a complex muffler with non-overlapping perforated tubes, baffles and area discontinuities, and finally an acoustically lined duct with annular airgap – the so-called combination muffler.

4.7.1 A Muffler with Non-Overlapping Perforated Ducts and a Baffle

Figure 4.16 shows a muffler that consists of both perforated pipes and baffles without any area discontinuities in between. This muffler can be divided into 11 sections with three acoustic elements in each section depending on the presence of perforated pipes and baffles. Transfer matrices are then written for each section relating the upstream state variables of each element of the section to the downstream state variables. Let the state variables of an element, acoustic pressure p_i and volume velocity v_i, be clubbed into a

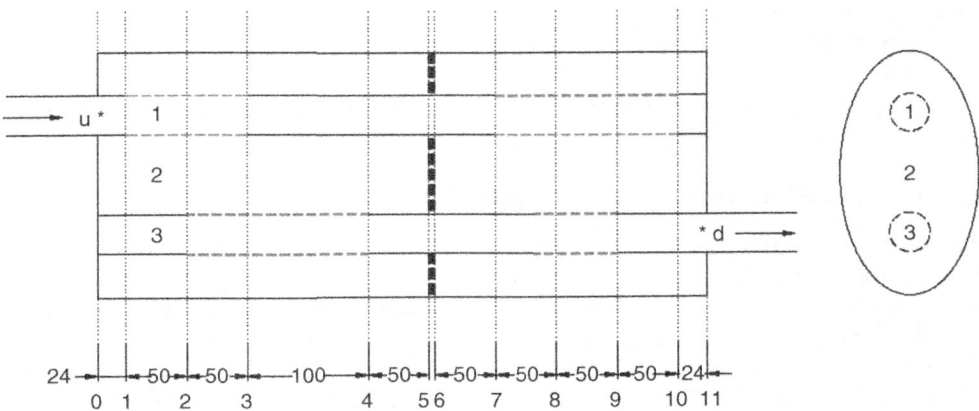

Figure 4.16 Schematic of a muffler with partially overlapping perforated ducts and a baffle, divided into sections to illustrate the integrated transfer matrix approach (points u and d represent the upstream and downstream points, respectively) (Adapted from [11])

state vector $[S_i]$. As is clear from Figure 4.16, there are three interacting ducts in each segment, the integrated state vector for which is given by [11]

$$\begin{bmatrix} p_1(0) \\ v_1(0) \\ p_2(0) \\ v_2(0) \\ p_3(0) \\ v_3(0) \end{bmatrix} \equiv \begin{bmatrix} S_{1,0} \\ S_{2,0} \\ S_{3,0} \end{bmatrix}. \qquad (4.116)$$

If $\mathbf{P_{i,j\_k}}$ be the 2×2 transfer matrix of uniform pipe of the ith acoustic element between the section (j_k), relating the state variables $\mathbf{S_{i,j}}$ and $\mathbf{S_{i,k}}$, then

$$[\mathbf{P_{i,j\_k}}] = \begin{bmatrix} \cos(kL_{j\_k}) & jY\sin(kL_{j\_k}) \\ \dfrac{j}{Y}\sin(kL_{j\_k}) & \cos(kL_{j\_k}) \end{bmatrix}, \qquad (4.117)$$

where k ($=\omega/c$) and Y ($=c_0/S$) denote wave number and characteristic impedance, respectively. The section (0_1) in Figure 4.16 can be viewed as a combination of three rigid-wall pipes, for which the individual transfer matrices are known. The transfer matrix for the whole section can be written by arranging the three known rigid-pipe transfer matrices. Then, the transfer matrix between the section (0_1) can be written in the following form [11]:

$$\begin{bmatrix} S_{1,0} \\ S_{2,0} \\ S_{3,0} \end{bmatrix} = \begin{bmatrix} p_{1,0} \\ v_{1,0} \\ p_{2,0} \\ v_{2,0}(=0) \\ p_{3,0} \\ v_{3,0}(=0) \end{bmatrix} = \begin{bmatrix} [\mathbf{P_{1,0\_1}}] & [\mathbf{O}]_{2\times 2} & [\mathbf{O}]_{2\times 2} \\ [\mathbf{O}]_{2\times 2} & [\mathbf{P_{2,0\_1}}] & [\mathbf{O}]_{2\times 2} \\ [\mathbf{O}]_{2\times 2} & [\mathbf{O}]_{2\times 2} & [\mathbf{P_{3,0\_1}}] \end{bmatrix} \begin{bmatrix} S_{1,1} \\ S_{2,1} \\ S_{3,1} \end{bmatrix}. \qquad (4.118)$$

In Equation 4.118 as per international convention, [**O**] represents a null matrix.

Similarly, the section (1_2) consists of elements 1 and 2 interacting through perforated surface with rigid-pipe annulus as the third element. Let $\mathbf{E_{i,m,j\_k}}$ be a 4×4 matrix representing the interaction of ducts i and m, between the sections j and k. The transfer matrix may be evaluated by means of the eigenanalysis of the coupled equations. Coefficients of the matrix may be evaluated from the coupled equations by means of Kar and Munjal's generalized algorithm [5]. Thus, the overall transfer matrix for the section (1_2) consists of a 4×4 interacting ducts matrix, [**E**], and a 2×2 straight uniform pipe matrix, [**P**], as follows:

$$\begin{bmatrix} S_{1,1} \\ S_{2,1} \\ S_{3,1} \end{bmatrix} = \begin{bmatrix} [\mathbf{E_{1,2,1\_2}}]_{4\times 4} & [\mathbf{O}]_{4\times 2} \\ [\mathbf{O}]_{2\times 4} & [\mathbf{P_{3,1\_2}}]_{2\times 2} \end{bmatrix} \begin{bmatrix} S_{1,2} \\ S_{2,2} \\ S_{3,2} \end{bmatrix}. \qquad (4.119)$$

Similarly let $\mathbf{F_{i,m,p,j\_k}}$ be a 6×6 matrix representing the interaction of ducts i and m across the annulus p, between the sections j and k. The matrix can be evaluated by using the generalized algorithm of [5], as indicated above. And for the cross baffle, the transfer matrix for the flow-acoustic resistance offered by it can be written as [9]

$$\mathbf{B_{j\_k}} = \begin{bmatrix} 1 & Z_{j\_k} \\ 0 & 1 \end{bmatrix}, \quad (4.120)$$

where Z_{j_k} is the lumped impedance offered by the baffle to the axially moving plane wave in a moving medium.

From Equations 4.117, 4.119 and 4.120, it follows that the transfer matrix can be defined between any section by a proper combination of the matrices of individual elements of the section. The transfer matrices of different sections of the muffler of Figure 4.16 are as follows:

$$\begin{bmatrix} S_{1,2} \\ S_{2,2} \\ S_{3,2} \end{bmatrix} = \begin{bmatrix} \mathbf{F}_{1,3,2,2\_3} \end{bmatrix}_{6 \times 6} \begin{bmatrix} S_{1,3} \\ S_{2,3} \\ S_{3,3} \end{bmatrix} \quad (4.121)$$

$$\begin{bmatrix} S_{1,3} \\ S_{2,3} \\ S_{3,3} \end{bmatrix} = \begin{bmatrix} [\mathbf{P}_{1,3\_4}]_{2 \times 2} & [\mathbf{O}]_{2 \times 4} \\ [\mathbf{O}]_{4 \times 2} & [\mathbf{E}_{2,3,3\_4}]_{4 \times 4} \end{bmatrix} \begin{bmatrix} S_{1,4} \\ S_{2,4} \\ S_{3,4} \end{bmatrix} \quad (4.122)$$

$$\begin{bmatrix} S_{1,4} \\ S_{2,4} \\ S_{3,4} \end{bmatrix} = \begin{bmatrix} [\mathbf{P}_{1,4\_5}] & [\mathbf{O}]_{2 \times 2} & [\mathbf{O}]_{2 \times 2} \\ [\mathbf{O}]_{2 \times 2} & [\mathbf{P}_{2,4\_5}] & [\mathbf{O}]_{2 \times 2} \\ [\mathbf{O}]_{2 \times 2} & [\mathbf{O}]_{2 \times 2} & [\mathbf{P}_{3,4\_5}] \end{bmatrix} \begin{bmatrix} S_{1,5} \\ S_{2,5} \\ S_{3,5} \end{bmatrix} \quad (4.123)$$

$$\begin{bmatrix} S_{1,5} \\ S_{2,5} \\ S_{3,5} \end{bmatrix} = \begin{bmatrix} [\mathbf{P}_{1,5\_6}] & [\mathbf{O}]_{2 \times 2} & [\mathbf{O}]_{2 \times 2} \\ [\mathbf{O}]_{2 \times 2} & [\mathbf{B}_{5\_6}] & [\mathbf{O}]_{2 \times 2} \\ [\mathbf{O}]_{2 \times 2} & [\mathbf{O}]_{2 \times 2} & [\mathbf{P}_{3,5\_6}] \end{bmatrix} \begin{bmatrix} S_{1,6} \\ S_{2,6} \\ S_{3,6} \end{bmatrix} \quad (4.124)$$

$$\begin{bmatrix} S_{1,6} \\ S_{2,6} \\ S_{3,6} \end{bmatrix} = \begin{bmatrix} [\mathbf{P}_{1,6\_7}] & [\mathbf{O}]_{2 \times 2} & [\mathbf{O}]_{2 \times 2} \\ [\mathbf{O}]_{2 \times 2} & [\mathbf{P}_{2,6\_7}] & [\mathbf{O}]_{2 \times 2} \\ [\mathbf{O}]_{2 \times 2} & [\mathbf{O}]_{2 \times 2} & [\mathbf{P}_{3,6\_7}] \end{bmatrix} \begin{bmatrix} S_{1,7} \\ S_{2,7} \\ S_{3,7} \end{bmatrix} \quad (4.125)$$

$$\begin{bmatrix} S_{1,7} \\ S_{2,7} \\ S_{3,7} \end{bmatrix} = \begin{bmatrix} [\mathbf{E}_{1,2,7\_8}]_{4 \times 4} & [\mathbf{O}]_{4 \times 2} \\ [\mathbf{O}]_{2 \times 4} & [\mathbf{P}_{3,7\_8}]_{2 \times 2} \end{bmatrix} \begin{bmatrix} S_{1,8} \\ S_{2,8} \\ S_{3,8} \end{bmatrix} \quad (4.126)$$

$$\begin{bmatrix} S_{1,8} \\ S_{2,8} \\ S_{3,8} \end{bmatrix} = \begin{bmatrix} \mathbf{F}_{1,3,2,8\_9} \end{bmatrix}_{6 \times 6} \begin{bmatrix} S_{1,9} \\ S_{2,9} \\ S_{3,9} \end{bmatrix} \quad (4.127)$$

$$\begin{bmatrix} S_{1,9} \\ S_{2,9} \\ S_{3,9} \end{bmatrix} = \begin{bmatrix} [\mathbf{E}_{1,2,9\_10}]_{4 \times 4} & [\mathbf{O}]_{4 \times 2} \\ [\mathbf{O}]_{2 \times 4} & [\mathbf{P}_{3,9\_10}]_{2 \times 2} \end{bmatrix} \begin{bmatrix} S_{1,10} \\ S_{2,10} \\ S_{3,10} \end{bmatrix} \quad (4.128)$$

$$\begin{bmatrix} S_{1,10} \\ S_{2,10} \\ S_{3,10} \end{bmatrix} = \begin{bmatrix} [P_{1,10\_11}] & [O]_{2\times 2} & [O]_{2\times 2} \\ [O]_{2\times 2} & [P_{2,10\_11}] & [O]_{2\times 2} \\ [O]_{2\times 2} & [O]_{2\times 2} & [P_{3,10\_11}] \end{bmatrix} \begin{bmatrix} S_{1,11} \\ S_{2,11} \\ S_{3,11} \end{bmatrix} \quad (4.129)$$

The overall six-port transfer matrix connecting the 0-section to the 11th section can be calculated by successive multiplication of the constituent matrices:

$$\begin{bmatrix} p_{1,0} \\ v_{1,0} \\ p_{2,0} \\ v_{2,0} \\ p_{3,0} \\ v_{3,0} \end{bmatrix} = \begin{bmatrix} T_{11} & T_{12} & T_{13} & T_{14} & T_{15} & T_{16} \\ T_{21} & T_{22} & T_{23} & T_{24} & T_{25} & T_{26} \\ T_{31} & T_{32} & T_{33} & T_{34} & T_{35} & T_{36} \\ T_{41} & T_{42} & T_{43} & T_{44} & T_{45} & T_{46} \\ T_{51} & T_{52} & T_{53} & T_{54} & T_{55} & T_{56} \\ T_{61} & T_{62} & T_{63} & T_{64} & T_{65} & T_{66} \end{bmatrix} \begin{bmatrix} p_{1,11} \\ v_{1,11} \\ p_{2,11} \\ v_{2,11} \\ p_{3,11} \\ v_{3,11} \end{bmatrix}, \quad (4.130)$$

or

$$\begin{bmatrix} p_{1,11} \\ v_{1,11} \\ p_{2,11} \\ v_{2,11} \\ p_{3,11} \\ v_{3,11} \end{bmatrix} = \begin{bmatrix} R_{11} & R_{12} & R_{13} & R_{14} & R_{15} & R_{16} \\ R_{21} & R_{22} & R_{23} & R_{24} & R_{25} & R_{26} \\ R_{31} & R_{32} & R_{33} & R_{34} & R_{35} & R_{36} \\ R_{41} & R_{42} & R_{43} & R_{44} & R_{45} & R_{46} \\ R_{51} & R_{52} & R_{53} & R_{54} & R_{55} & R_{56} \\ R_{61} & R_{62} & R_{63} & R_{64} & R_{65} & R_{66} \end{bmatrix} \begin{bmatrix} p_{1,0} \\ v_{1,0} \\ p_{2,0} \\ v_{2,0} \\ p_{3,0} \\ v_{3,0} \end{bmatrix}, [R] = [T]^{-1}. \quad (4.131)$$

Assumptions of rigid end plates yields four boundary conditions at sections 0 and 11,

$$v_{2,0} = 0, \quad v_{3,0} = 0, \quad v_{1,11} = 0 \quad \text{and} \quad v_{2,11} = 0. \quad (4.132\text{--}4.135)$$

Combining the above 6×6 matrix with the boundary conditions yields one single final matrix relating the inlet state variables, p_{1_0}, v_{1_0}, and outlet state variables, p_{3_11}, v_{3_11}.

This can alternatively be implemented by writing all the available Equations 4.131–4.135 of the muffler in the Gaussian form:

$$\begin{bmatrix} R_{11} & R_{12} & R_{13} & R_{14} & R_{15} & R_{16} & -1 & 0 & 0 & 0 \\ R_{21} & R_{22} & R_{23} & R_{24} & R_{25} & R_{26} & 0 & -1 & 0 & 0 \\ R_{31} & R_{32} & R_{33} & R_{34} & R_{35} & R_{36} & 0 & 0 & -1 & 0 \\ R_{41} & R_{42} & R_{43} & R_{44} & R_{45} & R_{46} & 0 & 0 & 0 & -1 \\ R_{51} & R_{52} & R_{53} & R_{54} & R_{55} & R_{56} & 0 & 0 & 0 & 0 \\ R_{61} & R_{62} & R_{63} & R_{64} & R_{65} & R_{66} & 0 & 0 & 0 & 0 \\ 0 & 0 & 0 & 1 & 0 & 0 & 0 & 0 & 0 & 0 \\ 0 & 0 & 0 & 0 & 0 & 1 & 0 & 0 & 0 & 0 \\ 0 & 0 & 0 & 0 & 0 & 0 & 1 & 0 & 0 & 0 \\ 0 & 0 & 0 & 0 & 0 & 0 & 0 & 1 & 0 & 0 \end{bmatrix} \begin{bmatrix} p_{1,0} \\ v_{1,0} \\ p_{2,0} \\ v_{2,0} \\ p_{3,0} \\ v_{3,0} \\ p_{1,11} \\ v_{1,11} \\ p_{2,11} \\ v_{2,11} \end{bmatrix} = \begin{bmatrix} 0 \\ 0 \\ 0 \\ 0 \\ p_{3,11} \\ v_{3,11} \\ 0 \\ 0 \\ 0 \\ 0 \end{bmatrix} \quad (4.136)$$

Inverting the coefficient matrix by means of Gaussian elimination with pivoting, using a standard subroutine/function, yields

$$\begin{bmatrix} p_{1,0} \\ v_{1,0} \\ p_{2,0} \\ v_{2,0} \\ p_{3,0} \\ v_{3,0} \\ p_{1,11} \\ v_{1,11} \\ p_{2,11} \\ v_{2,11} \end{bmatrix} = [\mathbf{A}] \begin{bmatrix} 0 \\ 0 \\ 0 \\ 0 \\ p_{3,11} \\ v_{3,11} \\ 0 \\ 0 \\ 0 \\ 0 \end{bmatrix} \quad (4.137)$$

whence the desired 2×2 transfer matrix **[T]** may be obtained as follows:

$$\begin{bmatrix} p_{1,0} \\ v_{1,0} \end{bmatrix} = \begin{bmatrix} A_{15} & A_{16} \\ A_{25} & A_{26} \end{bmatrix} \begin{bmatrix} p_{3,11} \\ v_{3,11} \end{bmatrix} = [\mathbf{T}] \begin{bmatrix} p_{3,11} \\ v_{3,11} \end{bmatrix} \quad (4.138)$$

The transmission loss can now be easily calculated by means of Equation 3.27. Thus,

$$TL = 20 \log \left[\frac{1+M_u}{1+M_d} \left(\frac{Y_d}{Y_u} \right)^{1/2} \left| \frac{T_{11} + T_{12}/Y_d + Y_u T_{21} + (Y_u/Y_d) T_{22}}{2} \right| \right] \quad (4.139)$$

4.7.2 Muffler with Non-Overlapping Perforated Elements, Baffles and Area Discontinuities

The muffler configuration of Figure 4.17 adopted from [9] consists of both perforated pipes and cross baffles. The total muffler can be divided into 17 sections with three interacting

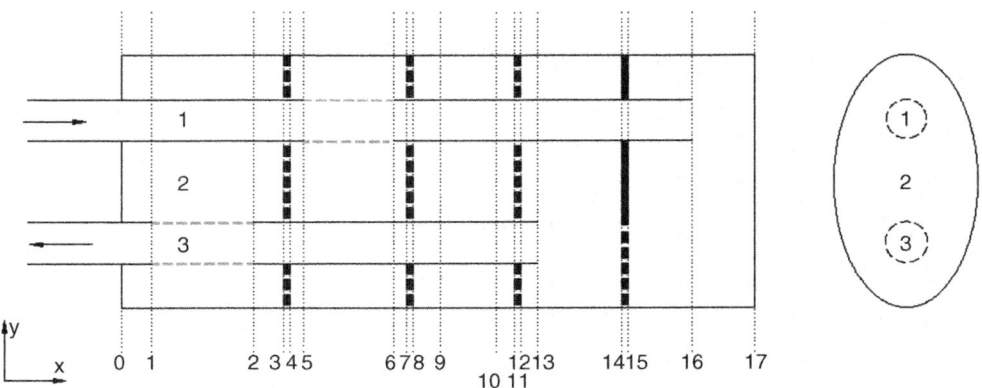

Figure 4.17 Schematic of a muffler with perforated elements, cross baffles and area discontinuities (Adapted from [9,11])

acoustic elements in each section. The muffler can be visualized as one with continuous acoustic elements up to section 13 and another with area discontinuities.

The transfer matrices for the sections 0 through to 13, for the muffler configuration of Figure 4.17 are given below, making use of the Integrated Transfer Matrix method illustrated previously for the muffler configuration of Figure 4.16:

$$\begin{bmatrix} S_{1,0} \\ S_{2,0} \\ S_{3,0} \end{bmatrix} = \begin{bmatrix} [P_{1,0\_1}] & [O]_{2\times 2} & [O]_{2\times 2} \\ [O]_{2\times 2} & [P_{2,0\_1}] & [O]_{2\times 2} \\ [O]_{2\times 2} & [O]_{2\times 2} & [P_{3,0\_1}] \end{bmatrix} \begin{bmatrix} S_{1,1} \\ S_{2,1} \\ S_{3,1} \end{bmatrix} \quad (4.140)$$

$$\begin{bmatrix} S_{1,1} \\ S_{2,1} \\ S_{3,1} \end{bmatrix} = \begin{bmatrix} [P_{1,1\_2}]_{2\times 2} & [O]_{2\times 4} \\ [O]_{4\times 2} & [E_{2,3,1\_2}]_{4\times 4} \end{bmatrix} \begin{bmatrix} S_{1,2} \\ S_{2,2} \\ S_{3,2} \end{bmatrix} \quad (4.141)$$

$$\begin{bmatrix} S_{1,2} \\ S_{2,2} \\ S_{3,2} \end{bmatrix} = \begin{bmatrix} [P_{1,2\_3}] & [O]_{2\times 2} & [O]_{2\times 2} \\ [O]_{2\times 2} & [P_{2,2\_3}] & [O]_{2\times 2} \\ [O]_{2\times 2} & [O]_{2\times 2} & [P_{3,2\_3}] \end{bmatrix} \begin{bmatrix} S_{1,3} \\ S_{2,3} \\ S_{3,3} \end{bmatrix} \quad (4.142)$$

$$\begin{bmatrix} S_{1,3} \\ S_{2,3} \\ S_{3,3} \end{bmatrix} = \begin{bmatrix} [P_{1,3\_4}] & [O]_{2\times 2} & [O]_{2\times 2} \\ [O]_{2\times 2} & [B_{3\_4}] & [O]_{2\times 2} \\ [O]_{2\times 2} & [O]_{2\times 2} & [P_{3,3\_4}] \end{bmatrix} \begin{bmatrix} S_{1,4} \\ S_{2,4} \\ S_{3,4} \end{bmatrix} \quad (4.143)$$

$$\begin{bmatrix} S_{1,4} \\ S_{2,4} \\ S_{3,4} \end{bmatrix} = \begin{bmatrix} [P_{1,4\_5}] & [O]_{2\times 2} & [O]_{2\times 2} \\ [O]_{2\times 2} & [P_{2,4\_5}] & [O]_{2\times 2} \\ [O]_{2\times 2} & [O]_{2\times 2} & [P_{3,4\_5}] \end{bmatrix} \begin{bmatrix} S_{1,5} \\ S_{2,5} \\ S_{3,5} \end{bmatrix} \quad (4.144)$$

$$\begin{bmatrix} S_{1,5} \\ S_{2,5} \\ S_{3,5} \end{bmatrix} = \begin{bmatrix} [E_{1,2,5\_6}]_{4\times 4} & [O]_{4\times 2} \\ [O]_{2\times 4} & [P_{3,5\_6}]_{2\times 2} \end{bmatrix} \begin{bmatrix} S_{1,6} \\ S_{2,6} \\ S_{3,6} \end{bmatrix} \quad (4.145)$$

$$\begin{bmatrix} S_{1,6} \\ S_{2,6} \\ S_{3,6} \end{bmatrix} = \begin{bmatrix} [P_{1,6\_7}] & [O]_{2\times 2} & [O]_{2\times 2} \\ [O]_{2\times 2} & [P_{2,6\_7}] & [O]_{2\times 2} \\ [O]_{2\times 2} & [O]_{2\times 2} & [P_{3,6\_7}] \end{bmatrix} \begin{bmatrix} S_{1,7} \\ S_{2,7} \\ S_{3,7} \end{bmatrix} \quad (4.146)$$

$$\begin{bmatrix} S_{1,7} \\ S_{2,7} \\ S_{3,7} \end{bmatrix} = \begin{bmatrix} [P_{1,7\_8}] & [O]_{2\times 2} & [O]_{2\times 2} \\ [O]_{2\times 2} & [B_{7\_8}] & [O]_{2\times 2} \\ [O]_{2\times 2} & [O]_{2\times 2} & [P_{3,7\_8}] \end{bmatrix} \begin{bmatrix} S_{1,8} \\ S_{2,8} \\ S_{3,8} \end{bmatrix} \quad (4.147)$$

$$\begin{bmatrix} S_{1,8} \\ S_{2,8} \\ S_{3,8} \end{bmatrix} = \begin{bmatrix} [P_{1,8\_9}] & [O]_{2\times 2} & [O]_{2\times 2} \\ [O]_{2\times 2} & [P_{2,8\_9}] & [O]_{2\times 2} \\ [O]_{2\times 2} & [O]_{2\times 2} & [P_{3,8\_9}] \end{bmatrix} \begin{bmatrix} S_{1,9} \\ S_{2,9} \\ S_{3,9} \end{bmatrix} \quad (4.148)$$

$$\begin{bmatrix} S_{1,9} \\ S_{2,9} \\ S_{3,9} \end{bmatrix} = \begin{bmatrix} [P_{1,9\_10}]_{2\times 2} & [O]_{2\times 4} \\ [O]_{4\times 2} & [E_{2,3,9\_10}]_{4\times 4} \end{bmatrix} \begin{bmatrix} S_{1,10} \\ S_{2,10} \\ S_{3,10} \end{bmatrix} \quad (4.149)$$

$$\begin{bmatrix} S_{1,10} \\ S_{2,10} \\ S_{3,10} \end{bmatrix} = \begin{bmatrix} [\mathbf{P}_{1,10\_11}] & [\mathbf{O}]_{2\times 2} & [\mathbf{O}]_{2\times 2} \\ [\mathbf{O}]_{2\times 2} & [\mathbf{P}_{2,10\_11}] & [\mathbf{O}]_{2\times 2} \\ [\mathbf{O}]_{2\times 2} & [\mathbf{O}]_{2\times 2} & [\mathbf{P}_{3,10\_11}] \end{bmatrix} \begin{bmatrix} S_{1,11} \\ S_{2,11} \\ S_{3,11} \end{bmatrix} \quad (4.150)$$

$$\begin{bmatrix} S_{1,11} \\ S_{2,11} \\ S_{3,11} \end{bmatrix} = \begin{bmatrix} [\mathbf{P}_{1,11\_12}] & [\mathbf{O}]_{2\times 2} & [\mathbf{O}]_{2\times 2} \\ [\mathbf{O}]_{2\times 2} & [\mathbf{B}_{11\_12}] & [\mathbf{O}]_{2\times 2} \\ [\mathbf{O}]_{2\times 2} & [\mathbf{O}]_{2\times 2} & [\mathbf{P}_{3,11\_12}] \end{bmatrix} \begin{bmatrix} S_{1,12} \\ S_{2,12} \\ S_{3,12} \end{bmatrix} \quad (4.151)$$

$$\begin{bmatrix} S_{1,12} \\ S_{2,12} \\ S_{3,12} \end{bmatrix} = \begin{bmatrix} [\mathbf{P}_{1,12\_13}] & [\mathbf{O}]_{2\times 2} & [\mathbf{O}]_{2\times 2} \\ [\mathbf{O}]_{2\times 2} & [\mathbf{P}_{2,12\_13}] & [\mathbf{O}]_{2\times 2} \\ [\mathbf{O}]_{2\times 2} & [\mathbf{O}]_{2\times 2} & [\mathbf{P}_{3,12\_13}] \end{bmatrix} \begin{bmatrix} S_{1,13} \\ S_{2,13} \\ S_{3,13} \end{bmatrix} \quad (4.152)$$

Successively multiplying these thirteen 6×6 matrices, we can calculate the 6-port transfer matrix connecting the 0-section to the 13th section as follows:

$$\begin{bmatrix} p_{1,0} \\ v_{1,0} \\ p_{2,0} \\ v_{2,0} \\ p_{3,0} \\ -v_{3,0} \end{bmatrix} = \begin{bmatrix} T_{11} & T_{12} & T_{13} & T_{14} & T_{15} & T_{16} \\ T_{21} & T_{22} & T_{23} & T_{24} & T_{25} & T_{26} \\ T_{31} & T_{32} & T_{33} & T_{34} & T_{35} & T_{36} \\ T_{41} & T_{42} & T_{43} & T_{44} & T_{45} & T_{46} \\ T_{51} & T_{52} & T_{53} & T_{54} & T_{55} & T_{56} \\ T_{61} & T_{62} & T_{63} & T_{64} & T_{65} & T_{66} \end{bmatrix} \begin{bmatrix} p_{1,13} \\ v_{1,13} \\ p_{2,13} \\ v_{2,13} \\ p_{3,13} \\ v_{3,13} \end{bmatrix}. \quad (4.153)$$

The two-port transfer matrices of the remaining elements may be used to close the model. For the uniform pipe section between Chapters 13 and 16,

$$\begin{bmatrix} p_{1,13} \\ v_{1,13} \end{bmatrix} = [\mathbf{P}_{1,13\_16}] \begin{bmatrix} p_{1,16} \\ v_{1,16} \end{bmatrix}. \quad (4.154)$$

Considering transverse plane waves in the end chambers as modelled by Elnady et al. [9], we have

$$\begin{bmatrix} p_{1,16} \\ v_{1,16} \end{bmatrix} = \begin{bmatrix} 1 & 0 \\ \dfrac{1}{Z_6} & 1 \end{bmatrix} \begin{bmatrix} \text{TM} \\ \text{(Variable} \\ \text{area duct)} \end{bmatrix} [\mathbf{B}_{14\_15}] [\mathbf{P}_{2,13\_14}] \begin{bmatrix} p'_{3,13} \\ v'_{3,13} \end{bmatrix} \quad (4.155)$$

For transverse plane wave propagation in an elliptical (variable-area) cross-section duct, the transfer matrix is given by Equations 2.185–2.189 as per [7,8], and inserted in Equation 4.155 above. Successive multiplication of the five 2×2 transfer matrices of Equations 4.154 and 4.155 yields

$$\begin{bmatrix} p_{1,13} \\ v_{1,13} \end{bmatrix} = \begin{bmatrix} T'_{11} & T'_{12} \\ T'_{21} & T'_{22} \end{bmatrix} \begin{bmatrix} p'_{3,13} \\ v'_{3,13} \end{bmatrix} \quad (4.156)$$

At the junction/section 13, it may be noted that

$$p_{2,13} = p_{3,13} = p'_{3,13} \qquad (4.157, 4.158)$$

$$v_{2,13} - v_{3,13} = -v'_{3,13} \qquad (4.159)$$

Equations 4.156–4.159, along with $v_{2,0} = 0$, provide six equations or boundary equations. However, they have also introduced, incidentally, two additional variables, $p'_{3,13}$ and $v'_{3,13}$. These may be combined with Equation 4.153 to reduce the 6×6 matrix in Equation 4.153 to the desired 2×2 matrix of the entire configuration as follows:

$$\begin{bmatrix} 1 & 0 & 0 & 0 & -T_{11} & -T_{12} & -T_{13} & -T_{14} & -T_{15} & -T_{16} & 0 & 0 \\ 0 & 1 & 0 & 0 & -T_{21} & -T_{22} & -T_{23} & -T_{24} & -T_{25} & -T_{26} & 0 & 0 \\ 0 & 0 & 1 & 0 & -T_{31} & -T_{32} & -T_{33} & -T_{34} & -T_{35} & -T_{36} & 0 & 0 \\ 0 & 0 & 0 & 1 & -T_{41} & -T_{42} & -T_{43} & -T_{44} & -T_{45} & -T_{46} & 0 & 0 \\ 0 & 0 & 0 & 0 & -T_{51} & -T_{52} & -T_{53} & -T_{54} & -T_{55} & -T_{56} & 0 & 0 \\ 0 & 0 & 0 & 0 & -T_{61} & -T_{62} & -T_{63} & -T_{64} & -T_{65} & -T_{66} & 0 & 0 \\ 0 & 0 & 0 & 0 & 1 & 0 & 0 & 0 & 0 & 0 & -T'_{11} & -T'_{12} \\ 0 & 0 & 0 & 0 & 0 & 1 & 0 & 0 & 0 & 0 & -T'_{21} & -T'_{22} \\ 0 & 0 & 0 & 0 & 0 & 0 & 1 & 0 & -1 & 0 & 1 \\ 0 & 0 & 0 & 0 & 0 & 0 & 1 & 0 & -1 & 0 & 0 & 0 \\ 0 & 0 & 0 & 0 & 0 & 0 & 0 & 1 & 0 & -1 & 0 \\ 0 & 0 & 0 & 1 & 0 & 0 & 0 & 0 & 0 & 0 & 0 \end{bmatrix} \begin{bmatrix} p_{1,0} \\ v_{1,0} \\ p_{2,0} \\ v_{2,0} \\ p_{1,13} \\ v_{1,13} \\ p_{2,13} \\ v_{2,13} \\ p_{3,13} \\ v_{3,13} \\ p'_{3,13} \\ v'_{3,13} \end{bmatrix} = \begin{bmatrix} 0 \\ 0 \\ 0 \\ 0 \\ p_{3,0} \\ -v_{3,0} \\ 0 \\ 0 \\ 0 \\ 0 \\ 0 \\ 0 \end{bmatrix}$$

(4.160)

Inverting the coefficient matrix by means of Gaussian elimination with pivoting, using a standard subroutine/function, yields the overall transfer matrix of the entire muffler configuration. Finally, Transmission Loss, Insertion Loss and Noise Reduction (or Level Difference) can be evaluated as explained before in Chapters 2 and 3.

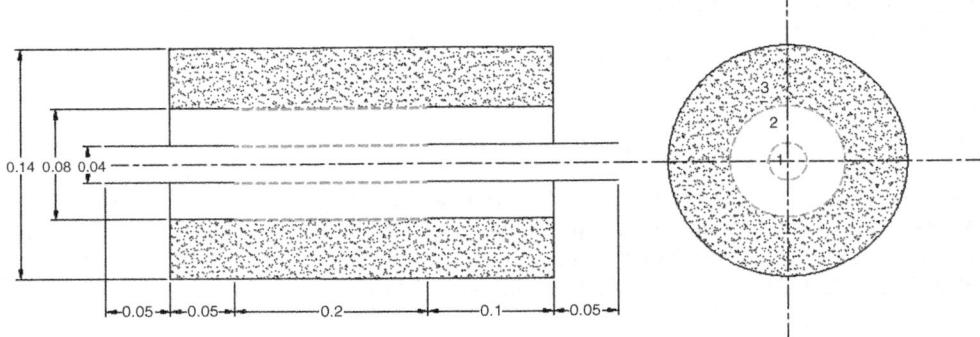

Figure 4.18 Schematic of lined duct with annular air-gap between the central airway and the absorptive material (all dimensions are in meters) (Adapted from [11])

4.7.3 A Combination Muffler

The muffler in Figure 4.18 is a combination muffler consisting of the reflective as well as absorptive elements. As can be observed from the figure, the sound waves get dispersed in the perforated tubes and then interact with the absorptive lining through the annular perforated tube.

Let us now illustrate the present ITM approach for this combination muffler. The muffler can be divided into three sections with a total of three interacting perforated elements. The elements 1 and 2 represent conventional unlined elements and hence the medium in these elements is considered as air or exhaust gas, as the case may be. The third element has absorptive lining, and therefore, complex characteristic impedance due to the lining is considered in the analysis.

Section 0_1 consists of three hard pipes wherein the outer duct is filled with absorptive material. Hence, the first and second elements are treated as simple hard walled pipes whereas the third element is considered as dissipative. Let $\mathbf{D}_{i,j\_k}$ be the 2×2 transfer matrix of uniform dissipative pipe of the ith acoustic element between the section (j_k), relating the state variables $\mathbf{S}_{i,j}$ and $\mathbf{S}_{i,k}$. Thus,

$$[\mathbf{D}_{i,j\_k}] = \begin{bmatrix} \cos(k_w L_{j\_k}) & jY_w \sin(k_w L_{j\_k}) \\ \dfrac{j}{Y_w} \sin(k_w L_{j\_k}) & \cos(k_w L_{j\_k}) \end{bmatrix} \quad (4.161)$$

where k_w and Y_w are the complex wave number and the characteristic impedance of the acoustic lining. Expressions for k_w and Y_w are given later in Chapter 6. Thus, the overall transfer matrix between sections 0_1 and 2_3 can be written as

$$\begin{bmatrix} \mathbf{S}_{1,0} \\ \mathbf{S}_{2,0} \\ \mathbf{S}_{3,0} \end{bmatrix} = \begin{bmatrix} p_{1,0} \\ v_{1,0} \\ p_{2,0} \\ v_{2,0}(=0) \\ p_{3,0} \\ v_{3,0}(=0) \end{bmatrix} = \begin{bmatrix} [\mathbf{P}_{1,0\_1}] & [\mathbf{O}]_{2\times2} & [\mathbf{O}]_{2\times2} \\ [\mathbf{O}]_{2\times2} & [\mathbf{P}_{2,0\_1}] & [\mathbf{O}]_{2\times2} \\ [\mathbf{O}]_{2\times2} & [\mathbf{O}]_{2\times2} & [\mathbf{D}_{3,0\_1}] \end{bmatrix} \begin{bmatrix} \mathbf{S}_{1,1} \\ \mathbf{S}_{2,1} \\ \mathbf{S}_{3,1} \end{bmatrix} \quad (4.162)$$

$$\begin{bmatrix} S_{1,2} \\ S_{2,2} \\ S_{3,2} \end{bmatrix} = \begin{bmatrix} [P_{1,2\_3}] & [O]_{2\times 2} & [O]_{2\times 2} \\ [O]_{2\times 2} & [P_{2,2\_3}] & [O]_{2\times 2} \\ [O]_{2\times 2} & [O]_{2\times 2} & [D_{3,2\_3}] \end{bmatrix} \begin{bmatrix} S_{1,3} \\ S_{2,3} \\ S_{3,3} \end{bmatrix}. \qquad (4.163)$$

The section 1_2 represents a three-duct interacting element with dissipative material in the third element. The transfer matrix is evaluated by means of the eigenanalysis of the coupled equations. Coefficients of the matrix may be evaluated from the coupled equations by means of Kar and Munjal's generalized algorithm, as done for normal interacting ducts [5], the only difference being the impedance across perforates. The modified impedance expressions for lined ducts is given by [12]

$$Z_p = \left[0.06 + jk_0\left\{t_w + 0.375d_h\left(1 + \frac{Y_w}{Y_0}\frac{k_0}{k_w}\right)\right\}\right]\bigg/\sigma, \qquad (4.164)$$

where t_w and d_h are the thickness of the perforated wall and the diameter of the holes or perforations, respectively. Using these impedance expressions appropriately in the coupled equations and making use of eigen analysis, the 6×6 transfer matrix across the section 1_2 can be obtained.

Once the transfer matrices are obtained, the ordinary procedure of the ITM method as described in the earlier sections (4.7.1 and 4.7.2) is followed in order to obtain the four-pole parameters relating the upstream and the downstream variables.

Results from the present approach and 3D FEM analysis have been shown in Figure 4.19. The absorptive lining has been assigned complex sound speed and density, calculated from

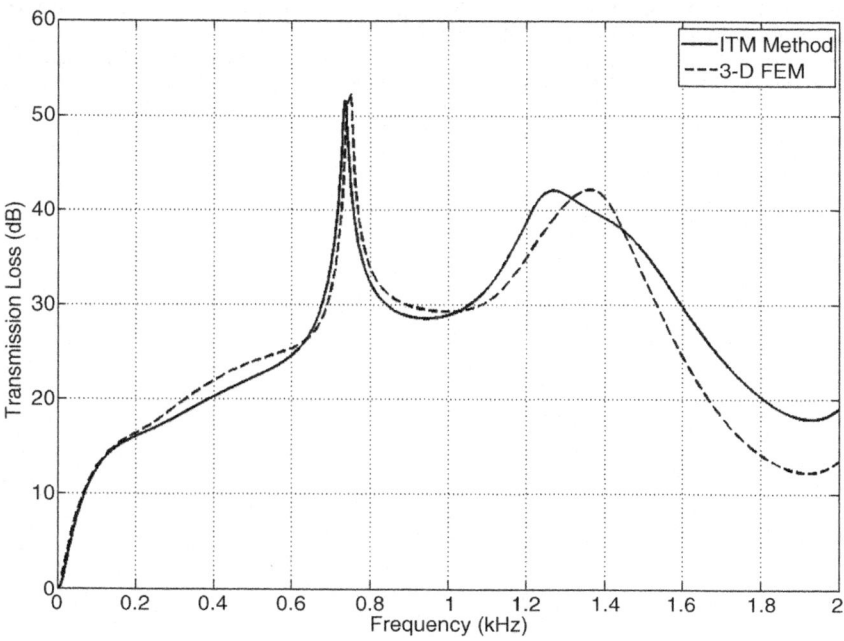

Figure 4.19 Comparison of TL spectrum from the 3D FEM analysis with that of the ITM method for the configuration of Figure 4.18 (Reproduced with permission from [11])

the material's complex characteristic impedance and wave number in 3D FEM analysis. The flow resistivity of the lining has been taken as 16 000 Pa.s/m^2. It is observed that the results agree reasonably well in the low frequency range, but deviate at higher frequencies owing to the cut on of three-dimensional modes.

References

1. Selamet, A., Dickey, N.S. and Novak, J.M. (1994) The Herschel-Quincke tube: A theoretical, computational and experimental investigation. *Journal of the Acoustical Society of America*, **96**(5), 3177–3185.
2. Selamet, A. and Easwaran, V. (1997) The Modified Herschel-Quincke tube: Attenuation and resonance for n-duct configuration. *Journal of the Acoustical Society of America*, **102**(1), 164–169.
3. Selamet, A., Easwaran, V. and Falkowski, A.G. (1995) Three-pass mufflers with uniform perforations. *Journal of the Acoustical Society of America*, **105**(3), 1548–1562.
4. Munjal, M.L. (1997) Analysis of a flush-tube three-pass perforated element muffler by means of transfer matrices. *International Journal of Acoustics and Vibration*, **2**(2), 63–68.
5. Kar, T. and Munjal, M.L. (2005) Generalized analysis of a muffler with any number of interacting ducts. *Journal of Sound and Vibration*, **285**(3), 585–596.
6. Singh, B. and Munjal, M.L. (2012) Flow-acoustic analysis of commercial automotive mufflers: matrizant approach. *Journal of the Acoustical Society of India*, **39**(3), 142–151.
7. Mimani, A. and Munjal, M.L. (2011) Transverse plane wave analysis of short elliptical chamber mufflers: An analytical approach. *Journal of Sound and Vibration*, **330**(7), 1472–1489. doi: 10.1016/j.jsv.2010.09.035
8. Mimani, A. and Munjal, M.L. (2010) Transverse plane-wave analysis of short elliptical end-chamber and expansion-chamber mufflers. *International Journal of Acoustics and Vibration*, **15**(1), 24–38.
9. Elnady, T., Abom, M. and Allam, S. (2010) Modeling perforates in mufflers using two-ports. *Journal of Vibration and Acoustics*, **132**, 061010-1-11.
10. Vijayasree, N.K. (2011) Flow acoustic analysis of complex muffler configurations M.Sc., (Engg.) thesis, *Indian Institute of Science*, Bangalore.
11. Vijayasree, N.K. and Munjal, M.L. (2012) On an integrated transfer matrix method for multiply connected mufflers. *Journal of Sound and Vibration*, **331**, 1926–1938.
12. Selamet, A., Lee, I., Ji, Z. and Huff, N. (2001) Acoustic attenuation performance of perforated absorbing silencers, *SAE Technical Paper 2001-01-1435*. doi: 10.4271/2001-01-1435

5

Flow-Acoustic Measurements

Measurements are required for supplementing the analysis by providing certain basic data or parameters that cannot be predicted precisely, for verifying the analytical/numerical predictions, and also for evaluating the overall performance of a system configuration so as to check if it satisfied the design requirements. In particular, in the field of exhaust systems, where mean flow introduces quite a few complications, measurements are required for evaluation of radiation impedance or reflection coefficient at the radiation end (tail pipe end), flow-acoustic attenuation constant, characteristics of the engine exhaust source, level difference across, or transmission loss of, one or more acoustic elements in order to verify the transfer matrices thereof, dissipation of acoustic energy emerging from the tail pipe end in the shear layer of the mean-flow jet, and, finally, the insertion loss of the exhaust muffler as required by the designer and user.

Measurement of insertion loss of a muffler is the easiest thing to do inasmuch as it requires a measurement of sound pressure in the far field without and with the muffler. The output of the microphone is fed through a preamplifier, to a spectrum analyzer, and from there to a measuring amplifier and/or level recorder. Of course, one requires a sufficiently anechoic environment so as to ensure that the two positions of the microphone (without and with the muffler) are subjected to almost the same reverberation sound, and this should be at least 10 dB less than the direct sound from the radiation end of the exhaust pipe or tail pipe over the entire frequency range of interest.

Evaluation of other parameters, however, requires picking up of the sound within the pipe, and this is where one comes across major difficulties because of hot and moving medium. Measurement of level difference or noise reduction is not so difficult in that one has to pick up sound from two discrete points across the muffler elements under consideration. The real problem is encountered in evaluation of impedance or reflection coefficient of a termination.

5.1 Impedance of a Passive Subsystem or Termination

The probe-tube method used for this purpose generally involves continuous traverse of the microphone in an impedance tube or probe tube to get a continuous trace of SPL variation and, in particular, the exact locations and amplitudes of SPL maxima and minima. This is very tricky because of certain unsteadiness resulting from the flow-probe interaction.

There have been a number of developments in the field of the probe-tube method for stationary medium [1–10] and moving medium [11–20]. In what follows, the most accurate and the most convenient of the probe-tube methods that are specially suited for moving media [18–20], are described. The subsequent sections describe some alternatives to this method, namely, the two-microphone method and certain variants thereof.

5.1.1 The Probe-Tube Method

Figure 5.1 shows an impedance tube with a (black box) passive termination at one end and an acoustic source at the other. The tube is filled with the required medium in the presence of which the impedance of the termination is to be determined. The tube is excited at the desired frequency and the sound pressures $p^{(r)}$ are measured at fixed positions $z^{(r)}$, $r = 1, 2, 3, \ldots$. From the plane wave theory discussed in Chapter 3, one gets for incompressible moving medium (Mach number $M < 0.25$),

$$p^{(r)} = p^+ e^{j\omega t} \left\{ e^{jk^+ z^{(r)}} e^{\alpha^+ z^{(r)}} + |R| e^{j\theta} e^{-jk^- z^{(r)}} e^{-\alpha^- z^{(r)}} \right\} \tag{5.1}$$

and

$$\left| p^{(r)} \right| = |p^+| \left[e^{2\alpha^+ z^{(r)}} + |R|^2 e^{-2\alpha^- z^{(r)}} + 2|R| e^{(\alpha^+ - \alpha^-) z^{(r)}} \times \cos\{\theta - (k^+ + k^-) z^{(r)}\} \right]^{1/2}, \tag{5.2}$$

where the distance $z^{(r)}$ is measured from the reflective surface ($z = 0$) in a direction opposite to that of the incident wave, Figure 5.1.

$|R|$ and θ, the amplitude and phase angle of the reflection coefficient, are to be determined. But, then, α is also generally unknown. This can be calculated by means of a rigid termination.

5.1.1.1 Evaluation of the Attenuation Constant and Wave Number

For the case of a rigid termination, Figure 5.2,

$$M = 0, \quad |R| = 1, \quad \theta = 0, \quad \alpha^+ = \alpha^- = \alpha, \quad k^+ = k^- = k. \tag{5.3}$$

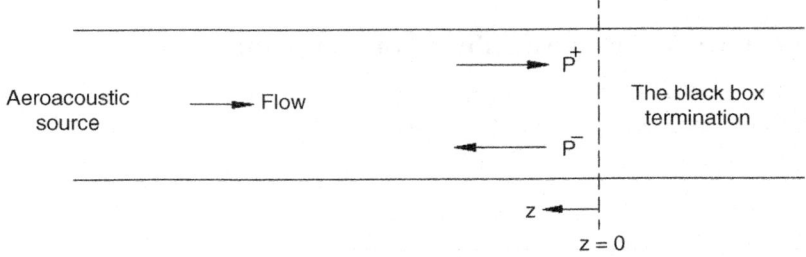

Figure 5.1 Schematic of an impedance tube

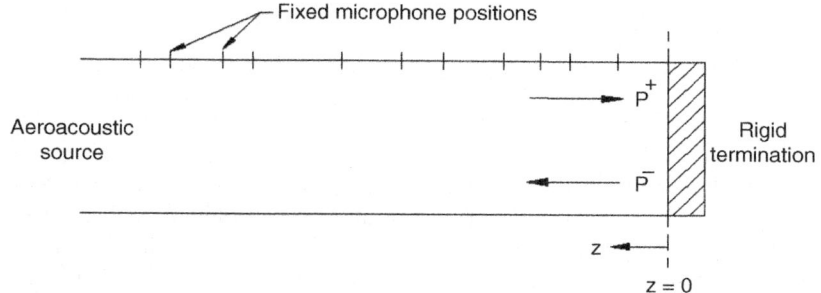

Figure 5.2 Schematic of an impedance tube with rigid termination for evaluation of α

Substituting the value of k from Equation 1.69, that is,

$$k = k_0 + \alpha = \omega/c_0 + \alpha \tag{5.4}$$

into Equation 5.2 and making use of Equation 5.3, Equation 5.2 reduces to

$$|p^{(r)}|^2 = |p^+|^2 \left[e^{2\alpha z^{(r)}} + e^{-2\alpha z^{(r)}} + 2\cos(2k_0 z^{(r)})\cos(2\alpha z^{(r)}) - 2\sin(2k_0 z^{(r)})\sin(2\alpha z^{(r)}) \right],$$
$$r = 1, 2, \ldots, \tag{5.5}$$

where k_0 is the free-medium wave number. As α is of the order of $0.05/m$ (or less), for a tube of 1 m or less, one may justly retain terms of degree two or less, and neglect all the higher degree terms of $2\alpha z^{(r)}$ in the series expansion of Equation 5.5.

Thus,

$$|p^{(r)}|^2 = 2|p^+|^2 \left[1 + 2\alpha^2 (z^{(r)})^2 \{1 - \cos(2k_0 z^{(r)})\} + \cos(2k_0 z^{(r)}) - 2\alpha z^{(r)} \sin(2k_0 z^{(r)}) \right]. \tag{5.6}$$

Defining $\delta_{i,j} = SPL_i - SPL_j = 10\log|p^{(i)}/p^{(j)}|^2 = 10\log \beta_{i,j}, \quad i,j = 1,2,3,\ldots,$ (5.7)

$\beta_{i,j}$ can be evaluated from the actual experimental values of level difference $\delta_{i,j}$.
From Equations 5.6 and 5.7 one gets

$$\beta_{i,j} = \frac{2\alpha^2 (z^{(i)})^2 [1 - \cos(2k_0 z^{(i)})] - 2\alpha z^{(i)} \sin(2k_0 z^{(i)}) + [1 + \cos(2k_0 z^{(i)})]}{2\alpha^2 (z^{(j)})^2 [1 - \cos(2k_0 z^{(j)})] - 2\alpha z^{(j)} \sin(2k_0 z^{(j)}) + [1 + \cos(2k_0 z^{(j)})]}, \tag{5.8}$$

which, on rearrangement, reduces to a quadratic in α:

$$\alpha^2 - b_{i,j}\alpha + \frac{1}{2}c_{i,j} = 0, \quad i,j = 1,2,3\ldots, \tag{5.9}$$

where

$$b_{i,j} = \frac{\beta_{i,j} z^{(j)} \sin(2k_0 z^{(j)}) - z^{(i)} \sin(2k_0 z^{(j)})}{\beta_{i,j}(z^{(j)})^2 [1 - \cos(2k_0 z^{(j)})] - (z^{(i)})^2 [1 - \cos(2k_0 z^{(i)})]}, \tag{5.10}$$

and

$$c_{i,j} = \frac{\beta_{i,j}\left[1 + \cos\left(2k_0 z^{(j)}\right)\right] - \left[1 + \cos\left(2k_0 z^{(j)}\right)\right]}{\beta_{i,j}(z^{(j)})^2[1 - \cos(2k_0 z^{(j)})] - (z^{(i)})^2[1 - \cos(2k_0 z^{(i)})]}. \quad (5.11)$$

For every pair of positions (i, j), this quadratic yields two possible roots for α. The positive root is the required one. If both the roots are positive, then solution of a number of such quadratic equations formed from SPL observations at a number of points arranged in pairs (i, j) would be necessary to pick out the correct value of α as the one that repeats itself consistently.

A number of observations from actual experiments [20] have revealed that the location of the measurement points with respect to the standing wave pattern influences the accuracy of observation very substantially. An analytical study of the behavior of α with $p(z)$ indicates that observations on the rising flank of the standing wave just after a pressure minimum would yield a relatively more accurate value of α. In fact, actual calculations have shown that greater accuracy can be obtained if the positions of observations are closer to the pressure minimum.

Tube attenuation α having been so determined, k, the wave number for stationary medium in a tube, can then be calculated from Equation 5.4, where ω is the source frequency in rad/s and

$$c_0 = (\gamma RT)^{1/2}, \quad (5.12)$$

where the ratio of specific heats γ, and R the gas constant, are calculated from the chemical composition of the medium, and T is the temperature in degrees Kelvin.

5.1.1.2 Evaluation of $\alpha^\pm$ and $k^\pm$

Standing wave measurements in a moving medium are extremely difficult, especially when the mean flow is turbulent. The disturbances in the flow interfere with the measurements, rendering them unsteady at high-flow Mach numbers. However, at the moderate flow values of M usually found in engine exhaust systems ($M < 0.25$), the standing wave pattern can be obtained from SPL measurements at discrete positions. $\alpha^\pm$ and $k^\pm$ can then be calculated from Equations 3.63–3.65), namely [20].

$$k^\pm = k/(1 \pm M), \quad (5.13)$$

$$\alpha^\pm = \alpha(M)/(1 \pm M), \quad (5.14)$$

where

$$\alpha(M) = \alpha + \xi M, \quad \xi = F/2D. \quad (5.15)$$

It is worth noting here that while in Equations 3.63 and 3.64, k could be approximated as k_0, it cannot be done in the impedance-tube calculations, where small errors in $k^\pm$ and $\alpha^\pm$ can cause large errors in the standing wave parameters. Unlike in the case of stationary medium, where α could be determined from closed-end experiments in the tube, one has to keep the tube end open to allow the mean flow, and this brings into play a coupling between $\alpha^{\pm}$, $|R|$ and θ. Simultaneous solution of a set of equations formulated from the observations at a number of points is rather involved and the results are not sufficiently consistent. Hence, it appears preferable to estimate the wave numbers $k^{\pm}$ and attenuation constants $\alpha^{\pm}$ from Equations 5.13 and 5.14, with α, the attenuation constant for a stationary medium, measured as in the preceding subsection, and F, the Froude's friction factor, evaluated from the steady flow pressure drop.

5.1.1.3 Evaluation of the Reflection Coefficient $|R| \exp(j\theta)$

From Equation 5.2 it can be shown [18] that for any termination, the positions of the rth and $(r+1)$th pressure minima are related to the phase angle θ by

$$\theta = \left[\frac{2z_{min}^{(r)}}{z_{min}^{(r+1)} - z_{min}^{(r)}} - (2r-1)\right]\pi. \tag{5.16}$$

Hence the phase angle can be evaluated if any two consecutive pressure nodes are located. As any two consecutive minima would occur at a distance of approximately $\lambda/2 (\lambda =$ wave length), the minimum length of an impedance tube has to be equal to λ, that is,

$$l_{min} = \lambda_{max} = \frac{c_0}{f_{min}}. \tag{5.17}$$

Thus if the minimum frequency of interest is 100 Hz, then for experiments at normal room temperature, the minimum length of the impedance tube has to be about 3.4 m.

With the values of $k^{\pm}$, $\alpha^{\pm}$, and θ thus calculated, it is possible to determine the amplitude of the reflection coefficient $|R|$ from SPL measurements at a number of discrete locations. With Equation 5.7, Equation 5.2 yields an equation that on readjustment becomes [18]

$$|R|^2 + 2(A_{i,j}\cos\theta + B_{i,j}\sin\theta)|R| + C_{i,j} = 0, \quad i,j = 1,2,3,\ldots \tag{5.18}$$

where

$$A_{i,j} = \left[\beta_{i,j}e^{(\alpha^+ - \alpha^-)z^{(j)}}\cos\{(k^+ + k^-)z^{(j)}\} - e^{(\alpha^+ - \alpha^-)z^{(j)}}\cos\{(k^+ + k^-)z^{(j)}\}\right]/D_{i,j}, \tag{5.19}$$

$$B_{i,j} = \left[\beta_{i,j}e^{(\alpha^+ - \alpha^-)z^{(j)}}\sin\{(k^+ + k^-)z^{(j)}\} - e^{(\alpha^+ - \alpha^-)z^{(j)}}\sin\{(k^+ + k^-)z^{(j)}\}\right]/D_{i,j}, \tag{5.20}$$

$$C_{i,j} = \left[\beta_{i,j}e^{2\alpha^+ z^{(j)}} - e^{2\alpha^+ z^{(i)}}\right]/D_{i,j}, \tag{5.21}$$

and

$$D_{i,j} = \left[\beta_{i,j}e^{-2\alpha^- z(j)} - e^{-2\alpha^- z(i)}\right]. \tag{5.22}$$

Quadratics in $|R|$ obtained from SPL measurements at a number of pairs of positions (i,j) can be solved and the approximate value of $|R|$ can be evaluated, recognizing the correct root in each case by keeping in mind the fact that the maximum value of $|R|$ is $(1+M)/(1-M)$.

In conclusion, evaluation of α (and hence $k, k^{\pm}$, and $\alpha^{\pm}$) involves SPL values at a few points on the rising flank of one or more pressure minima, θ requires exact locations of two consecutive minima, and $|R|$ requires SPLs at a few discrete points (located anywhere, including of course near the pressure minima, as required by α). So, one requires a pressure probe within the tube to locate pressure minima (not to measure the values of SPL at these minima, which are never sufficiently reliable). The same probe can be used to pick up SPLs at a few (say, four to six) discrete points.

5.1.1.4 Experimental Setup

Figure 5.3 shows a typical laboratory layout for the experimental setup for SPL measurements at discrete positions by a probe tube connected to a 6-mm Bruel and Kjaer (B & K) condenser microphone [20]. The equipment consists of a header of 129.4-mm inner diameter to which various sizes of impedance tubes can be connected. A probe traverse is mounted on a tube of 30.4-mm inner diameter connected the rear side of the header. Two types of probe tubes with different termination have been used. An open-ended tube was used for measurements without flow. For measurements of SPL in a moving medium, a closed-ended tube having four 0.8-mm-diameter communication holes at 50 mm from the closed tip was used. Each probe tube has a 3.5-mm outer diameter and a 2-mm inside diameter and is 1.8 m long. It is supported at regular intervals of 300 mm by means of 1.5-mm diameter steel wires introduced laterally to keep it from sagging. The probe tube does not extend throughout the test section; however, the duct area change, being about 1% or less, is not expected to lead to any significant errors. The distance of the end section of the probe tube from the plane of reflection is measured by a scale and vernier system to an accuracy of 0.1 mm. Of course, this

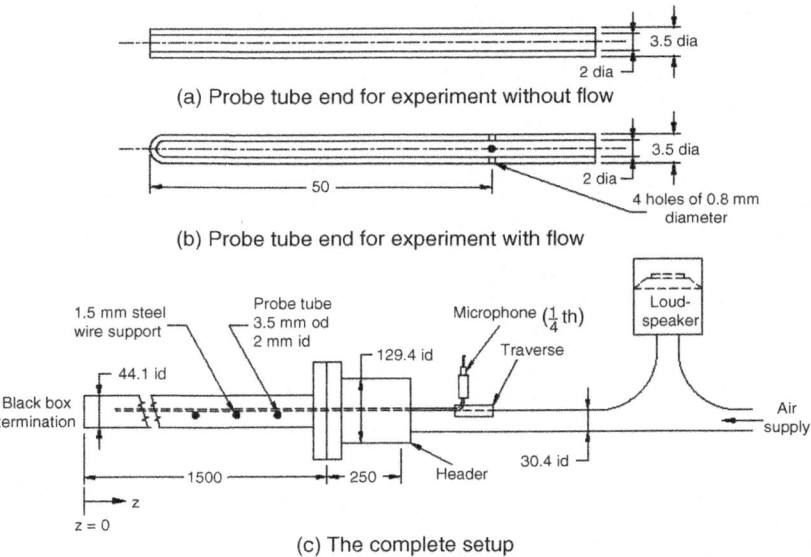

Figure 5.3 Schematic layout of the impedance-tube experimental setup (Reproduced with permission from [20])

accuracy is partially nullified by the fact that the pressure-sensing holes in the probe tube have a diameter of 0.8 mm. A loudspeaker is housed in a chamber that is connected to the inlet tube of the header and is excited by a B & K type-1023 beat frequency oscillator. The sound signal picked up by the microphone through the probe tube is filtered through a B & K type-2020 heterodyne slave filter at a bandwidth of 3.16 Hz and measured by a B & K measuring amplifier type 2606, which gives an accuracy of 0.2 dB. For experiments with a moving medium, compressed air from a compressor is throttled by a pressure control valve and is passed through the system. The flow rate is measured by an orifice plate meter incorporated in the air supply facility. The static pressure and temperature at the test section are measured. The Froude's friction factor is calculated from the rate of pressure drop in the impedance tube by a precision micromanometer. The Mach number of flow is estimated on the basis of the mean velocity averaged at the test section.

This layout has been used successfully for evaluation of α for tubes of various diameters and also for experimental evaluation of the radiation impedance for an unflanged tail pipe end with mean flow [22].

5.1.2 The Two-Microphone Method

This method, as its name indicates makes use of two microphones located at fixed positions. The excitation may be random signal (containing all frequencies of interest) or a discrete frequency signal, as used in the probe-tube method of the preceding section. A schematic diagram of the method originally proposed by Seybert and Ross [23] is shown in Figure 5.4, wherein two 6-mm microphones are located at fixed positions. A random-noise generator gives the required random-noise signal, which is passed through a filter so as to retain only the desired frequency range, and then power-amplified before it is fed to an electropneumatic exciter, which creates an acoustic pressure field on the moving medium in the impedance tube (also called the transmission tube). The signal picked up by each microphone is amplified by a preamplifier before it is fed to a two-channel Fourier analyzer cum correlator, which may be a digital computer system preceded by an analog-to-digital converter. The measured data are

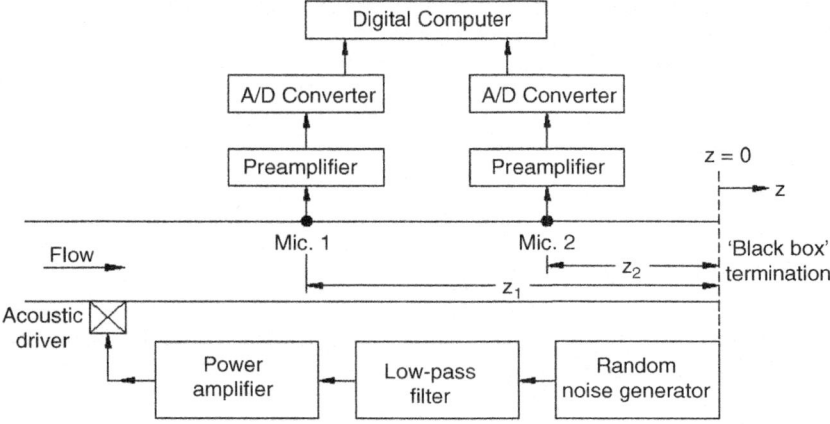

Figure 5.4 Experimental setup for evaluation of the reflection characteristics of a termination by means of the two-microphone random-excitation method (Reproduced with permission from [23])

autospectral densities of the signals at two microphone locations and the cross-spectral density between them. Making use of these measured data, the reflection coefficient of the termination is calculated as follows [23].

As shown in Chapter 3,

$$p(z,t) = A(f)e^{j(\omega t - k^+ z)} + B(f)e^{j(\omega t + k^- z)}, \tag{5.23}$$

$$v(z,t) = \frac{1}{Y}\left\{A(f)e^{j(\omega t - k^+ z)} - B(f)e^{j(\omega t + k^- z)}\right\}, \tag{5.24}$$

and therefore

$$p(0,t) = (A(f) + B(f))e^{j\omega t} \tag{5.25}$$

and

$$v(0,t) = \frac{1}{Y}(A(f) - B(f))e^{j\omega t}, \tag{5.26}$$

where A and B are functions of f and $k^{\pm} = k/(1 \pm M)$. Attenuation is obviously ignored:

If p and v are random functions of time, then the required impedance $Z(\omega)$ or $Z(f)$ at $z = 0$, normalized with respect to the characteristic impedance of the tube Y, is given by [25],

$$\frac{Z(f)}{Y} = Z_n(f) = \frac{S_{pv}(f)}{YS_{vv}(f)}, \tag{5.27}$$

where

$S_{pv}(f)$ = cross-spectral density between p and v at $z = 0$.
$S_{vv}(f)$ = autospectral density of the mass velocity v at $z = 0$.

and these, in turn, are related to the finite Fourier transforms of p and v at $z = 0$ by

$$S_{pv}(f) = \frac{1}{T}P_0(f,T)V_0^*(f,T), \tag{5.28}$$

$$S_{vv}(f) = \frac{1}{T}V_0(f,T)V_0^*(f,T). \tag{5.29}$$

Similarly,

$$S_{pp}(f) = \frac{1}{T}P_0(f,T)P_0^*(f,T), \tag{5.30}$$

where the superscript * as usual denotes complex conjugate and T is the finite time of the record used in the Fourier transforms as

$$P_0(f,T) = \frac{1}{T}\int_0^T P(0,t)e^{-j\omega t}dt \tag{5.31}$$

Equation 5.25 can be substituted in Equation 5.31 to obtain

$$P_0(f, T) = A(f, T) + B(f, T), \qquad (5.32)$$

where $A(f)$ and $B(f)$ of Equation 5.25 have been rewritten as $A(f, T)$ and $B(f, T)$ to indicate their being finite Fourier transforms.

Similarly, Equations 5.26 and 5.31 yield

$$V_0(f, T) = \frac{1}{Y}\{A(f, T) - B(f, T)\}. \qquad (5.33)$$

Substituting Equations 5.32 and 5.33 in Equations 5.28–5.30, and noting that

$$\frac{1}{T}(BA^* - AB^*) = -2jQ_{AB} \quad \text{and} \quad \frac{1}{T}(BA^* + AB^*) = 2C_{AB}, \qquad (5.34)$$

yields

$$S_{pv}(f) = \frac{1}{Y}\{S_{AA}(f) - S_{BB}(f) - j2Q_{AB}(f)\}, \qquad (5.35)$$

$$S_{vv}(f) = \frac{1}{Y^2}\{S_{AA}(f) - S_{BB}(f) - 2C_{AB}(f)\}, \qquad (5.36)$$

and

$$S_{pp}(f) = S_{AA}(f) + S_{BB}(f) + 2C_{AB}(f), \qquad (5.37)$$

where $S_{AA}(f)$ and $S_{BB}(f)$ are the autospectral densities of $A(f)$ and $B(f)$, respectively, and C_{AB} and Q_{AB} are the real and imaginary parts of S_{AB}, the cross-spectral density between A and B. Thus,

$$S_{AB}(f) = C_{AB}(f) + jQ_{AB}(f). \qquad (5.38)$$

Again, finite Fourier transforms of Equations 5.23 and 5.24 yield

$$P_1(f, T) = A(f, T)e^{-jk^+ z_1} + B(f, T)e^{+jk^- z_1}, \qquad (5.39)$$

$$P_2(f, T) = A(f, T)e^{-jk^+ z_2} + B(f, T)e^{+jk^- z_2}, \qquad (5.40)$$

$$V_1(f, T) = \frac{1}{Y}\left\{A(f, T)e^{-jk^+ z_1} - B(f, T)e^{+jk^- z_1}\right\}, \qquad (5.41)$$

and

$$V_2(f, T) = \frac{1}{Y}\left\{A(f, T)e^{-jk^+ z_2} - B(f, T)e^{+jk^- z_2}\right\}. \qquad (5.42)$$

Now,

$$S_{11}(f) = \frac{1}{T}\{P_1(f,T)P_1^*(f,T)\} = S_{AA}(f) + S_{BB}(f) \qquad (5.43)$$
$$+ 2\{C_{AB}(f)\cos(k^+ + k^-)z_1 + Q_{AB}(f)\sin(k^+ + k^-)z_1\},$$

$$S_{22}(f) = \frac{1}{T}\{P_2(f,T)P_2^*(f,T)\} = S_{AA}(f) + S_{BB}(f) \qquad (5.44)$$
$$+ 2\{C_{AB}(f)\cos(k^+ + k^-)z_2 + Q_{AB}(f)\sin(k^+ + k^-)z_z\},$$

$$C_{12}(f) = \text{Re}[S_{12}(f)] = \text{Re}\left[\frac{1}{T}P_1(f,T)P_2^*(f,T)\right]$$
$$= S_{AA}(f)\cos k^+(z_1 - z_2) + S_{BB}(f)\cos k^-(z_1 - z_2) \qquad (5.45)$$
$$+ C_{AB}(f)[\cos(k^-z_1 + k^+z_2) + \cos(k^+z_1 + k^-z_2)]$$
$$+ Q_{AB}(f)[\sin(k^-z_1 + k^+z_2) + \sin(k^+z_1 + k^-z_2)],$$

and

$$Q_{12}(f) = \text{Im}[S_{12}(f)] = \text{Im}\left[\frac{1}{T}P_1(f,T)P_2^*(f,T)\right]$$
$$= -S_{AA}(f)\sin k^+(z_1 - z_2) + S_{BB}(f)\sin k^-(z_1 - z_2) \qquad (5.46)$$
$$+ C_{AB}(f)[-\sin(k^+z_1 + k^-z_2) + \sin(k^-z_1 + k^+z_2)]$$
$$+ Q_{AB}(f)[\cos(k^+z_1 + k^-z_2) - \cos(k^-z_1 + k^+z_2)].$$

Equations 5.43–5.46 are four inhomogeneous algebraic linear equations that can be solved to obtain values of the four unknowns, $S_{AA}, S_{BB}, C_{AB},$ and Q_{AB}. Finally, these can be substituted in Equations 5.35 and 5.36, which can in turn be substituted in Equation 5.27 to obtain the normalized impedance

$$\frac{Z(f)}{Y} = Z_n(f) = \frac{S_{AA}(f) - S_{BB}(f) - j2Q_{AB}(f)}{S_{AA}(f) + S_{BB}(f) - 2C_{AB}(f)}. \qquad (5.47)$$

The reflection coefficient of the termination is now given by Equation 2.30, that is,

$$R(f) \equiv |R|e^{j\theta} = \frac{Z_n(f) - 1}{Z_n(f) + 1}$$
$$= \frac{-S_{BB} + C_{AB}(f) - jQ_{AB}(f)}{S_{AA} - C_{AB}(f) - jQ_{AB}(f)}. \qquad (5.48)$$

Thus, one needs to measure spectral densities $S_{11}(f), S_{22}(f)$, and the real and imaginary parts of $S_{12}(f)$, then solve Equations 5.43–5.46 simultaneously for $S_{AA}, S_{BB}, C_{AB},$ and Q_{AB}, and finally evaluate the reflection coefficient from Equation 5.48. Obviously, therefore, one needs a digital computer and also the intermediaries, like an FM tape recorder and A/D convertors.

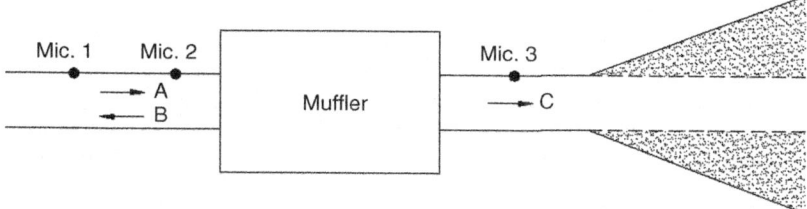

Figure 5.5 A setup for evaluation of the transmission loss of a muffler by means of the three-microphone random-excitation method (Reproduced with permission from [25])

Incidentally, this method can also be used for the evaluation of transmission loss of a subsystem (like a muffler) by using another microphone on the downstream side that ends in an anechoic termination, as shown in Figure 5.5.

Then,

$$TL = 10 \log \frac{S_{AA}(f)}{S_{CC}(f)}, \qquad (5.49)$$

where $S_{cc}(f)$ is the autospectral density of the acoustic pressure signal picked up from point 3, which is the pressure of the wave transmitted into the anechoic termination.

This method was used by Seybert and Ross for the stationary-medium case. Prasad [24] has used it successfully for the exhaust system of an internal combustion engine.

5.1.3 Transfer Function Method

Spectral density being energy in a very small frequency band $\Delta f (\Delta f = 1/T)$ around frequency f, the preceding analysis is indeed an analysis of energies associated with the incident wave and the reflected wave. The real thing that makes measurement of pressures at two discrete points sufficient is that the real part C_{12} as well as the imaginary part Q_{12} of the cross-spectral density S_{12} is also calculated, which accounts for the phase difference. This basic fact is made use of by Schmidt and Johnson, who used a discrete frequency technique to evaluate orifices, employing two wall-mounted microphones at different upstream positions along a tube [26]. By measuring the pressure amplitudes at the two points in the tube, as well as the phase shift between the points, they deduced the reflection coefficient of the sample. Thus, the real difference between the discrete frequency method of Section 5.1 and these methods is the fact that while the former consists in measuring amplitudes only, the latter measure amplitude as well as the phase difference.

A very useful variant of the two-microphone random-excitation method is Chung and Blaser's transfer function method [27]. The experimental setup is about the same as in Figure 5.4. However, the reflection coefficient of the termination is calculated from the acoustic transfer function H_{12} between the two signals rather than the spectral densities instead of working with the convolution integrals and their Fourier transforms (autocorrelation and cross-correlation).

Effect of acoustic absorption, which is neglected in references [23,27,28], has been incorporated in [29,30]. Chu [30] has, in fact, shown that acoustic absorption may play a

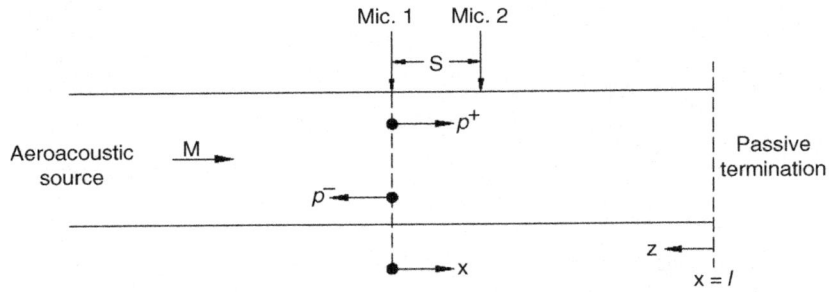

Figure 5.6 Schematic of the impedance tube and the notation adopted

significant role in some cases. However, he has neglected the mean flow effect which has been considered by others [23,27,28]. Both these effects are incorporated simultaneously in what follows.

Figure 5.6 shows a configuration adopted from Boden and Abom [31]. Comparison of the notations in Figures 5.4 and 5.6 indicates that $x = z + l, l = z_1$ and $s = z_1 - z_2$. For sinusoidal time dependence, the standing wave pressure at a distance x downstream of microphone 1 would be given by [21]

$$p(x) = p^+ e^{-jk^+ x} e^{-\alpha^+ x} + p^- e^{jk^- x} e^{\alpha^- x} \\ = e^{(jk_c + \alpha_c)Mx} \{ p^+ e^{-(jk_c + \alpha_c)x} + p^- e^{(jk_c + \alpha_c)x} \}, \quad (5.50)$$

where

$$k^\pm = k/(1 \pm M) = k_c(1 \mp M), \qquad \alpha^\pm = \alpha(M)/(1 \pm M) = \alpha_c(1 \mp M), \quad (5.51, 5.52)$$

$$k = k_0 + \alpha, \quad k_0 = \omega/c_0, \quad \alpha(M) = \alpha + MF/2D, \quad (5.53\text{--}5.55)$$

$$k_c = (k^+ + k^-)/2 = k(1 - M^2), \quad \alpha_c = (\alpha^+ + \alpha^-)/2 = \alpha(M)(1 - M^2), \quad (5.56, 5.57)$$

In Equations 5.50–5.57, M is the mean flow Mach number, c_0 is the sound speed in the free medium, F is Froude's friction factor, D is the diameter (or hydraulic diameter) of the duct, and α_0 is the stationary medium viscothermal attenuation constant of the duct.

Let us define

$$\beta \equiv jk_c + \alpha_c. \quad (5.58)$$

Then, Equation 5.50 can be written in the concise form

$$p(x) = e^{\beta M x} \{ p^+ e^{-\beta x} + p^- e^{\beta x} \} \quad (5.59)$$

Therefore, the standing wave pressures at the two microphones in Figure 5.6 would be given by

$$p_1 = p(0) = p^+ + p^-, \quad p_2 = p(s) = e^{\beta M s} \{ p^+ e^{-\beta s} + p^- e^{\beta s} \}. \quad (5.60, 5.61)$$

If we define the reflection coefficient at $z=0$ (i.e. the microphone 1 location) and the transfer function as

$$R(0) = p^-/p^+, \quad H_{21} = p_2/p_1, \quad (5.62, 5.63)$$

then, use of Equations 5.60–5.63 readily yields the following expression for reflection coefficient at $x=0$, the microphone 1 location:

$$R(0) = \left(H_{21}e^{-\beta Ms} - e^{-\beta s}\right) / \left(e^{\beta s} - H_{21}e^{-\beta Ms}\right). \quad (5.64)$$

Incidentally, dividing the numerator as well as the denominator of Equation 5.64 by $e^{-\beta Ms}$ and making use of Equations 5.51, 5.52 and 5.58 gives

$$R(0) = \left(H_{21} - e^{-[(jk+\alpha)/(1+M)]s}\right) / \left(e^{[(jk+\alpha)/(1-M)]s} - H_{21}\right), \quad (5.65)$$

Incidentally, for $\alpha = 0$, Equation 5.65 reduces to Equation 5.19 (read with Equations 5.33 and 5.34) by Chung and Blaser [33].

The reflection coefficient at the passive termination $(x = l)$ can now be calculated by means of Equations 5.59, 5.62 and 5.64 as follows:

$$R(l) = \frac{p^- e^{\beta l}}{p^+ e^{-\beta l}} = \left\{\frac{H_{21}e^{-\beta Ms} - e^{-\beta s}}{e^{\beta s} - H_{21}e^{-\beta Ms}}\right\} e^{2\beta l}. \quad (5.66)$$

For a consistency check, Equation 5.66 may be seen to reduce to Equation 5.9 of Chu [30] for $M = 0$, and to Equation 5.7 of Boden and Abom [31] for $M = 0$ and $\alpha = 0$.

The acoustic impedance of the passive termination may now be evaluated by means of Equation 5.66 and the expression (1.121) or (3.62) for the characteristic impedance as follows:

$$Z(l) = Y\frac{1 + R(l)}{1 - R(l)}, Y = Y_0\left\{1 - \frac{\alpha(M)}{k_0} + j\frac{\alpha(M)}{k_0}\right\}$$

or

$$Z(l) = Y_0\left(1 - \frac{\alpha(M)}{k_0} + j\frac{\alpha(M)}{k_0}\right)\frac{e^{\beta s} - H_{21}e^{-\beta Ms} + \left[H_{21}e^{-\beta Ms} - e^{-\beta s}\right]e^{2\beta l}}{e^{\beta s} - H_{21}e^{-\beta Ms} - \left[H_{21}e^{-\beta Ms} - e^{-\beta s}\right]e^{2\beta l}}. \quad (5.67)$$

It may be recalled that $Y_0 = c_0/S$ is the characteristic impedance of the duct for an ideal fluid. For the limiting case of an ideal stationary medium $(\alpha = 0, M = 0)$, Equation 5.67 simplifies to

$$Z(l) = jY_0\frac{H_{21}\sin k_0 l - \sin k_0(l-s)}{\cos k_0(l-s) - H_{21}\cos k_0 l}, \quad (5.68)$$

For a check, Equation 5.68 may be seen to be identical to Equation 5.8 of Boden and Abom [31].

Unfortunately, measurement of the transfer function H_{21} may be in error owing to possible non-equalization errors and/or discretization errors implicit in the data processing [31,32]. The consequent uncertainty in the terminal reflection coefficient $R(l)$ may be calculated by differentiating Equation 5.66 partially with respect to H_{21} as follows [34]:

$$\delta R(l)|_{H_{21}} = \frac{\partial R(l)}{\partial H_{21}} \delta H_{21} = \frac{e^{-\beta M s} e^{2\beta l}}{e^{\beta s} - H_{21} e^{-\beta M s}} \{1 + R(l)\} \delta H_{21} \qquad (5.69)$$

Equation 5.66 can be rearranged to obtain H_{21} in terms of $R(l)$:

$$H_{21} = \frac{R(l)e^{\beta s} + e^{\beta(2l-s)}}{e^{2\beta l} + R(l)} e^{\beta M s}. \qquad (5.70)$$

Obviously, the uncertainty in $R(l)$ would be maximum when the denominator of Equation 5.69 tends to zero.

Therefore, substitution of Equation 5.70 in the denominator of Equation 5.69 yields the instability condition $e^{2\beta l}(e^{\beta s} - e^{-\beta s}) = 0$, or

$$\sinh \beta s = 0, \qquad (5.71)$$

or, upon making use of the identity (5.58),

$$\sin k_c s \cosh \alpha_c s - j \cos k_c s \sinh \alpha_c s = 0. \qquad (5.72)$$

This condition will never occur precisely, because of α_c, which includes both damping and flow effects. Thus the convective attenuation constant α_c can be considered as a stabilizing factor. However, the uncertainty in $R(l)$ will still be largest when $\sinh \beta s$ becomes small. Physical significance of the instability condition (5.72) may be appreciated from its limiting cases as follows.

In the limiting case of zero damping, the instability Equation 5.72 would reduce to

$$\sin\{k_0 s/(1-M^2)\} = 0, or \quad k_0 = n\pi(1-M^2)/s, \quad n = 0, 1, 2, 3, \ldots, \qquad (5.73)$$

which is the same as obtained by Seybert [28].

For inviscid moving medium, Equation 5.71 would reduce to $\sin k_c s = 0$, implying $k_c s = m\pi, m = 1, 2, 3, \ldots$, or with reference to Figure 5.4,

$$(k^+ + k^-)(z_1 - z_2) = 2m\pi,$$

or

$$s \equiv z_1 - z_2 = (1 - M^2)m(\lambda/2), \quad m = 0, 1, 2, 3, \ldots \qquad (5.74)$$

This clearly indicates that the reflection coefficient cannot be determined from Equation 5.66 at discrete frequency points for which the microphone spacing is almost equal to an integer multiple of the half-wavelength of sound. In order to avoid these points up to a frequency f_{max}, the microphone spacing $z_1 - z_2$ must be chosen such that

$$z_1 - z_2 < \frac{c_0}{2f_{max}}(1 - M^2). \tag{5.75}$$

Transfer function can be measured easily by means of a spectral analyzer. However, a careful calibration is required for the gain factor as well as the phase factor of the transfer function. One of the ways of doing it is the sensor switching procedure [27,35]. In this procedure, the measurement of the transfer function is made with an initial microphone configuration and a second measurement is made with the locations 1 and 2 switched or interchanged. The mean of the two results is taken as the desired result. Thus, not only is the measurement error due to phase mismatch between the two microphone channels eliminated, but also the result becomes independent of the gain factor of the two measurement channels.

In this method, the reflection coefficient (and thence impedance) can be evaluated conveniently by a programmable digital spectral analyzer. However, in digital spectral computation the signal-to-noise ratio is one of the most important factors affecting computational accuracy. Signal interference can be minimized by means of Chung's signal-enhancement technique [36], which unfortunately requires an additional microphone channel and presumes that (flow) noise at the different microphones must be mutually independent, that is, not coherent.

Incidentally, the Ross-Seybert method and the Chung-Blaser method are conceptually the same, as one can be derived from the other. The Chung-Blaser method, however, is preferred, as it is computationally more efficient and easier to implement. For this reason, it has been adopted as the ASTM standard for the two-microphone impedance tube method [37].

The correction factors associated with the two microphone technique have been evaluated by four different calibration methods [38]. Unlike the microphone switching technique [27,37] and exposing the same sound field to both the microphones [23,39] methods, a Section Reversal method [38] is introduced below for obtaining the correction factors associated with the measured transfer functions. These correction factors can then be used for testing different acoustic materials, as long as the microphones are not disturbed from their respective microphone holders.

As there are no materials which have an absolutely defined refection coefficient or impedance, there are limited tests which can be performed to verify the accuracy of a particular method. The case of an infinitely rigid termination is theoretically possible, but for all practical purposes cannot be achieved or measured experimentally either due to the electrical noise or due to the losses through the tube walls or other factors [40]. The impedance of an ideally rigid termination would be infinite. It is therefore not practical to compare the measured impedance with the theoretical value. However, the reflection coefficient is well defined and is suitable for measurement and comparison purposes. An ideal rigid termination has a reflection coefficient $|R| = 1$. Similarly, it is impossible to make a totally anechoic termination which would have a reflection coefficient $|R| = 0$. Another termination which has a well-defined solution, other than the idealized rigid and anechoic terminations, is an open-ended termination.

Measurement of the transfer functions is made with an initial configuration of the symmetric section and a second measurement is made with the microphone sensing locations interchanged by reversing the symmetric section, thus avoiding the removal of the microphones from their respective holders. The correction factor is then obtained from the geometric mean of the two measured transfer functions as follows.

If h^{12} and h^{21} are the two measured transfer functions between the microphones 1 and 2, for the 'section forward' and 'section reversed' configurations respectively, then the correction factor k_{21} and the true transfer function H_{21} for the corresponding pair of microphones are given by [38]

$$K_{12} = \sqrt{h^{12}h^{21}}; \text{ and } H_{12} = \frac{h^{12}}{K_{12}} \qquad (5.76, 5.77)$$

This true transfer function can be used in Equation 5.66 to compute the reflection coefficient. The impedance tube technology presumes the tube to be a one-dimensional waveguide. This would require complete absence of higher-order modes and rigid walls. While one-dimensionality may be ensured by selecting the tube diameter so as to satisfy the criterion (1.48), the requirement of rigid wall would call for very thick impedance tube wall [41]. For practical values of wall thickness, the wall compliance would not be negligible, and the speed of sound in the duct would be a little less than that in the free-field [41,42].

5.1.4 Comparison of the Various Methods for a Passive Subsystem

The classical moving-probe, discrete-frequency excitation method described in Section 5.1 has a number of advantages over the two-microphone methods and the transient testing method in that

i. it needs very simple instrumentation because only SPLs are to be measured (the random-excitation and transient testing methods call for a dual-channel real time analyzer and a digital computer, or else an on-line data processing system);
ii. it is more accurate, because the usual errors that creep in with the data acquisition and retrieval systems are absent;
iii. it requires only a sinusoidal signal that can be provided with sufficient strength to ensure that SPL at even the pressure minima is sufficiently higher than the broadband ambient flow noise; and

On the other hand, it has a number of disadvantages in that

i. the experiment is very time-consuming in view of the fact that measurements are to be taken separately for each of the numerous frequencies of interests (for example, if one chooses a frequency step of 20 Hz, a frequency range of 20–2000 Hz (the usual range for typical exhaust mufflers) would involve measurements at as many as 100 different frequencies);
ii. the acoustic field is likely to be disturbed by the presence of the probe tube, while wall-mounted microphones used in the two-microphone and transient testing methods do not interfere with the acoustic field;
iii. the tube length required at low frequencies is large (at least one wave length at the lowest frequency of interest); and
iv. the test conditions (room temperature) may not remain constant over the time length of the measurement.

However, the transfer function method, which is the best type of random excitation method, makes use of fixed-position microphones and thereby suffers from certain inherent weaknesses. For example, at the frequency for which the microphone spacing $s = z_1 - z_2$ equals one-half a wavelength, and its multiples, the reflection coefficient in Equation 5.66 becomes indeterministic and the impedance cannot be evaluated. This effect has been noted for stationary medium [26] as well as moving medium [24,34]. Thus, one would need to have more than two sets of microphones, preamplifiers, and A/D convertors for covering the complete frequency range of interest.

Again, Seybert and Ross's scheme of calculation [23] involves simultaneous solution of four algebraic equations, the coefficients of which, being experimentally determined, are not exact. This is likely to result in numerical instabilities and hence very substantial errors at certain frequencies. This has been observed by Panicker and Munjal in connection with the simultaneous evaluation of amplitude and phase of the reflection coefficient from acoustic measurements through wall-mounted (that is, fixed-position) microphones [20].

Another limitation of Seybert and Ross's method, as well as Chung and Blaser's transfer function method, is that they neglect acoustic attenuation. This could result in significant errors at certain frequencies [34].

5.2 Four-Pole Parameters of a Flow-Acoustic Element or Subsystem

The transfer matrix representation for a passive system (or subsystem) of Figure 5.7 is

$$\begin{bmatrix} p_1 \\ v_1 \end{bmatrix} = \begin{bmatrix} A & B \\ C & D \end{bmatrix} \begin{bmatrix} p_2 \\ v_2 \end{bmatrix} \tag{5.78}$$

The four-pole parameters A, B, C and D were first measured for a uniform tube by To and Doige [39] making use of their transient testing method, for a stationary medium.

This technique was successfully extended to the direct measurement of the four-pole parameters of uniform tubes, flare tubes and sudden area changes, by Lung and Doige [43], for small mean flow velocities. They made use of a two-load, four-microphone technique.

The flow noise produced at sudden area discontinuities being random in nature, makes it very difficult to employ random excitation that is generally used with the four-microphone two-load method, particularly at low frequencies. The two-load method suffers from an additional disadvantage in that the two loads may not be 'sufficiently' different at all frequencies of interest.

Therefore an investigation was undertaken for development of a new measurement technique for the determination of the four-pole parameters or transfer matrices of pipeline elements with flow. This resulted in a new method, called the two source-location method, to replace the two-load method [39,43]. Besides, the use of a signal enhancement technique (ensemble averaging in the time domain) helped to filter out the flow noise generated by the element under investigation.

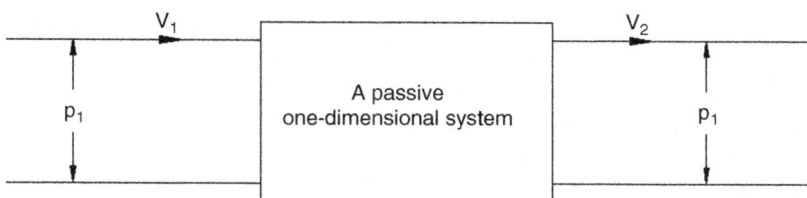

Figure 5.7 Four-pole representation of an acoustic system

5.2.1 Theory of the Two Source-Location Method

The two source-location method consists of the following:

a. Measuring acoustic pressures (or rather their ratios called transfer functions) at four fixed locations, two upstream and two downstream of the test element, as shown in Figure 5.8a, with a pseudo-random source on the left side;
b. Shifting the source to the right side, as shown in Figure 5.8b, and measuring acoustic pressures at the same four locations;
c. Calculating A, B, C and D (the four-pole parameters of the test element) by means of a dual-channel FFT analyzer, use of the time-domain ensemble averaging and certain relations that are derived below [44].

Referring to Figure 5.8a the state variables at various junctions are related as follows:

$$\begin{bmatrix} p_{1a} \\ v_{1a} \end{bmatrix} = \begin{bmatrix} A_{12} & B_{12} \\ C_{12} & D_{12} \end{bmatrix} \vdots \begin{bmatrix} A & B \\ C & D \end{bmatrix} \vdots \begin{bmatrix} A_{34} & B_{34} \\ C_{34} & D_{34} \end{bmatrix} \begin{bmatrix} p_{4a} \\ p_{4a}/Z_a \end{bmatrix}. \quad (5.79)$$

Carrying out successive multiplication of transfer matrices starting from the right end yields

$$p_{3a}/p_{4a} = A_{34} + B_{34}/Z_a, \quad (5.80)$$

$$\frac{p_{2a}}{p_{4a}} = AA_{34} + BC_{34} + \frac{AB_{34} + BD_{34}}{Z_a} = A\left\{\frac{p_{3a}}{p_{4a}}\right\} + B\left\{C_{34} + \frac{D_{34}}{Z_a}\right\} \quad (5.81)$$

$$\frac{p_{1a}}{p_{4a}} = A_{12}(AA_{34} + BC_{34}) + B_{12}(CA_{34} + DC_{34}) + \frac{A_{12}(AB_{34} + BD_{34}) + B_{12}(CB_{34} + DD_{34})}{Z_a} \quad (5.82)$$

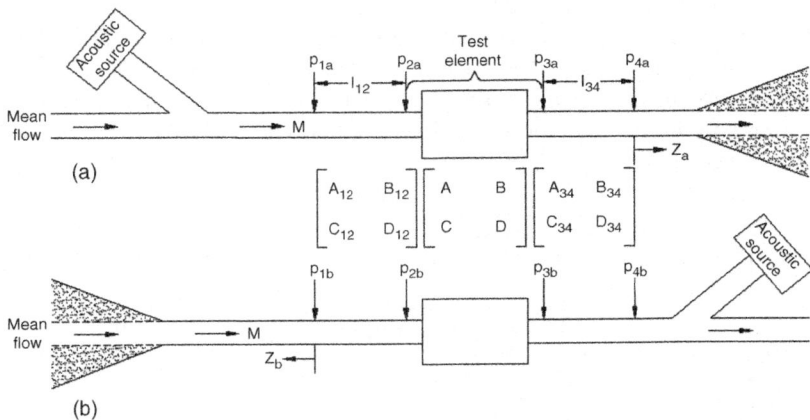

Figure 5.8 The two test configurations for the two source-location method (Adapted from [44])

or, upon making use of Equations 5.80 and 5.81,

$$\frac{p_{1a}}{p_{4a}} = A_{12}\left\{\frac{p_{2a}}{p_{4a}}\right\} + B_{12}\left\{C\frac{p_{3a}}{p_{4a}} + D\left(C_{34} + \frac{D_{34}}{Z_a}\right)\right\}. \quad (5.83)$$

Noting that the direction of the forward progressive wave and hence of the mass velocity v in Figure 5.8b is from right to left in contrast to that in Figure 5.7, it is evident that Equation 5.78 is to be replaced by

$$\begin{bmatrix} p_2 \\ -v_2 \end{bmatrix} = \frac{1}{\Delta}\begin{bmatrix} D & B \\ C & A \end{bmatrix}\begin{bmatrix} p_1 \\ -v_1 \end{bmatrix}, \quad (5.84)$$

where $\Delta = AD - BC$ is the determinant of the matrix.

Proceeding from the source to the load (impedance Z_b) in Figure 5.8b, yields

$$\begin{bmatrix} p_{4b} \\ v_{4b} \end{bmatrix} = \begin{bmatrix} \frac{D_{34}}{\Delta_{34}} & \frac{B_{34}}{\Delta_{34}} \\ \frac{C_{34}}{\Delta_{34}} & \frac{A_{34}}{\Delta_{34}} \end{bmatrix} \overset{p_{3b}}{\underset{v_{3b}}{\vdots}} \begin{bmatrix} \frac{D}{\Delta} & \frac{B}{\Delta} \\ \frac{C}{\Delta} & \frac{A}{\Delta} \end{bmatrix} \overset{p_{2b}}{\underset{v_{2b}}{\vdots}} \begin{bmatrix} \frac{D_{12}}{\Delta_{12}} & \frac{B_{12}}{\Delta_{12}} \\ \frac{C_{12}}{\Delta_{12}} & \frac{A_{12}}{\Delta_{12}} \end{bmatrix} \begin{bmatrix} p_{1b} \\ p_{1b}/Z_b \end{bmatrix}, \quad (5.85)$$

whence

$$p_{2b}/p_{1b} = (1/\Delta_{12})\{D_{12} + B_{12}/Z_b\}, \quad (5.86)$$

$$\frac{p_{3b}}{p_{1b}} = \frac{1}{\Delta\Delta_{12}}\left\{DD_{12} + BC_{12} + \frac{DB_{12} + BA_{12}}{Z_b}\right\} = \frac{1}{\Delta}\left\{D\frac{p_{2b}}{p_{1b}} + \frac{B}{\Delta_{12}}\left(C_{12} + \frac{A_{12}}{Z_b}\right)\right\}, \quad (5.87)$$

$$\frac{p_{4b}}{p_{1b}} = \frac{1}{\Delta_{34}\Delta\Delta_{12}}\left\{D_{34}(DD_{12} + BC_{12}) + B_{34}(CD_{12} + AC_{12}) + \frac{D_{34}(DB_{12}+BA_{12})+B_{34}(CB_{12}+AA_{12})}{Z_b}\right\},$$

or, upon making use of Equations 5.86 and 5.87,

$$\frac{p_{4b}}{p_{1b}} = \frac{1}{\Delta_{34}}\left[D_{34}\frac{p_{3b}}{p_{1b}} + \frac{B_{34}}{\Delta}\left\{C\frac{p_{2b}}{p_{1b}} + \frac{A}{\Delta_{12}}\left(C_{12} + \frac{A_{12}}{Z_b}\right)\right\}\right]. \quad (5.88)$$

It may be noted that Equations 5.80 and 5.86 yield the following expressions for the load impedances Z_a and Z_b, respectively:

$$Z_a = \frac{B_{34}}{(p_{3a}/p_{4a}) - A_{34}}, \quad Z_b = \frac{B_{12}}{\Delta_{12}(p_{2b}/p_{1b}) - D_{12}}. \quad (5.89, 5.90)$$

Simultaneous solution of Equations 5.81, 5.83, 5.87 and 5.88 would yield explicit expressions for A, B, C and D. Here it may be noted that determinants of the constituent matrices are given by

$$\Delta_{12} = A_{12}D_{12} - B_{12}C_{12}, \quad \Delta = AD - BC, \quad \Delta_{34} = A_{34}D_{34} - B_{34}C_{34}. \quad (5.91-5.93)$$

Determinants Δ_{12} and Δ_{34} would be known a priori, but Δ would not be. Equation 5.92 should therefore be used to write Δ explicitly in terms of A, B, C and D before attempting simultaneous solution of Equations 5.81, 5.83, 5.87 and 5.88, with Z_a and Z_b given by Equations 5.89 and 5.90.

For convenience of writing, one can define a transfer function H_{ij} as

$$H_{ij} \equiv p_i/p_j, \tag{5.94}$$

so that

$$p_{ia}/p_{ja} \equiv H_{ij,a} \quad \text{and} \quad p_{ib}/p_{jb} \equiv H_{ij,b}. \tag{5.95}$$

Substituting the value of Z_a from Equation 5.89 in Equations 5.81 and 5.83 and rearranging, yields

$$(H_{34,a})A + \{C_{34} + (D_{34}/B_{34})(H_{34,a} - A_{34})\}B = H_{24,a} \tag{5.96}$$

and

$$(H_{34,a})C + \{C_{34} + (D_{34}/B_{34})(H_{34,a} - A_{34})\}D = (H_{14,a} - A_{12}H_{24,a})/B_{12}. \tag{5.97}$$

respectively. Similarly, substituting the value of Z_b from Equation 5.90 in Equations 5.87 and 5.88, and rearranging, gives

$$\{C_{12} + (A_{12}/B_{12})(\Delta_{12}H_{21,b} - D_{12})\}B + (\Delta_{12}H_{21,b})D = (H_{31,b}\Delta_{12})\Delta \tag{5.98}$$

and

$$\{C_{12} + (A_{12}/B_{12})(\Delta_{12}H_{21,b} - D_{12})\}A + (\Delta_{12}H_{21,b})C$$
$$= \{(H_{41,b}\Delta_{34} - H_{31,b}D_{34})/B_{34}\}\Delta\Delta_{12} \tag{5.99}$$

By eliminating Δ from Equations 5.98 and 5.99, we get a single linear equation in A, B, C and D, which can be combined with Equations 5.96 and 5.97 to obtain expressions for three of the four variables in terms of the fourth. Substitution of these expressions in Equations 5.96 or 5.97, yields the value of this fourth variable and thence all the other three variables. These algebraic manipulations (omitted here) lead ultimately to the following expressions:

$$A = \frac{\Delta_{34}(H_{23,a}H_{43,b} - H_{23,b}H_{43,a}) + D(H_{23,b} - H_{23,a})}{\Delta_{34}(H_{43,b} - H_{43,a})}, \tag{5.100}$$

$$B = B_{34}(H_{23,a} - H_{23,b})/\Delta_{34}(H_{43,b}H_{43,a}), \tag{5.101}$$

$$C = \frac{(H_{13,a} - A_{12}H_{23,a})(\Delta_{34}H_{43,b} - D_{34}) - (H_{13,b} - A_{12}H_{23,b})(\Delta_{34}H_{43,a} - D_{34})}{B_{12}\Delta_{34}(H_{43,b} - H_{43,a})} \tag{5.102}$$

$$D = B_{34}\{(H_{13,a}H_{13,b}) + A_{12}(H_{23,b}H_{23,a})\}/B_{12}\Delta_{34}(H_{43,b} - H_{43,a}). \tag{5.103}$$

The determinant is given by

$$\Delta = B_{34}(H_{13,a}H_{23,b} - H_{13,b}H_{23,a})/B_{12}\Delta_{34}(H_{43,b} - H_{43,a}). \tag{5.104}$$

In these final expressions, making use of Equations 3.62–3.67 and 3.70, we get

$$\begin{bmatrix} A_{12} & B_{12} \\ C_{12} & D_{12} \end{bmatrix} = e^{-M\beta_{12}} \begin{bmatrix} \cosh\beta_{12} & Y\sinh\beta_{12} \\ \sinh\beta_{12}/Y & \cosh\beta_{12} \end{bmatrix}, \quad \Delta_{12} = e^{-2M\beta_{12}},$$

$$\begin{bmatrix} A_{34} & B_{34} \\ C_{34} & D_{34} \end{bmatrix} = e^{-M\beta_{34}} \begin{bmatrix} \cosh\beta_{34} & Y\sinh\beta_{34} \\ \sinh\beta_{34}/Y & \cosh\beta_{34} \end{bmatrix}, \quad \Delta_{34} = e^{-2M\beta_{34}},$$

$$\beta_{12} = (jk_c + \alpha_c)l_{12}, \quad \beta_{34} = (jk_c + \alpha_c)l_{34}, \quad k_c = k/(1 - M^2), \quad \alpha_c = \alpha/(1 - M^2),$$

$$k = k_0 + \alpha, \quad \alpha = \alpha_0 + MF/2D, \quad Y = Y_0\{1 - (\alpha/k_0) + j\alpha/k_0l\}.$$

Here α_0, M, F and D are the viscothermal pressure attenuation constant, the mean flow Mach number, Froude's friction factor, and the pipe diameter, respectively.

5.2.2 Theory of the Two-Load Method

In contrast to the two source-location method, the two-load method consists of conducting the test with two different loads with the source on the same side. These two configurations are shown in Figure 5.9. The first configuration of Figure 5.9 is of course the same as in Figure 5.8. Therefore, Equations 5.96 and 5.97 hold for this as well. Equations 5.98 and 5.99 will now be replaced by equations that are identical to Equations 5.96 and 5.97 except that the subscript a is replaced by b. Omitting the algebraic manipulations, it can be shown that simultaneous solution of these four equations yields Equations 5.100–5.103 for the two-load method as well as the two-source method. Incidentally, the same expressions can also be

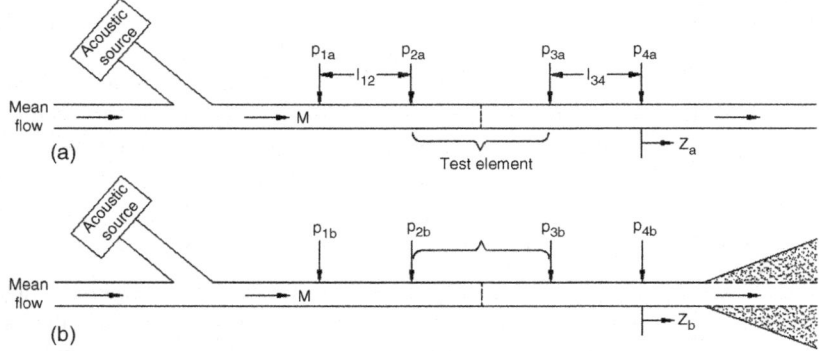

Figure 5.9 The two test configurations for the two-load method. Z_a and Z_b are the respective loads (Adapted from [44])

obtained by rearranging those of reference [43]. This is not surprising inasmuch as the measurement of the four-pole parameters should be independent of the methods one chooses to create the two-test states. Significantly, however, as shown below, the two-source location method is superior to the two-load method in that the former always creates two independent states.

5.2.3 Comparison of the Two Methods

In the two-load method (see Figure 5.9), if Z_b tends to Z_a at any frequency, there would be no difference whatsoever between the two configurations, so that $H_{23,b} \rightarrow H_{23,a}$, $H_{43,b} \rightarrow H_{43,a}$ and $H_{13,b} \rightarrow H_{13,a}$, and substituting these limiting equalities in Equations 5.100–5.103 indicates that A, B, C and D would then become indeterminate. In fact, even in the neighborhood of these frequencies, the uncertainty of the two-load method is much greater than that of the two-source method, which incidentally would not fail even in the hypothetical eventuality of the two impedances Z_a and Z_b being identical. It can be argued that the validity of the measurements would extend over the full range if l_{12} and l_{34} were less than $\pi/3.68$ times D. This is because one is making use of the plane wave theory which breaks down as $k_0 D/2$ approaches 1.84. In fact, the three-dimensional effects are known to become significant at still lower frequencies and therefore, reasonable design values l_{12} and l_{34} are given by the inequality.

$$L_{12}, L_{34} < 0.75\, D. \tag{5.105}$$

For logistic reasons or otherwise, if one uses l_{12} and/or l_{34} larger than D, one should keep the highest frequency proportionately down: that is the highest frequency should not be decided by the plane wave propagation criterion

$$k_0 D/2 < 1.84, \tag{5.106}$$

but by the reliability inequality [45,46]

$$0.1\pi < k_0 l_{12}, k_0 l_{34} < 0.8\pi. \tag{5.107}$$

As Equations 5.100–5.103 hold for the two-load method as well as the two-source location method, for both the methods the same expressions would hold for estimation of uncertainty in any of the four-pole parameters on account of a possible error in prediction or measurement of the velocity of wave propagation c_0 (and hence k_0), the pressure attenuation coefficient α, the lengths l_{12} and l_{34}, and the magnitude or phase of any of the measured transfer functions (resulting from the equalization errors or the data-processing errors).

5.2.4 Experimental Validation

Although the same expressions govern the four-pole parameters of the test element in both the schemes, yet the two-source location method is functionally much more stable than the two-load method. In fact, it is entirely independent of the loading terminations on either side. This has been amply demonstrated for orifice plates as well as uniform tubes, with typical mean flow velocities: see Figures 5.10–5.12. Details of the basic arrangement for laboratory test

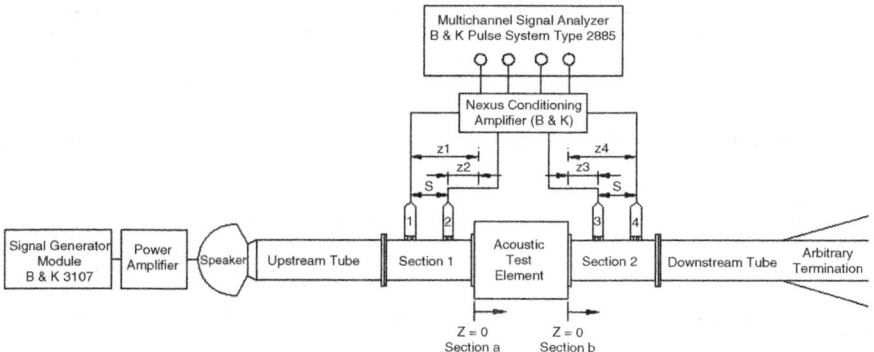

Figure 5.10 Schematic of the experimental set-up (Reproduced with permission from [47])

(Figure 5.10) are given in reference [47]. The broken line curves in Figure 5.11 indicate the analytically computed values (see Equation 3.70). Figures 5.11 and 5.12 are two typical graphs representing successful measurement when the two-source location method is used and the superiority thereof over the two-load method.

Incidentally, it may be observed from Figures 5.8 and 5.10 that the measured transfer matrix refers to the subsystem between microphones 2 and 3. Thus, it includes not only the test element but also uniform tubes of length z_2 upstream and z_3 downstream. Symbolically,

$$[T_m] = [T_u][T_{test}][T_d], \tag{5.108}$$

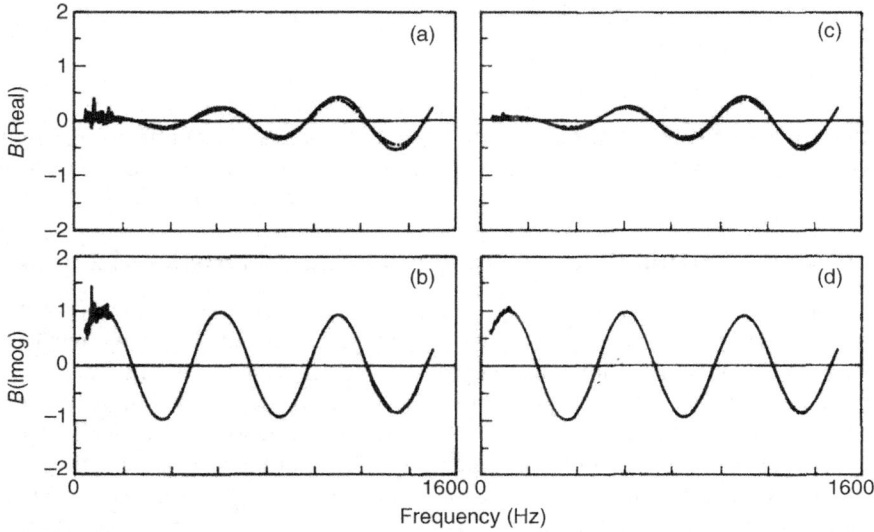

Figure 5.11 B parameter for a straight pipe, $M = 0.03$. (a) and (b), two load method; (c) and (d), two source-location method (Reproduced with permission from [44])

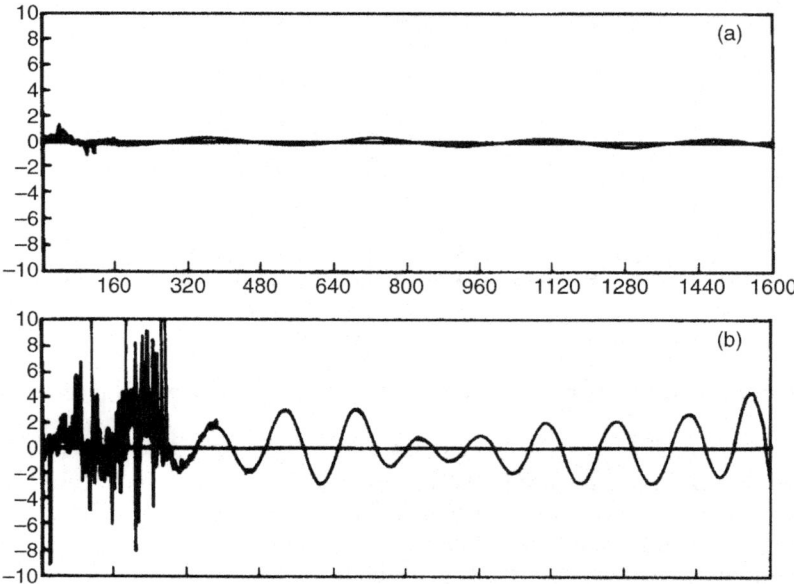

Figure 5.12 Percentage uncertainty in A due to a 1% error in $H_{23,b}$ (real part). (a) Two source-location method (b) two-load method (Reproduced with permission from [44])

where subscripts m, u, test and d denote measured, upstream, test element and downstream, respectively. As $[T_u]$ and $[T_d]$ are known from Equation 3.70, $[T_{test}]$ may be extracted from Equation 5.108 by premultiplying both sides of Equation 5.108 with $[T_u]^{-1}$ and then post multiplying it with $[T_d]^{-1}$ so as to obtain

$$[T_{test}] = [T_u]^{-1}[T_m][T_d]^{-1} \quad (5.109)$$

Significantly it has been shown in Ref [47] that higher-order mode effects are implicitly captured by the experimental technique even though it is based essentially on the plane wave theory.

5.3 An Active Termination – Aeroacoustic Characteristics of a Source

The foregoing sections have dealt with various methods for measuring the reflection characteristics and/or impedance of a passive termination (black box). As remarked in Chapters 2 and 3, however, flow-acoustic analysis of an exhaust muffler [evaluation of insertion loss – Equation 3.51] requires a prior knowledge of $Z_{c,n+1}$, the internal impedance (also called the output impedance) of the exhaust source. Prediction of sound radiated by an n-element exhaust system would further require knowledge of the source strength $v_{c,n+1}$ and hence $p_{c,n+1}$ (because $v_{c,n+1} \equiv p_{c,n+1}/Z_{c,n+1}$). Thus pressure $p_{c,n+1}$ (corresponding to the open-circuit voltage of an electrical source) and impedance $Z_{c,n+1}$ are the two characteristics of the exhaust source that need to be known a priori all over the frequency range of interest. Alternatively, $p_{c,n+1}$ and $Z_{c,n+1}$ could also be denoted by $p_{c,s}$ and $Z_{c,s}$.

Unlike the electroacoustic driver, a reciprocating internal combustion engine (or compressor, for the matter) is a peculiar source of acoustic signals inasmuch as its geometry changes rapidly with time because of

a. large piston displacement and velocities,
b. the presence of an exhaust valve (or port) that varies the communication passage very sharply, and
c. high cylinder pressure during the initial part of blow-down, which results in choked sonic flow at the valve throat and hence an acoustic diode during this phase of the exhaust process.

Because of these and fast changes in the geometry of the exhaust source, it is extremely difficult to model it acoustically for prediction of the source characteristics $p_{c,n+1}$ (or $p_{c,s}$) and $Z_{c,n+1}$ (or $Z_{c,s}$). In fact, at the time of writing of this book there is no analytical method available for the purpose!

A number of methods have, however, been developed for experimental evaluation of the source characteristics. These can be grouped under two headings:

i. Direct measurement $Z_{c,s}$ ($p_{c,s}$ cannot be measured directly).
ii. Indirect measurement of $p_{c,s}$ and $Z_{c,s}$.

5.3.1 Direct Measurement of Source Impedance

The method of direct measurement rests on the hypothesis that if 'sufficiently strong' acoustic waves are directed onto a running engine, the resulting pressure field would be more or less independent of the sound radiated by the engine, and analysis of the pressure field should yield impedance of the source Z_s, which could then be converted to $Z_{c,s}$. The signal sent in by the acoustic driver should predominate, all over the frequency range of interest, over the ambient pressure field in the pipe created by the running engine. In other words, the signal-to-noise ratio should be sufficiently large. Implications of this condition are as follows:

i. A normal electroacoustic driver (loudspeaker) would not do; one must use an electropneumatic driver, the output of which can be made an order higher than the strongest electroacoustic driver by adjusting the pressure and amount of the compressed air supply.
ii. In order to reduce the engine noise, one may like to motor it instead of firing. But then this would alter the flow conditions across the exhaust valve (for example, there would be no choked-sonic conditions) and hence the impedance of the exhaust source. So, the engine must be running normally (in the firing mode) at rated load and speed.
iii. It has been observed [24,48] that with the engine firing normally, even the pneumatic driver cannot produce a sound that would predominate over the engine's noise at low frequencies, that is, the firing frequency F and its first few harmonics at which the engine noise is maximum. For example, for an eight-cylinder, four-stroke-cycle engine running at 2000 RPM, F would amount to 133.3 Hz and the method of direct measurement would fail at frequencies lower than 400 Hz (3F). Unfortunately, however, these are the frequencies at which maximum insertion loss is called for and the muffler designer's need for a precise knowledge of the source impedance is the strongest.

iv. With the driver producing acoustic signals of the order of 140–150 dB, nonlinearity effects might become significant, resulting in substantial unpredictable errors in the measured impedance.

Direct measurement of the exhaust source impedance Z_s has been tried with a fair amount of success by Prasad and Crocker [24,48], making use of the wall mounted two-microphone random-excitation method (transfer function approach), on a 5.77 liter displacement, eight-cylinder engine exhaust system. Figure 5.13 shows a diagram of the experimental setup. In order to overcome the half-wavelength spacing limitation, three microphones were used, selecting a particular combination of two microphones for the transfer function approach [Equation 5.66]. The noise source used was an electropneumatic driver. When the engine was in operation, a sufficient signal-to-noise ratio was obtained with a current of 2.5 A and a supply air pressure of about 30 psi (about two bars), as shown in Figure 5.14. In view of high temperatures (of the order of 670° K near the manifold), piezoelectric pressure transducers with cooling jackets were used instead of the B and K condenser microphones. Impedance analysis was carried out using a dual-channel fast Fourier analyzer with a frequency resolution of 10 Hz and block size of 512 Hz.

Figure 5.14 clearly indicates that the driver signal was generally 20 dB (at least 10 dB) more than the engine noise at frequencies of 350 Hz or higher (firing frequency equals 133.3 Hz in this case of 2000 RPM). However, the combined SPL spectrum of the engine noise and acoustic driver is not equal to the antilogarithmic addition of the two except in an average sense. This is because of a general unsteadiness caused by the flow-acoustic interaction. The discrepancies are of the order of 5–10 dB, large enough to suggest that the impedance values resulting from the transfer function approach, shown in Figure 5.15, are valid only in an average sense. The corresponding values of source impedance obtained by Seybert and Ross [23], superimposed on the same, should generally be more reliable in view of a judicious time-averaging of the observed values by the observer in the case of standing wave method or the moving-probe discrete frequency excitation method. Automatic data reduction can be very misleading and unreliable at times, especially when the measurables are unsteady. Incidentally, this explains why there are large differences (0.25–1.0) in the values of

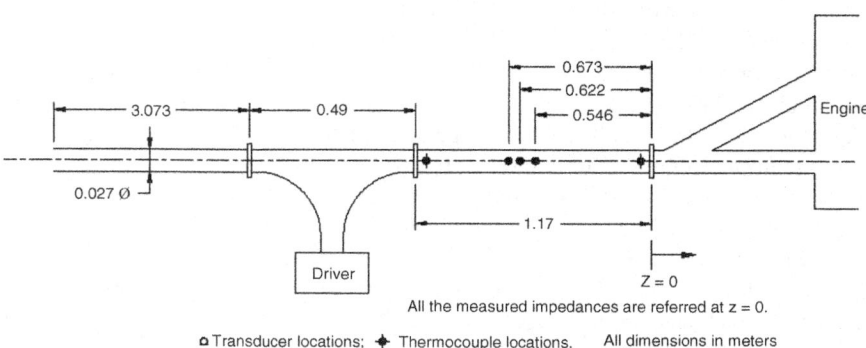

Figure 5.13 Diagram of the setup for engine impedance measurement (Reproduced with permission from [24])

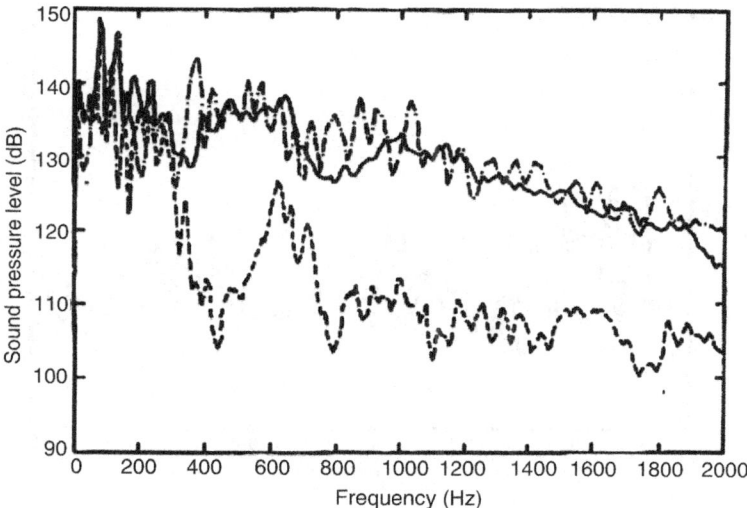

Figure 5.14 Sound pressure level spectra for various operating systems for the purpose of signal-to-noise ratio. ——, Engine and driver; — ··· —, driver only; – – –, engine noise only (Reproduced with permission from [48])

the normalized source impedance measured by the two methods. The agreement is observed to be only in an average sense.

Figure 5.16 shows the insertion loss of an expansion chamber on the same engine [48], where the measured values of IL are compared with those predicted using the measured engine impedance (Figure 5.15). The agreement is good in an average sense; local discrepancies of 5–10 dB are apparent – perhaps a direct consequence of the discrepancies in the measured engine impedance, more so at lower frequencies.

Sneckenberger [49] utilized the two-sensor, random-excitation, transfer-function technique for measurement of impedance of the engine, interpreted within a format of passive acoustic systems. It was a hypothesis of the experimentation that the operating engine can be modeled

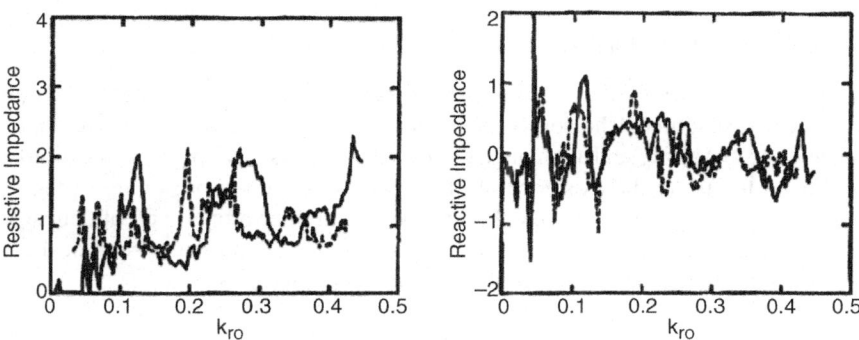

Figure 5.15 Measured dimensionless specific acoustic impedance of an engine operating at 2000 RPM. ——, Transfer function method; – – –, Standing wave method (Reproduced with permission from [48])

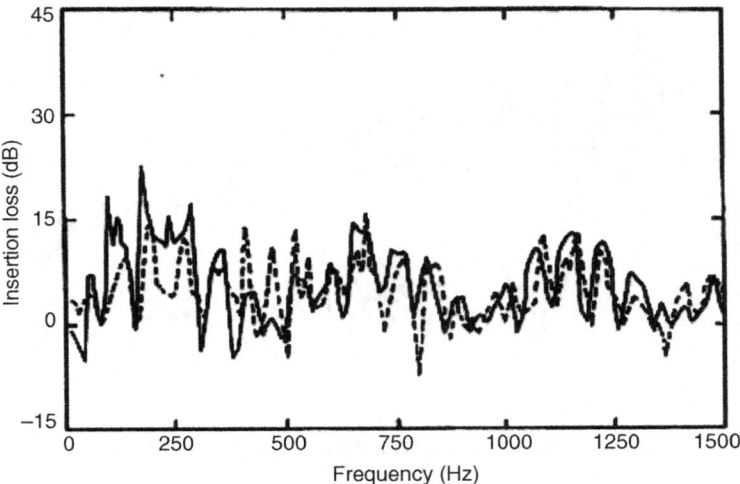

Figure 5.16 Insertion loss of an expansion chamber on an engine operating at 2000 RPM. ——, Predicted (using measured engine impedance); – – –, Measured (Reproduced with permission from [24])

as a series of cyclic passive systems. This hypothesis has been assessed from experimental results obtained from impedance studies of the motored engine. Modeling of the engine as a cyclic series of passive systems was found to be quite useful for interpreting experimental acoustic impedance for motored engines [49].

5.3.2 Indirect Measurement of Source Characteristics

This indirect approach makes use of the fact that when a source with internal characteristics $p_{c,s}$ and $Z_{c,s}$ is connected to a load impedance Z_L, the acoustic power W_L picked up by the load would be a function of not only Z_L but also the source characteristics $p_{c,s}$ and $Z_{c,s}$. Thus, by connecting the source to different loads and measuring the power output one could formulate the required number of equations that could be solved simultaneously to extract values of the source characteristics [15,18].

5.3.2.1 Impedance-Tube Method

Let $p_{c,s}$ and $Z_{c,s}$ be the two hypothetical aeroacoustic characteristics of the engine (or compressor) exhaust source. The source here is meant to include the core, giving rise to the pulsating gas (or air) flow, and a small length of the exhaust gas pipe, equal to about six diameters, so as to ensure the existence of plane waves before the beginning of the muffler system, which includes the rest of the exhaust pipe, the muffler proper, and of course the tail pipe. Let $Z_{c,l}$ be the equivalent aeroacoustic impedance of the muffler system at the source end. With reference to the analogous circuit of Figure 5.17, one gets

$$p_{c,s} = v_c(Z_{c,s} + Z_{c,l}) = v_c Z_{c,s} + p_c, \qquad (5.110)$$

where p_c, v_c, and $Z_{c,s}$ are all complex functions of ω.

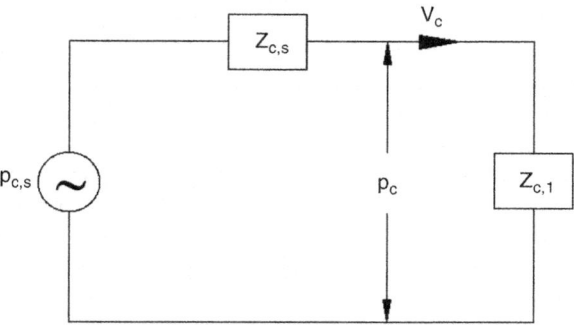

Figure 5.17 Electrical analogous circuit for the four-load model (Adapted from [56])

Let the experiments be conducted with two different muffler systems, connected to the source in turn, say, the first with a pipe alone and the next with an expansion chamber and a pipe, as shown in Figure 5.18a and b . In both these cases, the exhaust pipe comprises a simple tube and an impedance tube. Let the source, which could be an engine pump or compressor, be run at the same speed for the two configurations of Figure 5.18 so as to ensure that the basic source characteristics remain unchanged in both the cases.

The impedance-tube method is used here to measure the reflection characteristics of the atmosphere termination in the first case (Figure 5.18a) and those of an expansion chamber muffler in the second case (Figure 5.18b); both are passive terminations for which, indeed, the impedance tube method is intended. The exhaust source is used in its normal mode of

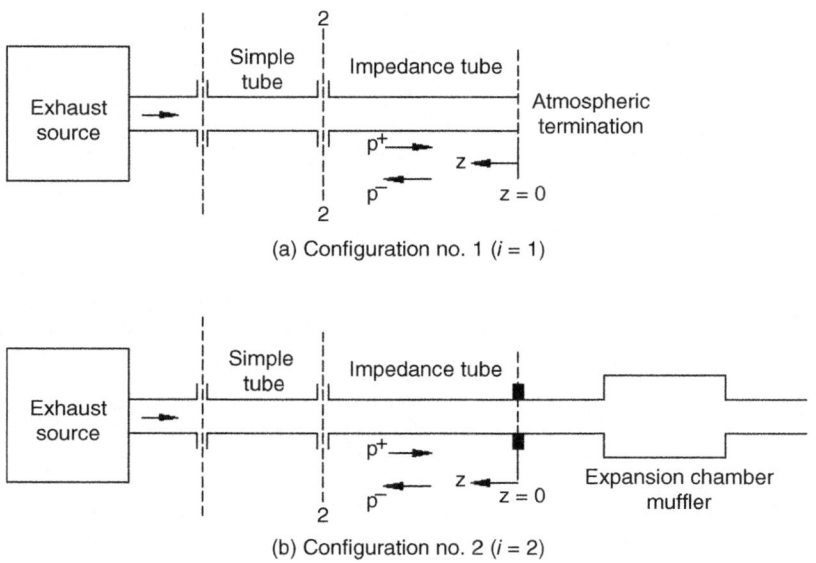

Figure 5.18 Measurement of source characteristics using the impedance-tube method (Reproduced with permission from [18])

functioning to produce the flux of gases and, of course, the noise. Sound picked up from the impedance tube by means of a moving probe is passed through a narrow-band heterodyne slave filter (with bandwidth of 3.18 or 10 Hz) with center frequency set at the firing frequency or one of its integral multiples, as required [18].

Let $|R|^{(i)}$ and $\theta^{(i)}$ be the amplitude and phase of the reflection coefficient at $z=0$ for the ith configuration of Figure 5.18 ($i=1, 2$). Using, the discrete frequency method described in Section 5.1, one can evaluate α, $\alpha^{\pm}$, $k^{\pm}$, $|R|$, and θ for both the configurations. Then $Y^{\pm}$ can be determined from Equation 1.121. If the length of the impedance tube in each case is, say, l, then, for acoustic pressure and mass velocity at the upstream end of the impedance tube (corresponding to section 2-2 in Figure 5.18), one gets

$$p_{2-2}^{(i)} = p^{+(i)} e^{j\omega t} \left[e^{jk^+(i)l + \alpha^+(i)l} + |R|^{(i)} e^{j\theta^{(i)} - jk^-(i)l - \alpha^-(i)l} \right], \quad i = 1, 2, \quad (5.111)$$

and

$$v_{2-2}^{(i)} = p^{+(i)} e^{j\omega t} \left[\frac{e^{jk^+(i)l + \alpha^+(i)l}}{Y^+} - \frac{|R|^{(i)} e^{j\theta^{(i)} - jk^-(i)l - \alpha^-(i)l}}{Y^-} \right], \quad i = 1, 2. \quad (5.112)$$

With $\alpha^{\pm}, k^{\pm}, R$, and θ being known, $p^{+(i)}$ can be determined from Equation 5.1 using complex pressure picked up from any point (z), and p and v at section 2-2 can be calculated from Equations 5.111 and 5.112 for both the configurations.

Now making use of Equations 3.32 and 3.33 yields

$$p_{c,2-2}^{(i)} = p_{2-2}^{(i)} + M_{2-2}^{(i)} Y_{2-2}^{(i)} v_{2-2}^{(i)}, \quad (5.113)$$

and

$$v_{c,2-2}^{(i)} = v_{2-2}^{(i)} + M_{2-2}^{(i)} p_{2-2}^{(i)} / Y_{2-2}^{(i)}, \quad (5.114)$$

where

$$Y_{2-2}^{(i)} = (c/S)_{2-2}^{(i)}. \quad (5.115)$$

Making use of the transfer matrix relation (3.72), these values of $p_{c,2-2}^{(i)}$ and $v_{c,2-2}^{(i)}$ can be transferred to the upstream end of the simple tube in Figure 5.18 (this corresponds to section 1-1). Thus,

$$\begin{bmatrix} p_{c,1-1} \\ v_{c,1-1} \end{bmatrix}^{(i)} = e^{-j(k_c Ml)_{1-2}} \begin{bmatrix} \cos(k_c l)_{1-2} & jY_{1-2} \sin(k_c l)_{1-2} \\ \dfrac{j}{Y_{1-2}} \sin(k_c l)_{1-2} & \cos(k_c l)_{1-2} \end{bmatrix}^{(i)} \times \begin{bmatrix} p_{c,2-2} \\ v_{c,2-2} \end{bmatrix}^{(i)} \quad (5.116)$$

Now Equation 5.110 implies

$$p_{c,s} = v_{c,1-1}^{(1)} Z_{c,s} + p_{c,1-1}^{(1)} = v_{c,1-1}^{(2)} Z_{c,s} + p_{c,1-1}^{(2)} \quad (5.117)$$

whence

$$Z_{c,s} = \frac{p_{c,1-1}^{(1)} - p_{c,1-1}^{(2)}}{v_{c,1-1}^{(2)} - v_{c,1-1}^{(1)}}. \qquad (5.118)$$

Thus, the source impedance $Z_{c,s}$, as seen at the upstream end of the simple tube can be calculated. Finally, the strength of the source $p_{c,s}$ can be calculated from either of Equation 5.117. It will in general be a complex number.

Incidentally, source characteristics $p_{c,s}$ and $Z_{c,s}$ can be transferred to the desired section of the exhaust pipe by making use of the relationships [50].

$$p_{c,s,d} = p_{c,s,u}/(T_{11} + T_{21}Z_{c,s,u}) \quad \text{or} \quad p_{c,s,u} = p_{c,s,d}\frac{T_{11}T_{22} - T_{12}T_{21}}{T_{22} - T_{21}Z_{c,s,d}}, \qquad (5.119)$$

$$Z_{c,s,d} = \frac{T_{12} + T_{22}Z_{c,s,u}}{T_{11} + T_{21}Z_{c,s,u}} \quad \text{or} \quad Z_{c,s,u} = \frac{T_{11}Z_{c,s,d} - T_{12}}{T_{22} - T_{21}Z_{c,s,d}}, \qquad (5.120)$$

where T_{ij} are the convective four-pole parameters of the pipe element (or for that matter, any subsystem of elements) between the upstream point u and the downstream point d. Relations (5.119) and (5.120) would also hold for the acoustic source characteristics p_s and Z_s provided the transfer matrix $[T]$ connecting points u and d is written out in terms of the classical state variables p and v (instead of the convective state variables p_c and v_c).

Similarly, the convective source impedance $Z_{c,s}$ and the conventional source impedance Z_s are related to each other as follows [51]:

$$Z_{c,s} = \frac{Z_s - M_n Y_n}{1 - M_n Z_s/Y_n} \quad \text{or} \quad Z_s = \frac{Z_{c,s} + M_n Y_n}{1 + M_n Z_{c,s}/Y_n}, \qquad (5.121)$$

This may be noted to be similar to Equation 3.58 except that the direction of the meanflow Mach number M is reversed because for source impedance we are looking into the source.

The corresponding relationships for the source strength are [51]

$$p_{c,s} = \frac{p_s(1 - M_n^2)}{1 - M_n Z_s/Y_n} \quad \text{and} \quad p_s = \frac{p_{c,s}}{1 + M_n Z_{c,s}/Y_n}, \qquad (5.122)$$

where subscript n refers to the exhaust pipe next to the engine manifold.

5.3.2.2 Method of External Measurements

To minimize the errors in the source characteristics consequent to the unavoidable measurement errors, a multi-load least-squares method was proposed by Desmons and Hardy [52] that makes use of complex pressure measurements. However, these measurements can be influenced by the standing wave effects. Besides, complex pressure measurements require a phase reference signal which can be hard to find in some machines. In the case of the exhaust systems of IC Engines, there are additional problems due to the turbulent flow and the hot and corrosive gases. Therefore, an indirect method making use of sound pressure level measurement outside the tail pipe in a free environment recommends itself.

The first attempt in this direction was made by Prasad [53] who suggested a four-load method making external measurements, and demonstrated it successfully for a loudspeaker source. However, an error analysis of this method [52,54,55] has revealed that Prasad's four-load method is very sensitive to errors in the input data. To avoid or minimize these errors, Boden [55] proposed what he called a Direct Least-Squares Multi-Load method. This improved the results considerably. For further improvement in this direction Jang and Ih [56] proposed a refined multi-load method that uses an error function based on the linear time invariant source model and demonstrated it successfully on the exhaust system of an IC Engine as well as a centrifugal blower and a loudspeaker. This is detailed below.

With respect to Figure 5.17 acoustic pressure at the junction between source and load, is given by

$$p_c = p_{c,s} Z_{c,l} / (Z_{c,l} + Z_{c,s}), \tag{5.123}$$

where convective variables p_c and v_c and impedance $Z_{c,l}$ are given by the equations

$$p_c = p + MY_0 v = p(1 + MY_0/Z_l), \tag{5.124a}$$

$$v_c = v + Mp/Y_0, \tag{5.124b}$$

$$Z_{c,l} = (Z_l + MY_0)/(1 + MZ_l/Y_0) \tag{5.124c}$$

Here, subscripts c, s and l denote convective, source and load, respectively.

Therefore, for N different acoustic loads we get

$$|P_{c,s}|^2 = |P_{c,n}|^2 \frac{|Z_{c,n} + Z_{c,s}|^2}{|Z_{c,n}|^2}, (n = 1, 2, \ldots, N) \tag{5.125}$$

Although the engine exhaust source is not time-invariant, yet we can make use of

$$F(Z) = \sum_{m<n} w_{mn} \left\{ \frac{|P_{c,m}|^2}{|Z_{c,m}|^2} |Z_{c,m} + Z|^2 - \frac{|P_{c,n}|^2}{|Z_{c,n}|^2} |Z_{c,n} + Z|^2 \right\}^2, \tag{5.126}$$

where w_{mn} is the weighting factor, to write the following equation that compares the magnitude of the source strength $p_{c,s}$ for two load applications m and n. Here $F(Z)$ is the error function which should be zero only when $Z = Z_{c,s}$. Obviously, this would not be zero due to finite precision in the evaluation of loads. Therefore, the error function $F(Z)$ should be minimized in the least squares sense in order to obtain $Z_{c,s}$. Thus, the necessary condition for minimizing $F(Z)$ is given by

$$dF = \sum_{m<n} 2w_{mn} \left\{ \frac{|P_{c,m}|^2}{|Z_{c,m}|^2} |Z_{c,m} + Z|^2 - \frac{|P_{c,n}|^2}{|Z_{c,n}|^2} |Z_{c,n} + Z|^2 \right\}$$

$$\times \text{Re} \left[\left(\frac{|P_{c,m}|^2}{|Z_{c,m}|^2} (Z_{c,m}^* + Z^*) - \frac{|P_{c,n}|^2}{|Z_{c,n}|^2} (Z_{c,n}^* + Z^*) \right) dZ \right] = 0. \tag{5.127}$$

By substituting $Z \equiv Z_{c,s} = x + jy$ in Equation 5.127, we can express dF as a function of Z and dZ. Differential increments dx and dy being arbitrary variables, Equation 5.127 can be written as

$$G(Z) = \sum_{m<n} w_{mn} \left\{ \frac{|P_{c,m}|^4}{|Z_{c,m}|^4} |Z_{c,m} + Z|^2 (Z_{c,m} + Z) - \frac{|P_{c,m}|^2 |P_{c,n}|^2}{|Z_{c,m}|^2 |Z_{c,n}|^2} |Z_{c,n} + Z|^2 (Z_{c,m} + Z) \right.$$
$$\left. + \frac{|P_{c,n}|^4}{|Z_{c,n}|^4} |Z_{c,n} + Z|^2 (Z_{c,n} + Z) - \frac{|P_{c,m}|^2 |P_{c,n}|^2}{|Z_{c,m}|^2 |Z_{c,n}|^2} |Z_{c,m} + Z|^2 (Z_{c,n} + Z) \right\} = 0.$$

(5.128)

Expressing

$$Z = x + jy = re^{j\theta} = r(\cos\theta + j\sin\theta) \qquad (5.129)$$

we can get, after considerable algebra, two separate cubic equations as follows:

$$\text{Re}[G(Z)] = A_1 r^3 + B_1 r^2 + C_1 r + D_1 = 0 \qquad (5.130)$$

$$\text{Im}[G(Z)] = A_2 r^3 + B_2 r^2 + C_2 r + D_2 = 0 \qquad (5.131)$$

Dividing the angle of 2π into N equal angular segments,

$$0 \le \theta < 2\pi; \quad \theta_j = 2\pi/N_\theta, \qquad (5.132)$$

equations (5.130) and (5.131) yield $2N_\theta$ equations, with A, B, C and D calculated at each of those N angular locations.

$$\text{Re}[G(Z)] = A_1 r^3 + B_1 r^2 + C_1 r + D_1 = 0 \qquad (5.133)$$

$$\text{Im}[G(Z)] = A_2 r^3 + B_2 r^2 + C_2 r + D_2 = 0 \qquad (5.134)$$

Here A_1, B_1, C_1, D_1, A_2, B_2, C_2, and D_2 are given by the following equations [56]:

$$A_1 = \sum_{m<n} \left(\frac{|p_{c,m}|^2}{|Z_{c,m}|^2} - \frac{|p_{c,n}|^2}{|Z_{c,n}|^2} \right)^2 (\cos^3\theta + \sin^2\theta \cos\theta), \qquad (5.135)$$

$$B_1 = \sum_{m<n} \left\{ \frac{|p_{c,m}|^4}{|Z_{c,m}|^4} (2R_m \cos^2\theta + 2X_m \cos\theta \sin\theta + R_m) \right.$$
$$+ \frac{|p_{c,n}|^4}{|Z_{c,n}|^4} (2R_n \cos^2\theta + 2X_n \cos\theta \sin\theta + R_n),$$
$$\left. - \frac{|p_{c,m}|^2 |p_{c,n}|^2}{|Z_{c,m}|^2 |Z_{c,n}|^2} [2(R_m + R_n) \cos^2\theta + 2(X_m + X_n) \cos\theta \sin\theta + R_m + R_n] \right\}$$

(5.136)

$$C_1 = \sum_{m<n} \left\{ \frac{|p_{c,m}|^4}{|Z_{c,m}|^4} (3R_m^2 \cos\theta + 2R_m X_m \sin\theta + X_m^2 \cos\theta) \right.$$

$$+ \frac{|p_{c,m}|^4}{|Z_{c,m}|^4} (3R_n^2 \cos\theta + 2R_n X_n \sin\theta + X_n^2 \cos\theta) - \frac{|p_{c,m}|^2}{|Z_{c,m}|^2} \frac{|p_{c,n}|^2}{|Z_{c,n}|^2} \left[(R_m^2 + R_n^2) \cos\theta \right.$$

$$\left. + 4R_m R_n \cos\theta + (X_m^2 + X_n^2) \cos\theta + 2(R_m X_n + R_n X_m) \sin\theta \right] \right\}, \quad (5.137)$$

$$D_1 = \sum_{m<n} \left\{ \frac{|p_{c,m}|^4}{|Z_{c,m}|^4} (R_m^3 + R_m X_m^2) + \frac{|p_{c,n}|^4}{|Z_{c,n}|^4} (R_n^3 + R_n X_n^2) \right.$$

$$\left. - \frac{|p_{c,m}|^2}{|Z_{c,m}|^2} \frac{|p_{c,n}|^2}{|Z_{c,n}|^2} (R_m R_n^2 + R_m X_n^2 + R_m^2 R_n + R_n X_m^2) \right\} \quad (5.138)$$

$$A_2 = \sum_{m<n} \left(\frac{|p_{c,m}|^2}{|Z_{c,m}|^2} - \frac{|p_{c,n}|^2}{|Z_{c,n}|^2} \right)^2 (\sin^3\theta + \cos^2\theta \sin\theta), \quad (5.139)$$

$$B_2 = \sum_{m<n} \left\{ \frac{|p_{c,m}|^4}{|Z_{c,m}|^4} (2X_m \sin^2\theta + 2R_m \cos\theta \sin\theta + X_m) \right.$$

$$+ \frac{|p_{c,n}|^4}{|Z_{c,n}|^4} (2X_n \sin^2\theta + 2R_n \cos\theta \sin\theta + X_n) \quad (5.140)$$

$$\left. - \frac{|p_{c,m}|^2}{|Z_{c,m}|^2} \frac{|p_{c,n}|^2}{|Z_{c,n}|^2} \left[2(X_m + X_n) \sin^2\theta + 2(R_m + R_n) \cos\theta \sin\theta + X_m + X_n \right] \right\}$$

$$C_2 = \sum_{m<n} \left\{ \frac{|p_{c,m}|^4}{|Z_{c,m}|^4} (3X_m^2 \sin\theta + 2R_m X_m \cos\theta + R_m^2 \sin\theta) + \frac{|p_{c,n}|^4}{|Z_{c,n}|^4} (3X_n^2 \sin\theta \right.$$

$$+ 2R_n X_n \cos\theta + R_n^2 \sin\theta) - \frac{|p_{c,m}|^2}{|Z_{c,m}|^2} \frac{|p_{c,n}|^2}{|Z_{c,n}|^2} \left[(X_m^2 + X_n^2) \sin\theta + 4X_m X_n \sin\theta \right.$$

$$\left. + (R_m^2 + R_n^2) \sin\theta + 2(R_m X_n + R_n X_m) \cos\theta \right] \right\} \quad (5.141)$$

$$D_2 = \sum_{m<n} \left\{ \frac{|p_{c,m}|^4}{|Z_{c,m}|^4} (X_m^3 + R_m^2 X_m) + \frac{|p_{c,n}|^4}{|Z_{c,n}|^4} (X_n^3 + R_n^2 X_n) \right.$$

$$\left. - \frac{|p_{c,m}|^2}{|Z_{c,m}|^2} \frac{|p_{c,n}|^2}{|Z_{c,n}|^2} (X_m X_n^2 + R_n^2 X_m + X_m^2 X_n + R_m^2 X_n) \right\}. \quad (5.142)$$

Both of these cubics have real co-efficients. Therefore, each will have one real root and one pair of conjugate roots. We need to select the values of r and θ that satisfy Equations 5.130

and 5.131 and minimize the error function $F(Z)$. Then,

$$Z_{c,s} = r_{c,s}e^{i\theta_{c,s}}. \tag{5.143}$$

Finally, the source strength can be obtained easily from Equation 5.123 as follows:

$$|p_{c,s}| = \frac{\sum_m |p_{c,m}| \left|\frac{Z_{c,m}}{Z_{c,m}+Z_{c,s}}\right|}{\sum_m \left|\frac{Z_{c,m}}{Z_{c,m}+Z_{c,s}}\right|^2}. \tag{5.144}$$

This Refined multi-load Method of Jiang and Ih [56], however, is not so robust as Boden's Direct Least-Squares method [55] described below.

Using the classical state variables and Figure 5.19 with Z_L being denoted by Z for convenience, we have

$$p_s(Z/p) = Z_s + Z, \tag{5.145}$$

where, following Boden, Z_s is the normalized source impedance and Z is the normalized acoustic impedance of the rest of the system seen from the source.

Taking the square magnitude of Equation 5.145 and rearranging the terms gives [55]

$$\left(|Z|^2/G_p\right)G_{ps} - 2\operatorname{Re}(Z_1)\operatorname{Re}(Z_s) - 2\operatorname{Im}(Z)\operatorname{Im}(Z_s) - |Z_s|^2 = |Z|^2. \tag{5.146}$$

where $G_{ps} = |p_s|^2$ is the auto-spectrum or autospectral density of the source pressure, p_s.

Use of four different loads with normalized acoustic impedances Z_1, Z_2, Z_3 and Z_4 in Equation 5.146 yields

$$\left(|Z_1|^2/G_{p1}\right)G_{ps} - 2\operatorname{Re}(Z_1)\operatorname{Re}(Z_s) - 2\operatorname{Im}(Z_1)\operatorname{Im}(Z_s) - |Z_s|^2 = |Z_1|^2,$$

$$\left(|Z_2|^2/G_{p2}\right)G_{ps} - 2\operatorname{Re}(Z_2)\operatorname{Re}(Z_s) - 2\operatorname{Im}(Z_2)\operatorname{Im}(Z_s) - |Z_s|^2 = |Z_2|^2,$$

$$\left(|Z_3|^2/G_{p3}\right)G_{ps} - 2\operatorname{Re}(Z_3)\operatorname{Re}(Z_s) - 2\operatorname{Im}(Z_3)\operatorname{Im}(Z_s) - |Z_s|^2 = |Z_3|^2,$$

$$\left(|Z_4|^2/G_{p4}\right)G_{ps} - 2\operatorname{Re}(Z_4)\operatorname{Re}(Z_s) - 2\operatorname{Im}(Z_4)\operatorname{Im}(Z_s) - |Z_s|^2 = |Z_4|^2, \tag{5.147}$$

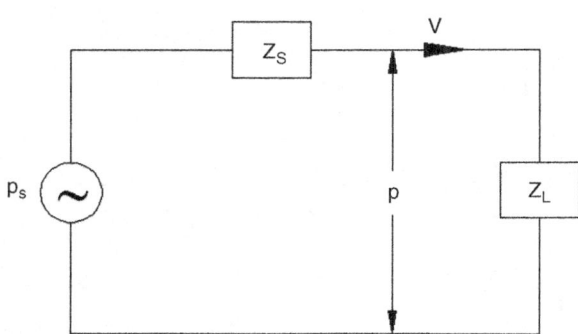

Figure 5.19 Electrical analogous circuit for the two-load method

or, in vector notation,

$$\left(|\mathbf{Z}|^2/\mathbf{G_p}\right)G_{ps} - 2\,\text{Re}(\mathbf{Z})\,\text{Re}(Z_s) - 2\,\text{Im}(\mathbf{Z})\,\text{Im}(Z_s) - \mathbf{I}|Z_s|^2 = |\mathbf{Z}|^2, \qquad (5.148)$$

where the vectors have dimensions 4×1, and $\mathbf{I}$ is a 4×1 vector with all elements equal to one.

It may be noted from Equations 5.147 and 5.148 that we are implicitly interpreting the nonlinear system of equations as a linear system of equations with $|Z_s|^2$ as an independent unknown quantity.

Following Boden [55], Equation 5.148 can be rewritten as

$$\mathbf{f}\left(G_{ps}, |Z_s|\cos(\varphi_s), |Z_s|\sin(\varphi_s)\right) = -\left(|\mathbf{Z}|^2/\mathbf{G_p}\right)G_{ps} + 2\,\text{Re}(\mathbf{Z})|Z_s|\cos(\varphi_s) \\ + 2\,\text{Im}(\mathbf{Z})|Z_s|\sin(\varphi_s) + \mathbf{I}|Z_s|^2 + |\mathbf{Z}|^2 = 0, \qquad (5.149)$$

where use has been made of the fact that $\text{Re}(Z_s) = |Z_s|\cos(\varphi_s)$ and $\text{Im}(Z_s) = |Z_s|\sin(\phi_s)$.

Equation 5.149 is then multiplied by the transpose of $\mathbf{f}$: i.e., $\mathbf{f}^T\mathbf{f}$. This function $F(G_{ps}, |Z_s|, \varphi_s)$ should be minimized with respect to, for example G_{ps}: that is

$$\frac{\delta(F)}{\delta(G_{ps})} = \frac{\delta(\mathbf{f}^T\mathbf{f})}{\delta(G_{ps})} = 2\frac{\delta(\mathbf{f}^T)}{\delta(G_{ps})}\mathbf{f} = -2\left\{-\left(\frac{|\mathbf{Z}|^2}{|\mathbf{P}|^2}\right)^T\mathbf{f}\right\} = -2\left\{-\left(\frac{|\mathbf{Z}|^2}{|\mathbf{P}|^2}\right)^T\left(\frac{|\mathbf{Z}|^2}{|\mathbf{P}|^2}\right)G_{ps}\right.$$

$$+ 2|Z_s|\cos(\varphi_s)\left(\frac{|\mathbf{Z}|^2}{|\mathbf{P}|^2}\right)^T\text{Re}(\mathbf{Z}) + 2|Z_s|\sin(\varphi_s)\left(\frac{|\mathbf{Z}|^2}{|\mathbf{P}|^2}\right)^T\text{Im}(\mathbf{Z}) \qquad (5.150)$$

$$\left. + |Z_s|^2\left(\frac{|\mathbf{Z}|^2}{|\mathbf{P}|^2}\right)^T\mathbf{I} + \left(\frac{|\mathbf{Z}|^2}{|\mathbf{P}|^2}\right)^T|\mathbf{Z}|^2\right\} = 0,$$

which gives

$$G_p = \left(2|Z_s|\cos(\varphi_s)\left(\frac{|\mathbf{Z}|^2}{|\mathbf{P}|^2}\right)^T\text{Re}(\mathbf{Z}) + 2|Z_s|\sin(\varphi_s)\left(\frac{|\mathbf{Z}|^2}{|\mathbf{P}|^2}\right)^T\text{Im}(\mathbf{Z})\right.$$

$$\left. + |Z_s|^2\left(\frac{|\mathbf{Z}|^2}{|\mathbf{P}|^2}\right)^T\mathbf{I} + \left(\frac{|\mathbf{Z}|^2}{|\mathbf{P}|^2}\right)^T|\mathbf{Z}|^2\right\}\Bigg/\left(\frac{|\mathbf{Z}|^2}{|\mathbf{P}|^2}\right)^T\left(\frac{|\mathbf{Z}|^2}{|\mathbf{P}|^2}\right). \qquad (5.151)$$

Inserting Equation 5.151 in Equation 5.149 one can calculate a remainder:

$$\mathbf{g}(|Z_s|, \varphi_s|) = 2|Z_s|\cos(\varphi_s)\left\{\text{Re}(\mathbf{Z}) - \frac{\left(|\mathbf{Z}|^2/|\mathbf{P}|^2\right)^T\text{Re}(\mathbf{Z})\left(|\mathbf{Z}|^2/|\mathbf{P}|^2\right)}{\left(|\mathbf{Z}|^2/|\mathbf{P}|^2\right)^T\left(|\mathbf{Z}|^2/|\mathbf{P}|^2\right)}\right\}$$

$$+ 2|Z_s|\sin(\varphi_s)\left\{\text{Im}(\mathbf{Z}) - \frac{\left(|\mathbf{Z}|^2/|\mathbf{P}|^2\right)^T\text{Im}(\mathbf{Z})\left(|\mathbf{Z}|^2/|\mathbf{P}|^2\right)}{\left(|\mathbf{Z}|^2/|\mathbf{P}|^2\right)^T\left(|\mathbf{Z}|^2/|\mathbf{P}|^2\right)}\right\}$$

$$+ |Z_s|^2\left\{\mathbf{I} - \frac{\left(|\mathbf{Z}|^2/|\mathbf{P}|^2\right)^T\mathbf{I}\left(|\mathbf{Z}|^2/|\mathbf{P}|^2\right)}{\left(|\mathbf{Z}|^2/|\mathbf{P}|^2\right)^T\left(|\mathbf{Z}|^2/|\mathbf{P}|^2\right)}\right\} + |\mathbf{Z}|^2 - \frac{\left(|\mathbf{Z}|^2/|\mathbf{P}|^2\right)^T|\mathbf{Z}^2|\left(|\mathbf{Z}|^2/|\mathbf{P}|^2\right)}{\left(|\mathbf{Z}|^2/|\mathbf{P}|^2\right)^T\left(|\mathbf{Z}|^2/|\mathbf{P}|^2\right)}.$$

$$(5.152)$$

Minimizing $G(|Z_s|, \varphi_s) = \mathbf{g}^T\mathbf{g}$ with respect to $|Z_s|$ gives [55]

$$\frac{\delta(G)}{\delta(|Z_s|)} = \left[(2\cos(\varphi_s)) \left\{ \operatorname{Re}(\mathbf{Z}) - \frac{\left(|\mathbf{Z}|^2/|\mathbf{P}|^2\right)^T \operatorname{Re}(\mathbf{Z})\left(|\mathbf{Z}|^2/|\mathbf{P}|^2\right)}{\left(|\mathbf{Z}|^2/|\mathbf{P}|^2\right)^T \left(|\mathbf{Z}|^2/|\mathbf{P}|^2\right)} \right\}^T \right.$$

$$+ 2\sin(\varphi_s) \left\{ \operatorname{Im}(\mathbf{Z}) - \frac{\left(|\mathbf{Z}|^2/|\mathbf{P}|^2\right)^T \operatorname{Im}(\mathbf{Z})\left(|\mathbf{Z}|^2/|\mathbf{P}|^2\right)}{\left(|\mathbf{Z}|^2/|\mathbf{P}|^2\right)^T \left(|\mathbf{Z}|^2/|\mathbf{P}|^2\right)} \right\}^T$$

$$\left. + 2|Z_s| \left\{ \mathbf{I} - \frac{\left(|\mathbf{Z}|^2/|\mathbf{P}|^2\right)^T \mathbf{I}\left(|\mathbf{Z}|^2/|\mathbf{P}|^2\right)}{\left(|\mathbf{Z}|^2/|\mathbf{P}|^2\right)^T \left(|\mathbf{Z}|^2/|\mathbf{P}|^2\right)} \right\}^T \right] \mathbf{g} = 0. \quad (5.153)$$

This leads to a cubic equation in $|Z_s|$, which can be written as [55]

$$|Z_s|^3 \mathbf{c}^T\mathbf{c} + |Z_s|^2 \{2\cos(\varphi_s)\mathbf{c}^T\mathbf{a} + 2\sin(\varphi_s)\mathbf{c}^T\mathbf{b} + \cos(\varphi_s)\mathbf{a}^T\mathbf{c} + \sin(\varphi_s)\mathbf{b}^T\mathbf{c}\}$$
$$+ |Z_s|\{\{\mathbf{c}^T\mathbf{d} + 2\cos^2(\varphi_s)\mathbf{a}^T\mathbf{a} + 2\sin^2(\varphi_s)\mathbf{b}^T\mathbf{b} + 2\sin(\varphi_s)\cos(\varphi_s)\mathbf{a}^T\mathbf{b} \quad (5.154)$$
$$+ 2\sin(\varphi_s)\cos(\varphi_s)\mathbf{b}^T\mathbf{a}\} + \cos(\varphi_s)\mathbf{a}^T\mathbf{d} + \sin(\varphi_s)\mathbf{b}^T\mathbf{d},$$

where

$$\mathbf{a} = \operatorname{Re}(\mathbf{Z}) - \frac{\left(|\mathbf{Z}|^2/|\mathbf{P}|^2\right)^T \operatorname{Re}(\mathbf{Z})\left(|\mathbf{Z}|^2/|\mathbf{P}|^2\right)}{\left(|\mathbf{Z}|^2/|\mathbf{P}|^2\right)^T \left(|\mathbf{Z}|^2/|\mathbf{P}|^2\right)}, \quad (5.155)$$

$$\mathbf{b} = \operatorname{Im}(\mathbf{Z}) - \frac{\left(|\mathbf{Z}|^2/|\mathbf{P}|^2\right)^T \operatorname{Re}(\mathbf{Z})\left(|\mathbf{Z}|^2/|\mathbf{P}|^2\right)}{\left(|\mathbf{Z}|^2/|\mathbf{P}|^2\right)^T \left(|\mathbf{Z}|^2/|\mathbf{P}|^2\right)}, \quad (5.156)$$

$$\mathbf{c} = \mathbf{I} - \frac{\left(|\mathbf{Z}|^2/|\mathbf{P}|^2\right)^T \mathbf{I}\left(|\mathbf{Z}|^2/|\mathbf{P}|^2\right)}{\left(|\mathbf{Z}|^2/|\mathbf{P}|^2\right)^T \left(|\mathbf{Z}|^2/|\mathbf{P}|^2\right)}, \quad (5.157)$$

$$\mathbf{d} = |\mathbf{Z}|^2 - \frac{\left(|\mathbf{Z}|^2/|\mathbf{P}|^2\right)^T |\mathbf{Z}|^2 \left(|\mathbf{Z}|^2/|\mathbf{P}|^2\right)}{\left(|\mathbf{Z}|^2/|\mathbf{P}|^2\right)^T \left(|\mathbf{Z}|^2/|\mathbf{P}|^2\right)}. \quad (5.158)$$

The cubic in Equation 5.154 can be solved readily on computer using an appropriate subroutine or library function in MATLAB®.

The procedure of calculating the source data using Boden's direct least-squares method may be itemized as follows [55]:

i. Calculate the source impedance $|Z_s|$ from Equation 5.154 for a sufficiently large number of values of the angle ϕ_s between 0 and 2π.
ii. Insert the $|Z_s|$ data in Equation 5.151 to obtain G_{ps} for all the different ϕ_s values.
iii. Insert the data in the function $F(G_{ps}, |Z_s|, \phi_s)$ and identify the value of ϕ_s which minimizes the function, making use of an appropriate routine from the MATLAB® which finds the minimum values of a vector.
iv. The resulting data for G_{ps}, $|Z_s|$ and ϕ_s constitutes the solution of the problem of fitting the original data to Equation 5.148 in the least squares sense.

As one does not manipulate the nonlinear equations in any way in this method (one only tries to fit the experimental data directly to the model), one can justly call Boden's method [55] the 'direct least squares method'. Boden has demonstrated this method successfully on a number of sources, including a fan, and internal combustion engines. He has shown his direct least-squares method to be better (less prone to error) than the conventional four-load method. However, it is not so efficient as the two-load method that makes use of flush-mounted microphones picking up complex values of acoustic pressures within the tube.

5.3.3 Numerical Evaluation of the Engine Source Characteristics

On the face of it, this section title does not appear to belong to this chapter on flow-acoustic measurements. However, while discussing various methods for indirect evaluation of source characteristics, it may not be irrelevant to include a numerical method that makes use of the scheme of calculation given in the foregoing paragraphs for the method of external measurements, with the vital difference that the far-field SPLs for different exhaust pipe lengths are not measured; they are instead calculated by means of the time-domain analysis.

The existence of unique source characteristics is practically not achievable for the engine exhaust source [50]. Nevertheless, nonlinear fluid dynamic equations can be solved by means of the method of characteristics [57–59] or the 3D finite volume method [60]. This method has the advantage of being self-sufficient inasmuch as it does not require prior knowledge of the source characteristics.

Yet another approach is the hybrid approach where one combines the time-domain analysis of the exhaust/intake source with the frequency domain analysis of the muffler downstream. Satyanarayana and Munjal proposed a simple hybrid approach making use of an interrelationship between progressive wave variables of the linear acoustic theory and Riemann variables of the method of characteristics [58]. This approach serves well for the free radiation condition but fails in the case of complex mufflers. This limitation was removed by Hota and Munjal [59] while retaining the associated simplicity of Ref. [58]. They incorporated the reflection of the forward wave at the exhaust valve at each of the harmonics. With this approach they were able to predict SPL for any complex commercial muffler downstream for a single cylinder engine. However, this approach proved to be inadequate to predict the SPL of a multi-cylinder engine, particularly for the case of a turbocharged engine.

Prasad and Crocker, based on their direct measurements of source impedance of a multi-cylinder inline CI engine [48], proposed the anechoic source approximation: $Z_s = Y_0$. Callow and Peat came out with a relatively more realistic expression [61]:

$$Z_s(\text{exhaust}) = Y_0(0.707 - j0.707), \qquad (5.159)$$

where Y_0 is the characteristic impedance of the exhaust pipe, c_o/S. Here, S is the area of cross-section of the exhaust pipe, and c_o is the sound speed of the exhaust gases.

Fairbrother et al. and Boden et al. tried to extract the linear source characteristics data from nonlinear finite-volume CFD simulation [62,63], using the two-load method [18]. Knutsson and Boden then tried to extract the intake source data from the one-dimensional CFD simulation using the commercial software Ricardo-WAVE which uses the finite volume approach to solve the 1D compressible gas dynamics equations for mass, energy and momentum [64]. Their results were corroborated with the measured values quite satisfactorily. Hota and Munjal extended the work of Fairbrother et al. [55,56] to formulate source characteristics of a compression-ignition (CI) engine as functions of the engine's physical and thermodynamic parameters and incorporated them as empirical formulas into the scheme to predict the un-muffled noise using a multi-load method [65]. Again, inspired by the work of Knutsson and Boden [57], the investigation of Ref. [65] was extended to the intake source characterization of C.I. engines by Hota and Munjal [66].

The engine exhaust or intake source can be characterized in accordance with the electrical analogy as can be seen in Figure 5.19 for the un-muffled and muffled system, respectively. Here acoustic pressure p and mass velocity v are analogous to voltage (or electromotive force) and current in the electrical network theory, respectively.

As per the electrical analogous circuits of the un-muffled system depicted in Figure 5.19, for two different acoustic loads (impedances) Z_{L1} and Z_{L2}, one can write [62]:

$$p_s Z_{L1} - p_1 Z_s = p_1 Z_{L1} \quad \text{and} \quad p_s Z_{L2} - p_2 Z_s = p_2 Z_{L2} \qquad (5.160, 5.161)$$

These two equations may be solved simultaneously to obtain:

$$p_s = p_1 p_2 \frac{Z_{L1} - Z_{L2}}{p_2 Z_{L1} - p_1 Z_{L2}} \quad \text{and} \quad Z_s = Z_{L1} Z_{L2} \frac{p_1 - p_2}{p_2 Z_{L1} - p_1 Z_{L2}} \qquad (5.162, 5.163)$$

The following steps are adopted to calculate the source characteristics of the intake system or the exhaust system of the engine:

i. Store two arrays of pressure histories corresponding to two loads, predicted by means of AVL-BOOST [67]. These are computed just upstream of air filter for the intake side and just downstream of the exhaust runners for the exhaust side. These arrays are in the time domain.
ii. Compute the discrete Fourier transform (DFT) of the above arrays to find pressure in the frequency domain for different speed orders corresponding to the firing frequency and its integral multiples. For example, for a single cylinder four-stroke-cycle engine, the speed orders of interest would be: 0.5, 1.0, 1.5, 2.0, . . . , 49.5, 50.

iii. With Z_0 as the radiation impedance, calculate Z_{L1} and Z_{L2} of the pipe as:

$$Z_{Ln} = \frac{Z_0 \cos(kl_n) + jY_0 \sin(kl_n)}{(j/Y_0)Z_0 \sin(kl_n) + \cos(kl_n)}, n = 1, 2 \tag{5.164}$$

where $k = \omega/c_0$ is the wave number and $Y_0 = c_0/S$ is the characteristic impedance of the load pipe of length l_n and area of cross-section S.

iv. Find out p_s and Z_s from Equations 5.162 and 5.163 for the chosen pair of pipe lengths.

v. Repeat steps i to iv above for different pairs of pipe lengths.

vi. Average of all calculated values represents the source characteristics p_s and Z_s. The source pressure p_s has been re-defined in terms of the Source Strength Level, SSL, as:

$$SSL = 20 \log|p_s/p_{th}|, \tag{5.165}$$

where, threshold pressure $p_{th} = 2.0 \times 10^{-5}$ Pa.

It has been observed by Hota and Munjal for the CI engines that if a least square fit is done on the SSL spectrum at different frequencies or speed orders, the curve goes down more or less exponentially [65,66]. Hence the generalized formula for the SSL can be defined as:

$$SSL = A \times \left(\frac{\text{speed order}}{N_{cyl}/2}\right)^B \text{ dB}, \tag{5.166}$$

where N_{cyl} is the number of cylinders in the four-stroke cycle engine, $N_{cyl}/2$ represents the speed order of the firing frequency of a four-stroke cycle engine, and constant A represent SSL at the firing frequency.

Speed order, n, of frequency f_n is defined as

$$n = \frac{f_n}{RPM/60} \tag{5.167}$$

where RPM denotes the shaft speed in revolutions per minute.

The firing frequency of a multi-cylinder engine is given by

$$\text{firing frequency} = \frac{RPM}{60} \times \frac{2}{N_{st}} \times N_{cyl} \tag{5.168}$$

As there is one firing in two revolutions of a four-stroke ($N_{st} = 4$) cycle engine, the speed order of the firing frequency of a four-stroke cycle engine becomes $N_{cyl}/2$.

This kind of least square fit has been done to discount sharp peaks and troughs because computations have been made by assuming that speed of the engine remains absolutely constant. But in reality there may be around 1 to 5 percent variation in speed because the pressure *vs* crank angle diagrams of successive cycles would never be identical.

A parametric study has been conducted for the following parameters, varying one at a time, and keeping other parameters constant at their default (underlined) values.

Turbocharged diesel engines

Air fuel ratio, AFR = 18.0, **23.7**, 29.2, 38.0
Engine speed in RPM = 1000, 1300, 1600, 2100, 2400, 3000, 3500, **4000**, 4500
Engine capacity (displacement), V (in liters) = 1.0, 1.5, 2.0, **2.5**, 3.0, 4.0

Number of cylinders, $N_{cyl} = 1, 2, 3, \underline{\mathbf{4}}, 6$
So the default turbocharged engine is: 4 cylinders, 2.5 liters, running at 4000 rpm, with the air-fuel ratio 23.7.

Naturally aspirated diesel engines

Air fuel ratio, AFR = 14.5, **17.0**, 29.0, 39.6
Values of RPM, V and N_{cyl} are the same as for the turbocharged engine above.
The resultant values of constants A and B of Equation 5.166 are as follows [65,66,68].

Exhaust System of CI Engine [65]

Turbocharged diesel engines:

$$A = 173.4 \times (1 - 0.0019\, AFR)(1 + 0.12\, NS - 0.016\, NS^2)(1 - 0.0023\, V)(1 - 0.021\, N_{cyl}) \tag{5.169}$$

$$B = -0.093 \times (1 + 0.016\, AFR)(1 + 0.31\, NS - 7.62\, NS^2)(1 - 0.03\, V)(1 + 0.026\, N_{cyl}) \tag{5.170}$$

Naturally aspirated diesel engines:

$$A = 167 \times (1 - 0.0015\, AFR)(1 + 0.125\, NS - 0.002\, NS^2)(1 + 0.0018\, V) \\ \times (1 - 0.0233\, N_{cyl}) \tag{5.171}$$

$$B = -0.13 \times (1 - 0.123\, AFR)(1 + 0.19\, NS - 0.053\, NS^2)(1 - 0.007\, V)(1 - 0.026\, N_{cyl}) \tag{5.172}$$

where $NS = $ *engine speed in RPM*$/1000$.

Intake System of CI Engine [66]

Turbocharged diesel engines:

$$A = 214 \times (1 + 0.0018\, AFR)(1 - 0.08\, NS + 0.01\, NS^2)(1 - 0.0021\, V)(1 - 0.05\, N_{cyl}) \tag{5.173}$$

$$B = -0.318 \times (1 - 0.0033\, AFR)(1 - 0.039\, NS)(1 - 0.173\, V)(1 + 0.022\, N_{cyl}) \tag{5.174}$$

Naturally aspirated diesel engines:

$$A = 198 \times (1 + 0.00075\, AFR)(1 - 0.1\, NS + 0.016\, NS^2)(1 - 0.001\, V)(1 - 0.028\, N_{cyl}) \tag{5.175}$$

$$B = -0.15 \times (1 + 0.0012\, AFR)(1 + 0.005\, NS)(1 - 0.0064\, V)(1 + 0.109\, N_{cyl}) \tag{5.176}$$

where $NS = $ *engine speed in RPM*$/1000$.

Naturally aspirated SI engines [68]:

An acoustic parametric study has been conducted for the following basic parameters, varying one at a time, keeping other parameters constant at their default (underlined) values for both intake and exhaust.

Air fuel ratio, AFR = 12.0, **15.0**, 20.0, 24.0, 28.0

Engine speed in RPM = 1000, 1500, 2000, 2500, 3000, 3500, 4000, 4500, 5000, 5500, 6000, 6500, 7000

Engine swept volume or displacement volume, V (in liters) = 0.5, 1.0, 1.5, **2.0**, 2.5, 3.0,

Number of cylinders, N_{cyl} = 1, 2, 3, **4**, 6

Here, default parameters of the naturally aspirated SI engine are: 4 cylinders, 2.0 liters, 5500 rpm, and the air-fuel ratio of 15. The resultant values of A and B of SSL are as follows:

Exhaust System:

$$A = 198.7 \times (1 - 0.0015\,AFR)(1 + 0.0445\,NS - 0.00765\,NS^2)(1 + 0.0021\,V)$$
$$\times (1 - 0.0374\,N_{cyl}) \tag{5.177}$$

$$B = -0.53 \times (1 + 0.008\,AFR)(1 + 0.246\,NS - 0.0314\,NS^2)(1 - 0.028\,V)(1 - 0.287\,N_{cyl}) \tag{5.178}$$

Intake System:

$$A = 176.36 \times (1 - 0.00022\,AFR)(1 + 0.0517\,NS - 0.00866\,NS^2)(1 - 0.00184\,V)$$
$$\times (1 - 0.03336\,N_{cyl}) \tag{5.179}$$

$$B = -0.106 \times (1 + 0.0021\,AFR)(1 + 0.256\,NS - 0.032\,NS^2)(1 + 0.0075\,V)(1 + 0.122\,N_{cyl}) \tag{5.180}$$

where $NS = engine\ speed\ in\ RPM/1000$.

The resultant source characteristics are used with the transfer matrix based muffler program [69] to predict the intake as well as exhaust sound pressure level of a diesel or gasoline engine. Thus, the designer would be able to make a rough estimate of the exhaust sound pressure level, and thereby synthesize the required muffler configuration for an engine under development, and the rather cumbersome measurements of source characteristics would be avoided.

5.3.4 A Comparison of the Various Methods for Measuring Source Characteristics

The method of direct measurements is suitable for only one of the two source characteristics, that is, source impedance Z_s (and hence $Z_{c,s}$). It fails at lower frequencies where the engine noise is comparable to the acoustic signal generated by the external source (electropneumatic driver). Unfortunately, these are the frequencies that interest muffler designers the most. Besides, there are errors resulting from flow-acoustic unsteadiness, but these are not larger than those appearing in the indirect methods that will now be compared.

With the impedance-tube two-load method, the measurements are made inside the tube and, as such, the laboratory does not have to be acoustically anechoic. But the setup requires precision in fabrication. Measurement of complex pressure requires a reference source as well as a real time analyzer. Turbulent and hot meanflow create additional challenges.

The impedance-tube two-load method, described earlier in this chapter, accounts for meanflow convection only; the temperature gradients are ignored. Although Munjal and Prasad [70] have derived the four-pole parameters for a tube with mean flow and a linear temperature gradient, there is still no easy way of incorporating temperature gradient in the impedance-tube relations.

The method of external measurements of noise involves SPL measurements in free atmosphere. This is definitely easier than the ones involving in-pipe measurements. However, it requires reflection-free environments, although the ground reflections can be taken into account if the ground is known to be hard enough.

One possible source of error in the finite volume analysis and the resultant Equation 5.166 is that it presumes the SSL and thence SPL to peak at the firing frequency, whereas the mixing in some manifolds results in the exhaust/intake noise to peak at a multiple of the firing frequency.

The finite volume analysis, or for that matter, method of characteristics, is, however, free from all experimental difficulties and errors, as it does not call for any measurements except the valve timings and valve time history, which are generally known from the design data of the engine. But this method involves many computations, which are both time-consuming and costly.

One particular weakness in the indirect (two-load or four-load) methods involving in-duct measurements is that considerable errors will occur whenever a pressure node exists in the vicinity of the measuring microphone. Here again, the finite volume analysis method would be better because it does not depend on any measurements.

References

1. Davies, A.H. and Evans, E.J. (1930) Measurements of absorbing power of materials by the standing wave method. *Proceedings of the Royal Society of London*, **127**(A), 89–110.
2. White, W.M. (1939) An acoustic transmission line for impedance tube measurements. *Journal of the Acoustical Society of America*, **11**, 140–146.
3. Beranek, L.L. (1940) Precision measurement of acoustic impedance. *Journal of the Acoustical Society of America*, **12**, 3–13.
4. Scott, R.A. (1946) An apparatus for accurate measurement of the acoustic impedance of sound absorbing materials. *Proceedings of the Physical Society of London*, **58**, 253–264.
5. Lippert, W.K.R. (1953) The practical representation of standing waves in acoustic impedance tube. *Acustica*, **3**, 153–160.
6. Ando, Y. (1969) An extrapolation of measuring the reflection coefficient of an acoustic tube. *Applied Acoustics*, **2**, 95–99.
7. ASTM 384-58 (1972) *Impedance and Absorption of Acoustical Material by the Tube Method – Reapproved*, American Society for Testing Materials, Philadelphia, PA.
8. Melling, T.H. (1973) An impedance tube for precise measurement of acoustic impedance and insertion loss at high speed pressure levels. *Journal of Sound and Vibration*, **28**, 23–54.
9. Yaniv, S.L. (1973) Impedance tube measurements of propagation constant and characteristic impedance of porous acoustical materials. *Journal of the Acoustical Society of America*, **54**, 1138–1142.
10. Kathuriya, M.L. and Munjal, M.L. (1975) Accurate method for the experimental evaluation of the acoustical impedance of black box. *Journal of the Acoustical Society of America*, **58**, 451–454.

11. Alfredson, R.J. and Davies, P.O.A.L. (1979) The radiation of sound from an engine exhaust. *Journal of Sound and Vibration*, **13**(4), 389–408.
12. Ahrems, C. and Romneberger, D. (1971) Acoustic attenuation in rigid and rough tubes with turbulent air flow. *Acustica*, **25**, 150–157.
13. Ingard, U. and Singhal, V.K. (1974) Sound attenuation in turbulent pipe flow. *Journal of the Acoustical Society of America*, **55**, 535–538.
14. Ingard, U. and Singhal, V.K. (1975) Effect of flow on the acoustical resonance of an open ended duct. *Journal of the Acoustical Society of America*, **55**, 778–793.
15. Kathuriya, M.L. and Munjal, M.L. (1976) A method for the experimental evaluation of the acoustic characteristics of an engine exhaust system in the presence of mean flow. *Journal of the Acoustical Society of America*, **60**(3), 745–751.
16. Kathuriya, M.L. and Munjal, M.L. (1977) Measurement of he acoustical impedance of a black box at low frequencies using a shorter impedance tube. *Journal of the Acoustical Society of America*, **62**(3), 751–754.
17. Kathuriya, M.L. and Munjal, M.L. (1977) A method for evaluation of the acoustical impedance of a black box with or without mean flow. *Journal of the Acoustical Society of America*, **62**, 755–759.
18. Kathuriya, M.L. and Munjal, M.L. (1979) Experimental evaluation of the aeroacoustic characteristics of a source of pulsating gas flow. *Journal of the Acoustical Society of America*, **65**(1), 240–248.
19. Davies, P.O.A.L., Bhattacharya, M. and Bento Coelho, J.L. (1980) Measurement of plane wave acoustic field in flow ducts. *Journal of Sound and Vibration*, **72**, 539–542.
20. Panicker, V.B. and Munjal, M.L. (1981) Impedance tube technology for flow acoustics. *Journal of Sound and Vibration*, **77**(4), 573–577.
21. Panicker, V.B. and Munjal, M.L. (1981) Acoustic dissipation in a uniform tube with moving medium. *Journal of the Acoustical Society of India*, **IX**(3), 95–101.
22. Panicker, V.B. and Munjal, M.L. (1982) Radiation impedance of an unflanged pipe with mean flow. *Noise Control Engineering Journal*, **18**(2), 48–51.
23. Seybert, A.F. and Ross, D.F. (1977) Experimental determination of acoustic properties using a two-microphone random-excitation technique. *Journal of the Acoustical Society of America*, **61**(5), 1362–1370.
24. Prasad, M.G. and Crocker, M.J. (1983) Studies on acoustical modeling of a multi-cylinder engine exhaust system. *Journal of Sound and Vibration*, **90**(4), 491–508.
25. Bendat, J.S. and Piersol, A.G. (1971) *Random Data: Analysis and Measurement Procedures*, Wiley-Interscience, New York.
26. Schmidt, W.E. and Johnson, J.P. (1975) Measurement of acoustic reflection from obstruction in a pipe with flow, NSF Rep. PD-20.
27. Chung, J.Y. and Blaser, D.A. (1980) Tranfer function method of measuring in-duct acoustic properties I: Theory, II experiment. *Journal of the Acoustical Society of America*, **68**(3), 907–913, 914–921.
28. Seybert, A.F. (1988) Two sensor methods for the measurement of sound intensity and acoustic properties in ducts. *Journal of the Acoustical Society of America*, **83**(6), 2233–2239.
29. To, C.W.S. and Doige, A.G. (1980) The application of a transient technique to the determination of acoustic properties of unknown systems. *Journal of Sound and Vibration*, **71**, 545–554.
30. Chu, W.T. (1986) Extension of the two-microphone transfer function method for impedance tube measurements. *Journal of the Acoustical Society of America*, **80**, 347–348.
31. Boden, M. and Abom, M. (1986) Influence of errors on the two-microphone method for measuring acoustic properties in a duct. *Journal of the Acoustical Society of America*, **79**, 541–549.
32. Seybert, A.F. and Soenarko, B. (1981) Errors analysis of spectral estimates with application to the measurement of acoustic parameters using random sound fields in ducts. *Journal of the Acoustical Society of America*, **69**, 1190–1198.
33. Chung, J.Y. and Blaser, D.A. (1980) Transfer function method of measuring acoustic intensity in a duct system with flow. *Journal of the Acoustical Society of America*, **68**, 1570–1577.
34. Munjal, M.L. and Doige, A.G. (1990) The two-microphone method incorporating the effects of mean flow and acoustic damping. *Journal of Sound and Vibration*, **137**(1), 135–138.
35. Chung, J.Y. (1978) Cross-spectral method of measuring acoustic intensity without error caused by instrument phase mismatch. *Journal of the Acoustical Society of America*, **64**, 1613–1616.
36. Chung, J.Y. (1977) The rejection of flow noise using a coherent function method. *Journal of the Acoustical Society of America*, **62**, 388–395.

37. ASTM Standard Designation: E 1050-98 (1998) Standard test method for impedance and absorption materials using a tube, two microphones, and a digital frequency analysis system.
38. Narayana, T.S.S. and Munjal, M.L. (2005) Comparison of different calibration techniques for estimation of transfer function using the two-microphone method. *Journal of Acoustical Society of India*, **33**, 464–469.
39. To, C.W.S. and Doige, A.G. (1979) A transient testing technique for the determination of matrix parameters of acoustic systems I: Theory and principles, II: Experimental procedure and results. *Journal of Sound and Vibration*, **62**, 207–233.
40. Katz, Brian F.G. (2000) Method to resolve sample location errors in the two microphone duct measurement method. *Journal of the Acoustical Society of America*, **80**, 2231–2237.
41. Easwaran, V. and Munjal, M.L. (1995) A note on the effect of wall compliance on lowest-order mode propagation in fluid-filled/submerged impedance tube. *Journal of the Acoustical Society of America*, **97**(6), 3494–3501.
42. Munjal, M.L. and Thawani, P.T. (1996) Acoustic performance of hoses – a parametric study. *Noise Control Engineering Journal*, **44**(6), 274–280.
43. Lung, T.Y. and Doige, A.G. (1983) A time-averaging transient testing method for acoustic properties of piping systems and mufflers with flow. *Noise Control Engineering Journal*, **73**(3), 867–876.
44. Munjal, M.L. and Doige, A.G. (1990) Theory of a two source-location method for direct experimental evaluation of the four-pole parameters of an aeroacoustic element. *Journal of Sound and Vibration*, **141**(2), 323–333.
45. Boden, H. and Abom, M. (1986) Influence of errors in the two-microphone method for measuring acoustic properties in ducts. *Journal of the Acoustical Society of America*, **79**, 541–549.
46. Abom, M. and Boden, H. (1988) Error analysis of two-microphone measurements in ducts with flow. *Journal of the Acoustical Society of America*, **83**, 2429–2438.
47. Narayana, T.S.S. and Munjal, M.L. (2005) Prediction and measurement of the four-pole parameters of a muffler including higher-order mode effects. *Noise Control Engineering Journal*, **53**(6), 270–276.
48. Prasad, M.G. and Crocker, M.J. (1983) On the measurement of the internal source impedance of a multi-cylinder engine exhaust system. *Journal of Sound and Vibration*, **90**(4), 479–490.
49. Sneckenberger, J.E. (1984) Experimental source impedance study of a single cylinder engine, Nelson Acoustic Conference, Madison, WI.
50. Munjal, M.L. and Doige, A.G. (1988) On uniqueness, transfer and combination of acoustic sources in one-dimensional systems. *Journal of Sound and Vibration*, **121**(1), 25–35.
51. Munjal, M.L. and Doige, A.G. (1990) On the relation between convective source characteristics and their acoustic counterparts. *Journal of Sound and Vibration*, **136**(2), 343–346.
52. Desmons, L. and Hardy, J. (1994) A least squares method for evaluation of characteristics of acoustic sources. *Journal of Sound and Vibration*, **175**, 365–376.
53. Prasad, M.G. (1987) A four-load method for evaluation of acoustical source impedance in a duct. *Journal of Sound and Vibration*, **114**, 347–356.
54. Sridhara, B.S. and Crocker, M.J. (1992) Error analysis for the four-load method used to measure the source impedance in ducts. *Journal of the Acoustical Society of America*, **92**, 2924–2931.
55. Boden, H. (1995) On multi-load methods for determination of the source data of acoustic one-port sources. *Journal of Sound and Vibration*, **180**, 725–743.
56. Jang, S.-H. and Ih, J.-G. (2000) Refined multiload method for measuring acoustical source characteristics of an intake or exhaust system. *Journal of the Acoustical Society of America*, **107**(6), 3217–3225.
57. Gupta, V.H. and Munjal, M.L. (1992) On numerical prediction of the acoustical source characteristics of an engine exhaust system. *Journal of the Acoustical Society of America*, **92**(5), 2716–2725.
58. Satyanarayana, Y. and Munjal, M.L. (2000) A hybrid approach for aero-acoustic analysis of the engine exhaust system. *Applied Acoustics*, **60**, 425–450.
59. Hota, R.N. and Munjal, M.L. (2004) A new hybrid approach for thermo-acoustic modeling of engine exhaust system. *International Journal of Acoustics and Vibration*, **9**(3), 129–138.
60. Harten, A., Engquist, B., Osher, S. and Chakravarthy, S.R. (1987) Uniformly high order accurate essentially non-oscillatory schemes, III. *Journal of Computational Physics*, **71**(2), 231–303.
61. Callow, G.D. and Peat, K.S. (1988) Insertion loss of engine inflow and exhaust silencers, *I Mech. E C19/88*, 39–46.
62. Fairbrother, R., Boden, H. and Glav, R. (2005) Linear acoustic exhaust system simulation using source aata from linear simulation, *SAE Technical Paper series*, 2005-01-2358.

63. Boden, H., Tonse, M. and Fairbrother, R. (2004) On extraction of IC-engine acoustic source data from non-linear simulations, Proceedings of the Eleventh International Congress on Sound and Vibration (ICSVII), St. Petersburg, Russia.
64. Knutsson, M. and Boden, H. (2007) IC-Engine intake source data from non-linear Simulation, *SAE Technical Paper series*, 2007-01-2209.
65. Hota, R.N. and Munjal, M.L. (2008) Approximate empirical expressions for the aeroacoustic source strength level of the exhaust system of compression ignition engines. *International Journal of Aeroacoustics*, **7**(34), 349–371.
66. Hota, R.N. and Munjal, M.L. (2008) Intake source characterization of a compression ignition engine: empirical expressions. *Noise Control Engineering Journal*, **56**(2), 92–106.
67. BOOST Version 5.0.2 (2007) AVL LIST GmbH, Graz, Austria.
68. Munjal, M.L. and Hota, R.N. (June 2010) Acoustic source characteristics of the exhaust and intake systems of a spark ignition engine, *Internoise* 2010, Lisbon.
69. Munjal, M.L., Panigrahi, S.N. and Hota, R.N. (2007) FRITAmuff: A comprehensive platform for prediction of un-muffled and muffled exhaust noise of IC engines, *ICSV14*, Cairns, Australia.
70. Munjal, M.L. and Prasad, M.G. (1986) On plane wave propagation in a uniform pipe in the presence of mean flow and a temperature gradient. *Journal of Acoustical Society of America*, **80**(5), 1501–1506.

6

Dissipative Ducts and Parallel Baffle Mufflers

Mufflers used in air-conditioning and ventilation ducts, industrial fans, ventilation and access openings of acoustic enclosures, intake and exhaust ducts of power stations, cooling tower installations, gas turbines, and jet-engines test cells, do not act only by muffling the sources through successive reflections of sound by means of impedance mismatching, but also by dissipating the incident sound energy as heat. These mufflers, are in fact, primarily dissipative or absorptive mufflers with the advantage of providing definite attenuation of sound over a wide range of frequencies. Unfortunately, their performance is poor at lower frequencies, the actual limit being governed by the cross dimensions of the duct, lining thickness, sound-absorptive properties of the lining, and so forth.

Dissipative mufflers were used earlier for automotive exhaust silencing as well, but eventually they had to be replaced by reflective mufflers owing to the problem of

i. the unburnt carbon particles tending to close the pores of sound-absorbing materials lining the walls of the mufflers,
ii. the high-velocity unsteady flow of exhaust gases blowing out the fibers of the absorptive lining,
iii. thermal cracking of the linings,
iv. poor attenuation at low frequencies (of the order of the firing frequency) where most of the exhaust noise is concentrated, and
v. relatively higher costs.

Nevertheless, absorptive mufflers (often called silencers to differentiate them from reflective mufflers) do find application where the flow is steady, not so hot, and clean, or else where there is no flow, as in the slits of windows and doors. Besides, of late, long strand fibers have been developed that are deployed within a very thin unperforated plastic layer without any binder. On exposure to heat these materials expand to fill up the available annular space. Often the binder material (resin) melts at higher temperatures (of the order of 250°C) and limits the use of absorptive materials at high temperatures that are typical of the exhaust systems of the reciprocating internal combustion (IC) engines and gas turbines. The present-day long-strand fibers that are deployed in exhaust mufflers without any binder have no such limitations. Therefore, combination mufflers are being introduced extensively in the present-day automotive exhaust systems.

It is, of course, very rare that an exhaust system or intake system consists of a lined duct alone. Invariably there is a chamber or receiver on the upstream side and a length of the unlined duct (having more cross-sectional area than the clear flow area of the lined portion of the duct) on either side of the lined portion of the duct, creating reflections of sound at the inlet and outlet. Thus, aeroacoustic analysis of an air duct system should also be done according to the theory outlined in Chapters 2–4. Of course, one would require the transfer matrix of an acoustically lined duct in terms of its attenuation constant.

Therefore, in the next section, we revisit the analysis of Sections 1.7.1 and 1.7.2 in order to incorporate the convective effect of incompressible moving medium.

Here, we are interested in air ducts where air, carrying noise (from a fan or compressor or a rotary engine), moves at small velocities (Mach number of the order of 0.1 or less). The duct may be circular or rectangular in cross section. It may be lined on all the sides, or on two of the four sides (in the case of rectangular ducts), or may contain parallel baffles made up of acoustically absorptive material covered with thin, perforated metallic or plastic sheets for protection against flow, or sometimes an impermeable, very thin layer of a synthetic material like Mylar. This protective layer plays a nontrivial role in acoustics of the lined ducts and parallel baffle mufflers. This is attempted in the following section. Subsequent sections deal with evaluation of attenuation constant from the physical properties of the porous material, design of the lining (or baffles) for maximum attenuation of a progressive plane wave, effect of mean flow on attenuation, and static pressure drop across an acoustically absorptive duct without or with baffles.

6.1 Acoustically Lined Rectangular Duct with Moving Medium

Mean flow affects the propagation of waves not only by convection but also by altering the wall impedance of an acoustically lined duct. The latter effect is determined only by extensive experimentation and its discussion is deferred. Here it should suffice to write wall impedances in the x and y directions as

$$Z_{w,x} = Z_{w,x}(M) \quad \text{and} \quad Z_{w,y} = Z_{w,y}(M), \tag{6.1}$$

respectively, where M is the average Mach number of the grazing mean flow.

The acoustic pressure field in the acoustically lined duct of Figure 6.1 would be governed by the convected wave Equation 1.91. One may expect the general solutions to be

$$p(x,y,z,t) = \left(C_1 e^{-jk_z^+ z} + C_2 e^{+jk_z^- z}\right)\left(e^{-jk_x x} + C_3 e^{+jk_x x}\right) e^{j\omega t} \times \left(e^{-jk_y y} + C_4 e^{+jk_y y}\right), \tag{6.2}$$

which retains the x and y components of the solution for stationary medium (Equation 1.25), as there is no mean flow in the transverse direction. Wave numbers $k_z^\pm$ would then be given by relation (1.93) or (1.94) in its general form

$$k_z^\pm = \frac{\pm M k_0 + \left[k_0^2 - \left(1 - M^2\right)\left(k_x^2 + k_y^2\right)\right]^{1/2}}{1 - M^2}. \tag{6.3}$$

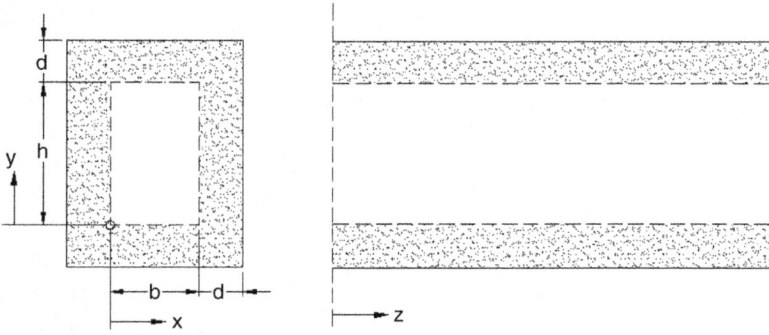

Figure 6.1 Schematic views of an acoustically lined rectangular duct with clear dimensions b and h (cf. Figure 1.1)

The particle velocity components u_x, u_y and u_z can be obtained by writing each of them in the form of Equation 6.2 with four different constants, substituting the same in the three components of the momentum Equation 1.90, namely,

$$\rho_0 \frac{Du_x}{Dt} + \frac{\partial p}{\partial x} = 0, \qquad (6.4a)$$

$$\rho_0 \frac{Du_y}{Dt} + \frac{\partial p}{\partial y} = 0, \qquad (6.4b)$$

$$\rho_0 \frac{Du_z}{Dt} + \frac{\partial p}{\partial z} = 0, \qquad (6.4c)$$

and equating the coefficients of the different exponentials separately to zero. Thus one would get

$$u_x(x,y,z,t) = \frac{1}{\rho_0 c_0} \frac{k_x}{k_0} \left(\frac{C_1}{1 - Mk_z^+/k_0} e^{-k_z^+ z} + \frac{C_2}{1 + Mk_z^-/k_0} e^{+k_z^- z} \right) \\ \times \left(e^{-jk_x x} - C_3 e^{+jk_x x} \right) \left(e^{-jk_y y} + C_4 e^{+jk_y y} \right) e^{j\omega t}, \qquad (6.5)$$

and similar expressions for u_y and u_z.

The boundary condition at the duct wall is based on the assumption that at the duct wall the fluid particle displacement and the wall particle displacement are the same. Let it be noted by η. Now, the wall impedance is related to the radial velocity within the lining; that is,

$$p/Z_{w,x} = \partial \eta / \partial t, \qquad (6.6)$$

where the radial velocity in the propagating medium is given by

$$u_x = D\eta/Dt. \qquad (6.7)$$

Eliminating the displacement η from these equations, the boundary conditions at the walls become

$$\frac{Dp(0,y,z,t)/Dt}{-\partial u_x(0,y,z,t)/\partial t} = \frac{Dp(b,y,z,t)/Dt}{\partial u_x(b,y,z,t)/\partial t} = Z_{w,x}(M), \qquad (6.8)$$

$$\frac{Dp(x,0,z,t)/Dt}{-\partial u_y(x,0,z,t)/\partial t} = \frac{Dp(x,h,z,t)/Dt}{\partial u_y(x,h,z,t)/\partial t} = Z_{w,y}(M), \qquad (6.9)$$

[cf. Equations 1.125 and 1.126]. Now, substituting Equations 6.2 and 6.5 in Equation 6.8 brings into play a coupling between k_x and k_z. Thus, in the case of a moving medium, there would be a k_x^+ corresponding to k_z^+ and k_x^- corresponding to k_z^-. That being the case, Equations 6.2 and 6.5 are incorrect; k_x^+ and k_z^+ have to be determined from the forward-moving parts of Equations 6.2 and 6.5, and k_x^- and k_z^- from the reflected parts thereof. Thus, substituting

$$p^+(x,y,z,t) = C_1 \left(e^{-jk_x^+ x} + C_3^+ e^{+jk_x^+ x} \right) \left(e^{-jk_y^+ y} + C_4^+ e^{+jk_y^+ y} \right) e^{-jk_z^+ z} e^{j\omega t} \qquad (6.10)$$

and

$$u_x^+(x,y,z,t) = \frac{1}{\rho_0 c_0} \frac{k_x^+}{k_0(1 - Mk_z^+/k_0)} C_1 \left(e^{-jk_x^+ x} - C_3^+ e^{+jk_x^+ x} \right) \\ \times \left(e^{-jk_y^+ x} + C_4^+ e^{+jk_y^+ y} \right) e^{-jk_z^+ z} e^{j\omega t} \qquad (6.11)$$

in the first of Equation 6.8 yields

$$-Z_{w,x}(M) = \rho_0 c_0 \frac{k_0}{k_x^+} \left(1 - Mk_z^+/k_0 \right)^2 \frac{1 + C_3^+}{1 - C_3^+}, \qquad (6.12)$$

and in the latter of Equation 6.8 yields

$$Z_{w,x}(M) = \rho_0 c_0 \frac{k_0}{k_x^+} \left(1 - Mk_z^+/k_0 \right)^2 \frac{e^{-jk_x^+ b} + C_3^+ e^{+jk_x^+ b}}{e^{-jk_x^+ b} - C_3^+ e^{+jk_x^+ b}}. \qquad (6.13)$$

Equation 6.12 can be rearranged in the form

$$C_3^+ = \frac{F_x^+ + G^+}{F_x^+ - G^+}, \qquad (6.14)$$

where

$$F_x^+ = \frac{Z_{w,x}(M) k_x^+}{\rho_0 c_0 k_0} \qquad (6.15)$$

and

$$G^+ = \left\{1 - \frac{Mk_z^+}{k_0}\right\}^2 \tag{6.16}$$

Substituting for C_3^+ from Equation 6.14 in Equation 6.13 and rearranging gives a quadratic in F_x^+, which in turn yields.

$$F_x^+ = \frac{j(\cos k_x b \pm 1)}{\sin k_x b} G^+. \tag{6.17}$$

These two equations can be rewritten in the expanded form

$$\frac{Z_{w,x}(M)}{\rho_0 c_0} \frac{k_x^+}{k_0} = j \cot\left(\frac{k_x^+ b}{2}\right)\left(1 - \frac{Mk_z^+}{k_0}\right)^2, \tag{6.18a}$$

$$\frac{Z_{w,x}(M)}{\rho_0 c_0} \frac{k_x^+}{k_0} = -j \tan\left(\frac{k_x^+ b}{2}\right)\left(1 - \frac{Mk_z^+}{k_0}\right)^2. \tag{6.18b}$$

The coupling between k_x^+ and k_z^+ is obvious from the transcendental Equations 6.18 and 6.3. Identical relations would hold for k_y^+ and C_4^+, with $Z_{w,x}$ and b being replaced by $Z_{w,y}$ and h, respectively; that is,

$$\frac{Z_{w,y}(M)}{\rho_0 c_0} \frac{k_y^+}{k_0} = j \cot\left(\frac{k_y^+ h}{2}\right)\left(1 - \frac{Mk_z^+}{k_0}\right)^2, \tag{6.19a}$$

$$\frac{Z_{w,y}(M)}{\rho_0 c_0} \frac{k_y^+}{k_0} = -j \tan\left(\frac{k_y^+ h}{2}\right)\left(1 - \frac{Mk_z^+}{k_0}\right)^2. \tag{6.19b}$$

For the (m,n) order mode, $k_{x,m}^+, k_{y,n}^+$ and $k_{z,m,n}^+$ have to be gotten from simultaneous solutions of Equations 6.18, 6.19 and 6.3. This can be done on a digital computer by means of the Newton-Raphson iteration scheme, making use of the no-flow values of the three variables (gotten from the procedure described in Section 1.7.1) for the start of the iteration. One can also evaluate $k_{x,m}^+, k_{y,n}^+$ and $k_{z,m,n}^+$ by iterating between Equations 6.3, 6.18a and 6.19a. First, Equations 6.18a and 6.19a are solved for $M = 0$. The resulting values of $k_{x,m}^+$ and $k_{y,n}^+$ are substituted in Equation 6.3 to evaluate $k_{z,m,n}^+$. This is now used again in Equations 6.18a and 6.19a, which are solved for the new values of $k_{x,m}^+$ and $k_{y,n}^+$. These are now substituted in Equation 6.3 for the new value of $k_{z,m,n}^+$. This is repeated until all the three variables are evaluated to the required accuracy.

After one has evaluated $k_{x,m}^+, k_{y,n}^+$ and $k_{z,m,n}^+$, $C_{3,m}^+$ is evaluated from Equations 6.14–6.16:

$$C_{3,m}^+ = \frac{Z_{w,x}(M)k_{x,m}^+/\rho_0 c_0 k_0 + \left(1 - Mk_{z,m,n}^+/k_0\right)^2}{Z_{w,x}(M)k_{x,m}^+/\rho_0 c_0 k_0 - \left(1 - Mk_{z,m,n}^+/k_0\right)^2}. \tag{6.20}$$

Similarly,

$$C_{4,n}^+ = \frac{Z_{w,y}(M)k_{y,n}^+/\rho_0 c_0 k_0 + \left(1 - Mk_{z,m,n}^+/k_0\right)^2}{Z_{w,y}(M)k_{y,n}^+/\rho_0 c_0 k_0 - \left(1 - Mk_{z,m,n}^+/k_0\right)^2}. \qquad (6.21)$$

For the reflected wave, the foregoing procedure, outlined in Equations 6.8–6.21, can be repeated, replacing $C_{1,m,n}$ by $C_{2,m,n}$, M by $-M$, and superscript $+$ by $-$. In this way, $k_{x,m}^-, k_{y,n}^-$ and $k_{z,m,n}^-$, $C_{3,m}^-$ and $C_{4,n}^-$ can be determined.

Finally, for standing waves, the preceding solutions can be combined to obtain the general solution

$$p(x,y,z,t) = \sum_{m=0}^{\infty}\sum_{n=0}^{\infty}\left[C_{1,m,n}\left(e^{-jk_{x,m}^+ x} + C_{3,m}^+ e^{+jk_{x,m}^+ x}\right)\left(e^{-jk_{y,n}^+ y} + C_{4,n}^+ e^{+jk_{y,n}^+ y}\right)e^{-jk_{z,m,n}^+ z}\right.$$
$$\left. + C_{2,m,n}\left(e^{-jk_{x,m}^- x} + C_{3,m}^- e^{+jk_{x,m}^- x}\right)\left(e^{-jk_{y,n}^- y} + C_{4,n}^- e^{+jk_{y,n}^- y}\right)e^{+jk_{z,m,n}^- z}\right]e^{j\omega t}. \qquad (6.22)$$

For convenience, it can be written in the form

$$p(x,y,z,t) = \sum_n \sum_m \left[C_{1,m,n} F_m^+(x) F_n^+(y) e^{-jk_{z,m,n}^+ z} + C_{2,m,n} F_m^-(x) F_n^-(y) e^{+jk_{z,m,n}^- z}\right] e^{j\omega t}, \qquad (6.23)$$

where

$$F_m^\pm(x) = e^{-jk_{x,m}^\pm x} + C_{3,m}^\pm e^{+jk_{x,m}^\pm x} \qquad (6.24a)$$

and

$$F_n^\pm(y) = e^{-jk_{y,n}^\pm y} + C_{4,n}^\pm e^{+jk_{y,n}^\pm y}. \qquad (6.24b)$$

In this notation, the axial particle velocity is given by

$$u(x,y,z,t) = \sum_n \sum_m \frac{1}{\rho_0 c_0}\left[\frac{k_{z,m,n}^+}{k_0 - Mk_{z,m,n}^+} C_{1,m,n} F_m^+(x) F_n^+(y) e^{-jk_{z,m,n}^+ z}\right.$$
$$\left. - \frac{k_{z,m,n}^-}{k_0 + Mk_{z,m,n}^-} C_{2,m,n} F_m^-(x) F_n^-(y) e^{+jk_{z,m,n}^- z}\right]e^{j\omega t}, \qquad (6.25)$$

The progressive wave of a given frequency (or a particular mode) would propagate unattenuated if $k_z^\pm$ are real (not complex or imaginary). This would be so if the term under the radical sign in Equation 6.3 is greater than or equal to zero; that is,

$$k_0^2 \geq \left(1 - M^2\right)\left(k_{x,m}^2 + k_{y,n}^2\right). \qquad (6.26)$$

Here values of $k_{x,m}$ and $k_{y,n}$ are selected for the particular direction ($+$ or $-$). Thus, the plane wave [i.e. the (0, 0) mode] of any frequency can always propagate unattenuated in a rigid unlined duct, because $k_x = k_y = 0$ for the (0, 0), mode.

For a duct with lined walls, even the lowest k_x and k_y, (i.e. $k_{x,1}$ and $k_{y,1}$) would be greater than zero. Therefore, all modes (including the lowest) would be attenuated at frequencies given by

$$k_0 = \frac{2\pi f}{c_0} < \left\{ (1 - M^2) \left(k_{x,m}^2 + k_{y,n}^2 \right) \right\}^{1/2}. \tag{6.27}$$

Thus, the cut-off frequencies for a moving medium are lower than $[(1 - M^2)^{1/2}$ times] those for a stationary medium.

It is worth noting from Equation 6.3 that at frequencies given by the inequality (6.27), a given mode would exponentially decay without oscillation along the axis if the medium is stationary, and with oscillation if the medium is moving. In other words, the convective effect of mean flow ensures propagation of every mode, albeit with decreasing amplitude.

6.2 Acoustically Lined Circular Duct with Moving Medium

As mentioned before, mean flow alters the wall impedance and convects the waves downstream. The acoustic pressure field in the circular flow duct would be governed by the convected wave equation

$$\left[\frac{D^2}{Dt^2} - c_0^2 \left(\frac{\partial^2}{\partial r^2} + \frac{1}{r} \frac{\partial}{\partial r} + \frac{1}{r^2} \frac{\partial^2}{\partial \theta^2} + \frac{\partial^2}{\partial z^2} \right) \right] p = 0. \tag{6.28}$$

The general solution of this equation for compliant walls (see Figure 6.2) can be constructed from solution (1.143), building into it the convective effect of mean flow and keeping in mind the coupling between k_r and k_z because of mean-flow convection. Thus,

$$p(r, \theta, z, t) = \sum_{m=0}^{\infty} \sum_{n=0}^{\infty} \left\{ C_{1,m,n} J_m \left(k_{r,m,n}^+ r \right) e^{-jk_{z,m,n}^+ z} + C_{2,m,n} J_m \left(k_{r,m,n}^- r \right) e^{+jk_{z,m,n}^- z} \right\} e^{jm\theta} e^{j\omega t}, \tag{6.29}$$

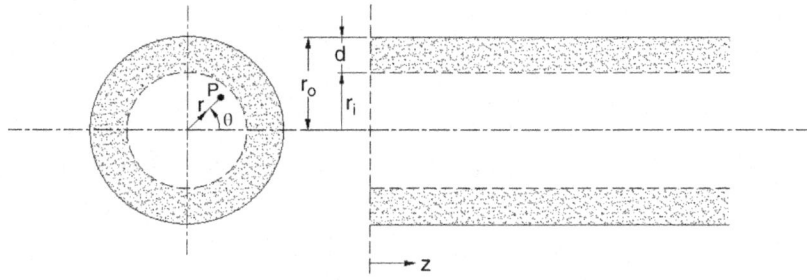

Figure 6.2 Schematic views of an acoustically lined circular duct with clear radius r_i (cf. Figure 1.2)

where $k_{r,m,n}$, $n = 0, 1, 2, \ldots$, are obtained by applying the wall-boundary condition [cf. Equations 6.8 and 6.9 for rectangular ducts]

$$\frac{Dp/Dt}{\partial u_r/\partial t} = Z_w(M) \text{ at } r = r_0, \tag{6.30}$$

where u_r is given by the momentum equation in the radial direction

$$\rho_0 \frac{Du_r}{Dt} + \frac{\partial p}{\partial r} = 0 \tag{6.31}$$

or

$$\rho_0 \left(\frac{\partial u_r}{\partial t} + U \frac{\partial u_r}{\partial z} \right) + \frac{\partial p}{\partial r} = 0. \tag{6.32}$$

Clearly, u_r would be of the type

$$u_r(r, \theta, z, t) = \sum_{m=0}^{\infty} \sum_{n=0}^{\infty} \left\{ C_{3,m,n} J'_m \left(k^+_{r,m,n} r \right) e^{-jk^+_{z,m,n} z} + C_{4,m,n} J'_m \left(k^-_{r,m,n} r \right) e^{+jk^-_{z,m,n} z} \right\} e^{jm\theta} e^{j\omega t}. \tag{6.33}$$

Substituting expressions (6.29) and (6.33) in Equation 6.32 and equating the coefficients of $e^{-jk^+_{z,m,n} z}$ and $e^{+jk^-_{z,m,n} z}$ separately to zero yields

$$C_{3,m,n} = \frac{j}{\rho_0 c_0} \frac{k^+_{r,m,n}}{k_0} \frac{C_{1,m,n}}{1 - Mk^+_{z,m,n}/k_0} \tag{6.34}$$

$$C_{4,m,n} = \frac{-j}{\rho_0 c_0} \frac{k^-_{r,m,n}}{k_0} \frac{C_{2,m,n}}{1 + Mk^-_{z,m,n}/k_0}. \tag{6.35}$$

As illustrated earlier for rectangular ducts, a relation between $k^+_{z,m,n}$ and $k^+_{r,m,n}$ can be got from the wall-boundary condition (6.30) by substituting in it the forward wave components of p and u_r. This gives

$$\frac{J_m \left(k^+_{r,m,n} r_0 \right)}{\left(k^+_{r,m,n} r_0 \right) J'_m \left(k^+_{r,m,n} r_0 \right)} \left(1 - \frac{Mk^+_{z,m,n}}{k_0} \right) = j \frac{Z_w(M)}{\rho_0 c_0} \frac{1}{k_0 r_0}. \tag{6.36}$$

For the (m, n) order $k^+_{r,m,n}$ and $k^+_{z,m,n}$ can be evaluated by simultaneous solution of Equation 6.36 and equation

$$k^{\pm}_{z,m,n} = \frac{\mp M k_0 + \left[k_0^2 - (1 - M^2) k^{\pm 2}_{r,m,n} \right]^{1/2}}{1 - M^2}. \tag{6.37}$$

This can be done on a digital computer by means of the Newton–Raphson iteration scheme, making use of the no-flow values of the two variables for the start of the iteration.

Another way of evaluating $k_{r,m,n}^+$ and $k_{z,m,n}^+$ is by the successive iteration method. One starts with solving Equation 6.36 for $k_{r,m,n}^+$ for $M = 0$, substitutes it in Equation 6.37 to find $k_{z,m,n}^+$, substitutes it in Equation 6.36 to find a new value of $k_{r,m,n}^+$, then evaluates the new value of $k_{z,m,n}^+$ from Equation 6.37, and so on.

For the reflected wave, the foregoing procedure is repeated by working with reflected wave components of p and u_r.

Finally, for standing waves, these two solutions can be combined to obtain the general solution

$$p(r,\theta,z,t) = \sum_{m=0}^{\infty}\sum_{n=0}^{\infty} \left\{ C_{1,m,n} J_m\left(k_{r,m,n}^+ r\right) e^{-jk_{z,m,n}^+ z} + C_{2,m,n} J_m\left(k_{r,m,n}^- r\right) e^{+jk_{z,m,n}^- z} \right\} e^{jm\theta} e^{j\omega t}, \tag{6.38}$$

and

$$u_z(r,\theta,z,t) = \sum_{m=0}^{\infty}\sum_{n=0}^{\infty} \frac{1}{\rho_0 c_0} \left[\frac{k_{z,m,n}^+}{k_0 - M k_{z,m,n}^+} C_{1,m,n} J_m\left(k_{r,m,n}^+ r\right) e^{-jk_{z,m,n}^+ z} \right.$$
$$\left. - \frac{k_{z,m,n}^-}{k_0 + M k_{z,m,n}^-} C_{2,m,n} J_m\left(k_{r,m,n}^- r\right) e^{+jk_{z,m,n}^- z} \right] e^{jm\theta} e^{j\omega t}. \tag{6.39}$$

6.3 Transfer Matrix Relation for a Dissipative Duct

The final results of Sections 6.1 and 6.2 are repeated here for propagation (with the attenuation of course) of the lowest mode or the least attenuated mode, in a form required for derivation of the desired transfer matrix relation in the axial (z) direction, absorbing the dependence on the transverse coordinates and time in the two constants A and B. The least naturally attenuated mode corresponds to the first (smallest) root of Equation 6.18 for rectangular ducts and (6.36) with $m=0$ (indicating axial symmetry) for circular ducts. Adopting the meaning of m and n as indicated in Figures 1.3 and 1.4 of Chapter 1 (i.e. the number of pressure nodal lines) yields

$$m = 0 \quad \text{and} \quad n = 0 \tag{6.40}$$

for rectangular as well as circular ducts.

These values for m and n are understood in the following analysis; therefore, the subscripts m and n are dropped henceforth. Thus, Equations 6.23 and 6.38 yield

$$p(z) = A e^{-jk_z^+ z} + B e^{+jk_z^- z} \tag{6.41}$$

where $k_z^\pm$ are given by

$$k_z^\pm = \frac{\mp M k_0 + \left[k_0^2 - (1 - M^2)\{(k_x^\pm)^2 + (k_y^\pm)^2\} \right]^{1/2}}{1 - M^2} \tag{6.42a}$$

and

$$k_z^\pm = \frac{\mp M k_0 + \left[k_0^2 - (1-M^2)(k_r^\pm)^2\right]^{1/2}}{1-M^2}, \tag{6.42b}$$

for rectangular and cylindrical ducts, respectively. $k_x^\pm$, $k_y^\pm$ and $k_r^\pm$ are the first (or the smallest) roots of Equations 6.18a, 6.19a and 6.36, that is,

$$\frac{Z_{w,x}}{\rho_0 c_0} \frac{k_x^\pm}{k_0} = j \cot\left(\frac{k_x^\pm b}{2}\right)\left(1 \mp \frac{M k_z^\pm}{k_0}\right)^2, \tag{6.43a}$$

$$\frac{Z_{w,y}}{\rho_0 c_0} \frac{k_y^\pm}{k_0} = j \cot\left(\frac{k_y^\pm h}{2}\right)\left(1 \mp \frac{M k_z^\pm}{k_0}\right)^2, \tag{6.43b}$$

and

$$\frac{Z_w}{\rho_0 c_0} \frac{1}{k_0 r_o} = -j \frac{J_0(k_r^\pm r_o)}{(k_r^\pm r_o) J_0'(k_r^\pm r_o)}\left(1 \mp \frac{M k_z^\pm}{k_0}\right)^2, \tag{6.43c}$$

respectively. Coupled Equation 6.42 and 6.43 have to be solved simultaneously, as indicated above, by means of one of the iteration methods.

The corresponding expression for axial particle velocity is obtained from Equation 6.25 and 6.39, that is,

$$u_z(z) = \frac{1}{\rho_o c_o}\left\{\frac{k_z^+}{k_0 - M k_z^+} A e^{-j k_z^+ z} - \frac{k_z^-}{k_0 + M k_z^-} B e^{+j k_z^- z}\right\} \tag{6.44}$$

for rectangular as well as circular ducts.

Dependence of both the acoustic pressure p and particle velocity u_z on the transverse coordinates is the same for all z. Thus, integrating p and u_z over the cross section would yield p and u_z multiplied by the same factor. This common factor would be identical for the forward wave as well as the reflected wave, and therefore can be absorbed in A and B in Equation 6.41 and 6.44. Thus Equation 6.44 would yield the following expression for axial mass velocity $v(z)$:

$$v(z) = \frac{A}{Y^+} e^{-j k_z^+ z} - \frac{B}{Y^-} e^{+j k_z^- z} \tag{6.45}$$

where

$$Y^\pm = Y_0 \frac{k_0 \mp M k_z^\pm}{k_z^\pm} \tag{6.46}$$

and

$$Y_0 = \frac{c_0}{S}. \tag{6.47}$$

If Equations 6.45 and 6.41 are combined, the transfer matrix for an acoustically absorptive duct for the lowest or least attenuated mode can be obtained as follows.

$$p(0) = A + B, \tag{6.48}$$

$$v_z(0) = \frac{A}{Y^+} - \frac{B}{Y^-}, \tag{6.49}$$

whence

$$A = \frac{p(0)/Y^- + v_z(0)}{1/Y^- + 1/Y^+} = \frac{Y^+\{p(0) + Y^- v_z(0)\}}{Y^+ + Y^-} \tag{6.50}$$

and

$$B = \frac{p(0)/Y^+ - v_z(0)}{1/Y^- + 1/Y^+} = \frac{Y^-\{p(0) - Y^+ v_z(0)\}}{Y^+ + Y^-}. \tag{6.51}$$

Now,

$$p(l) = Ae^{-jk_z^+ l} + Be^{+jk_z^- l}$$

$$= [1/(Y^+ + Y^-)]\left[Y^+ e^{-jk_z^+ l}\{p(0) + Y^- v_z(0)\} + Y^- e^{+jk_z^- l}\{p(0) - Y^+ v_z(0)\}\right] \tag{6.52}$$

$$= [1/(Y^+ + Y^-)]\left[\{Y^+ e^{-jk_z^+ l} + Y^- e^{+jk_z^- l}\}p(0) + Y^+ Y^- (e^{-jk_z^+ l} - e^{+jk_z^- l})v_z(0)\right]. \tag{6.53}$$

Similarly,

$$v(l) = [1/(Y^+ + Y^-)]\left[\{e^{-jk_z^+ l} - e^{+jk_z^- l}\}p(0) + \{Y^- e^{-jk_z^+ l} + Y^+ e^{+jk_z^- l}\}v_z(0)\right]. \tag{6.54}$$

Equations 6.53 and 6.54 can be arranged in the matrix form

$$\begin{bmatrix} p(l) \\ v_z(l) \end{bmatrix} = \begin{bmatrix} \dfrac{Y^+ e^{-jk_z^+ l} + Y^- e^{+jk_z^- l}}{Y^+ + Y^-} & \dfrac{Y^+ Y^- (e^{-jk_z^+ l} - e^{+jk_z^- l})}{Y^+ + Y^-} \\ \dfrac{e^{-jk_z^+ l} - e^{+jk_z^- l}}{Y^+ + Y^-} & \dfrac{Y^- e^{-jk_z^+ l} + Y^+ e^{+jk_z^- l}}{Y^+ + Y^-} \end{bmatrix} \begin{bmatrix} p(0) \\ v_z(0) \end{bmatrix}, \tag{6.55}$$

which can be inverted to yield the desired transfer matrix relation

$$\begin{bmatrix} p(0) \\ v_z(0) \end{bmatrix} = \frac{e^{j(k_z^+ - k_z^-)l}}{Y^+ + Y^-} \begin{bmatrix} Y^- e^{-jk_z^+ l} + Y^+ e^{+jk_z^- l} & Y^+ Y^- (e^{+jk_z^- l} - e^{-jk_z^+ l}) \\ e^{+jk_z^- l} - e^{-jk_z^+ l} & Y^+ e^{-jk_z^+ l} + Y^- e^{+jk_z^- l} \end{bmatrix} \begin{bmatrix} p(l) \\ v(l) \end{bmatrix}. \tag{6.56}$$

It can easily be verified that for the case of rigid-walled (unlined) duct with mean flow,

$$k_z^{\pm} = \frac{k_0}{1 \pm M}, \quad Y^{\pm} = Y_0, \tag{6.57}$$

and then Equation 6.56 reduces to Equation 3.70 (for $\alpha = \zeta = 0$). Incidentally, Equation 3.70 came into that relatively simpler form because of Equation 3.62, where Y^+ and Y^- were assumed to be equal as an approximation. This kind of approximation is permissible for a rigid-walled pipe where wall friction effect is small. However, in the case of lined duct, $k_z^{\pm}$ are substantially different from k_0 and Equation 6.46 cannot be reduced to $Y^+ \cong Y^-$ even as an approximation.

With the transfer matrix relation (6.56) (which is in a general form), an acoustically lined duct can be integrated into the rest of the exhaust system for aeroacoustic analysis with $Y^{\pm}$ being calculated from Equation 6.46 and $k_z^{\pm}$ from Equation 6.42. However $k_z^{\pm}$ depend on $k_{x,o,o}$ and $k_{y,o,o}$ for a rectangular duct and $k_{r,o,o}$ for a circular duct, which represent the first (or the smallest) roots of the transcendental Equations 6.43a and 6.43b, or 6.43c, which involve normal impedance of lined wall $Z_{w,x}$ and $Z_{w,y}$ or Z_w.

6.4 Transverse Wave Numbers for a Stationary Medium

Transverse wave number $k_{x,0}$ and $k_{y,0}$ for a stationary medium (which is used extensively for design purposes) are first roots of Equations 6.18a and 6.19a for $M = 0$. Earlier, the real and imaginary parts of the first root of such a transcendental equation used to be obtained from nomograms. However, a reasonable and convenient approximation (there are many such approximations in vogue!) can be obtained by writing [1]

$$\tan x \simeq 0.1875x - \frac{1.0047}{x + \pi/2} - \frac{1.0047}{x - \pi/2}, \tag{6.58}$$

which, when used for $M = 0$ in Equation 6.18a, yields the quadratic

$$\left(\frac{k_x b}{2}\right)^2 \simeq \frac{2.47 + Q \pm \sqrt{(2.47 + Q)^2 - 1.87Q}}{0.38}, \tag{6.59}$$

where

$$Q = jk_0 \frac{b}{2} \frac{\rho_0 c_0}{Z_{w,x}}. \tag{6.60}$$

Equation 6.59 gives two complex values for k_x. Of particular importance is the one that gives lower attenuation, that is, lower imaginary part of $k_{z,0}$ when $k_{x,0}$ is substituted in Equation 6.43a.

Replacing b by h and $Z_{w,x}$ and $Z_{w,y}$, Equations 6.59 and 6.60 also yield $k_{y,0}$, the desired root of Equation 6.43b.

If only two opposite sides of a rectangular duct are lined (say, the ones normal to the x axis), then the wave number in the other directions (k_y) would be zero because $Z_{w,y}$ tends to infinity. In this case, negative sign is appropriate in Equation 6.59.

For $M = 0$ and $m = 0$, on making use of Equation A1.7, Equation 6.43c becomes

$$\frac{(k_r r_0) J_1(k_r r_0)}{J_0(k_r r_0)} = j \frac{(\rho_0 c_0) k_0 r_0}{Z_w}. \quad (6.61)$$

This equation can also be solved by means of nomograms. However, approximately [1],

$$(k_{r,0} r_0)^2 \simeq \frac{96 + 36jQ \pm \sqrt{9216 + 2304jQ - 912Q^2}}{12 + jQ}, \quad (6.62)$$

where

$$Q = (k_0 r_0) \frac{\rho_0 c_0}{Z_w}. \quad (6.63)$$

Here, again, the appropriate sign before the radical corresponds to the value of $k_{r,0,0}$ that yields lesser attenuation, that is, lower imaginary part of $k_{z,0,0}$ when $k_{r,0,0}$ is substituted in Equation 6.42b with $M = 0$. Now, k_z can be written as

$$k_z^{\pm} = \beta^{\pm} - j\alpha^{\pm}, \quad (6.64)$$

where $\beta^{\pm}$ are propagation constants in the two directions and $\alpha^{\pm}$ are the corresponding attenuation constants. The difference between α^+ and α^- (and, for that matter, β^+ and β^-) is the result of the convective effect of mean flow in the positive direction.

All the effort in the design of dissipative ducts is directed toward maximizing the part of $\alpha^{\pm}$ that is independent of Mach number M, as one can do little about the convective effect, which, fortunately is not very significant anyway for $M^2 \ll 1$.

Let this M-independent part of $\alpha^{\pm}$ be denoted by α_0. Then it obvious from Equation 6.42 that

$$\alpha_0 = -\text{Im}\left\{k_0^2 - \left(k_{x,0}^2 + k_{y,0}^2\right)\right\}^{1/2} \quad (6.65)$$

for rectangular ducts and

$$\alpha_0 = -\text{Im}\left\{k_0^2 - k_{r,0,0}^2\right\}^{1/2} \quad (6.66)$$

for circular ducts. α_0 is nothing but the attenuation constant for stationary medium, for the determination of which much theoretical as well as experimental work has been done and reported in the published literature (see e.g. [2–6]).

6.5 Normal Impedance of the Lining

For the idealized case of plane wave incident on a locally reacting lining of uniform thickness d backed by a rigid wall, the impedance encountered by the plane wave, Z_w, is given by Equation 2.26 that is,

$$Z_w = -jY_w \cot(k_w d), \quad (6.67)$$

where Y_w and k_w are the complex characteristic impedance and wave number of the absorptive lining.

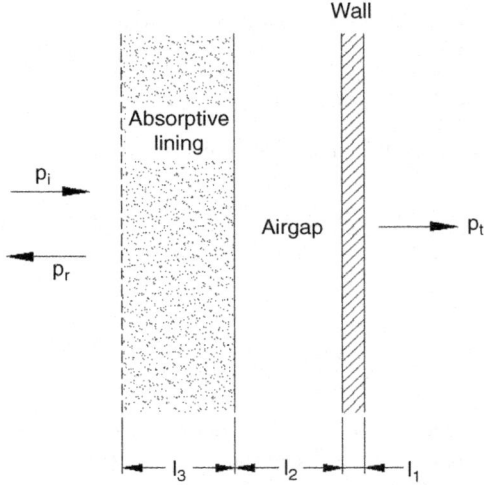

Figure 6.3 Normal impedance of an acoustically lined wall

Equation 6.67 is used very widely, but not wisely, inasmuch as it assumes that the metallic wall of the duct has very high (infinite) impedance (as compared to Y_w), and that there is no airgap between the acoustic lining and the metallic wall. These assumptions are often not true. In the general case shown in Figure 6.3, one could evaluate Z_w by means of the transfer matrix approach as follows:

$$\begin{bmatrix} p_3 \\ v_3 \end{bmatrix} = \begin{bmatrix} \cos k_3 l_3 & jY_3 \sin k_3 l_3 \\ j/Y_3 \sin k_3 l_3 & \cos k_3 l_3 \end{bmatrix} \begin{bmatrix} \cos k_2 l_2 & jY_2 \sin k_2 l_2 \\ j/Y_2 \sin k_2 l_2 & \cos k_2 l_2 \end{bmatrix}$$
$$\times \begin{bmatrix} \cos k_1 l_1 & jY_1 \sin k_1 l_1 \\ j/Y_1 \sin k_1 l_1 & \cos k_1 l_1 \end{bmatrix} \begin{bmatrix} p_0 \\ v_0 \end{bmatrix}, \qquad (6.68)$$

where

k_3, Y_3, and l_3 are the complex wave number, characteristic impedance, and thickness of the acoustic lining,

k_2, Y_2, and l_2 are the wave number, characteristic impedance, and thickness of the airgap (if any) between the lining and the metallic duct wall, and

k_1, Y_1, and l_1 are the wave number, characteristic impedance, and thickness of the metallic wall.

In particular,

$$k_2 = k_0 = \omega/c_0, \quad Y_2 = Y_0 = \rho_0 c_0, \qquad (6.69)$$

$$k_1 = \omega(\rho_1/E_1)^{1/2}, \quad Y_1 = (\rho_1 E_1)^{1/2}, \qquad (6.70)$$

$$E_1 = \frac{E(1-v)}{(1+v)(1-2v)} \qquad (6.71)$$

where ρ and E are the density and the elastic modulus, respectively, and ν is the Poisson's ratio. Writing the matrix Equation 6.68 in the form

$$\begin{bmatrix} p_3 \\ v_3 \end{bmatrix} = \begin{bmatrix} A_{11} & A_{12} \\ A_{21} & A_{22} \end{bmatrix} \begin{bmatrix} Y_0 v_0 \\ v_0 \end{bmatrix}, \tag{6.72}$$

we get

$$Z_w \equiv \frac{p_3}{v_3} = \frac{A_{11} Y_0 + A_{12}}{A_{21} Y_0 + A_{22}}. \tag{6.73}$$

This expression (Equation 6.73) should replace Equation 6.67.

For fiber-based porous sound-absorbing materials, which are used most often in mufflers, k_3 and Y_3, are given by [7]

$$\frac{k_w}{k_0} = (\chi)^{1/2} \left\{ 1 - j \frac{E}{\rho_0 \omega \chi} \right\}^{1/2}, \tag{6.74}$$

$$\frac{Y_w}{Y_0} = \frac{1}{\sigma} \frac{k_w}{k_0}, \tag{6.75}$$

where

ρ_0 = density of the gas (at ambient temperature),
c_0 = speed of sound in the gas (at ambient temperature),
E = flow resistance of the unit thickness of porous bulk material (at ambient temperature), called flow resistivity.
σ = porosity of the bulk material (usually $0.9 < \sigma < 1$),
χ = structural-factor, which is in the range of 1 to 3 in most porous materials,

$$k_0 = \omega/c_0 \quad \text{and} \quad Y_0 = \rho_0 c_0.$$

According to Ver [7], Y_w and k_w are more accurately described by the empirical formulae of Delany and Bazley [5] as modified and improved by Mechel [6]. These are

$$\frac{Y_w}{\rho_0 c_0} = \begin{bmatrix} 1 + 0.0485(A)^{0.754} - j0.087(A)^{0.73} & \text{for } A < 60, \\ \dfrac{0.5A/\pi + j1.4}{\{-1.466 + j0.212A\}^{1/2}} & \text{for } A > 60, \end{bmatrix} \tag{6.76}$$

$$\frac{k_w}{k_0} = \begin{bmatrix} -j0.189(A)^{0.6185} + 1 + 0.0978(A)^{0.6929} & \text{for } A < 60, \\ \{1.466 - j0.212A\}^{1/2} & \text{for } A > 60, \end{bmatrix} \tag{6.77}$$

where A is the normalized flow resistance of a λ-deep layer; that is

$$A = \frac{E\lambda}{\rho_0 c_0}. \tag{6.78}$$

Here λ is the wavelength of the air (at ambient temperature). In order to have some idea of the numbers involved, the real and imaginary parts of the characteristic impedance Y_w are plotted in Figure 6.4 as functions of the loss parameter A.

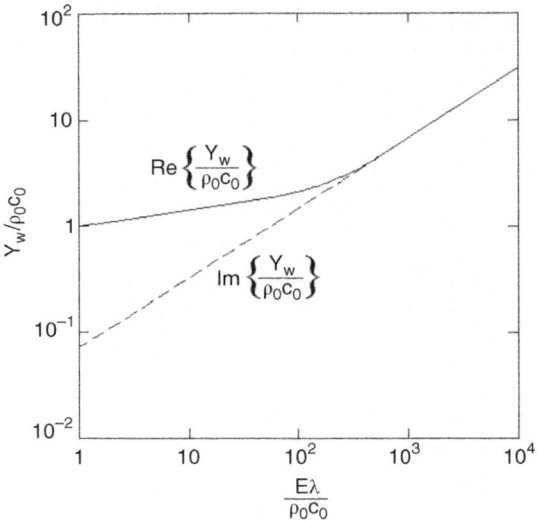

Figure 6.4 Normalized characteristic impedance of sound-absorbing materials as a function of loss parameter. ___, Real component; -----, Imaginary component (Reproduced with permission from [7])

For consolidated granular materials like porous rigid tiles, the complex wave number k_w is given by [2]

$$\frac{k_w}{k_0} = (\chi\sigma)^{1/2}\left\{1 - j\frac{E}{\rho_0\chi\omega}\right\}^{1/2} \tag{6.79}$$

[compare Equation 6.74], and Y_w may then be calculated from Equation 6.75. A more accurate modeling of such materials has been provided by Attenborough [3], according to which

$$\frac{k_w}{k_0} = q\left[\frac{1 + (\gamma - 1)T(C)}{1 - T(B)}\right]^{1/2} \tag{6.80}$$

$$\frac{Y_w}{Y_0} = \frac{q^2}{\sigma}\frac{1}{1 - T(B)}\frac{k_0}{k_w}, \tag{6.81}$$

where

q is a tortuosity factor,
γ is the ratio of specific heats for the gaseous medium,
$T(x) = 2J_1(x)/\{xJ_0(x)\}$, $x = B$ or C,
$B = \lambda_p(1-j)^{1/2}$,
$C_2 = N_{pr}^{1/2}B$,
$\lambda_p^2 = 8\rho_0 q^2 S\omega/(n^2\sigma E)$,
S is the steady flow shape factor,
n is the dynamic shaper factor $= 2 - S$, and
N_{pr} is the Prandtl number.

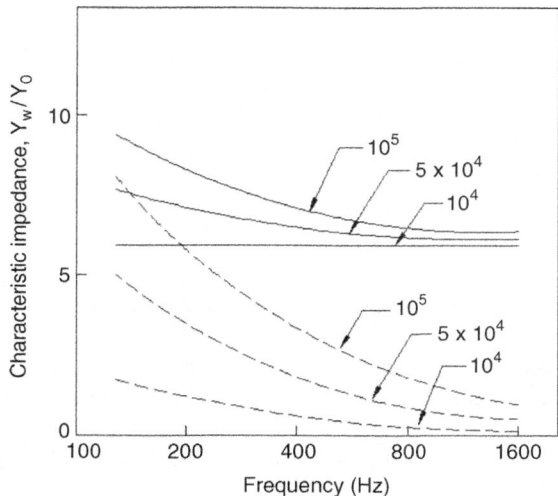

Figure 6.5 Normalized characteristic impedance of granular ceramic tiles with specific flow resistance E (in N-s/m^4) as a parameter. ___, Real component; -----, Imaginary component (Reproduced with permission from [3])

Figure 6.5 shows the plot of Equation 6.81 for porous ceramic absorber tiles with tortuosity factor $q = 2.0$, shape factor $S = 0.9$, porosity $\sigma = 0.37$, and with the flow resistivity E as a parameter [4].

Thus, from the knowledge of flow resistivity and porosity (generally provided by the manufacturer of the porous material or measured in the laboratory), one can evaluate Y_w and k_w from Equations 6.76 and 6.77, respectively, and, working backward Z_w from Equation 6.73, Q from Equation 6.60 or 6.63, $k_{x,0}$ (and $k_{y,0}$) or $k_{r,0,0}$ from Equations 6.59 or 6.62, $Y^{\pm}$ from Equation 6.46, $k_{z,0,0}$ from Equations 6.42a and 6.42b, and finally the transfer matrix from Equation 6.56, to be integrated into the complete aeroacoustic analysis of the duct system according to the theory given in Chapter 3.

However, sometimes the duct system consists of a single acoustically absorptive duct, and one is simply interested in the transmission loss of the duct. This can be calculated in a relatively much simpler way as follows.

6.6 Transmission Loss

Figure 6.6 shows an isolated acoustically absorptive duct of length l followed by nonreflective termination required for TL. For an unlined, rigid-walled, uniform area duct, $TL = 0$. On comparing Figure 6.6 with the unlined duct, it can be noticed that TL of a lined duct would be a function of reflections at the sudden area changes at the entry as well as at the exit, and acoustic dissipation in the lined portion of length l. But it can be proved readily that the total TL is not an algebraic sum of the three components as popularly believed; that is, TL is not equal to the sum of TL_{ent}, Tl_l, and TL_{ex},

where

TL_{ent} is TL due to the area change at the entrance,
Tl_l is TL due to the absorptive section of length l, and
TL_{ex} is TL due to the area change at the exit of the absorptive section.

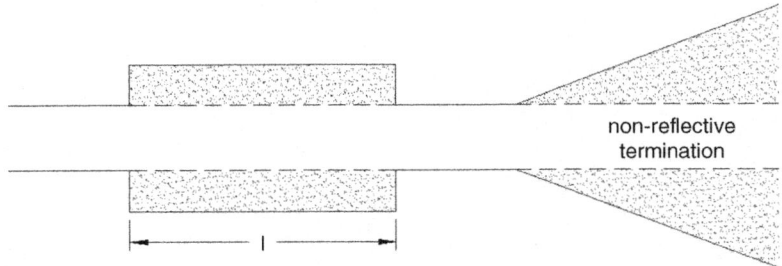

Figure 6.6 Measurement of the transmission loss of an acoustically absorptive duct

In any case, for plane waves, which are of primary concern in this monograph, TL_{ent} and TL_{ex} are quite small and can therefore be neglected. Thus, for design purposes

$$TL \simeq TL_l. \tag{6.82}$$

TL_l, the transmission loss across the lined portion of length l would, according to the definition of transmission loss, be equal to the attenuation of a forward moving progressive wave in decibels, that is,

$$TL_l = 20 \log \left| \frac{p(0)}{p(l)} \right| \simeq 20 \log \left(e^{\alpha_0 l} \right) = 8.68 \alpha_0 l, \tag{6.83}$$

where, as indicated earlier, α_0 is the pressure attenuation constant of the lined duct for stationary medium.

α_0 can be evaluated from empirical formulae or by means of the analytical procedure discussed in the foregoing sections. It is not necessary that $k_{x,0}$ (and $k_{y,0}$) be calculated from Equation 6.59. In fact, one can solve Equation 6.43 by computerized iteration methods.

There are other ways, too. One can solve the coupled-wave equation in the passage and in the porous material of the lining, imposing the conditions of the same propagation constant and the same normal impedance at the interface, by means of computerized iteration methods, to get the common propagation constant, one part of which represents the attenuation constant α_0 [7].

Sometimes a designer or consultant needs to do some quick hand calculations of the effectiveness (in terms of transmission loss) of a lined duct. The datum available is $\bar{\alpha}$, the absorption coefficient of the material of the lining, defined as the fraction of the normally incident plane-wave energy absorbed by the given thickness of the lining, backed by a rigid wall. The value of the absorption coefficient $\bar{\alpha}$ supplied by the manufacturer is an average value over a certain frequency range. There are a number of empirical formulae for quick hand calculations. One popular example is Pienings empirical formula [1], according to which

$$TL_l \approx 1.5 \frac{P}{S} \bar{\alpha} l \quad (dB), \tag{6.84}$$

where

$\bar{\alpha}$ is the absorption coefficient of the material,
P is the lined perimeter, and
S is the free-flow area of the cross section.

Thus, for a circular duct of radius r_0 or a square duct with each side $2r_0$ long, lined all over the periphery,

$$TL_l \approx 3\overline{\alpha}(l/r_0). \tag{6.85}$$

Formula (6.85) is indeed very useful for a quick estimate of the effectiveness of an acoustically lined duct.

For example, it indicates that if a material with $\overline{\alpha} = 0.5$ were used to line a circular or square duct, it would yield a 3-dB attenuation across a length equal to one diameter or side length. Equation 6.84 incidentally implies that if only two of the four sides are lined, one would get only half as much attenuation.

Incidentally, equating Equations 6.83 and 6.84 yields a rough value for α_0, the attenuation of acoustic pressure per length:

$$\alpha_0 = 0.173 \frac{P}{S} \overline{\alpha} \quad \text{(nepers/m)}. \tag{6.86}$$

Rigorous analysis and optimization of absorptive ducts is dealt with in [8–18], among others.

6.7 Effect of Protective Layer

An analysis of the waves in tubes with compliant walls has been given above for a locally reacting lining, neglecting the effect of the protective layer (perforated or impervious), which however is invariably used for holding the absorptive material in place as well as for mechanical protection. In this section, the effect of protective layer is investigated for waves in stationary medium for bulk reacting lining as well as locally reacting lining [19].

Figure 6.7 shows a rigid-walled circular duct lined on the inside with an absorptive layer which is protected in turn by a protective layer depicted by a dotted line at $r = r_i$.

The thickness of the lining is $d = r_o - r_i$ where subscripts i and o denote inner and outer radii of the lining, the protective layer thickness at $r = r_i$ being too small to be reckoned.

Here Z_p is the impedance of the protective layer, which may be impervious or perforated. The impedance of the thin protective plate is given by

$$Z_p = j\omega\rho_p t_p \tag{6.87}$$

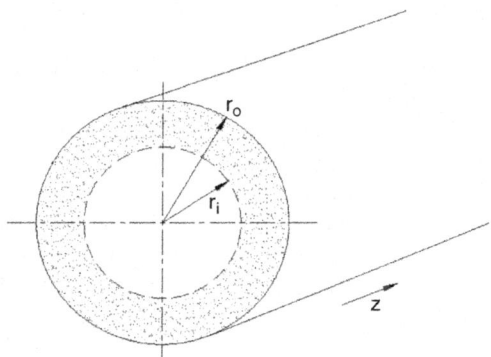

Figure 6.7 A lined circular duct

for an impervious or impermeable layer. For a perforated plate in stationary medium, the corresponding expression is given by Equation 3.129:

$$Z_p = \rho_0 c_0 \left[0.006 + jk_0\left(t_p + 0.75 d_h\right)\right]/\sigma \equiv Z_{pp}/\sigma \qquad (6.88)$$

where

$\rho_0 c_0$ is the characteristic impedance of the fluid medium inside the duct in $kg/m^2 - s$,
k_0 is the wavenumber, ω/c_0,
t_p is thickness of the perforated plate in m,
d_h is diameter of the holes or orifices,
σ is porosity of the perforated plate, defined as the ratio of open area to total surface area.

Selamet et al. [25], however, suggest a more realistic expression for Z_p, taking into account the fact that the perforated plate has air on one side and absorptive material on the other:

$$Z_p = \frac{\rho_0 c_0}{\sigma}\left[0.006 + jk_0\left\{t_p + 0.375 d_h\left(1 + \frac{Y_w k_0}{Y_0 k_w}\right)\right\}\right] \qquad (6.89)$$

where $Y_w = \rho_w c_w$, is the characteristic impedance of the absorptive layer.

The effect of this protective layer can be incorporated in the model of Section 1.7.4 as follows:

Equations 1.170, 1.179 and 1.181 remain unaltered. Only Equation 1.180 is altered to reflect the acoustic pressure drop across the lining impedance $Z_p(\omega)$

$$p(z, r_i, t) = p_w(z, r_i, t) + Z_p(\omega) u_{r,w}(z, r_i, t) \qquad (6.90)$$

Dividing both sides of this equation by the corresponding sides of Equation 1.181, and making use of Equations 1.70, 1.172, 1.173, 1.175 and 1.179 yields [19]

$$j\frac{\omega \rho_0}{k_{r,0}}\frac{J_0(k_{r,0} r_i)}{J_1(k_{r,0} r_i)} = Z_p(\omega) + j\frac{\omega \rho_w}{k_{r,w}}\frac{J_0(k_{r,w} r_i) + C_3 N_0(k_{r,w} r_i)}{J_1(k_{r,w} r_i) + C_3 N_1(k_{r,w} r_i)} \qquad (6.91)$$

(compare Equation 1.183). Here, C_3 is given by Equation 1.179, and Equations 1.184 and 1.185 hold as well.

For the limiting case of the locally reacting lining, $k_{r,w}$ in Equation 6.91 may be replaced by k_w, as explained before in Section 1.7.4.

Figure 6.8 shows a rectangular duct lined on two sides (generally, the longer sides of the duct are lined with acoustically absorptive lining). The other two sides being rigid (unlined), the wave number in that direction would be zero. Thus, referring to Figure 6.1, $k_x = 0$. Modeling two-dimensional waves with bulk-reacting lining with a thin protective layer would make use of Equations 1.153–1.163 and 1.165 of Section 1.7.3 without alteration. Only Equation 1.164 would now be rewritten to incorporate the pressure drop across the protective layer. Thus, for harmonic time dependence ($e^{j\omega t}$), Equation 1.164 would become

$$p(z, h, t) - p_w(z, d, t) = u_{y,0}(z, h, t) Z_p \qquad (6.92)$$

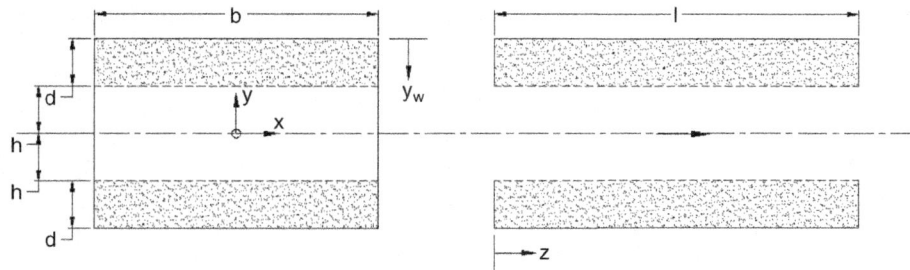

Figure 6.8 Schematic of a bulk reacting rectangular duct lined on two sides

or

$$p(z, h, t) = p_w(z, d, t) - u_{y,w}(z, h, t) Z_p \qquad (6.93)$$

Making use of Equations 1.153, 1.155, 1.156, 1.162 and 1.163 in Equation 6.93 we get

$$\cos(k_{y,0}h) = \cos(k_{y,w}d) + j \sin(k_{y,w}d) \frac{k_{y,w}}{k_w Y_w} Z_p \qquad (6.94)$$

Dividing the two sides of Equation 6.94 by the corresponding sides of Equation 1.167 yields [19]

$$\frac{k_0 Y_0}{k_{y,0}} \cot(k_{y,0}h) = -jZ_p - \frac{k_w Y_w}{k_{y,w}} \cot(k_{y,w}d)$$

or

$$j \frac{k_0 Y_0}{k_{y,0}} \cot(k_{y,0}h) = Z_p - j \frac{Y_w k_w}{k_{y,w}} \cot(k_{y,w}d) \qquad (6.95)$$

For the limiting case of the locally reacting lining, $k_{y,w} = k_w$, and then Equation 6.95 would reduce to

$$j \frac{k_0 Y_0}{k_{y,0}} \cot(k_{y,0}h) = Z_p - j Y_w \cot(k_w d) \qquad (6.96)$$

The second term on the right hand side of Equation 6.96 may be recognized as impedance of the rigid plate transferred across the lining of thickness d (compare Equation 2.26).

Either of the transcendental Equations 6.95 and 6.96 can be solved for axial wave number k_z by means of a Newton–Raphson iteration scheme or any other numerical scheme.

Figures 6.9 and 6.10 compare the Transmission Loss curves for local reaction and bulk reaction for a lined circular duct with the absorptive lining protected by a perforated plate and impervious Mylar layer, respectively, for the following dimensions (refer to Figure 6.2):

$$h \equiv \frac{\text{cross-sectional area of the air passage}}{\text{lined perimeter}} = \frac{\pi r_i^2}{2\pi r_i} = \frac{r_i}{2} = 50 \text{ mm},$$

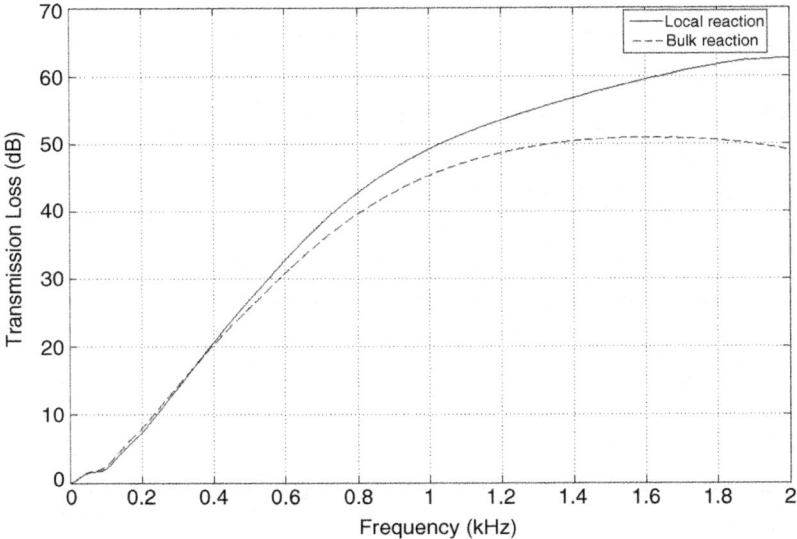

Figure 6.9 Effect of model on transmission loss of a 1-m lined circular duct with d = h = 50 mm, protected by a steel plate with porosity of 0.349

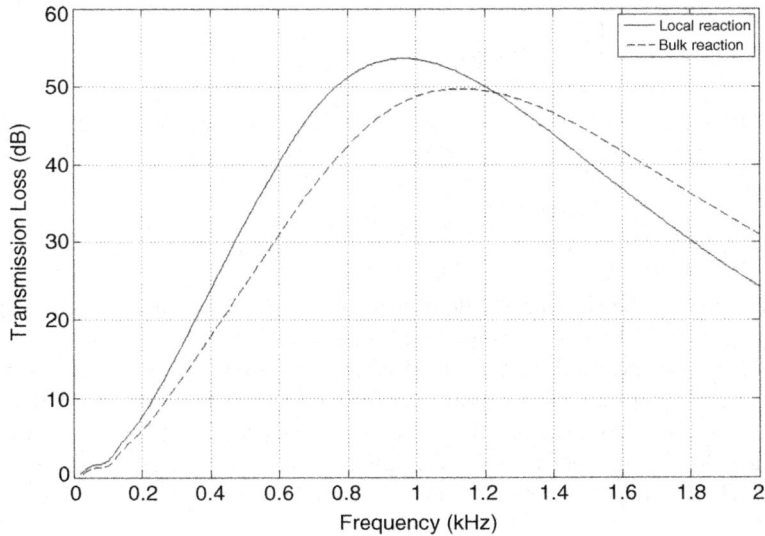

Figure 6.10 Effect of model on transmission loss of a 1-m lined circular duct with d = h = 50 mm, protected by a 0.04 mm thick Mylar layer

Flow resistivity of the fibrous lining material $= 40000 \dfrac{Pa \cdot s}{m^2}$,

Length of duct, $l = 1$ m,
internal radius of the lining, $r_i = 2h = 100$ mm,
outer radius of the lining, $r_0 = r_i + d = 150$ mm,
diameter of the holes constituting perforations, $d_h = 3$ mm,
thickness of the perforated steel plate, $t_p = 1$ mm
density of the impervious Mylar layer, $\rho_p = 1200$ kg/m^3

Incidentally, the cut-off frequency for the circular duct works out to be 1406 Hz and that for the rectangular duct amounts to 1730 Hz. Parts of the TL curves beyond these frequencies should be disregarded because then higher order modes would start propagating.

It is obvious from Figures 6.9 and 6.10 that for these typical ducts and absorptive material, the local reaction model is good enough for practical purposes, although strictly one should use the bulk reaction model.

Figure 6.11 shows the effect of porosity of the perforated protective plate on the TL curve. It may be seen that for the case of 34.9% porosity, there is hardly any deterioration in TL except at higher frequencies. However, in the case of a perforated plate with 4.9% porosity, it makes substantial difference right from 500 Hz onwards. In practice, one can use perforated plate of about 20% porosity and disregard its effect on TL.

Often, the absorptive material blankets are available in the market covered with a very thin impervious layer of Mylar, or the like. Figure 6.12 shows the effect of such an impervious (or impermeable) layer on TL of an absorptive rectangular duct, or a parallel baffle muffler consisting of several such ducts in parallel, as shown in Figure 6.13. It can be seen that even a

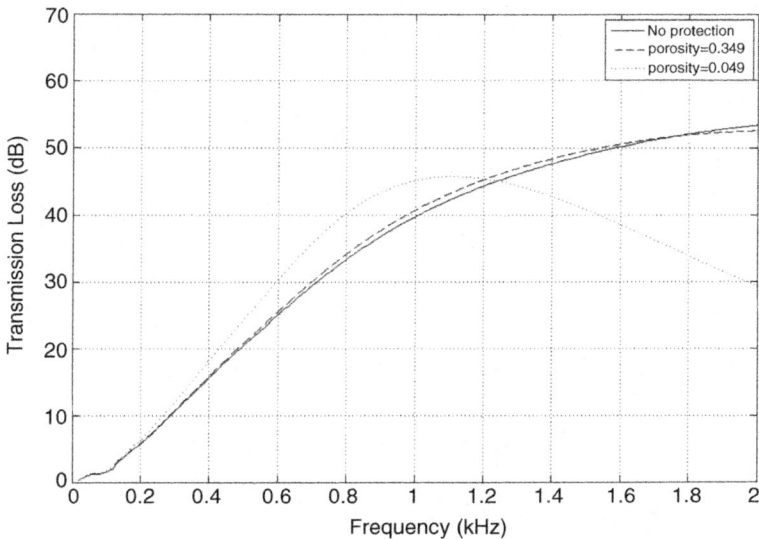

Figure 6.11 Effect of porosity of the perforated protective plate on transmission loss of a 1-m long parallel baffle muffler with d = h = 50 mm

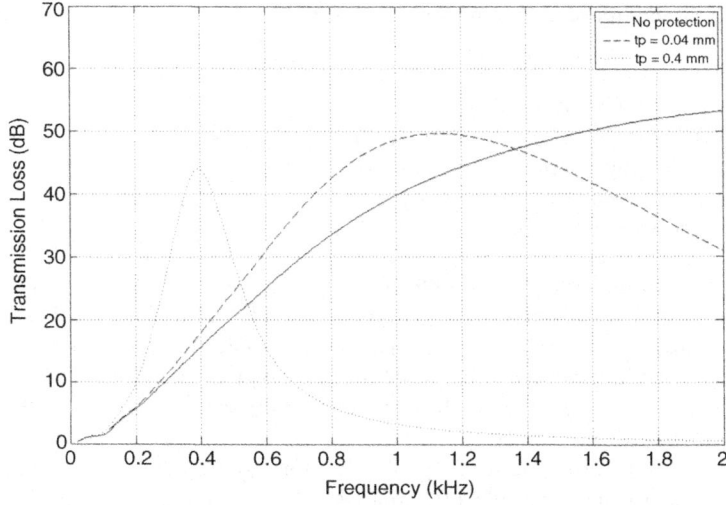

Figure 6.12 Effect of thickness of the protective Mylar layer on transmission loss of a 1-m long parallel baffle muffler with d = h = 50 mm

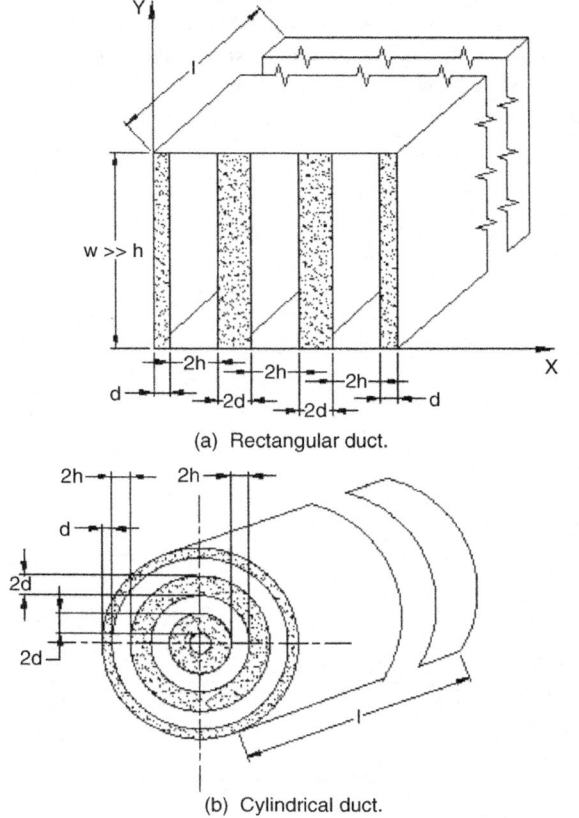

Figure 6.13 Parallel-baffle ducts or mufflers

layer of 40 microns (0.04 mm) makes a substantial difference; it increases TL at lower frequencies but reduces it drastically at higher frequencies. This effect is further accentuated with increase in thickness of the impervious layer. It may also be noticed that, covered with a thin impervious layer, the highly porous absorptive material gets trapped in a cavity, as it were. This results in a simple expansion chamber kind of behavior at higher frequencies, as is obvious from Figure 6.12.

Another important effect is that of impedance mismatch at the entrance and exit. This shows up at very low frequencies, where the absorptive effect of the lining is very weak.

6.8 Parallel Baffle Muffler

Single-passage ducts shown in Figures 6.2 and 6.8 are not sufficiently absorptive when the cross dimensions b and h, or r_0, are large, as would happen for large ducts required for the intake and exhaust systems of gas turbines, large industrial fans, cooling tower installations, and so forth. Flow passage in ducts with large transverse dimensions is subdivided through a parallel baffle system, shown in Figure 6.13a, for rectangular ducts. In the figure only three baffles are shown. Usually, however, there can be many baffles in parallel, each of thickness $2d$, with interbaffle spacing equal to $2h$. The other transverse dimension, W, is generally much larger than h, so that

$$\frac{S}{P} = \frac{W \times 2h}{2W} = h. \tag{6.97}$$

Now acoustic attenuation is known to be proportional to P/S, the ratio of the lined perimeter and flow area, and, of course, length l. Thus one could write

$$TL_l = TL_h \cdot l/h \tag{6.98}$$

where TL_h is the attenuation in a length equal to h, which is half the transverse dimension of the flow passage as shown in Figure 6.13a. TL_h may therefore be called specific transmission loss. It depends in a complex manner on the geometry of the passage and the baffle, the acoustic characteristics of the porous sound-absorbing material filling the baffles, the frequency, the temperature (and thence sound speed and wavelength), and the mean-flow velocity in the passage.

It is clear from Figure 6.13a that in parallel baffle mufflers where one of the cross dimensions (W in this case) is very much larger than half the passage width, h, wave propagation would be governed by a two-dimensional wave equation (x and z); derivatives of the state variables in the y direction would be negligible.

In the literature, the theory of parallel-baffle mufflers consists in evaluation of their attenuation characteristics for a stationary medium and then applying a correction for the mean flow.

For a stationary medium, Ver [7] has computed the normalized attenuation TL_h for various percentages of open area of the muffler cross section, $h/(h+d)$, and flow resistivity of the porous sound-absorbing material in the baffles, assuming $\sigma = 1$ and $\chi = 1$ in Equations 6.74 and 6.75. Ver's values of TL_h have been computed by means of the bulk-reaction model

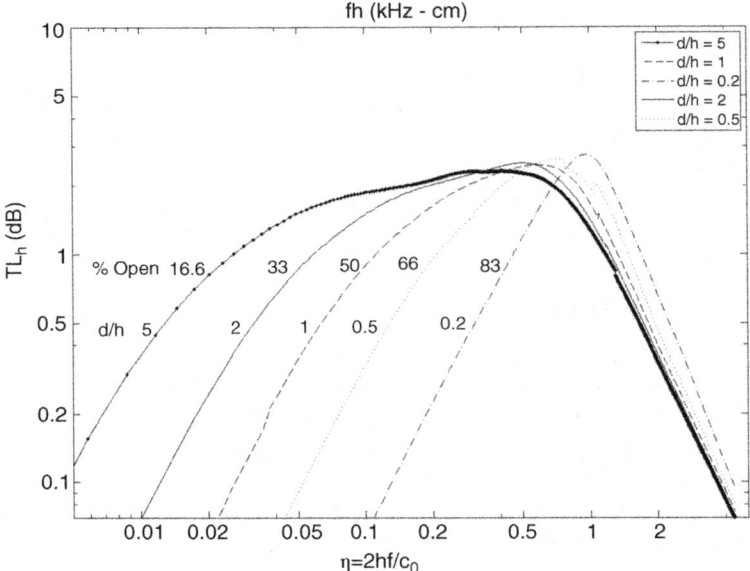

Figure 6.14 Normalized attenuation versus frequency curves for parallel-baffle mufflers, illustrating the effect of percentage open area on attenuation bandwidth for $R = Ed/(\rho_0 c_0) = 5$ (Adapted from [7])

outlined in Sections 1.7.3, 6.3 and 6.6, and the same are plotted in Figure 6.14 as a function of the normalized frequency

$$\eta \equiv 2hf/c_0 \qquad (6.99)$$

for parallel-baffle mufflers, with baffles of a normalized flow resistance,

$$R \equiv Ed/\rho_0 c_0 = 5, \qquad (6.100)$$

with the percentage of open area $h/(h+d)$ as parameter. The upper horizontal scale, which is valid only for air at room temperature, represents the product of h in centimeters and the frequency f in kilohertz. This has been omitted here.

It can be seen that an increase in d/h (thicker baffles and/or smaller flow passage) flattens the curve, thereby increasing the frequency range for which the mufflers would be sufficiently effective, and the increase in the frequency range is primarily on the lower-frequency side. Thus, low-frequency attenuation would require very thick baffles, which would increase the cost as well as the size of the muffler. The drop in attenuation at higher frequencies is due to propagation of higher-order modes.

The effect of flow resistance on TL_h for a practical baffle muffler with 50% open area ($d = h$) is shown in Figure 6.15. It is noted that the effect of variation in flow resistance is relatively insignificant (as compared to the effect of percentage of open area in Figure 6.14). This is indeed fortuitous inasmuch as our knowledge of the material characteristics represents the weakest link in the prediction process, and a strong dependence of attenuation on flow resistance parameter R would make table design of parallel baffle mufflers very inaccurate and hence unreliable.

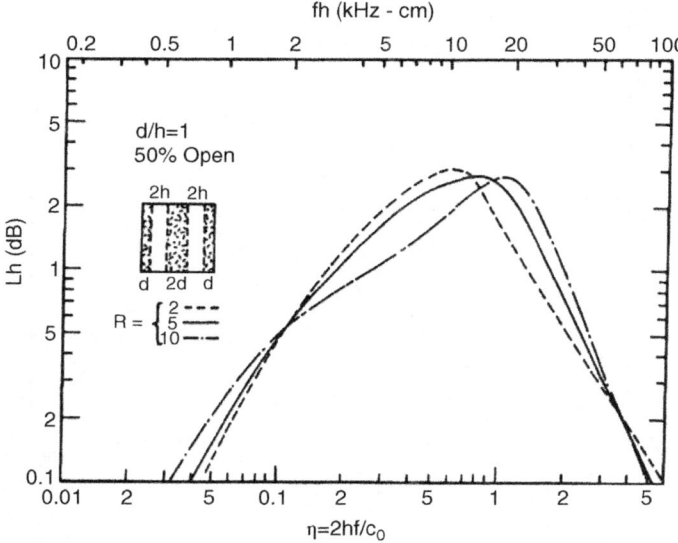

Figure 6.15 Normalized attenuation versus frequency curves for parallel-baffle muffler with 50% open area, illustrating the effect of baffle flow resistance R (Reproduced with permission from [7])

6.9 The Effect of Mean Flow

The convective as well as diffractive effects of mean flow on attenuation at high frequencies can be accounted for simply by appropriately shifting the attenuation curves of Figures 6.14 and 6.15 according to [20]. This is illustrated in [7] for which the mean flow usually increases the attenuation at higher frequencies and decreases it at lower frequencies if the flow direction coincides with the propagation direction of the sound. The trends reverse in the opposite direction (that is, for the reflected wave). The validity of this empirical procedure of accounting for mean flow is limited to fibrous materials.

The method described in Sections 6.1 and 6.2 is, however, quite general and can be used at all frequencies that are low enough to permit propagation of the lowest mode only. Figures 6.16 and 6.17 show the predicted effect of mean flow on TL_h of a square duct for progressive waves moving in the direction of mean flow and against mean flow, respectively. These figures confirm the popular view that mean flow decreases TL_h for the forward moving wave and increases it for the reflected or backward moving wave.

However, mean flow affects the acoustic propagation not only by downstream convection and sideways diffraction, but also by changing the absorption properties of the material by scattering through vortices and other nonlinear effects. These effects are generally ignored in the analysis because there is no simple way of accounting for them. They have to be determined experimentally if the gap between theory and practice must be filled up further. Besides, there is the flow-generated noise (at the entrance as well as the exit of a lined section) that can be evaluated by means of Ver's empirical formulae [21] or the ASHRAE formulae [23] given later in Section 6.13.

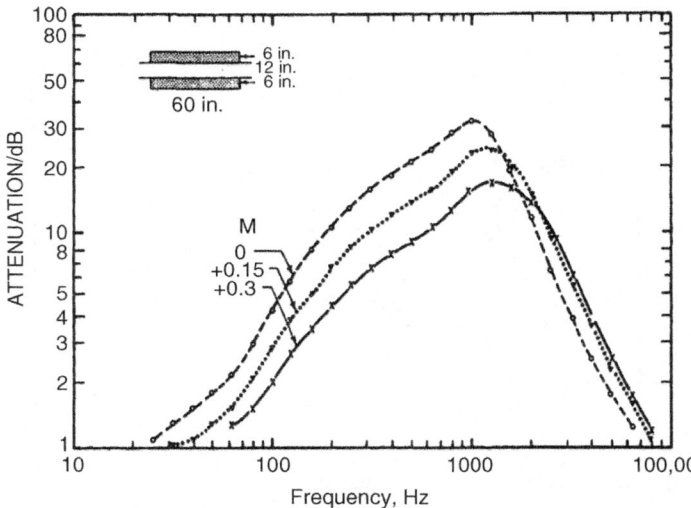

Figure 6.16 Effect of flow on sound attenuation for Mach numbers M = 0, 0.15, 0.3: sound propagation in flow direction (Reproduced with permission from [24])

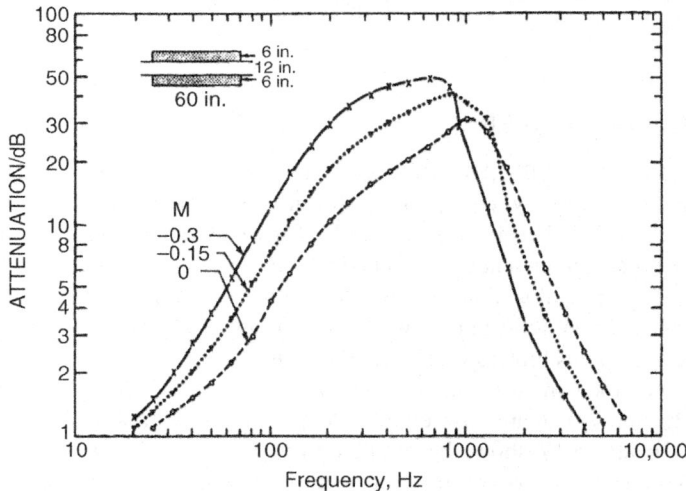

Figure 6.17 Effect of flow on sound attenuation for Mach numbers M = 0, 0.15, 0.3: sound propagation against flow direction (Reproduced with permission from [24])

6.10 The Effect of Terminations on the Performance of Dissipative Ducts

The preceding three sections of this chapter have dealt with the acoustic attenuation of a progressive wave as it travels down a dissipative duct in a stationary or moving medium. However, in practice, no duct has an anechoic termination. Usually it ends up in a bigger pipe or a chamber, or it radiates to the atmosphere. Thus, invariably, a reflected wave would be generated that would travel upstream against the flow. In the steady state, a standing wave would exist in the duct. The methods discussed in Sections 6.7 and 6.8 cannot be applied to standing waves, as they deal only with the attenuation constant, which is the imaginary part of the propagation constant. A more rigorous analysis indicated in Sections 6.1–6.3 yields a complex propagation constant for either of the two progressive waves, and therefore can be used to derive transfer matrices of the type of Equation 6.56 for the lowest-order mode.

The other termination is an active one – the source. The sound produced at the source is a function of how the equivalent impedance of the muffler matches with the source impedance, and also specifically on the resistive part of the equivalent impedance, as explained at length in Chapter 2 for a stationary medium, and in Chapter 3 for a moving medium. This requires transfer matrices (or equivalent relations) for all muffler elements including dissipative ducts (if any) and radiation impedance. And it also requires prior knowledge of the internal impedance of the flow source, which, in the case of a fan or compressor, is as difficult as that of an internal combustion engine, as discussed at length in Chapter 5.

6.11 Lined Bends

In ventilation systems there must be bends in ducts. At the bend, a part of the incoming acoustic energy flux is reflected back. The remainder, which passes through, is subjected to multiple reflections downstream of the bend. The bend, therefore, should be acoustically lined. In fact, a lined bend yields greater attenuation than the equivalent section of the lined duct. Typically, insertion loss due to 90° lined bend varies linearly from about 1 dB at 63 Hz to about 10 dB at 4 kHz and remains at about the same level in the next higher octave band [22]. As for lined ducts, the low-frequency performance of a lined bend can be improved by increasing thickness of the lining.

A 180° bend would, of course, yield higher insertion loss, but then it would cause a larger pressure drop. Considerations of pressure drop would suggest use of smoother bends instead of sharp ones and use of guide vanes in large bends.

6.12 Plenum Chambers

A plenum chamber is a large expansion chamber, used generally upstream of a lined duct or parallel-baffle muffler. It is lined all around with an absorptive material, leaving openings for the inlet and outlet ducts. As shown in Figure 6.18, the two openings are staggered so that a

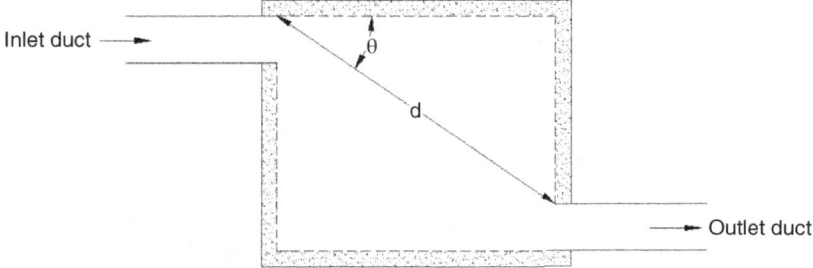

Figure 6.18 Layout of a single-chamber plenum

part of the acoustic energy is absorbed, owing to multiple reflections in the chamber. For this reason, it is classified as a dissipative element. However, it is also reflective, for a substantial part of the energy flux is reflected back to the source because of area discontinuities. The reflective part of the performance of the plenum chamber cannot be assessed in isolation from the rest of the system. However, attenuation due to dissipation may be estimated from the following approximate expression, which takes into account spatial expansion as well as dissipation of acoustic flux [22,23]:

$$\text{attenuation} = -10 \log \left[S \left\{ \frac{\cos \theta}{2\pi d^2} + \frac{1-\overline{\alpha}}{\overline{\alpha} S_w} \right\} \right] (dB). \tag{6.101}$$

where

$\overline{\alpha}$ is the average absorption coefficient of the plenum lining,
S is the area of cross section of the inlet pipe or outlet pipe (assumed to be equal),
S_w is the plenum wall area,
d is the slant distance from input to output, and
θ is the angle made by the line connecting inlet and outlet with the axis of the inlet pipe, as shown in Figure 6.18.

Equation 6.101 holds only for a progressive wave. It can, however, be integrated with the standing wave equations of the rest of the system if one uses this equation to calculate the equivalent, lumped, in-line acoustic resistance R and combines it with lumped shunt compliance C due to the volume of the chamber. While C is given by

$$C = V/c_o^2, \tag{6.102}$$

where V is volume of the plenum chamber, R may be seen to be

$$\frac{R}{Y} = \left[S \left\{ \frac{\cos \theta}{2\pi d^2} + \frac{1-\overline{\alpha}}{\overline{\alpha} S_w} \right\} \right]^{-1/2} - 1, \tag{6.103}$$

where Y is the characteristic impedance of the inlet pipe or outlet pipe (assumed to be equal).

6.13 Flow-Generated Noise

Insertion loss of a lined duct or muffler is limited by the noise generated by the flow of air (or gas) as it comes out of the muffler as a high-velocity jet. This jet noise is augmented by the noise generated at the area discontinuities within the muffler. Analytical estimation of the latter is very difficult. Therefore, the empirical predictive scheme presented here is based on a broad range of experimental data on flow-generated noise of duct silencers and is reproduced here from ISO 14163:1998(E) [26]. According to this source, an estimate for the octave-band sound power level of regenerated sound can be obtained from the following equation:

$$L_{W,oct} = B + \left\{ 10 \lg \frac{PcSn}{W_0} + 60 \lg M + 10 \lg \left[1 + \left(\frac{c}{2fH}\right)^2 \right] - 10 \lg \left[1 + \left(\frac{f\delta}{v}\right)^2 \right] \right\} \quad (6.104)$$

where

B = a value depending on type of silencer and frequency, dB
v = flow velocity in narrowest cross section of silencer, m/s
c = speed of sound in medium, m/s
M = Mach number ($M = v/c$)
P = static pressure in duct, Pa
S = area of narrowest cross section of passage, m^2
n = number of passages
f = octave-band center frequency, Hz
H = maximum dimension of duct perpendicular to baffles, m
δ = length scale characterizing high-frequency spectral content of regenerated noise, m
$W_0 = 1 \, W = 1 N \cdot m/s$

For smooth-walled dissipative splitter silencers used in heating, ventilation, and air-conditioning (HVAC) equipment, $B = 58$ and $\delta = 0.02$ m.

The maximum possible insertion loss obtainable, if one increases the length of the muffler, is equal to the difference between the upstream power level (incident on the muffler) and the flow-generated noise level. Insertion loss beyond this limit calls for not only increased length of the dissipative section but also increased transverse dimensions of the section so that flow passages could be increased and the flow velocities could be decreased.

6.14 Insertion Loss of Parallel Baffle Mufflers

Insertion loss is defined as the difference in the radiated power levels without and with a muffler. For a parallel baffle muffler (PBM) involving sudden area changes at the entrance and exit, IL may be written as

$$IL \simeq TL_{entrance} + TL_{PBM} + TL_{exit} \; (dB) \quad (6.105)$$

$TL_{entrance}$ is not only due to the sudden area changes but also due to the fact that parallel baffle mufflers are often fitted into plenums or acoustic enclosures, where lot of higher order modes would constitute the acoustic pressure field. The transformation from this highly three-dimensional random incidence field into a near plane wave in the louvers would result in considerable dissipation at the entrance. This may be evaluated from Figure 6.19 [24].

The transmission loss at the exit, is approximately given by Equation 2.106:

$$TL_{exit} = 10 \log \frac{(S_2 + S_1)^2}{4 S_2 S_1} = 10 \log \frac{\{(d+h)+h\}^2}{4(d+h)h} = 10 \log \frac{(d+2h)^2}{4(d+h)h} \quad (6.106)$$

Thus, for $d = h$ (50% open passage), $TL_{exit} = 0.5$ dB, and for $d = 2h$, $TL_{exit} = 1.2$ dB. TL_{exit} is clearly small, but $TL_{entrance}$ is substantial. Thus, insertion loss of a parallel baffle

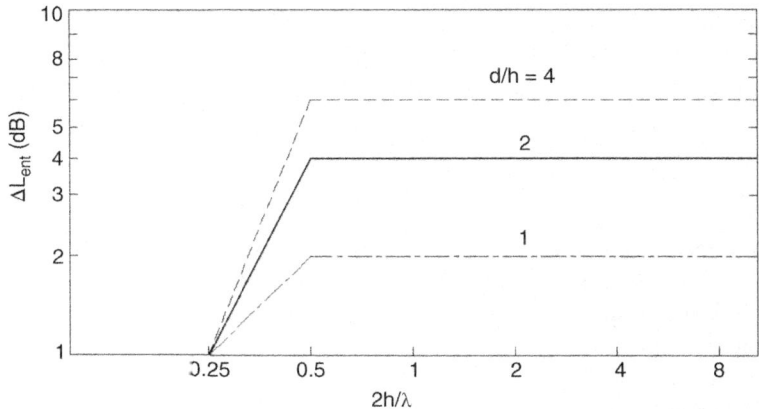

Figure 6.19 Acoustical entrance loss coefficient, ΔL_{ENT}, of silencers in a large duct with a semireverberant sound field in the entrance duct: $2h$ = silencer passage cross dimension (Reproduced with permission from [24])

muffler (PBM) will be considerably more than the corresponding TL calculated by means of Equation 6.98, particularly at higher frequencies.

References

1. Heckl, M. (1978) Foundations of noise control, Series of lectures delivered at I.I.T. Madras.
2. Beranek, L.L. (1947) Acoustical properties of homogeneous, isentropic rigid tiles and flexible blankets. *Journal of the Acoustical Society of America*, **19**(3), 556–568.
3. Attenborough, K. (1983) Acoustical properties of rigid fibrous absorbents and granular materials. *Journal of the Acoustical Society of America*, **73**(3), 785–799.
4. Shirahatti, U.S. (1985) Acoustic characterization of porous ceramic tiles, M. Sc. (Engg.) Thesis, *Indian Institute of Science*, Bangalore.
5. Delany, M.E. and Bazley, B.N. (1970) Acoustical characteristics of fibrous absorbent materials. *Applied Acoustics*, **3**, 106–116.
6. Mechel, F.P. (1976) Extension of low frequencies of the formulae of Delany and Bazley for absorbing materials (in German). *Acustica*, **35**, 210–213.
7. Ver, I.L. (July 1981) Acoustical design of parallel baffle mufflers, Proc. Nelson Acoustics Conference.
8. Schauer, J.J., Hoffman, E.P. and Guyton, R.P. (May 1978) Sound transmission through ducts, Report No. AFAPL-TR-78-25.
9. Mungur, P. and Plumblee, H.E. (1969) Propagation and attenuation of sound in soft walled annular duct containing a sheared flow. *NASA*, **SP-207**, 305–327.
10. Eversman, W. (1971) Effect of boundary layer on the transmission and attenuation of sound in an acoustically treated circular duct. *Journal of the Acoustical Society of America*, **49**(5), 1372–1380.
11. Rice, E.J. (1969) Propagation of waves in an acoustically lined duct with a mean flow. *NASA*, **SP-207**, 345–355.
12. Tester, B.J. (1973) The propagation and attenuation of sound in lined ducts containing uniform or "plug" flow. *Journal of Sound and Vibration*, **28**, 153–203.
13. Lansing, D.L. and Zorumski, W.E. (1973) Effects of wall admittance changes on duct transmission and radiation of sound. *Journal of Sound and Vibration*, **27**, 85–100.
14. Baumeister, K.J. and Rice, E.J. (1975) A difference theory for noise propagation in an acoustically lined duct with mean flow. *Progress in Astronautics and Aeronautics Series*, **37**, 435–453.
15. Kagawa, Y., Yambuchi, T. and Mori, A. (1977) Finite element simulation of an axi-symmetric acoustic transmission system with a sound absorbing wall. *Journal Sound and Vibration*, **53**(3), 357–374.

16. Craggs, A. (1977) A finite element method for modeling dissipative mufflers with a locally reactive lining. *Journal of Sound and Vibration*, **54**(2), 285–296.
17. Astley, R.J. and Eversman, W. (1979) A finite element formulation of the eigen value problem in lined ducts with flow. *Journal of Sound and Vibration*, **65**(1), 61–74.
18. Baumeister, K.J. (1979) Evaluation of optimized multisectioned acoustic liners. *AIAA Journal*, **17**(II), 1185–1192.
19. Munjal, M.L. and Thawani, P.T. (1997) Effect of protective layer on the performance of absorptive ducts. *Noise Control Engineering Journal*, **45**(1), 14–18.
20. Mechel, F.P. and Mertens, P. (1964) Sound attenuation and amplification in lined flow ducts. *Acustica*, **13**, 154–165.
21. Ver, I.L. (1972) Prediction scheme for self generated noise of silencers. *Proceedings Inter-Noise*, **72**, 294–298.
22. Irwin, J.D. and Graf, E.R. (1979) *Industrial Noise and Vibration Control*, Prentice-Hall, Englewood cliffs, NJ.
23. ANONYMOUS (1970) *Ashrae Guide and Data Book*, American Society for Heating, Refrigeration and Air,-Conditioning Engineers, New York, Chap. 33, Systems.
24. Ver, I.L. and Beranek, L.L. (2006) *Noise and Vibration Control*, 2nd edn, John Wiley & Sons, New York.
25. Selamet, A., Lee, I.J., Ji, Z.L. and Huff, N.T. (2001) Acoustic attenuation performance of perforated absorbing silencers, SAE Noise and Vibration Conference, Paper 2001-01-1435, Traverse City, Michigan, USA.
26. ISO14163-1998(E) (1998) Acoustics – Guidelines for Noise Control by Silencers, *International Organization for Standardization*, Geneva, Switzerland.

7

Three-Dimensional Analysis of Mufflers

Often one has to deal with high frequencies and/or chambers with large transverse dimensions, such that one or more of the higher modes gets cut-on in the chamber, even though the inlet and outlet pipes may permit plane waves only. This calls for a three-dimensional (3D) analysis. In this chapter we discuss a couple of analytical techniques as well as the finite element method.

The propagation of three-dimensional (higher-order) modes excited at an isolated, symmetric (coaxial) area discontinuity was first described by Miles [1], making use of known boundary (or junction) conditions and the fact that the infinite series of functions constituting acoustic pressure and particle velocity represented an orthogonal expansion. Decades later, El-Sharkawy and Nayfeh [4] extended this theory to three-dimensional (3D) analysis of a symmetric expansion chamber. They got four sets of equations containing an infinite number of unknowns for each spinning mode and each input. Fortunately, however, their computations indicated that each series could be truncated to the first five terms to yield a 0.1% accuracy for the whole range of parameters presented there. This analysis was extended to asymmetric expansion chambers by Jayaraman [3] and experimental corroboration was provided by Eriksson et al. [4]. A somewhat similar analysis has been presented by Ih and Lee for prediction of the transmission loss for simple chamber [5] and reversing chamber mufflers [6] of any circular geometry, replacing the inlet and outlet tubes by two hypothetical pistons, and superimposing the velocity potentials resulting therefrom. These methods are basically exact methods, notwithstanding the unavoidable truncation of the infinite series for numerical computation. However, the algebra involved there, particularly for asymmetric chambers, is very complicated and cumbersome.

Described below is a simple numerical approach for deriving the transfer matrix for the chamber, which then can be combined with plane-wave (one-dimensional) transfer matrices of the upstream and downstream elements to evaluate the overall performance of the muffler, as indicated earlier in this chapter.

7.1 Collocation Method for Simple Expansion Chambers

Let us consider a rectangular duct system as shown in Figure 7.1. Points 1 and 3 are just downstream and just upstream of the chamber, denoted as element number 2. Let b_i and h_i denote the breadth and height of the ith tube, $i = 1, 2$ and 3. In order to limit three-dimensional analysis to the chamber, let us assume that inlet pipe 3 and outlet pipe 1 permit only plane waves (this is being done only for algebraic simplicity; in fact, the method described here does not impose any such restriction). The two end plates of the chamber are notionally divided into a number of points as shown in the end views in Figure 7.1. The number of points on each end plate (including that in the inlet or outlet pipe) has to be equal to the total number of modes that would be cut-on in the frequency range of interest. The cut-on frequency of the (m, n) mode would be given by Equation 1.30, that is,

$$k_{z,m,n}^2 = k_0^2 - \left(\frac{m\pi}{b_2}\right)^2 - \left(\frac{n\pi}{h_2}\right)^2 = 0, \quad k_0 = \frac{\omega}{c_0} = \frac{2\pi f_{m,n}}{c_0},$$

whence

$$f_{m,n} = \frac{c_0}{2}\left\{\left(\frac{m}{b_2}\right)^2 - \left(\frac{n}{h_2}\right)^2\right\}^{1/2}. \tag{7.1}$$

For example, if

$$c_0 = 340 \text{ m/s}, \quad b_2 = h_2 = 15 \text{ cm},$$

then the cut-on frequencies would be those given in Table 7.1. It is obvious from this table that for a frequency range of 0–3000 Hz, it would suffice to take $m = 0, 1,$ and 2; $n = 0, 1$ and 2. Thus there would be nine modes (including the plane mode) and division of the end-plate regions into nine points (as shown in Figure 7.1) would be adequate.

Now, according to Equations 1.29 and 1.32, for sinusoidal time dependence and a rigid shell, we have

$$p_2(z, x, y) = \sum_{m=0}^{2}\sum_{n=0}^{2}\left\{A_{m,n}e^{-jk_{z,m,n}z} + B_{m,n}e^{+jk_{z,m,n}z}\right\} \times \cos\left(\frac{m\pi x}{b_2}\right)\cos\left(\frac{n\pi y}{h_2}\right); \tag{7.2}$$

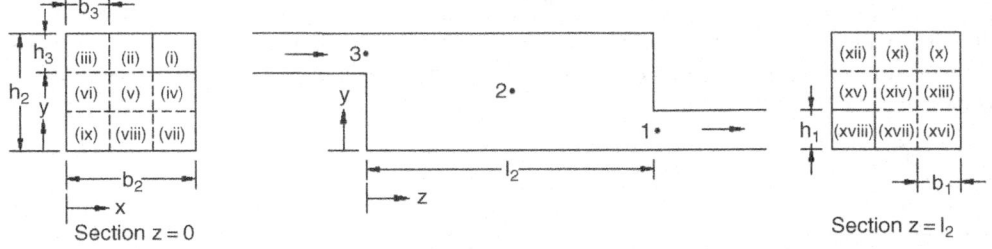

Figure 7.1 3D analysis of an expansion chamber (Adapted from [7])

Table 7.1 Cut-on frequencies $f_{m,n}$ in Hertz for a 15 cm × 15 cm duct

			m			
n	0	1	2	3	4	5
0	0	1133	2267	3400	4533	5667
1	1133	1603	2534	3584	4673	5779
2	2267	2534	3205	4086	5068	6103
3	3400	3584	4086	4808	5667	6608
4	4533	4673	5068	5667	6411	7257
5	5667	5779	6103	6608	7257	8014

$$u_2(z,x,y) = \sum_{m=0}^{2}\sum_{n=0}^{2} \frac{k_{z,m,n}}{\omega\rho_0}\{A_{m,n}e^{-jk_{z,m,n}z} - B_{m,n}e^{+jk_{z,m,n}z}\} \times \cos\left(\frac{m\pi x}{b_2}\right)\cos\left(\frac{n\pi y}{h_2}\right). \quad (7.3)$$

7.1.1 Compatibility Conditions at Area Discontinuities

For an incoming plane wave with amplitude A_3, there would be $2(M+1)(N+1)+2$ unknowns, namely, B_3, $A_{m,n}$, $B_{m,n}$ ($m=0,1,2,\ldots\ldots,M; n=0,1,2,\ldots\ldots,N$), and A_1. Here, $M = N = 2$ (i.e. m, $n = 0$, 1, and 2), requiring 20 compatibility equations for the system of Figure 7.1. These are provided by the physical requirements that acoustic pressure p and particle velocity u be equal at the junctions of inlet pipe and outlet pipe, and that axial particle velocity normal to the rigid part of the two end plates be zero. Symbolically,

$$p_2(0,x_q,y_q) = A_3 + B_3, \quad q = \text{i}; \quad (7.4)$$

$$u_2(0,x_q,y_q) = \frac{k_0}{\omega\rho_0}(A_3 - B_3), \quad q = \text{i}; \quad (7.5)$$

$$u_2(0,x_q,y_q) = 0, \quad q = \text{ii to ix}; \quad (7.6)$$

$$u_2(l_2,x_q,y_q) = 0, \quad q = \text{x to xvii}; \quad (7.7)$$

$$p_2(l_2,x_q,y_q) = A_1 + B_1, \quad q = \text{xviii}; \quad (7.8)$$

$$u_2(l_2,x_q,y_q) = \frac{k_0}{\omega\rho_0}(A_1 - B_1), \quad q = \text{xviii}; \quad (7.9)$$

where p_2 and u_2 are given by Equations 7.2 and 7.3, respectively.

Incidentally, the numbering scheme is entirely arbitrary. Thus there are 20 equations for as many variables, which can be solved for an arbitrary value of A_3 (say, unity).

Note that the reflected wave amplitude B_1 would depend on the value of A_1 and the termination of the outlet pipe or the subsystem downstream of point 1, and therefore is not an independent variable. This becomes clear from the following.

Writing the desired transfer matrix relation as

$$\begin{bmatrix} p_3 \\ v_3 \end{bmatrix} = \begin{bmatrix} T_{11} & T_{12} \\ T_{21} & T_{22} \end{bmatrix} \begin{bmatrix} p_1 \\ v_1 \end{bmatrix}, \qquad v = \rho_0 S u, \qquad (7.10)$$

we get

$$T_{11} = \left.\frac{p_3}{p_1}\right|_{v_1=0} = \left.\frac{A_3 + B_3}{A_1 + B_1}\right|_{B_1=A_1}, \qquad (7.11)$$

$$T_{12} = \left.\frac{p_3}{v_1}\right|_{p_1=0} = Y_1 \left.\frac{A_3 + B_3}{A_1 - B_1}\right|_{B_1=-A_1}, \qquad (7.12)$$

$$T_{21} = \left.\frac{v_3}{p_1}\right|_{v_1=0} = \frac{1}{Y_3}\left.\frac{A_3 - B_3}{A_1 + B_1}\right|_{B_1=A_1}, \qquad (7.13)$$

$$T_{22} = \left.\frac{v_3}{v_1}\right|_{p_1=0} = \frac{Y_1}{Y_3}\left.\frac{A_3 - B_3}{A_1 - B_1}\right|_{B_1=-A_1}. \qquad (7.14)$$

Thus, the preceding set of 20 inhomogeneous equations has to be solved two times with different values of B_1, and hence p_1 and v_1, to calculate all the four-pole parameters. It has been seen that the determinant of the transfer matrix works out to be unity at all frequencies.

Incidentally, transmission loss of the chamber may be evaluated from Equation 2.203 or from the relation

$$TL = 10\log\left[\frac{S_3}{S_1}\frac{|A_3|^2}{|A_1|^2}\right]_{B_1=0}, \qquad (7.15)$$

But then the anechoic boundary condition ($B_1 = 0$) would require another round of simultaneous solutions of 20 equations.

Figure 7.2 shows a comparison of the computed TL values for three dimensional waves with those for plane-wave propagation alone for the single chamber muffler of Figure 7.2. Vertical lines indicate the cut-on frequencies of higher-order modes (m and/or $n > 0$). It is obvious that for the dimensions considered here, three-dimensional effects start even earlier (at a lower frequency) than the cut-on frequency of the first higher-order mode, and at higher frequencies the two curves are wide apart, indicating predominant three-dimensional effects. Similar trends are obvious from Figure 7.3 where the performance of a symmetrical inlet-outlet chamber is considered.

Three-Dimensional Analysis of Mufflers

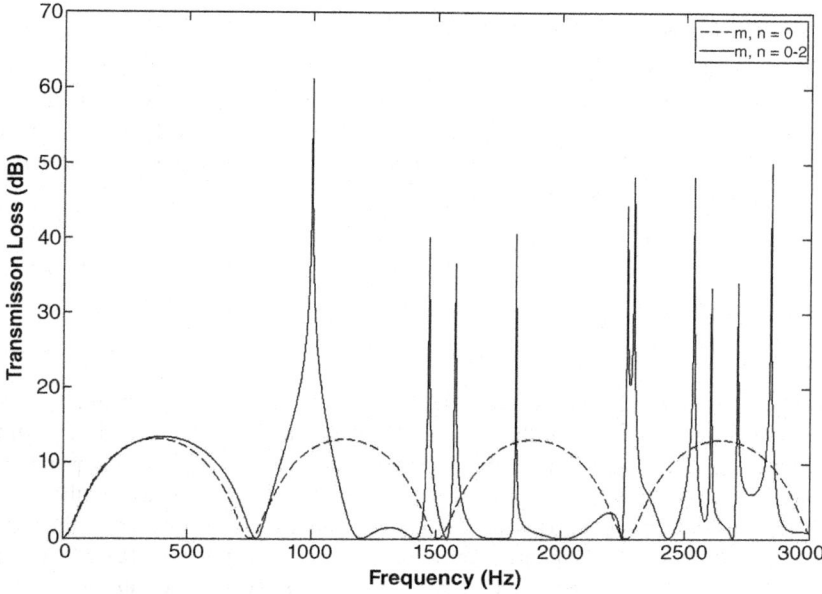

Figure 7.2 TL for the square duct chamber with staggered inlet and outlet (Adapted from [7])

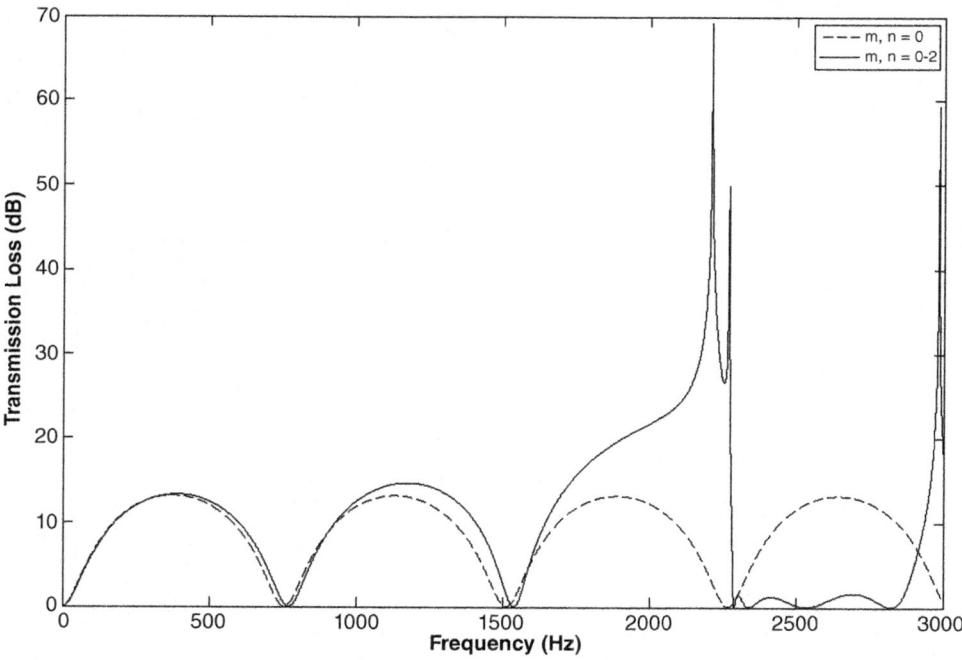

Figure 7.3 TL for the square duct chamber with symmetric inlet and outlet (Adapted from [7])

7.1.2 Extending the Frequency Range

If we want to increase the frequency range of Figures 7.2 and 7.3, we have to consider still higher modes that would get cut-on. That would require a corresponding increase in the number of points across the section of the chamber. This is shown in Figure 7.4, where, by comparison with Figure 7.1, it may be noted that the number of points has been increased from 9 to 36. That is four times. This would permit

$$m_1, n_1 = 0 \text{ and } 1;$$
$$m_1, n_2 = 0, 1, 2, 3, 4, \text{ and } 5; \quad (7.16)$$
$$m_3, n_3 = 0 \text{ and } 1.$$

Incidentally, the number of points in the inlet and outlet ducts has also increased from 1 to 4. The corresponding four modes in these ducts are (0, 0), (0, 1), (1, 0), and (1, 1). Thus, apart from the plane wave, propagation of three higher-order modes is also incorporated in the analysis, thereby removing the assumption of pure plane-wave propagation in these ducts, as implied in the preceding section. This new situation would do more justice to the physics of the situation, which does not preclude generation of higher-order modes in the inlet/outlet ducts at the junctions. This was indicated first by Miles [1] and is incorporated in the exact method [1–4]. However, the incident (or incoming) wave is a plane wave with amplitude $A_{3,0,0}$.

Equations 7.4–7.9 may now be rewritten as

$$A_{3,0,0} + \sum_{n_3=0}^{1} \sum_{m_3=0}^{1} \left[B_{3,m_3,n_3} \cos \frac{m_3 \pi x_q}{b_3} \cos \frac{n_3 \pi y_q}{h_3} \right]$$
$$= \sum_{n_2=0}^{5} \sum_{m_2=0}^{5} \left[\{A_{2,m_2,n_2} + B_{2,m_2,n_2}\} \cos \frac{m_2 \pi x_q}{b_2} \cos \frac{n_2 \pi y_q}{h_2} \right], \quad q = \text{i to iv}; \quad (7.17)$$

o x	o ix	o vi	o v	o ii	o i	o XLvi	o XLv	o XLii	o XLi	o xxxviii	o xxxvii
o xii	o xi	o viii	o vii	o iv	o iii	o XLviii	o XLvii	o XLiv	o XLiii	o XL	o xxxix
o xxii	o xxi	o xviii	o xvii	o xiv	o xiii	o Lviii	o Lvii	o Liv	o Liii	o L	o iL
o xxiv	o xxiii	o xx	o xix	o xvi	o xv	o LX	o Lix	o Lvi	o Lv	o Lii	o Li
o xxxiv	o xxxiii	o xxx	o xxix	o xxvi	o xxv	o LXX	o LXix	o LXvi	o LXV	o Lxii	o Lxi
o xxxvi	o xxxv	o xxxii	o xxxi	o xxviii	o xxvii	o LXXii	o LXXi	o LXviii	o LXvii	o LXiv	o LXiii
Section z = 0						Section z = l_2					

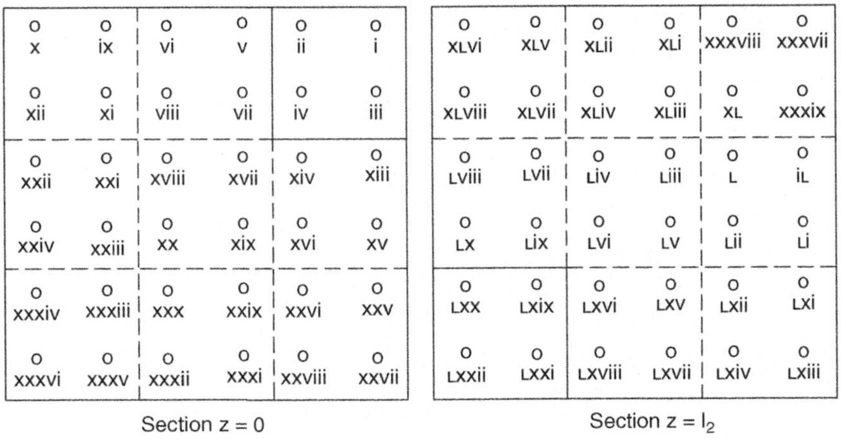

Figure 7.4 Division of a section and numbering of points for m, n = 0–5 (cf. Figure 7.1) (Adapted from [7])

$$k_{3,z,0,0}A_{3,0,0} + \sum_{n_3=0}^{1}\sum_{m_3=0}^{1}\left[k_{3,z,m_3,n_3}\{B_{3,m_3,n_3}\}\cos\frac{m_3\pi x_q}{b_3}\cos\frac{n_3\pi y_q}{h_3}\right]$$
$$= \sum_{n_2=0}^{5}\sum_{m_2=0}^{5}\left[k_{2,z,m_2,n_2}\{A_{2,m_2,n_2} - B_{2,m_2,n_2}\}\cos\frac{m_2\pi x_q}{b_2}\cos\frac{n_2\pi y_q}{h_2}\right], \quad q = \text{i to iv};$$
(7.18)

$$\sum_{n_2=0}^{5}\sum_{m_2=0}^{5}\left[k_{2,z,m_2,n_2}\{A_{2,m_2,n_2} - B_{2,m_2,n_2}\}\cos\frac{m_2\pi x_q}{b_2}\cos\frac{n_2\pi y_q}{h_2}\right] = 0, \quad q = \text{v to xxxvi}; \quad (7.19)$$

$$\sum_{n_2=0}^{5}\sum_{m_2=0}^{5}\left[k_{2,z,m_2,n_2}\{A_{2,m_2,n_2}\exp(-jk_{2,z,m_2,n_2}l_2) - B_{2,m_2,n_2}\exp(+jk_{2,z,m_2,n_2}l_2)\}\right.$$
$$\left. \times \cos\frac{m_2\pi x_q}{b_2}\cos\frac{n_2\pi y_q}{h_2}\right] = 0, \quad q = \text{xxxvii to lxviii};$$
(7.20)

$$\sum_{n_2=0}^{5}\sum_{m_2=0}^{5}\left[k_{2,z,m_2,n_2}\{A_{2,m_2,n_2}\exp(-jk_{2,,z,m_2,n_2}l_2) - B_{2,m_2,n_2}\exp(+jk_{2,,z,m_2,n_2}l_2)\}\right.$$
$$\left. \times \cos\frac{m_2\pi x_q}{b_2}\cos\frac{n_2\pi y_q}{h_2}\right] = \sum_{n_1=0}^{1}\sum_{m_1=0}^{1}\left[k_{1,z,m_1,n_1}A_{1,m_1,n_1}\cos\frac{m_1\pi x_q}{b_1}\cos\frac{n_1\pi y_q}{h_1}\right]$$
$$- B_{1,0,0}k_{1,z,0,0}, \quad q = \text{lxix to lxxii};$$
(7.21)

$$\sum_{n_2=0}^{5}\sum_{m_2=0}^{5}\left[\{A_{2,m_2,n_2}\exp(-jk_{2,,z,m_2,n_2}l_2) + B_{2,m_2,n_2}\exp(+jk_{2,,z,m_2,n_2}l_2)\} \times \cos\frac{m_2\pi x_q}{b_2}\cos\frac{n_2\pi y_q}{h_2}\right]$$
$$= \sum_{n_2=0}^{1}\sum_{m_2=0}^{1}\left[A_{1,m_1,n_1}\cos\frac{m_1\pi x_q}{b_1}\cos\frac{n_1\pi y_q}{h_1}\right] + B_{1,0,0}, \quad q = \text{lxix to lxxii};$$
(7.22)

Thus there are 80 equations for as many unknowns, namely,

$$\begin{array}{llll} B_{3,0,0} & B_{3,0,1} & B_{3,1,0} & B_{3,1,1}; \quad (4) \\ A_{2,m_2,n_2}, & B_{2,m_2,n_2}; & m_2, n_2 = 0 \text{ to } 5; & (36+36) \\ A_{1,0,0} & A_{1,0,1} & A_{1,1,0} & A_{1,1,1}. \end{array} \quad (7.23)$$

In Equations 7.17–7.22 it is assumed that the waves coming into the muffler from the inlet pipe as well as outlet pipe are plane waves, while those going out or away from the muffler (that is, generated by the area discontinuities) could contain higher-order modes as well. The former assumption would limit the applicability of the predicted TL to frequencies that are low enough not to let any higher-order modes propagate in the inlet and outlet pipes. For the muffler under consideration, this frequency limit would be 3400 Hz for (0, 1) and (1, 0) modes and 4810 Hz for the symmetric (1, 1) mode.

Figures 7.5 and 7.6 show plots of the computed values of TL up to 6000 Hz, involving simultaneous solution of an 80 × 80 matrix with complex coefficients. The corresponding

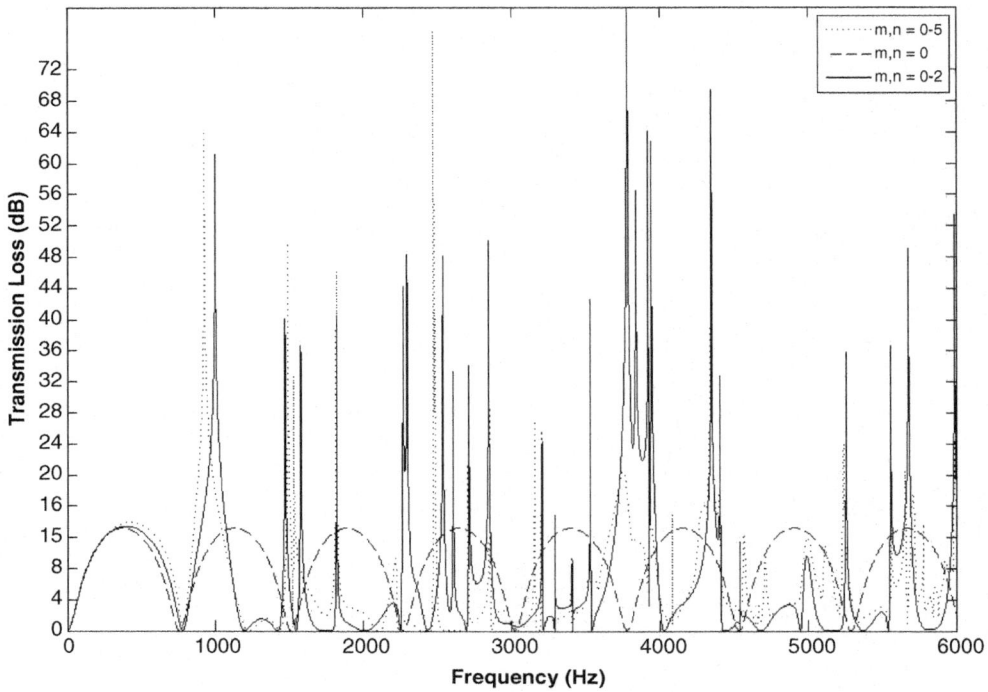

Figure 7.5 TL for the offset-inlet and offset-outlet square chamber muffler over extended frequency range (Adapted from [7])

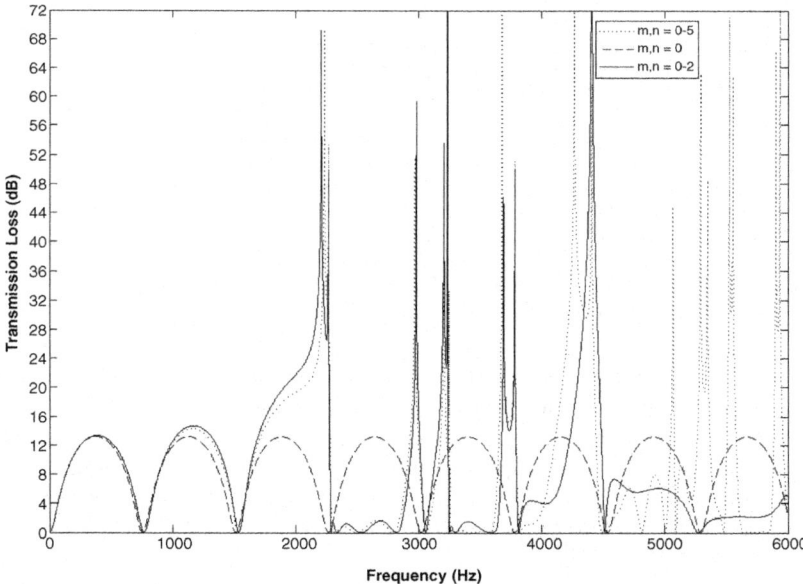

Figure 7.6 TL for symmetric square chamber muffler over extended frequency range (Adapted from [7])

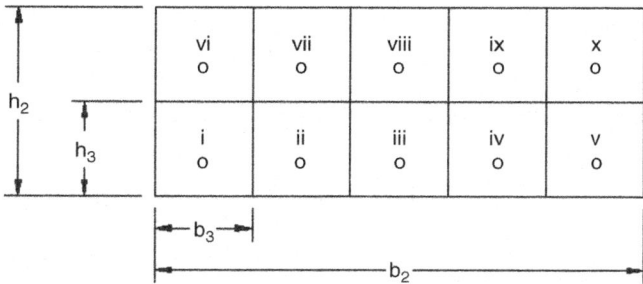

Figure 7.7 Division of a rectangular (nonsquare) chamber section. m = 0–4, n = 0 and 1 (Adapted from [7])

values of Figures 7.2 and 7.3 (computed from simultaneous solution of a 20 × 20 matrix) are superimposed for comparison. It is obvious that the latter are substantially in error at frequencies higher than 3300 Hz, where modes higher than the (2, 2) mode get cut-on. At lower frequencies, the two curves match sufficiently to suggest that, across the section, nine points (equal to the area ratio) are good enough for evaluation of the TL despite the existence of higher-order modes in the inlet/outlet ducts near the junction.

Incidentally, the computation (CPU) times on a digital computer compare as follows with the corresponding CPU time on the same computer for the finite element method (44 nodes) with banded matrix approach:

Nine points across the section (20 unknowns): 2%
Thirty-six points across the section (80 unknowns): 80%.

The computed values of the TL were almost identical for the two methods. In any case, the validity of TL beyond 3400 Hz is questionable on account of the assumption indicated above. Therefore, it would be appropriate to continue to use the configurations of Figure 7.1, where inlet and outlet pipes contain only one point, only plane waves are implicitly admitted, and there are nine points (equal to the area ratio), including the one at the junction, across the chamber.

For a rectangular (nonsquare) expansion chamber, the maximum limits of m_2 and n_2 would be proportional to b and h, respectively, as would the number of points (divisions) in the respective directions. This is illustrated in Figure 7.7.

7.1.3 Extension to Other Muffler Configurations

It may be observed that for a muffler with a simple circular tube chamber the compatibility conditions would remain the same, except that Equations 1.29, 1.30 and 1.32 would be replaced by Equations 1.46, 1.47 and 1.51 because of the change from Cartesian coordinates to cylindrical polar coordinates [7].

Similarly, a simple modification of the governing equations would enable us to incorporate the convective effect of mean flow.

7.2 Finite Element Methods for Mufflers

Developed originally as a tool for structural analysis (see, for example, [8,9]), the method was extended to acoustic analysis by [10–13] and Craggs [14,15]. The problem of acoustic propagation in mufflers, however, was tackled through FEM (albeit for stationary medium) by Young and Crocker [16,17]. Since then or concurrently, a large number of papers have appeared on the topic [18–49].

The finite element method has a number of advantages over the analytical as well as other numerical approaches.

a. It is completely general inasmuch as it has no limitation with respect to geometry of the muffler components and properties of the medium.
b. The boundary conditions in terms of pressure or velocity may be specified anywhere in the system.
c. Any desired degree of accuracy may be obtained by increasing the number of elements into which the system is subdivided.
d. For sinusoidal oscillation (i.e. working in the frequency domain), FEM produces symmetric positive definite band-structured matrices that can be solved by means of standard subroutines, effecting very valuable reduction of core memory and computation time.
e. Over the years, very efficient softwares have been developed that are commercially available with a graphic user interface (GUI).

On the other hand, the comparison with exact theories like the aeroacoustic theory (where applicable), FEM is much more cumbersome, time-consuming, and costly. Therefore, it is used only for those muffler configurations where three-dimensional effects are sure to dominate and analytical methods described elsewhere in this chapter are not readily applicable.

The finite element analysis procedure consists in discretizing the continuum by dividing it into an equivalent system of finite elements, selecting (or assuming) a field function (or model) to approximate the actual or exact field within an element, deriving the element matrices by means of a variational principle or one of the residual methods, preparing the algebraic equations (or matrices) for the overall finite element system, solving the same for the unknowns (pressure or velocities) at the nodes, and, finally computing the acoustic performance of the system in the form of four-pole parameters and thence the transmission loss, and so forth. These steps are discussed in the following sub-sections.

7.2.1 A Single Element

A finite element may be regarded as a piece of the continuum or as a region of the muffler. The shape or configuration of the basic element depends upon the geometry of the muffler and the number of independent space coordinates necessary to describe the problem. Thus, one could have one-dimensional elements in 'small'-diameter tubes [45], two-dimensional elements in elliptical chambers where the minor axis is sufficiently small (as compared to wavelength) [16,17,21], and three-dimensional elements in large mufflers and/or for analysis at high frequencies (see e.g. [19,43,44,46]).

Typical commercial mufflers, being round or elliptical, cannot be represented adequately by straight-edged elements; elements with curved-line boundaries have to be used [19]. A one-dimensional element is just a line connecting the two nodes; two-dimensional elements may be in the form of a triangle, rectangle, quadrilateral, and so on; and three-dimensional elements may have shapes of a tetrahedron, rectangular prism, arbitrary hexahedron, and combinations thereof like a hexahedron (cuboid) composed of five tetrahedra. Often, an assemblage of only one type of elements is not able to represent all components of a muffler adequately (or economically). Then two or more types of elements are employed simultaneously.

The corners of an element are called nodes or nodal points. Values of the field variable (pressure and its gradients) at these nodes are the unknowns. Values of the field variables at any point within an element can be written out as a function of the nodal values. These functions of space variables in acoustics are called pressure models, pressure functions, or pressure patterns (corresponding to displacement models, functions, fields, or patterns, in solid mechanics).

The pressure models must be continuous within the elements. This is ensured by choosing polynomial models which are inherently continuous. In addition, the pressures must be compatible between adjacent elements. This implies that adjacent elements must deform without causing openings, overlaps, or discontinuities between the elements. It can be shown [8,9] that inter element compatibility must be enforced for pressures and their derivatives up to the order $n-1$, where n is the highest-order derivative in the energy functional. Thus, for linear wave propagation, for which $n=2$, inter element compatibility has to be enforced for pressures and their first derivatives (proportional to, and hence representing, particle velocity). Patterns (or formulations or models or functions) that satisfy this condition are called compatible or conforming. Additionally, the pattern should be independent of the orientation of the local coordinate system. This property of the model is known as geometric isotropy, spatial isotropy, or geometric invariance.

The nodal pressures and their space derivatives (or velocity components) necessary to specify completely the deformation of the finite element are called the degrees of freedom of the element. Sometimes one makes use of secondary nodes or higher-order derivatives of pressures at the primary nodes. The degree of freedom occurring at external nodes is distinguished from those at internal nodes by referring to the former as joint or nodal degrees of freedom and the latter as internal degrees of freedom. Elements with these additional degrees of freedom are called higher-order elements.

The field variable at any point within an element p, can be related to the vector of the field variables at all the nodes of the element $\{p_n\}$ as

$$p = \{N\}^T \{p_n\}, \qquad (7.24)$$

where the vector $\{N\}$ consists of interpolation functions or shape functions. The entire vector is called an interpolation model.

If the shape functions or interpolation functions are selected in such a way that the coordinates of the control point inside the element x can also be related to the coordinates of the nodal point vector $\{x_n\}$ through the same interpolation model, that is,

$$x = \{N\}^T \{x_n\}, \qquad (7.25)$$

(in other words, if the geometry and displacements of the element are described in terms of the same parameters and are of the same order), then the elements are said to be isoparametric. Such elements satisfy the compatibility requirements and have isotropic field models [9]. The desired shape functions are in a natural coordinate system which, by definition, permits the specification of a point within the element by a set of dimensionless numbers whose magnitudes never exceed unity.

For example, for the line element shown in Figure 7.8a, with the natural coordinate r taking values as shown there,

$$x = x_1 + \frac{x_2 - x_1}{r_2 - r_1}(r - r_1)$$

$$= x_1 + \frac{x_2 - x_1}{2}(r + 1) \quad (7.26)$$

$$= \frac{1}{2}(1 - r)x_1 + \frac{1}{2}(1 + r)x_2.$$

Similarly,

$$p = \frac{1}{2}(1 - r)p_1 + \frac{1}{2}(1 + r)p_2. \quad (7.27)$$

Comparing these two equations with Equations 7.24 and 7.25 one gets

$$\{N\}^T = \left[\frac{1}{2}(1 - r) \quad \frac{1}{2}(1 + r)\right]. \quad (7.28)$$

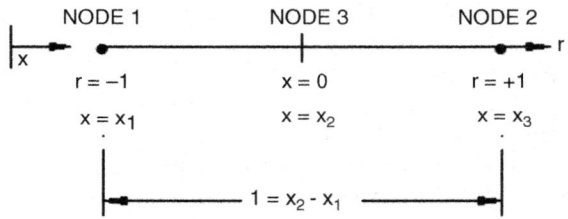

(a) A simple natural coordinate for line element.

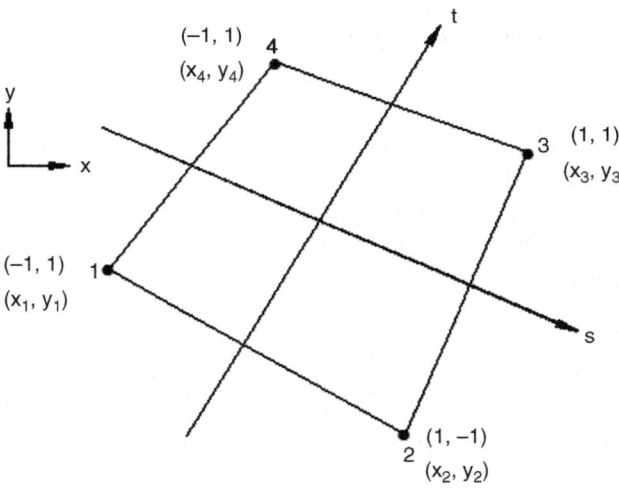

(b) Quadrilateral coordinates.

Figure 7.8 Natural coordinate systems (Reproduced with permission from [9])

Thus, the two interpolation functions of this linear interpolation model are

$$N_{12} = (1-r)/2 \quad \text{and} \quad N_2 = (1+r)/2. \tag{7.29}$$

Similarly, for a quadrilateral element shown in Figure 7.8b, the Cartesian coordinates (x, y) of any point within the quadrilateral can be related to the natural coordinates (s, t) through a linear interpolation model as follows [9]:

$$x = \{N\}^T \{x_n\}, \tag{7.30}$$

$$y = \{N\}^T \{y_n\}, \tag{7.31}$$

where

$$\begin{aligned} \{N\}^T &= \frac{1}{4}[(1-s)(1-t), (1+s)(1-t), (1+s)(1+t), (1-s)(1+t)], \\ \{x_n\}^T &= [x_1\, x_2\, x_3\, x_4], \\ \{y_n\}^T &= [y_1\, y_2\, y_3\, y_4], \end{aligned} \tag{7.32}$$

where x_1 and y_1 are the Cartesian coordinates of node 1, and so on.

The preceding examples illustrate the linear interpolation model. An interpolation function, also known as a shape function, is a function that has unit value at one nodal point and zero value at other nodal points. Figure 7.9 shows linear interpolation functions and quadratic interpolation functions for a one-dimensional element.

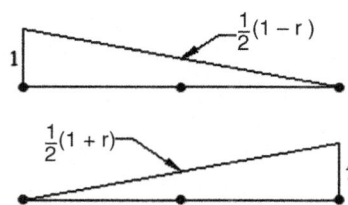

(a) One dimensional element.

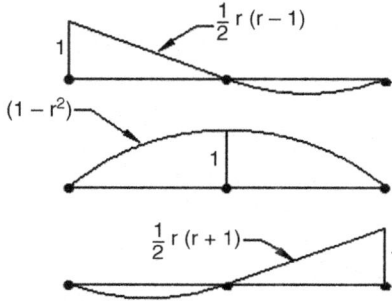

(b) Linear interpolation functions (two nodes).

(c) Quadratic interpolation functions (three nodes).

Figure 7.9 Interpolation functions for one-dimensional element (Reproduced with permission from [9])

Table 7.2 Typical shape functions for various elements

Order of model	Node I	Shape functions for various elements		
		Line	Quadrilateral	Hexahedron
Linear	Primary external (corner)	$\frac{1}{2}(1+rr_i)$	$\frac{1}{4}(1+ss_i)(1+tt_i)$	$\frac{1}{8}(1+rr_i)(1+ss_i)(1+tt_i)$
Quadratic	Primary external (corner)	$\frac{1}{2}(1+rr_i)rr_i$	$\frac{1}{4}(1+ss_i)(1+tt_i)(ss_i+tt_i-1)$	$\frac{1}{8}(1+rr_i)(1+ss_i)(1+tt_i)$ $\times (rr_i+ss_i+tt_i-2)$
	Secondary external (Midedge)	$(1-r^2)$	$\frac{1}{2}(1-s^2)(1+tt_i)$ for $s_i=0$ $\frac{1}{2}(1+ss_i)(1-t^2)$ for $t_i=0$	$\frac{1}{4}(1-r^2)(1+ss_i)(1+tt_i)$ for $r_i=0$, etc.

Similar relations hold for a hexahedral element as well. All these relations are assembled in Table 7.2. Higher-order models, which are used rather rarely, have been omitted from this table.

For use in mufflers, where axisymmetry is encountered often, an eight degree-of-freedom hexahedral element can be converted to a four degree-of-freedom axisymmetric ring element (Figure 7.10) by means of the following constraints applied to the nodal coordinates and the nodal pressures [19]:

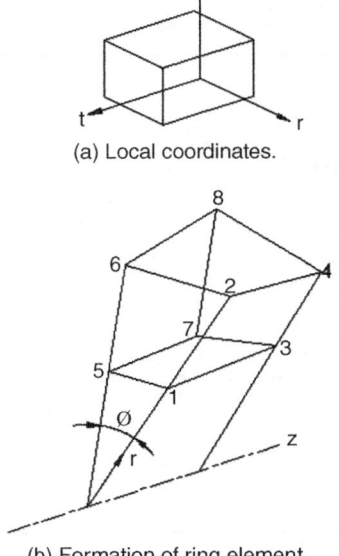

(a) Local coordinates.

(b) Formation of ring element.

Figure 7.10 The isoparametric element (Reproduced with permission from [19])

$$\begin{aligned}
&x_1 = x_2 = x_3 = x_4 = 0\\
&x_5 = r_1 \sin\phi, \quad x_6 = r_2 \sin\phi, \quad x_7 = r_3 \sin\phi, \quad x_8 = r_4 \sin\phi;\\
&y_1 = r_1, \quad y_2 = r_2, \quad y_3 = r_3, \quad y_4 = r_4,\\
&y_5 = r_1 \cos\phi, \quad y_6 = r_2 \cos\phi, \quad y_7 = r_3 \cos\phi, \quad y_8 = r_4 \cos\phi;\\
&z_1 = z_5 = \zeta_1, \quad z_2 = z_6 = \zeta_2, \quad z_3 = z_7 = \zeta_3, \quad z_4 = z_8 = \zeta_4,\\
&p_5 = p_1, \quad p_6 = p_2, \quad p_7 = r_3, \quad p_8 = p_4.
\end{aligned} \quad (7.33)$$

These relationships are obvious from Figure 7.10.

7.2.2 Variational Formulation of Finite Element Equations

There are a number of principles that are used in variational methods. The variational principle useful for problems in dynamics is Hamilton's principle enunciated as follows [9].

Among all possible time histories of displacement (or velocity) configurations that satisfy compatibility and the constraints or kinematic boundary conditions and that also satisfy conditions at time t_1 and t_2, the history representing the actual solution makes the Lagrangian functional L a minimum. Symbolically,

$$\delta \int_{t_1}^{t_2} L\, dt = 0, \Rightarrow \delta L = 0 \quad \text{for steady state solutions,} \tag{7.34}$$

where the Lagrangian L is defined as

$$L = \overline{U} - \overline{K} - \overline{W}, \tag{7.35a}$$

where, for a stationary medium,

$$\begin{aligned}
\overline{U} &= \text{the potential energy} = \frac{1}{2} \int_V \rho_0 c_0^2 (\mathrm{div}\,\xi)^2 dV,\\
\overline{K} &= \text{the kinetic energy} = \frac{1}{2} \int_V \rho_0 (\dot\xi)^2 dV,\\
\overline{W} &= \text{the external work function} = \int_{S_p} (-p)\xi_n\, dS,\\
\xi &= \text{the displacement,}\\
V &= \text{the volume,}
\end{aligned} \tag{7.35b}$$

and

$$S_p = \text{the surface on which pressure is defined.}$$

For sinusoidal time variations for all field variables, making use of the basic relations between velocity and pressure, the Lagrangian functional may be written in terms of pressure as [16]

$$L = \frac{1}{2\rho_0 c_0^2} \int_V p^2 dV - \frac{1}{2\rho_0 \omega^2} \int_V (\mathrm{grad}\,p)^2 dV + \frac{1}{j\omega} \int_{S_u} u_n p\, dS, \tag{7.36}$$

where

S_u = the surface over which velocity u is defined,
u_n = particle velocity (amplitude) normal to and away from the surface, $j\omega\xi$, and
p = (complex) amplitude of acoustic pressure.

In this form, L may be termed the complementary Lagrangian functional, and the corresponding version of Hamilton's principle as the complementary variational principle [11,16].

If surface S_u is made up of m portions with different (complex) impedances, then Equation 7.36 may be rewritten as

$$L = \int_V \left[\frac{pp^*}{2\rho_0 c_0^2} - \frac{(\text{grad } p)\cdot(\text{grad } p^*)}{2\rho_0 \omega^2} \right] dV + \sum_{n=1}^{m} \int_{S_n} \frac{pp^*}{j\omega Z_n} dS, \quad (7.37)$$

where the asterisk stands for the conjugate. Here terms like p^2 have been replaced by pp^* to make all the energy components real.

Rigid boundaries ($u_n \to 0$, and hence $Z_n \to \infty$) and the hypothetical boundaries where $Z_n \to 0$ and hence $p \to 0$, make no contributions to the functional L.

Reverting to Equation 7.36, substituting Equation 7.24 into it, noting that

$$p^2 = \{p_n\}_e^T \{N\}\{N\}^T \{p_n\}_e, \quad (7.38)$$

$$(\text{grad } p)^2 = \{p_n\}_e^T \{\text{grad } N\}\{\text{grad } N\}^T \{p_n\}_e, \quad (7.39)$$

and making use of the variational Equation 7.34 yields

$$\left([M]_e - k_0^2 [P]_e\right)\{p_n\}_e = -j\rho_0 \omega \{F\}_e, \quad (7.40)$$

where $[P]_e$ and $[M]_e$ are the stiffness and inertia matrices of the element, given by

$$[P]_e = \int_{V_e} [N][N]^T dV_e, \quad (7.41)$$

$$[M]_e = \int_{V_e} [\nabla N][\nabla N]^T dV_e, \quad (7.42)$$

where $[F]_e$ is the forcing vector given by

$$[F]_e = \int_{S_u} \{u_n\}\{N\}^T dS, \quad (7.43)$$

and subscript 'e' denotes element. For a rigid boundary and for internal elements, the forcing vector would reduce to a null vector.

In actual numerical analysis, we deal with scalar numbers, not vectors. Therefore, for three-dimensional elements,

$$[\nabla N] = [N_x N_y N_z]. \quad (7.44)$$

If $\{N\}$ is an 8×1 (column) matrix, then $\{N_x\}$, $\{N_y\}$, and $\{N_z\}$ are also 8×1 (column) matrices representing partial derivatives of N with respect to x, y, and z, respectively. Thus, $[N]$ is an 8×3 matrix with all three columns identically equal to $\{N\}$.

The relationships between partial derivatives in the Cartesian coordinate plane and the natural coordinate plane can be written in the usual way:

$$\begin{bmatrix} \partial/\partial r \\ \partial/\partial s \\ \partial/\partial t \end{bmatrix} = [J] \begin{bmatrix} \partial/\partial x \\ \partial/\partial y \\ \partial/\partial z \end{bmatrix}, \qquad (7.45)$$

where $[J]$ is the Jacobian

$$[J] \equiv \begin{bmatrix} \partial x/\partial r & \partial y/\partial r & \partial z/\partial r \\ \partial x/\partial s & \partial y/\partial s & \partial z/\partial s \\ \partial x/\partial t & \partial y/\partial t & \partial z/\partial t \end{bmatrix}, \qquad (7.46)$$

and

$$dV \equiv dx\,dy\,dz = |J|\,dr\,ds\,dt. \qquad (7.47)$$

Thus, the stiffness matrix of an element [Equation 7.41] is given by

$$[P]_e = \int_{V_e} [N][N]^T |J|\,dr\,ds\,dt. \qquad (7.48)$$

and the inertia matrix [Equation 7.42] is given by

$$\begin{aligned} [M]_e &= \int_{V_e} [N_x\,N_y\,N_z][N_x\,N_y\,N_z]^T dx\,dy\,dz \\ &= \int_{V_e} [N_r\,N_s\,N_t][J]^{-1} \left([J]^{-1}\right)^T [N_r\,N_s\,N_r]^T dr\,ds\,dt, \end{aligned} \qquad (7.49)$$

where

$$[J]^{-1} \equiv \begin{bmatrix} \partial r/\partial x & \partial s/\partial x & \partial t/\partial x \\ \partial r/\partial y & \partial s/\partial y & \partial t/\partial y \\ \partial r/\partial z & \partial s/\partial z & \partial t/\partial z \end{bmatrix}. \qquad (7.50)$$

Craggs integrated kernels in Equations 7.48 and 7.49 by using a 27-point Gauss scheme [19].

Equations of all finite elements constituting the muffler may finally be assembled into the matrix form (see Equation 7.40)

$$([M] - k_0^2[P])\{p_n\} = -j\rho_0\omega\{F\}, \qquad (7.51)$$

where the vector p_n represents the nodal pressures.

For dissipative mufflers, following Gladwell [12] and Morse and Ingard [51], use is made of an adjoint system which gains energy that the original system loses. The adjoint system may be looked upon as an image system of the original, having negative damping. In this case, a suitable functional, corresponding to the Lagrangian (7.37) is [12,19]

$$L = \int_V \left[\frac{pq^*}{2\rho_0 c_0^2} - \frac{1}{2\rho_0 \omega^2} (\text{grad } p) \cdot (\text{grad } q^*) \right] dV + \frac{1}{j\omega} \int_{S_u} A_n p q^* dS_n, \quad (7.52)$$

where

A_n is admittance, $1/Z_n$,
p is the acoustic pressure within the system, and
q is the acoustic pressure within the adjoint system.

This leads to the equation [12]

$$([M] - k_0^2[P] + j\rho_0\omega[C])\{p_n\} = -j\rho_0\omega\{F\} \quad (7.53)$$

(see Equation 7.51). Here $[C]$ is the damping matrix resulting from complex impedances Z_n, and hence admittances A_n, at one (or more) of the surfaces. Thus for an element e,

$$[C]_e = \sum_n \int_{S_e} A_n [N][N]^T dS_e. \quad (7.54)$$

The equations for the adjoint system are identical to Equation 7.53, except that the signs of the real components of A_n, and hence of $[C]$ and $\{F\}$, are reversed. However, because there is no link between p and q, the adjoint system equations do not have to be solved and computations need only be carried out on Equation 7.53.

Craggs [12] follows this approach of Gladwell but formulates his problem in terms of real and imaginary parts of the acoustic pressure rather than the complex pressure and its complex conjugate. Earlier he used it for flexible membranes [15] and Joppa and Fyfe used it for permeable membranes [25].

Ross [43] has extended this adjoint system approach to analysis of perforated-component acoustic systems, which includes concentric-tube resonators and plug mufflers. He considers the mean-flow effect on perforate impedance, but neglects the convective effect of mean flow. Neglecting (for convenience) the existence of other boundaries, for schematic of Figure 7.11, the functional of Equation 7.52 can be expanded to include both the enclosures as follows [43]:

$$L = \frac{1}{2\rho_0 c_0^2} \left[\int_{V_1} (p_1 q_1^* + p_1^* q_1) dV_1 + \int_{V_2} (p_2 q_2^* + p_2^* q_2) dV_2 \right]$$

$$- \frac{1}{2\rho_0 \omega^2} \left[\int_{V_1} (\text{grad } p_1 \cdot \text{grad } q_1^* + \text{grad } p_1^* \cdot \text{grad } q_1) dV_1 \right.$$

$$\left. + \int_{V_2} (\text{grad } p_2 \cdot \text{grad } q_2^* + \text{grad } p_2^* \cdot \text{grad } q_2) dV_2 \right] + \frac{1}{j\omega} \int_{S_{12}} [(A_{12} p_1 q_1^* - A_{12}^* p_1^* q_1)$$

$$- (A_{12} p_1 q_2^* - A_{12}^* p_1^* q_2) + (A_{12} p_2 q_2^* - A_{12}^* p_2^* q_2) - (A_{12} p_2 q_1^* - A_{12}^* p_2^* q_1)] dS_{12}, \quad (7.55)$$

where A_{12} is the perforate admittance $u/(p_1 - p_2)$.

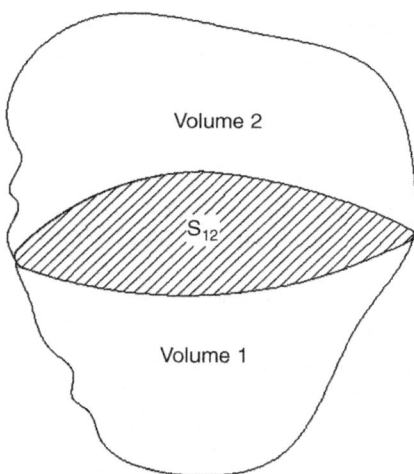

Figure 7.11 Schematic of two general acoustic volume enclosures with a common perforated boundary S_{12} (Reproduced with permission from [43])

Applying Hamilton's principle to Equation 7.55, varying q^* arbitrarily, discretizing all the terms of the Lagrangian by means of Equation 7.24 for every finite element, and combining all element equations, we get

$$([M] - k_0^2[P] + j\rho_0 \omega A_{12}([C]_{11} + [C]_{22} - [C]_{12} + [C]_{21}))\{p_n\} = \{0\}, \quad (7.56)$$

where, for an element e, $[P]_e$ and $[M]_e$ are given by Equations 7.41 and 7.42. For the elements connected along with perforated boundary S_{12}, the $[C]_{e,ij}$ matrices are defined by [43]

$$[C]_{e,ij} = \int_{S_{12}} [N]_i [N]_j^T dS_{12}, \quad i,j = 1,2. \quad (7.57)$$

The similarity (conceptual as well as formal) of Equations 7.56 and 7.57 with Equations 7.53 and 7.54 is obvious.

It may be noted that the variational formulation does not account for the convective effect of mean flow. This is better taken into account through the Galerkin approach discussed in the following section. This approach requires a governing equation, instead of energy functional, and is therefore more general in application.

7.2.3 The Galerkin Formulation of Finite Element Equations

The Galerkin method is the most popular of the residual methods, the others being the collocation method and the least squares method. It starts with the governing differential equation

$$\mathcal{L}p = f, \quad (7.58)$$

where p is a field variable, $\mathcal{L}$ is some differential operator, and f is a known forcing function. Then the residual R for an approximate trial solution p_n is defined as

$$R = \mathcal{L}p_a - f. \tag{7.59}$$

In finite element formulation, p_a would be a field variable. The various residual methods are based upon different techniques for minimizing the residual. The Galerkin method commences with the initial assumption of a trial solution of the type of Equation 7.24, that is,

$$p_a = \{N\}^T \{p_n\}, \tag{7.60}$$

where $\{p_n\}$ represents the nodal values of pressure and $\{N\}$ is a basic function vector.

This trial solution is substituted into Equation 7.59 to form a residual, which, when orthogonalized with respect to a complete set of weighting functions $\{W\}$, yields a set of n equations, where n is the total number of nodal degrees of freedom of a finite element, as follows:

$$\sum \int_{V_e} \{W\} R \, dV_e = \{0\}. \tag{7.61}$$

The integration over volume of the element in Equation 7.59 is carried out by parts to obtain weighted equations that would include both field and boundary residuals. The substitutions of the basic functions and weighting functions then yield a set of n linear algebraic equations that may be arranged in the usual matrix form.

In the conventional form of Galerkin formulation, the basic function N_i and the weighting functions W_i are selected to be the standard global finite element shape functions as defined in Equation 7.24, preferably in a natural coordinate system.

This method is illustrated hereunder for deriving the finite element equations for a duct with mean flow. Following Peat [44], equations have been written in terms of nondimensionalized velocity potential ϕ given by

$$\phi = \overline{\phi} + \phi' e^{j\omega t}, \quad \phi'/\overline{\phi} = 0(\alpha), \quad \alpha \ll 1, \tag{7.62}$$

where, for the linear case, the mean potential $\overline{\phi}$ and acoustic perturbation ϕ' are governed by the equations [32,44]

$$\nabla^2 \overline{\phi} = 0 \tag{7.63}$$

and

$$c_o^2 \nabla^2 \phi' + \omega^2 \phi' - 2j\omega \nabla \overline{\phi} \cdot \nabla \phi' = 0, \tag{7.64}$$

respectively. Here ∇, ω, $\overline{\phi}$ and ϕ' have their usual dimensions, unlike their nondimensional counterparts, made use of by Peat.

Because the mean-flow Mach number would not be uniform, in general, Equation 7.63 is solved first and the results are then incorporated into Equation 7.64 to determine ϕ', the acoustic perturbation. $\overline{\phi}$ and ϕ' are replaced by their finite element approximations in terms of their modal values,

$$\overline{\phi} = \{N\}^T\{\overline{\phi}_n\} \tag{7.65}$$

and

$$\phi' = \{N\}^T\{\phi'_n\}. \tag{7.66}$$

These are then substituted in Equations 7.63 and 7.64 and use is made of Equation 7.61 to obtain

$$\int_V (\{\nabla N\} \cdot \{\nabla N\}^T \{\overline{\phi}_n\}) dV = \{0\} \tag{7.67}$$

and

$$\int_V \left(c_0^2\{\nabla N\} \cdot \{\nabla N\}^T + \{N\}(\omega)^2\{N\}^T - 2j\omega\{\nabla N\}^T\{\overline{\phi}_n\}\{N\}\{\nabla N\}^T\right) \times \{\phi'_n\} dV = \{0\}. \tag{7.68}$$

Incidentally, it may be noticed that for the case of a stationary medium, Equation 7.68 reduces to Equation 7.40. Now Green's theorem can be used to reduce the volume integrals in both these equations to surface integrals [44]

$$\int_S \{N\}(\{\nabla N\}^T \cdot n dS)\{\overline{\phi}_n\} = \{0\}, \tag{7.69}$$

$$\int_S \{N\}(\{\nabla N\}^T \cdot n dS)\{\phi'_n\} = \{0\}, \tag{7.70}$$

where S is the surface enclosing the volume V and n is the unit outward normal to surface S.

For axisymmetric ducts, the surface integrals in Equations 7.69 and 7.70 may be replaced by line integrals [14]

$$\int_\Gamma r\{N\}\frac{\partial \overline{\phi}}{\partial n} d\Gamma = \{0\}, \tag{7.71}$$

and

$$\int_\Gamma r\{N\}\frac{\partial \phi'}{\partial n} d\Gamma = \{0\}, \tag{7.72}$$

where Γ is the contour of the duct and r is the distance from the line of axisymmetry. The physical state variables are related to the velocity potential by

$$u = \nabla\phi, \tag{7.73}$$

$$p = -\rho_0 \frac{D\phi}{Dt}. \tag{7.74}$$

Thus,

Mean velocity

$$\bar{u} = Mc_0 = \nabla\bar{\phi}; \tag{7.75}$$

Acoustic velocity

$$u' = \nabla\phi'; \tag{7.76}$$

Acoustic pressure

$$p' = -\rho_0 \frac{D\phi'}{Dt} = -\rho_0 \left(j\omega\phi' + \nabla\bar{\phi} \cdot \nabla\phi' \right). \tag{7.77}$$

As indicated earlier, Equations 7.67 or 7.69 or 7.71 is solved for mean-flow velocity distribution, and then Equations 7.68 or 7.70 or 7.72 is solved for evaluation of acoustic perturbation on velocity potential and then acoustic pressure and particle velocity.

7.2.4 Evaluation of Overall Performance of a Muffler

As discussed in Chapters 2 and 3, the overall performance of a muffler is evaluated in terms of insertion loss or transmission loss or level difference. All of these parameters can be evaluated from the overall transfer matrix of the muffler and radiation impedance and source impedance (relevant to insertion loss only).

State variables for the transfer matrix are acoustic pressure and mass (or volume) velocity. The latter implies plane wave propagation to the exclusion of higher modes or three-dimensional waves. This is indeed the case in the case of reciprocating-engine mufflers; the exhaust pipe and tail pipe (the two terminal elements) are generally too small in diameter to allow propagation of anything but plane (one-dimensional) waves at the frequencies of interest. This feature allows us to define an overall transfer matrix, notwithstanding the existence of three-dimensional waves in the chamber of the muffler (elsewhere called the muffler proper).

The overall transfer matrix relation can be written as

$$\begin{bmatrix} p_n \\ v_n \end{bmatrix} = \begin{bmatrix} A_{11} & A_{12} \\ A_{21} & A_{22} \end{bmatrix} \begin{bmatrix} p_1 \\ v_1 \end{bmatrix}, \tag{7.78}$$

where p and v are, respectively, acoustic pressure and mass velocity, subscripts n and 1 stands for exhaust pipe and tail pipe, respectively (Figure 7.12), and four-pole parameters are given by

$$A_{11} = \left.\frac{p_n}{p_1}\right|_{v_1=0} = \left.\frac{\phi'_n}{\phi'_1}\right|_{\frac{\partial \phi'_1}{\partial n}=0} \tag{7.79}$$

$$A_{12} = \left.\frac{p_n}{v_1}\right|_{p_1=0} = j\left.\frac{\omega p_n}{S_1 \partial p_1/\partial n}\right|_{p_1=0} = -j\left.\frac{\omega \phi'_n}{S_1 \partial \phi'_1/\partial n}\right|_{\phi'_1 = -\frac{j}{\omega}\frac{\partial \bar{\phi}_1}{\partial n}\frac{\partial \phi'_1}{\partial n}} \tag{7.80}$$

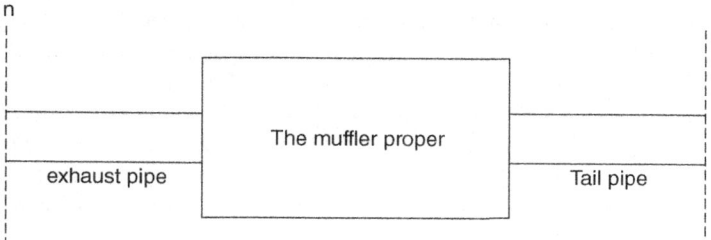

Figure 7.12 Schematic of a muffler

$$A_{21} = \left.\frac{v_n}{p_1}\right|_{v_1=0} = \left.j\frac{S_n \partial p_n/\partial n}{\omega p_1}\right|_{v_1=0} = \left.-j\frac{S_n \partial \phi'_n/\partial n}{\omega \phi'_1}\right|_{\partial \phi'_1/\partial n=0} \quad (7.81)$$

$$A_{22} = \left.\frac{v_n}{v_1}\right|_{p_1=0} = \left.j\frac{S_n}{S_1}\frac{\partial p_n/\partial n}{\partial p_1/\partial n}\right|_{p_1=0} = \left.\frac{S_n}{S_1}\frac{\partial \phi'_n/\partial n}{\partial \phi'_1/\partial n}\right|_{\phi'_1=\frac{j}{\omega}\frac{\partial \bar{\phi}_1}{\partial n}\frac{j\phi'_1}{\partial n}} \quad (7.82)$$

In all of the Equations 7.79–7.82 for the four-pole parameters,

$$\frac{\partial \bar{\phi}_1}{\partial n} = \bar{u}_1 = U_1, \quad (7.83)$$

which means that $\bar{u}_1$ represents the average value of mean velocity across the cross section of the tail pipe.

All the boundary conditions indicated in Equations 7.79–7.83 are indicated together in Figure 7.13. The additional conditions relevant to an axisymmetric hard-walled duct are also included there. The finite element equations are formulated and solved separately for evaluation of each of the four-pole parameters, A_{11}, A_{12}, A_{21} and A_{22} incorporating the relevant boundary conditions.

This kind of simplified analysis of the radiation condition would not be adequate in other applications, like arbitrary nacelle geometries of turbofan engines, where the length scale of disturbances is much smaller than the transverse dimension of the nacelle. An alternative approach is to match the numerical solution at the boundary of relatively small inner domain

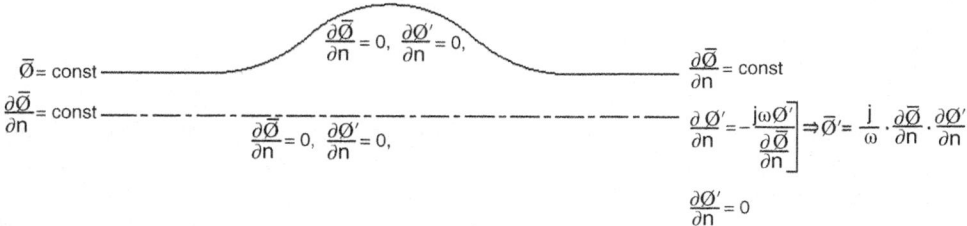

Figure 7.13 Boundary conditions for an axisymmetric hard-walled duct (Reproduced with permission from [37])

to an analytical far-field solution through an appropriate eigenfunction expansion at the matching surface [52] or via source distributions or related boundary integral equation techniques [18,27]. Astley and Eversman match a conventional finite element solution in the inner region to a 'wave envelope' finite element solution [42] in a large but finite outer region [46]. Fortunately however, none of these special cumbersome, and time-consuming techniques are necessary for the muffler of a reciprocating engine, the tail pipe of which is invariably of a small enough diameter to allow propagation of nothing but plane waves.

7.2.5 Numerical Computation

The first step in preparing the problem for numerical computation is to divide the total volume into an 'appropriate' number of finite elements, noting in the process that the greater the number of elements, the greater the precomputation effort, core memory requirements, and computation time. But what really matters is the number of nodes. In general, the more nodes per element (the higher the order of the element), the smaller will be the number of elements needed to resolve the acoustic field. On the other hand, since the size of the overall (global) matrix is proportional to the number of nodes, considerably more computational time and computer core memory are required for the higher-order elements. If the number of dependent variables equals N, the number of nodes, then

$$\text{core memory requirements} \propto N^2, \quad (7.84)$$

$$\text{solution time} \propto N^3; \quad (7.85)$$

hence the need for the least 'appropriate' value of N for required accuracy at the highest frequency of interest.

In this connection it is worth nothing that the C^1 continuity problem (in which continuity of the field variable p and its partial derivatives $\partial p/\partial x, \partial p/\partial y$, and $\partial p/\partial z$ is ensured) increases the unknowns to four times those of the C^0 continuity problem (in which continuity of only p is ensured), thereby increasing the core memory requirements 16 times and solution time 64 times. The continuity of $\partial p/\partial x, \partial p/\partial y$, and $\partial p/\partial z$ can be affected by means of Hermite polynomials for interpolation [9,10,14,29,40] instead of the usual Lagrange polynomials that are used for the C^0 continuity problem [38].

A quantitative idea is provided by Ross, who used eight-noded isoparametric elements. In this mesh configuration, he incorporated elements, the largest dimension of which was of the order of one-third of a wavelength at 2000 Hz. The arrangement (with measurements) was good up to 1300 Hz [43]. This seems to suggest the following general guideline:

$$\text{maximum typical dimension of a finite element} \leq \lambda_{\min}/6. \quad (7.86)$$

A major reduction in the number of nodes may be obtained by reducing the dimensionality of the elements from three to two or one, where applicable. For example, as indicated earlier in the chapter. Young and Crocker used, with commendable success, two-dimensional elements for elliptical chamber (Figure 7.14), the minor axis of which was considerably less than the smallest wavelength of interest [9,10,14]; and Craggs used one-dimensional (simple pipe) elements for analysis of acoustic propagation in hard-walled curved pipes with 'small'

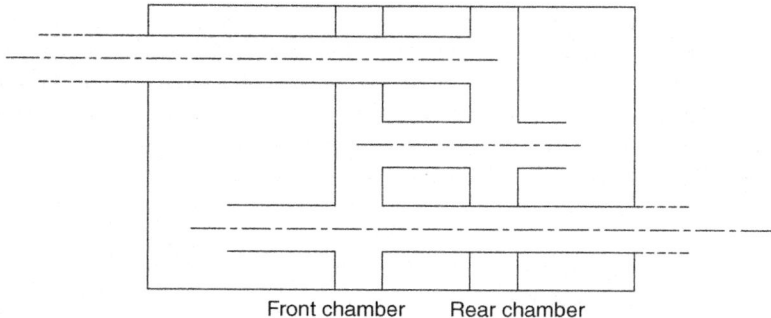

Figure 7.14 A double reversing end chamber-resonator combination (Reproduced with permission from [17])

transverse dimensions and in branched systems [38]. One may also employ with advantage finite elements of different dimensionalities for different parts of a given muffler.

The size of the solution (global) matrix may also be reduced by making use of the zero boundary conditions (even at the cost of destroying the numbering scheme). Young and Crocker were able to reduce the solution time to one-third of its original value in this way. A further 50% reduction in solution time was obtained by partitioning of the global matrix as follows.

Let there be n nodes out of which n_1 may be touching an active boundary and the remaining n_2 may be entirely 'internal' to the system. Thus, the global matrix equation may be rearranged and partitioned as

$$\begin{bmatrix} A_{11} & A_{12} \\ A_{21} & A_{22} \end{bmatrix} \begin{bmatrix} p^{(1)} \\ p^{(2)} \end{bmatrix} = \begin{bmatrix} F \\ 0 \end{bmatrix}, \quad (7.87)$$

where only $\{p^{(1)}\}^T = [p_1 p_2 \ldots p_{n1}]$ are of interest, constituting the external set of nodes, and generally

$$[A_{21}] = [A_{12}]^T.$$

The matrix Equation 7.87 may be solved to obtain

$$\{p^{(2)}\} = -[A_{22}]^{-1}[A_{12}]^T\{p^{(1)}\} \quad (7.88)$$

and thence

$$\left([A_{11}] - [A_{12}][A_{22}]^{-1}[A_{12}]^T\right)\{p^{(1)}\} = \{F\}. \quad (7.89)$$

This saves a great deal of solution time indeed.

In conclusion, the computation sequence involves six steps, namely:

i. Division of the system continuum into an appropriate number of finite elements and nodes as discussed in Section 7.2.1.

ii. Generation of element matrices by means of the variational method (if the energy functional is known) discussed in Section 7.2.2, or by the more general Galerkin method (if the governing differential equations and boundary conditions are available) discussed in Section 7.2.3.
iii. Formation of the total dynamic equations and matrix (also called the global matrix) from the element matrices and their modification by means of the appropriate boundary conditions.
iv. Reduction of the system matrix equation by partitioning.
v. Solution of the reduced system for nodal unknowns by the usual matrix method using standard subroutines.
vi. Evaluation of muffler performance from the values of the field variables at the nodes, as discussed in Section 7.2.4 above.

Computationally efficient and user-friendly softwares are now commercially available for machine meshing of complex muffler configurations, fast numerical solutions and post-processing of the results. However, large pre-computation effort makes the finite element analysis a last resort. In the following sections, we discuss a couple of analytical techniques.

7.3 Green's Function Method for a Rectangular Cavity

Rectangular expansion chamber silencers are extensively used in HVAC ducts as a plenum chamber and other industrial applications. Four-pole parameters are very useful for the analysis of complex HVAC system with different duct configurations.

Kim and Kang [53] gave a general formulation to derive the transfer matrix based on Green's functions for a circular expansion chamber with arbitrary locations of inlet, outlet port, termination conditions, and validated the results with experiments. Ih [54] discussed the transfer matrix formulation for different configurations of rectangular plenum chamber using the modal expansion method and validated the same with available literature. Venkatesham et al. [55] discussed the same problem as in [54] using Green's function to calculate the inside pressure. It is similar to that for a circular expansion chamber [53]. The assumptions involved in the formulation are rigid walls, no mean flow, and no acoustic source inside the plenum chamber. The Green's function approach has an advantage in modeling the absorptive boundary conditions and also flexible walls. This section deals with the rigid-wall chambers though. It follows [55] closely.

The rectangular expansion chamber is modeled as a piston-driven rectangular cavity where the pistons are fluctuating harmonically. The four-pole parameters are explicitly given in a very simple form using Green's function of rectangular cavity in terms of a finite number of modes. Initially, the four-pole parameters of the offset inlet and outlet configuration are derived. The four-pole parameters of other inlet/outlet configurations are rewritten from the offset configuration without further calculations. The derived transfer matrix may be combined with the transfer matrices of the muffler or system elements upstream and downstream in order to predict the overall performance of the system in terms of TL, IL or LD.

7.3.1 Derivation of the Green's Function

Green's function, $G(\vec{x}|\vec{x}_0)$, is the field at the observer's point $\vec{x}$ caused by a unit point source at the source point $\vec{x}_0$. For harmonic time dependence, it is governed by the inhomogeneous wave equation

$$\nabla^2 G(\vec{x}|\vec{x}_0) + k^2 G(\vec{x}|\vec{x}_0) = -4\pi\delta(\vec{x}|\vec{x}_0) \qquad (7.90)$$

Green's function G satisfies the reciprocity relation:

$$G(\vec{x}_0|\vec{x}) = G(\vec{x}|\vec{x}_0) \tag{7.91}$$

Figure 7.15 shows the schematic diagrams of a simple rectangular expansion chamber with different inlet and outlet configurations. If the dimensions of the inlet and outlet port were small compared to wavelength, then the uniform velocity assumption would be valid. Green's function for rectangular cavity with rigid walls in terms of the cavity mode shapes is given by [56]

$$G(\vec{x}|\vec{x}_0) = \sum_{mpn} \frac{\overline{\psi}_{mpn}(\vec{x}_0)\psi_{mpn}(\vec{x})}{k_{mpn}^2 - k^2} \tag{7.92a}$$

Alternatively [37],

$$G(\vec{x}|\vec{x}_0) = \sum_{mpn} \frac{\psi_{mpn}(\vec{x}_0)\psi_{mpn}(\vec{x})}{\left(k_{mpn}^2 - k^2\right)\int_V \{\psi_{mpn}(\vec{x})\}^2 dV} \tag{7.92b}$$

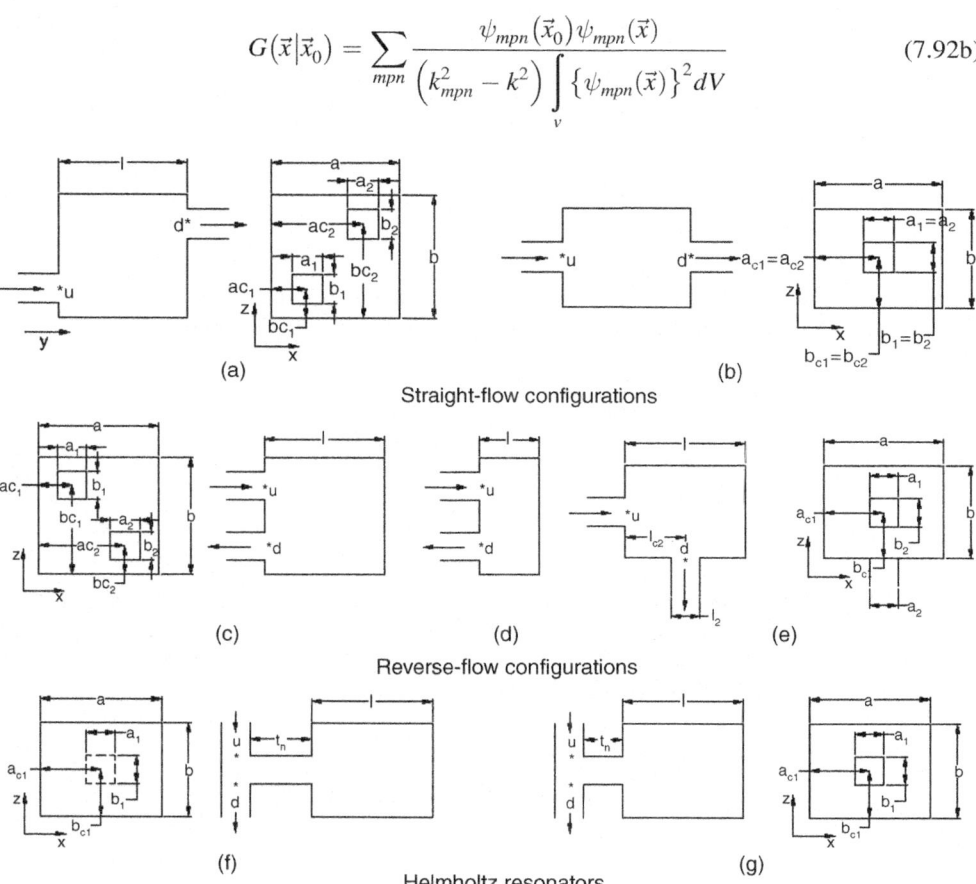

Figure 7.15 Schematic diagrams of rectangular expansion chamber with different inlet and outlet configurations (a) Offset inlet, outlet (two views); (b) Centered inlet, outlet; (c) Reverse flow expansion with long end-chamber; (d) Reverse flow expansion with short end-chamber; (e) End-in, side-out chamber; (f) Long neck Helmholtz-resonator; (g) Short neck Helmholtz-resonator (Reproduced with permission from [55])

Here, $k = \omega/c_0$ is the wave number, and the mode shape $\psi_{mpn}(\vec{x})$ of a rectangular cavity is given by [56]

$$\psi_{mpn}(\vec{x}) = \cos\frac{m\pi x}{a}\cos\frac{p\pi y}{l}\cos\frac{n\pi z}{b}, \tag{7.93}$$

with the wave number given by

$$k_{mpn} = \left[\left(\frac{m\pi}{a}\right)^2 + \left(\frac{p\pi}{l}\right)^2 + \left(\frac{n\pi}{b}\right)^2\right]^{\frac{1}{2}} \tag{7.94a}$$

and the corresponding natural frequency by

$$\omega_{mpn} = \pi c_0 \left[\left(\frac{m}{a}\right)^2 + \left(\frac{p}{l}\right)^2 + \left(\frac{n}{b}\right)^2\right]^{\frac{1}{2}} \tag{7.94b}$$

The orthonormal function $\overline{\psi}_{mpn}(\vec{x})$ is related to the mode shape function $\psi_{mpn}(\vec{x})$ as follows [56]:

$$\iiint_V \overline{\psi}_{q'}(\vec{x})\psi_q(\vec{x})dV = \delta_{qq'} \tag{7.95}$$

where q and q' are generic subscripts representing m, p and n as a set.

Thus, it can readily be proved that the orthonormal function $\overline{\psi}_{mpn}(\vec{x})$ is given by

$$\overline{\psi}_{mpn}(\vec{x}) = \frac{\psi_{mpn}(\vec{x})}{\int_V \{\psi_{mpn}(\vec{x})\}^2 dV}$$

$$= \frac{e_m e_p e_n}{V}\psi_{mpn}(\vec{x}). \tag{7.96}$$

Here, $\vec{x}$ and $\vec{x}_0$ are the response coordinates vector and source coordinates vector, respectively; n, p and m are integers; a, l and b are dimensions (in meters) of the rectangular expansion chamber in the x, y and z coordinate directions, as shown in Figure 7.15a; and V is volume of the chamber ($V = alb$). The normalization factors are given by

$e_m = 1$ if $m = 0$ and $e_m = 2$ if $m \geq 1$,

$e_n = 1$ if $n = 0$ and $e_n = 2$ if $n \geq 1$,

$e_p = 1$ if $p = 0$ and $e_p = 2$ if $p \geq 1$.

Substituting Equation 7.96 into Equation 7.92 yields

$$G(\vec{x}|\vec{x}_0) = \sum_{mpn} \frac{e_m e_p e_n}{V} \frac{\psi_{mpn}(\vec{x})\psi_{mpn}(\vec{x}_0)}{k_{mpn}^2 - k^2}. \tag{7.97}$$

The Helmholtz equation that determines the acoustic field in the rectangular chamber can be expressed as [56]

$$\nabla^2 \phi + k^2 \phi = -Q(\vec{x}), \tag{7.98}$$

where Q is the embedded source, if any, for the sake of generality.

For harmonic excitation, the velocity potential ϕ is related to the state variables acoustic pressure p and particle velocity u as follows:

$$p(\vec{x}) = -\rho \frac{\partial \phi(\vec{x}, t)}{\partial t} = -\rho j \omega \phi(\vec{x}) \tag{7.99}$$

and

$$u(\vec{x}) = \text{grad } \phi(\vec{x}) = \nabla \phi(\vec{x}) \tag{7.100}$$

General solution to Equation 7.98 in terms of Green's function is given by [56]

$$\phi(\vec{x}) = -\iint_S \{G(\vec{x}|\vec{x}_0)\nabla\phi(\vec{x}_0) - \phi(\vec{x}_0)\nabla_0 G(\vec{x}|\vec{x}_0)\} \cdot dS_0 + 4\pi \iiint_V Q(\vec{x}_0) G(\vec{x}|\vec{x}_0) dV_0 \tag{7.101}$$

where subscript 0 denotes the source.

In the present problem, $Q(\vec{x}_0) = 0$ and hypothetical pistons at the two ports in Figure 7.15 represent the two surface sources.

7.3.2 Derivation of the Velocity Potential

The principle of linear superposition is applicable here. Therefore, total velocity potential inside the chamber is expressed by superimposing the velocity potentials generated due to the inlet and outlet piston sources. Thus, the expression for total velocity potential can be deduced directly from Equation 7.101. Thus,

$$\phi(\vec{x}) = \phi_1(\vec{x}) + \phi_2(\vec{x}) \tag{7.102}$$

where, referring to Figure 7.15a,

$$\phi_1(\vec{x}) = -\iint_{piston1} G(\vec{x}|\vec{x}_0) N_1(x_0, z_0) dx dz \tag{7.103}$$

and

$$\phi_2(\vec{x}) = -\iint_{piston2} G(\vec{x}|\vec{x}_0) N_2(x_0, z_0) dx dz \tag{7.104}$$

In Equations 7.103 and 7.104, functions N_1 and N_2 may be shown to be [53]

$$N_1(x_0, z_0) = u_1 \left[H\left\{ x - \left(a_{c1} - \frac{a_1}{2} \right) \right\} - H\left\{ x - \left(a_{c1} + \frac{a_1}{2} \right) \right\} \right]$$
$$\times \left[H\left\{ z - \left(b_{c1} - \frac{b_1}{2} \right) \right\} - H\left\{ z - \left(b_{c1} + \frac{b_1}{2} \right) \right\} \right] \equiv u_1 f_1(x_0, z_0) \quad (7.105)$$

$$N_2(x_0, z_0) = u_2 \left[H\left\{ x - \left(a_{c2} - \frac{a_2}{2} \right) \right\} - H\left\{ x - \left(a_{c2} + \frac{a_2}{2} \right) \right\} \right]$$
$$\times \left[H\left\{ z - \left(b_{c2} - \frac{b_2}{2} \right) \right\} - H\left\{ z - \left(b_{c2} + \frac{b_2}{2} \right) \right\} \right] \equiv u_2 f_2(x_0, z_0) \quad (7.106)$$

where $H(\cdot)$ is Heaviside function (or the Unit Step function), and u_1 and u_2 are complex amplitudes of velocities of the hypothetical pistons at the inlet and outlet ports (or junctions), respectively.

Substituting the Green's function expression (7.97) into Equation 7.103 gives

$$\phi_1(x, y, z) = -\int_{b_{c1} - \frac{b_1}{2}}^{b_{c1} + \frac{b_1}{2}} \int_{a_{c1} - \frac{a_1}{2}}^{a_{c1} + \frac{a_1}{2}} \sum_{mpn} \frac{e_m e_p e_n}{V} \frac{\psi_{mpn}(x, y, z)}{k_{mpn}^2 - k^2} \psi_{mpn}(x_0, y_0, z_0) u_1 f_1(x_0, z_0) dx_0 dz_0 \quad (7.107)$$

or

$$\phi_1(x, y, z) = -u_1 \sum_{mpn} \frac{e_m e_p e_n}{V} \frac{\psi_{mpn}(\vec{x})}{k_{mpn}^2 - k^2} \left[\cos \frac{p \pi y_0}{l} C_{m1} C_{n1} \right], \quad (7.108)$$

where

$$C_{m1} = \begin{cases} a_1 & \text{for } m = 0 \\ \dfrac{2a}{m\pi} \cos\left(\dfrac{m\pi a_{c1}}{a}\right) \sin\left(\dfrac{m\pi a_1}{2a}\right) & \text{for } m \neq 0 \end{cases} \quad (7.109)$$

and

$$C_{n1} = \begin{cases} b_1 & \text{for } n = 0 \\ \dfrac{2b}{n\pi} \cos\left(\dfrac{n\pi b_{c1}}{b}\right) \sin\left(\dfrac{n\pi b_1}{2b}\right) & \text{for } n \neq 0 \end{cases} \quad (7.110)$$

Similar expressions would hold for velocity potential ϕ_2 due to the piston source at the outlet. Thus, in general, the velocity potential caused by port i can be written as

$$\varphi_i(\vec{x}) = \sum_{mpn} S^i_{mpn} \psi_{mpn}(\vec{x}), \quad (7.111)$$

where

$$S^i_{mpn} = \frac{(-1)^i e_m e_p e_n \cos(p\pi y_{io}/l)}{k^2_{mpn} - k^2} \frac{C_{mi} C_n u_i}{V} \quad (7.112)$$

and ψ_{mpn} is given by Equation 7.93, and C_{mi}, C_{ni} are given by Equations 7.109 and 7.110, with subscript 1 being replaced by i; where, $i = 1$ and 2, representing the two hypothetical pistons.

7.3.3 Derivation of the Transfer Matrix

The four-pole parameters may be derived from the velocity potentials by making use of Equations 7.99 and 7.100 that relate velocity potential to acoustic pressure and particle velocity.

The average sound pressure acting on the piston 'i' with cross-sectional area 'A_i' by the velocity field, which is generated by the rectangular piston 'i'', can be written as

$$\bar{p}_{ii'} = \frac{-j\omega\rho_0}{A_i} \int_{a_{ci}-\frac{a_i}{2}}^{a_{ci}+\frac{a_i}{2}} \int_{b_{ci}-\frac{b_i}{2}}^{b_{ci}+\frac{b_i}{2}} \phi_{i'} \, dz \, dx, \quad (7.113)$$

where overbar (¯) on p denotes average over the piston area.

Substituting Equations 7.111 and 7.112 into Equation 7.113 yields [55]

$$\bar{p}_{ii'} = \frac{(-1)^{i'}}{A_i A_{i'}} \cdot jY_0 v_{i'} \sum_{mpn} \frac{e_m e_p e_n}{l} \cdot \left(\frac{-k}{k^2_{mpn} - k^2}\right) \cos\left(\frac{p\pi y_{i'o}}{l}\right) \cos\left(\frac{p\pi y_{io}}{l}\right) C_{mi'} C_{ni'} C_{mi} C_{ni} \quad (7.114)$$

Here, $Y_0 = c_0/S$ is the characteristic impedance of the chamber, ρ_0 is mass density of the fluid (medium), c_0 is speed of sound, and $S = ab$ is the cross-sectional area of the chamber.

Similarly, defining the mass velocity $v_i = \rho_0 A_i u_i$, the averaged acoustic pressures on the surfaces of the two 'pistons' may be written as

$$\bar{p}_{ii'} = (-1)^{i'} jY_0 v_{i'} E_{ii'} \quad (7.115)$$

where [55]

$$E_{ii'} \equiv \frac{-1}{A_i A_{i'}} \sum_{mpn} \frac{e_m e_p e_n}{l} \cdot \left(\frac{k}{k^2_{mpn} - k^2}\right) \cos\left(\frac{p\pi y_{i'o}}{l}\right) \cos\left(\frac{p\pi y_{io}}{l}\right) C_{mi'} C_{ni'} C_{mi} C_{ni}, \quad (7.116a)$$

$$E_{ii} = \frac{-1}{A_i^2} \sum_{mpn} \frac{e_m e_p e_n}{l} \cdot \left(\frac{k}{k^2_{mpn} - k^2}\right) \cos^2\left(\frac{p\pi y_{io}}{l}\right) C_{mi}^2 C_{ni}^2, \quad i = 1 \text{ or } 2 \quad (7.116b)$$

Equation (7.116) indicates that $E_{ii'} = E_{i'i}$, which shows that these values are independent of the inlet and outlet port (or piston) positions. In other words, the E-functions satisfy the reciprocity principle.

The total sound pressure acting on the inlet and outlet ports or pistons can be evaluated readily by application of the principle of linear superposition. Thus,

$$\overline{p}_1 = \overline{p}_{11} + \overline{p}_{12} = -jv_1 Y_0 E_{11} + jv_2 Y_0 E_{12} \tag{7.117}$$

and

$$\overline{p}_2 = \overline{p}_{22} + \overline{p}_{21} = jv_2 Y_0 E_{22} - jv_1 Y_0 E_{21} \tag{7.118}$$

where $E_{21} = E_{12}$.

The transfer matrix of the rectangular expansion chamber of Figure 7.15a, relating the acoustic state variables at the inlet and outlet may now be written as

$$\begin{bmatrix} \overline{p}_1 \\ v_1 \end{bmatrix} = \begin{bmatrix} T_{11} & T_{12} \\ T_{21} & T_{22} \end{bmatrix} \begin{bmatrix} \overline{p}_2 \\ v_2 \end{bmatrix} \tag{7.119}$$

Use of Equations 7.117–7.119 yields the following expressions for the four-pole parameters $T_{ii'}$:

$$T_{11} = \left(\frac{\overline{p}_1}{\overline{p}_2}\right)_{v_2=0} = \frac{E_{11}}{E_{12}}; \quad T_{12} = \left(\frac{\overline{p}_1}{v_2}\right)_{\overline{p}_2=0} = jY_0\left(E_{12} - \frac{E_{11}E_{22}}{E_{12}}\right); \tag{7.120a,b}$$

$$T_{21} = \left(\frac{v_1}{\overline{p}_2}\right)_{v_2=0} = j(Y_0 E_{12})^{-1}; \quad T_{22} = \left(\frac{v_1}{v_2}\right)_{\overline{p}_2=0} = \frac{E_{22}}{E_{12}}. \tag{7.120c,d}$$

Finally, the transmission loss can be expressed in terms of the four-pole parameters as

$$TL = 20\log_{10}\left\{(Y_2/Y_1)^{1/2}|T_{11}+T_{12}/Y_2+T_{21}Y_1+T_{22}(Y_1/Y_2)|/2\right\}, \tag{7.121}$$

where Y_1 and Y_2 are characteristic impedances of the inlet duct and outlet duct, respectively: $Y_1 = c_0/S_1$ and $Y_2 = c_0/S_2$, where $S_1 = a_1 b_1$ and $S_2 = a_2 b_2$.

A special advantage of the Green's function method is that for different configurations shown in Figure 7.15 one has only to insert the new coordinates of the inlet/outlet ports (pistons), as shown below.

Straight-flow configurations:

The terms required to calculate the four-pole parameters of the straight-flow configurations shown in Figure 7.15a, b can be deduced from Equation 7.116 as:

$$E_{11} = \frac{-1}{(a_1 b_1)^2} \sum_{mpn} \frac{e_m e_p e_n}{l} \cdot \left(\frac{k}{k_{mpn}^2 - k^2}\right) C_{m1}^2 C_{n1}^2 \tag{7.122}$$

$$E_{12} = E_{21} = \frac{-1}{(a_1 b_1)(a_2 b_2)} \sum_{mpn} \frac{e_m e_p e_n}{l} \cdot \left(\frac{k}{k_{mpn}^2 - k^2}\right) C_{m1} C_{m2} C_{n1} C_{n2}(-1)^p \tag{7.123}$$

$$E_{22} = \frac{-1}{(a_2 b_2)^2} \sum_{mpn} \frac{e_m e_p e_n}{l} \cdot \left(\frac{k}{k_{mpn}^2 - k^2}\right) C_{m2}^2 C_{n2}^2 \qquad (7.124)$$

Reverse-flow configuration:

Generally in this configuration the inlet and outlet ports are on the same face and the flow direction is reversed as shown in Figure 7.15c and d. The quantities required to calculate the four-pole parameters in the reverse flow are E_{11}, E_{12} and E_{22}. Expressions for E_{11} and E_{22} are the same as for the straight-flow configurations. E_{12} is obtained from Equation 7.116a as

$$E_{12} = \frac{-1}{(a_1 b_1)(a_2 b_2)} \sum_{mpn} \frac{e_m e_p e_n}{l} \cdot \left(\frac{k}{k_{mpn}^2 - k^2}\right) C_{m1} C_{m2} C_{n1} C_{n2} \qquad (7.125)$$

The End-in Side-out configuration (EISO):

Figure 7.15e shows the end-in side-out cross-flow configuration. Generally in this configuration, inlet and outlet port are in the mutually perpendicular faces. The quantity required to calculate the four-pole parameter, E_{11}, is the same as for the straight-flow configuration, and the remaining quantities E_{12} and E_{22} are expressed by replacing b_2 with rectangular piston width l_2 in the reverse-flow configuration E_{12} and E_{22}:

$$E_{12} = \sum_{mpn} \frac{-1}{(a_1 b_1 a_2 l_2)} \frac{e_m e_p e_n}{l} \left(\frac{k}{k_{mpn}^2 - k^2}\right) (C_{m1} C_{m2} C_{n1} C_{n2}) \qquad (7.126)$$

$$E_{22} = \sum_{mpn} \frac{-1}{(a_2 l_2)^2} \frac{e_m e_p e_n}{l} \left(\frac{k}{k_{mpn}^2 - k^2}\right) (C_{m2} C_{n2})^2 \qquad (7.127)$$

Helmholtz resonators:

The four-pole parameters for Helmholtz resonator (Figure 7.15f and g) are expressed as [53]

$$T_{11} = 1, \quad T_{12} = 0, \quad T_{21} = \frac{j}{\left(Y_0 E_{11} - \frac{S\omega t_n}{a_1 b_1}\right)}, \quad T_{22} = 1 \qquad (7.128\text{a--d})$$

Here, acoustic inertance of the neck $S\omega t_n / a_1 b_1$ is added to the compliance of chamber volume in the four-pole parameters T_{21} of Helmholtz resonator. $S = ab$, and a_1 and b_1 are cross-dimensions of the neck. E_{11} represents the self-inertance and it is identical to E_{11} in the straight-flow configuration (see Equation 7.122).

7.3.4 Validation Against FEM

An FEM based numerical model has been used here to validate and compare the computational time between the analytical model and the numerical model. The 'FEM fluid' analysis has been performed in the commercial package named SYSNOISE [8]. A simple rectangular expansion chamber as shown in Figure 7.15b with dimensions of $a = b = 0.15$ m,

$a_1 = b_1 = a_2 = b_2 = 0.05$ m, $a_{c1} = b_{c1} = a_{c2} = b_{c2} = 0.125$ m. The inlet is provided with a uniform velocity source, and at the end termination impedance '$\rho_0 c_0 = 416.5$ kgm^{-2}s' is applied to simulate the anechoic termination condition. The transmission loss is calculated using the inlet and outlet pressure and velocities, and the resulting expression is given as [1].

$$TL = 20 \log \left| \frac{p_{in} + \rho_0 c_0 u_{in}}{2 \rho_0 c_0 u_{out}} \right| \quad (7.129)$$

A mesh of 4750 elements with 5736 nodes was used for these calculations. Two times the maximum transverse dimension of the piston has been used as the length of the inlet and outlet duct in order to ensure that evanescent waves generated at the inlet and outlet decay out and only the plane waves exist beyond that point. The existing mesh is valid up to a frequency of 5037 Hz, based on the assumption of a minimum of six elements per wavelength. The analysis is conducted from 10 Hz to 3000 Hz in steps of 10 Hz. Computations are carried out by using laptop (Dell latitude D620 with 1 GB ROM). The computational time for analytical model for 300 frequency steps is 13 seconds including I/O, whereas the numerical model computational time is 105 seconds, excluding the pre- and post-processing time.

Figure 7.16 shows the transmission loss of a rectangular chamber calculated by the analytical method presented here. It tallies so closely with that of the FE model that the comparison has not been shown there. Besides, the Green's function method captures the higher-order modes very well. It has been validated in Ref. [55] for all configurations shown in Figure 7.15. This analytical method has the additional advantage that a general subroutine or function can be prepared in FORTRAN or MATLAB® and the same can be incorporated in a general transfer matrix based muffler program (TMMP).

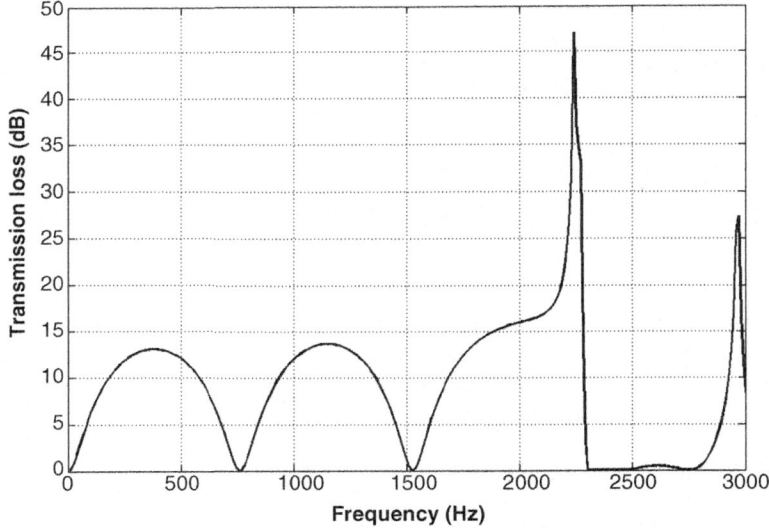

Figure 7.16 Prediction of the TL spectrum by means of the Green's function method for the centered inlet/outlet rectangular chamber of Figure 7.15(b) with $a = b = 0.15$ m, and $a_1 = b_1 = a_2 = b_2 = 0.05$ m (Adapted from [55])

7.4 Green's Function Method for Circular Cylindrical Chambers

The Green's function method presented in the preceding section for rectangular plenum can readily be extended to circular cylindrical chambers with circular inlet/outlet ports, that are used extensively in automotive mufflers. The main feature of this extension is the changeover from the Cartesian coordinates (x, y, z) to the cylindrical polar coordinates (r, θ, z).

Comparing Figure 7.17 with 7.15 and making use of the basic relationships developed in Sections 1.2 and 7.3, it can readily be seen that Equations 7.90–7.92 would apply here as well, but Equations 7.93 and 7.94 would get replaced with the following relationships [53]:

$$\phi_{mnp}(r) = J_m(k_{rmn}r)\cos(m\theta)\cos(k_{zp}z), \qquad (7.130)$$

$$k_{rmn} = \lambda_{mn}/R_c, k_{zp} = p\pi/L, k_{mnp}^2 = k_{mnp}^2 + k_{zp}^2. \qquad (7.131)$$

where m, n, and p represent the circumferential, radial, and longitudinal mode, respectively, λ_{mn} is a zero of $J'_m(\lambda_{mn})$, R_c and L are radius and length of the circular chamber shown in Figure 7.17, ϕ_{mnp} is the eigenfunction and k_{mnp} is the corresponding eigenfrequency or wave number for the (m, n, p) mode.

Use of orthogonality of the Bessel functions J_m (see Equations A.25 and A.26 of Appendix 1) and trigonometric (or circular) function in Equation 7.95 yields the following expression for the orthonormal set of eigenfunctions [53]:

$$\overline{\phi}_{mnp}(\vec{x}) = (N_r N_\theta N_z)^{-1} \phi_{mnp}(\vec{x}) \qquad (7.132)$$

where the normalizing factors N_r, N_θ, N_z for respective coordinates can be obtained as

$$N_r = \frac{R_c^2}{2}\left(1 - \frac{m^2}{(k_{rmn} + \mu_{mn})^2 R_c^2}\right) J_m^2(k_{rmn}R_c), \qquad (7.133a)$$

$$N_\theta = 2\pi/\varepsilon_m, \qquad (7.133b)$$

$$N_z = L/\varepsilon_p, \qquad (7.133c)$$

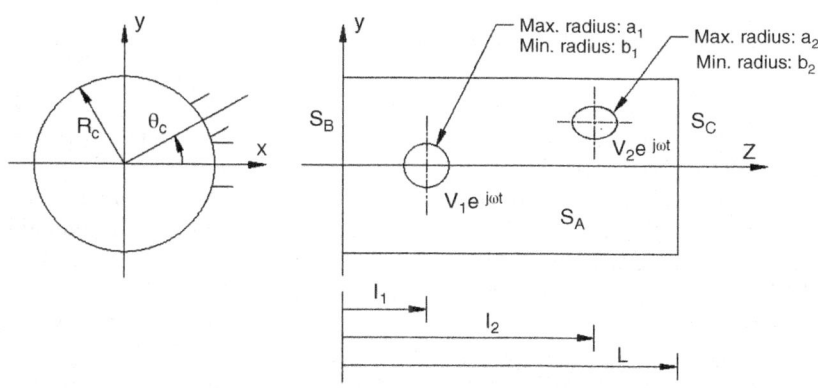

Figure 7.17 Circular expansion chamber with side-inlet/side-outlet ports (Adapted from [53])

where

$$\mu_{mn} = \begin{vmatrix} 1, & m=0, \ n=0, \\ 0, & \text{otherwise,} \end{vmatrix} \quad (7.134a)$$

$$\varepsilon_m, \varepsilon_p : \text{Neumann factor} = \begin{vmatrix} 1, & m=0, \ p=0, \\ 2, & \text{otherwise.} \end{vmatrix} \quad (7.134b)$$

Substituting Equation 7.133 into Equation 7.132 gives

$$\overline{\phi}_{mnp}(r) = \frac{\varepsilon_m \varepsilon_p (k_{rmn} + \mu_{mn})^2}{\pi L \left[(k_{rmn} + \mu_{mn})^2 R_c^2 - m^2 \right] J_m^2(k_{rmn} R_c)} \overline{\phi}_{mnp}(r). \quad (7.135)$$

The Green's function for an acoustic field in a circular chamber can now be obtained by substituting Equations 7.130 and 7.135 into Equation 7.92, that is [53],

$$G_k(\vec{x}/\vec{x}_0) = 4\pi \sum_{\substack{m \\ n \\ p}} G_{mnp}(k) H_{mnp}(\vec{x}) H_{mnp}(\vec{x}_0), \quad (7.136)$$

where

$$G_{mnp}(k) = \frac{\varepsilon_m \varepsilon_p (k_{rmn} + \mu_{mn})^2}{\pi L \left[(k_{rmn} + \mu_{mn})^2 R_c^2 - m^2 \right] J_m^2(k_{mn} R_c) \left(k_{mnp}^2 - k^2 \right)}, \quad (7.137a)$$

$$H_{mnp}(\vec{x}) = J_m(k_{rmn} r) \cos(m\theta) \cos(k_{zp} z), \quad (7.137b)$$

$$H_{mnp}(\vec{x}_0) = J_m(k_{rmn} r_0) \cos(m\theta_0) \cos(k_{zp} z_0). \quad (7.137c)$$

It may be noted that Green's function is the product of $H_{mnp}(\vec{x})$ and $H_{mnp}(\vec{x}_o)$, which are relevant to the observation point and source point, respectively, and $G_{mnp}(k)$, which is dependent on exciting frequency $k (k = \omega/c)$.

The velocity potential generated by the inlet and outlet ports (ports 1 and 2, respectively) can be obtained by substituting Green's function [Equation 7.136 along with Equation 7.137 into Equation 7.102 along with the cylindrical polar coordinates version of Equations 7.103 and 7.104. The resulting expressions and their integrations over the respective piston surface are, however, complex because when a circular tube enters the side of a circular cylinder, the junction port would be elliptical. Kim and Kang [53] have derived exact expressions for the velocity potential, transfer matrix and TL of the chamber configurations shown in Figures 7.17–7.19.

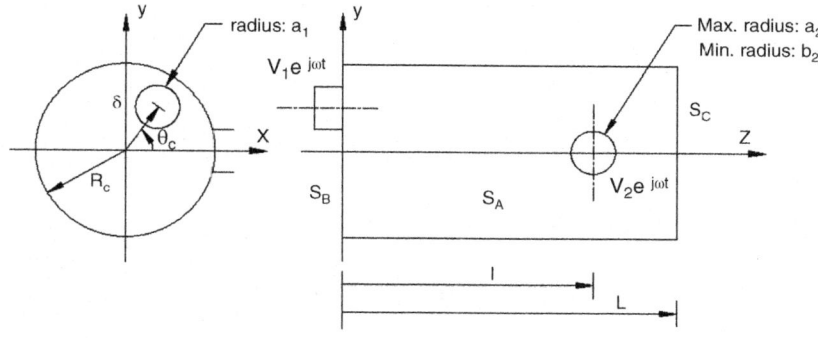

Figure 7.18 Circular expansion chamber with face-inlet/side-outlet ports (Adapted from [53])

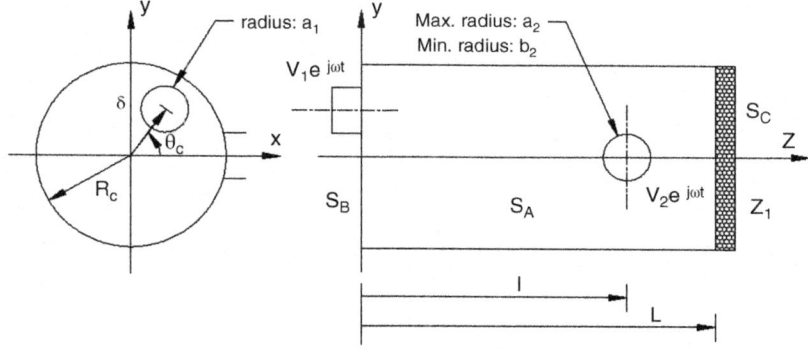

Figure 7.19 Circular expansion chamber with arbitrary termination impedance (Adapted from [53])

7.5 Green's Function Method for Elliptical Cylindrical Chambers

As shown in Section 1.8, 3D waves along elliptical cylindrical chambers are governed by the Helmholtz Equation 1.188 along with Equation 1.189 where ξ and η are elliptical coordinates. The general solution of Equation 1.188 is given by [57]

$$\phi(\vec{x}) = \sum_{m=0}^{\infty}\sum_{n=1}^{\infty}\sum_{p=0}^{\infty} \alpha_{mp} Se_m(q_{mn},\eta) Je_m(q_{mn},\xi)\cos\frac{p\pi z}{l} \\ + \sum_{m=0}^{\infty}\sum_{n=1}^{\infty}\sum_{p=0}^{\infty} \beta_{mp} So_m(\bar{q}_{mn},\eta) Jo_m(\bar{q}_{mn},\xi)\cos\frac{p\pi z}{l}. \qquad (7.138)$$

It may be recalled that Se_m and So_m are the even and odd Mathieu functions of the first kind, Je_m and Jo_m are the Modified Mathieu functions of the first kind [58].

Eigen-functions ϕ_1 and ϕ_2 of the elliptic systems are given by

$$\phi_1(\eta, \xi, z, q_{mn}) = Se_m(q_{mn}, \eta) Je_m(q_{mn}, \xi) \cos\frac{p\pi z}{l}, \qquad (7.139)$$

$$\phi_1(\eta, \xi, z, q_{mn}) = So_m(\bar{q}_{mn}, \eta) Jo_m(\bar{q}_{mn}, \xi) \cos\frac{p\pi z}{l}, \qquad (7.140)$$

and the eigen-frequencies k_{mnp} and $\bar{k}_{mnp}$ are given by

$$k_{mnp}^2 = \frac{4 q_{mn}}{h^2} + \left(\frac{p\pi}{l}\right)^2, \qquad (7.141)$$

$$\bar{k}_{mnp}^2 = \frac{4\bar{q}_{mn}}{h^2} + \left(\frac{p\pi}{l}\right)^2, \qquad (7.142)$$

where the parameters q and $\bar{q}$ are roots of the following equations arising from the rigid boundary condition

$$\left.\frac{\partial \phi}{\partial \xi}\right|_{\xi=\xi_1} = 0, \qquad (7.143)$$

that is,

$$Se'_m(q, \xi_1) = 0, \quad m = 0, 1, 2, \ldots \qquad (7.144)$$

$$So'_m(\bar{q}, \xi_1) = 0, \quad m = 0, 1, 2, \ldots \qquad (7.145)$$

Corresponding to Equation 7.92b in Section 7.3, Green's function for the elliptical configurations can be written as

$$G(\vec{x}/\vec{x}_0) = \sum_{mnp} \frac{\phi_1(\vec{x})\phi_1(\vec{x}_0)}{\left(k_{mnp}^2 - k^2\right) \int_v \phi_1(\vec{x})^2 dV} + \sum_{mnp} \frac{\phi_2(\vec{x})\phi_2(\vec{x}_0)}{\left(\bar{k}_{mnp}^2 - k^2\right) \int_v \phi_2(\vec{x})^2 dV}, \qquad (7.146)$$

and $\phi(\vec{x})$ would combine contributions from the two hypothetical pistons as follows:

$$\phi(\vec{x}) = \iint_{\text{piston 1}} G\langle \vec{x}|\vec{x}_0\rangle \nabla \phi(\vec{x}_0) d\vec{S}_0 + \iint_{\text{piston 2}} G(\vec{x}/\vec{x}_0) \nabla \phi(\vec{x}_0) d\vec{S}_0 \qquad (7.147)$$

The average sound pressure on piston $i (i = 1, 2)$ may be evaluated from the velocity potential as

$$p_i = -\frac{j\rho_0 \omega}{S_i} \int_{S_i} \phi(\vec{x}) dS_i \qquad (7.148)$$

Then p_1 and p_2 are given by Equations 7.117 and 7.118, where E-parameters can be evaluated numerically; analytical evaluation would be much too cumbersome!

Finally, the four-pole parameters are given by Equation 7.120. Banerjee and Jacobi present the transmission loss curves so computed for typical elliptical cylindrical chambers shown in Figures 7.20–7.23.

Three-Dimensional Analysis of Mufflers

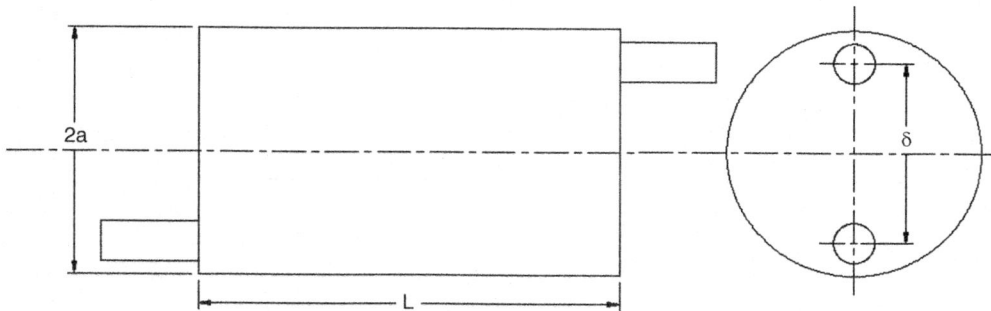

Figure 7.20 Geometry of the muffler: Length-L = 0.166 m, diameter-2a = 0.123 m, offset-δ = 0.091 m, diameter of inlet and outlet duct-d = 0.0205 m. (Adapted from [57])

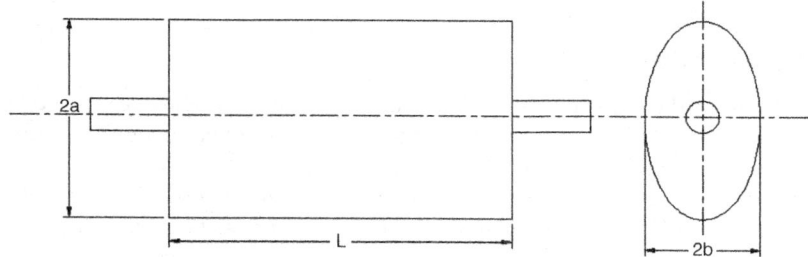

Figure 7.21 Geometry of the muffler: Length-L = 0.250 m, major axis-2a = 0.23 m, minor axis-2b = 0.13 m, diameter of inlet and outlet duct-d = 0.033 m (Adapted from [57])

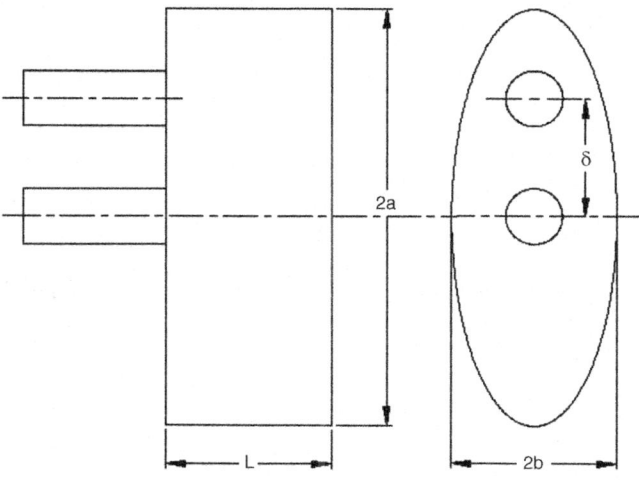

Figure 7.22 Geometry of the muffler: Length-L = 0.10 m, major axis-2a = 0.23 m, minor axis-2b = 0.13 m, diameter of inlet and outlet duct-d = 0.033 m, offset-δ = 0.0479 m (Adapted from [57])

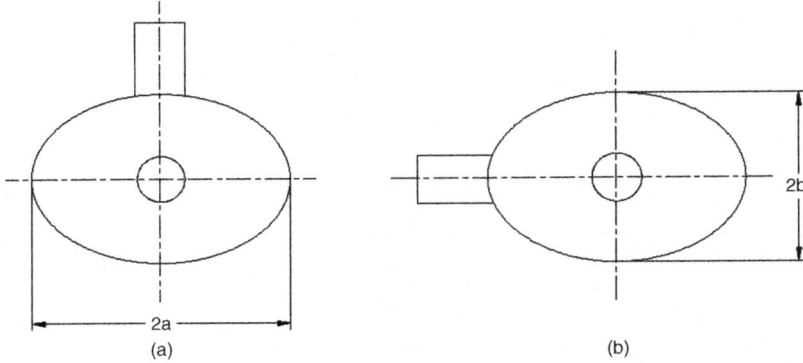

Figure 7.23 Geometry of the muffler with (a) outlet on the minor axis (b) outlet on the major axis: Length-L = 0.10 m, major axis-2a = 0.25 m, minor axis-2b = 0.15 m, diameter of inlet and outlet duct-d = 0.04859 m (Adapted from [57])

7.6 Breakout Noise

Acoustic pressure field inside a duct or chamber excites the walls of the duct (or shell and endplates of a chamber) into vibration which in turn radiates secondary sound outside. This is called breakout noise (or shell noise).

Basic physics of the break-out noise evaluation can be understood by means of the one-dimensional analysis of hoses (compliant-wall tubes) making use of radial admittance of the hose wall, as illustrated below.

7.6.1 Breakout Noise from Hoses

By virtue of its compliance, hose wall is a good representative of the shell of a cylindrical chamber. Assuming plane waves in an inviscid stationary medium, we can write the mass continuity and momentum equations as [60]

$$\frac{\partial \rho}{\partial t} + \rho_i \frac{\partial u}{\partial z} + \frac{2\rho_i}{r_i} u_w = 0, \tag{7.149}$$

$$\rho_i \frac{\partial u}{\partial t} + \frac{\partial p}{\partial z} = 0, \tag{7.150}$$

where

ρ_i is the density of the fluid medium inside the uniform hose (often a liquid),
u_w is radial velocity of the hose wall:

$$u_w = p/Z_w, \tag{7.151}$$

Z_w is the radial impedance of the hose wall:

$$Z_w = j\omega I + \frac{1}{j\omega C} + Z_0, \tag{7.152}$$

I is inertance, equal to mass density multiplied by a curvature factor [60]:

$$I = \rho_w t_w \left\{ 1 + (0.025/r_i)^2 \right\}, \qquad (7.153)$$

C is compliance, equal to radial deflection per unit pressure:

$$C = \frac{r_i}{E} \left\{ \frac{r_0^2 + r_i^2}{r_0^2 - r_i^2 r} + v \right\}, \qquad (7.154)$$

r_0 is outer radius ($r_0 = r_i + t_w$),
ρ_w and t_w are density and thickness of the hose wall.

Combining Equations 7.149 and 7.150 with the equation of isentropicity

$$dp = c_i^2 d\rho, \qquad (7.155)$$

eliminating ρ and u, assuming harmonic time dependence $\exp(j\omega t)$, yields the Helmholtz equation [60]:

$$\frac{d^2 p}{dz^2} + \left\{ k_i^2 - 2j \frac{k_i}{r_i} X \right\} p = 0, \qquad (7.156)$$

where X is the normalized wall admittance, defined as

$$X \equiv \frac{\rho_i c_i}{Z_w}. \qquad (7.157)$$

If we seek general solution of Equation 7.156 of the type

$$p(z) = A e^{-j k_z z} + B e^{j k_z z}, \qquad (7.158)$$

then it can readily be seen that

$$k_z = k_i \left[1 - 2j \frac{X}{k_i r_i} \right]^{1/2} \qquad (7.159)$$

Writing axial wave number as $k_z = \omega/c_{eq}$, where c_{eq} is the effective sound speed, gives

$$c_{eq} = c_i \frac{k_i}{k_z} = \frac{c_i}{\left\{ 1 - 2j \frac{X}{k_i r_i} \right\}^{1/2}} \qquad (7.160)$$

Incidentally, if we ignore the transverse radiation impedance Z_0, then combining Equations 7.157 and 7.152–7.154 reduces Equation 7.160 to that found in the literature (see [61] for example):

$$c_{eq} = c_i \bigg/ \left[1 + 2\frac{\rho_i c_i^2}{\rho_w c_w^2}\left\{\frac{r_0^2 + r_i^2}{r_0^2 - r_i^2} + v\right\}\right]^{1/2}, \quad c_w^2 = \frac{E}{\rho_w}. \tag{7.161}$$

It may be observed from Equation 7.161 that the equivalent speed of sound for plane waves inside the hose, c_{eq} is invariably less than c_i, the corresponding sound speed in the free medium.

The impedance model described above may be used to evaluate transverse transmission loss, TL_t, across the hose wall. As per the ASHRAE definition [62],

$$TL_t(\text{intensity}) = 10\log\left[\frac{1/2\text{Re}(p_i u_{zi}^*)}{1/2\text{Re}(p_0 u_w^*)}\right], \tag{7.162}$$

where * indicates complex conjugate. For a forward moving progressive wave, this yields

$$TL_t(\text{intensity}) = 10\log\left[\frac{\text{Re}\{pp^*/\rho_i c_{eq}\}}{\text{Re}\{(u_w Z_0)u_w^*\}}\right] = 10\log\left[\frac{|Z_w|^2}{\rho_i c_{eq} R_0}\right], \tag{7.163}$$

where R_0 is the radiation resistance or real part of the radiation impedance.

Velocity potential for cylindrical waves in the fluid outside the cylindrical hose is of the form [63]

$$\psi = AH_0^{(2)}(k_{r0}r)e^{-jkz}e^{j\omega t}, \quad k_{r0} = (k_0^2 - k^2)^{1/2}, k_0 = \omega/c_0, \tag{7.164}$$

where $H_0^{(2)}$ is the Hankel function of the second kind and zero order, and k_{r0} is the effective radial wave number. This yields the following expression for radiation impedance Z_0 at the outer boundary [63]:

$$Z_0 = \frac{p(r_0)}{u_w(r_0)} = \frac{-\rho_0(\partial\psi/\partial t)}{\partial\psi/\partial r} = j\frac{\omega\rho_0 H_0^{(2)}(k_{r0}r_0)}{k_{r0}H_1^{(2)}(k_{r0}r_0)}. \tag{7.165}$$

ASHRAE defines TL_t as the difference of intensity levels. It would be more practical, however, to define it as difference of power levels [64,65]. Then, areas of cross section πr_i^2 and radiating surface $2\pi r_0 l$ would be included. Thus,

$$TL_{tp}(\text{power}) = TL_t(\text{intensity}) + 10\log\left[\frac{\pi r_i^2}{2\pi r_0 l}\right], \tag{7.166}$$

where $TL_t(\text{intensity})$ is given by Equation 7.163 above.

Equation (7.166) by definition is a function of length. However, the theory presented here for prediction of TL_{tp} holds for a progressive wave only. It would not be valid for standing waves generated by a finite length hose.

Denoting axial TL by TL_a and transverse TL (power) by TL_{tp}, it may be noticed that

$$TL_a \equiv 10 \log \left(\frac{W_{i,a}}{W_{t,a}} \right) \qquad (7.167)$$

and

$$TL_{tp} \equiv 10 \log \left(\frac{W_{i,a}}{W_{t,t}} \right). \qquad (7.168)$$

Then, net TL will be given by

$$\begin{aligned} TL_{net} &\equiv 10 \log \frac{W_{i,a}}{W_{t,a} + W_{t,t}} \\ &= -10 \log \left(10^{-0.1 TL_a} + 10^{-0.1 TL_{tp}} \right). \end{aligned} \qquad (7.169)$$

The concept of TL_{net} has significance in the far field only because breakout noise would seem to have an entirely different effective source location than noise transmitted axially.

Equation (7.169) indicates that TL_{net} will be less than or equal to the lesser of TL_a and TL_{tp} at any frequency:

$$TL_{net} \leq \min \left(TL_a, TL_{tp} \right) \qquad (7.170)$$

Incidentally, hoses carrying air/gas act more or less like rigid pipes for acoustic waves, while playing a crucial role for vibration isolation. However, hoses carrying liquid/oil are good mufflers apart from being efficient vibration isolators as shown in Figure 7.24.

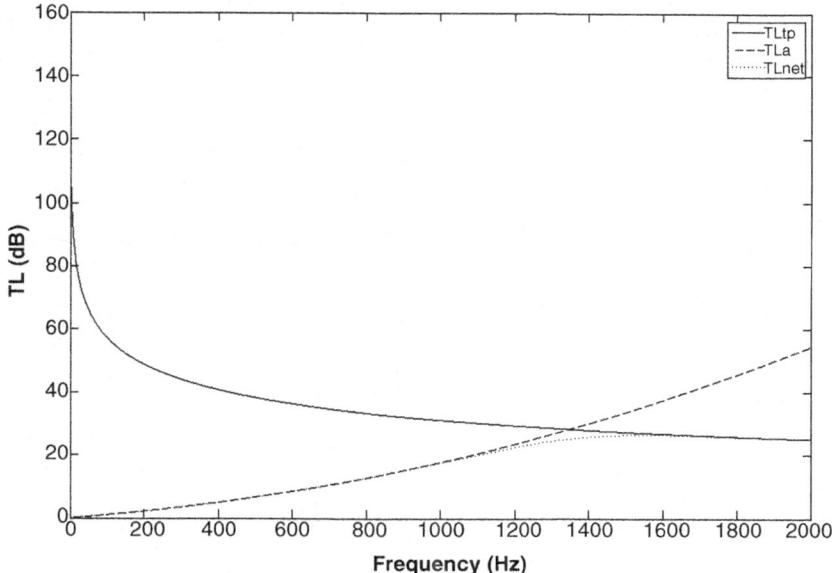

Figure 7.24 Acoustic performance of a 5.4-mm-radius hose with power steering oil medium and the default values of storage modulus, loss factor, and wall thickness (Adapted from [60])

It may be observed from Figure 7.24 that at lower frequencies, the TL_{net} curve overlaps with TL_a, and at higher frequencies, it coincides with TL_{tp}. At and around the frequency where $TL_a = TL_{tp}$, TL_{net} is up to 3 dB lower than the lesser of the two, governed by Equation 7.169.

7.6.2 Breakout Noise from Rectangular Ducts

Thin walled rectangular ducts used in the heating, ventilation and air-conditioning (HVAC) systems have very low transverse transmission loss, TL_{tp}. The resulting break-in and breakout of noise often leads to the phenomenon of cross-talk or loss of privacy [66]. At low frequencies of the order of the blade passing frequency of the fan or blower of typical air handling units of the HVAC systems, the coupling phenomenon is strong. Cummings modeled the coupling phenomena between the inside duct acoustic field and flexural waves along flexible duct wall by using a common axial wave number. He used the following three types of boundary conditions on a flexible duct wall [67,68]:

 i. The normal wall displacement at the edge of the duct is zero.
 ii. The duct edges remain right-angled.
 iii. Transverse bending moments on two different adjacent walls are equal.

The external acoustic radiation from the duct walls has been estimated in the literature: namely, a finite length line source model [67,68], equivalent cylindrical source of finite length [70], and a finite element method [71]. All these references [66–71] make use of progressive wave approach which implies an anechoic termination. Venkatesham et al. [72] studied the effect of different acoustic boundary conditions while making use of the same essentially one-dimensional approach. They observed that the progressive wave approach tends to overestimate the transverse transmission loss (or underestimate the breakout noise) although the overall mean pattern is captured, as shown for the duct of Figure 7.25 in Figures 7.26 and 7.27.

Recently, Venkatesham et al. [73] have developed a more rigorous method for calculation of breakout noise for a rectangular duct with compliant walls, incorporating three-dimensional effects along with the acoustical and structured wave coupling phenomena. First, the inside

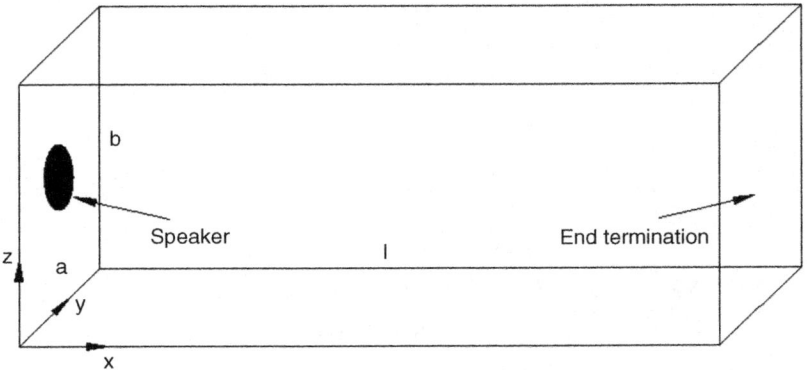

Figure 7.25 Rectangular duct with end termination excited by a speaker (Adapted from [72])

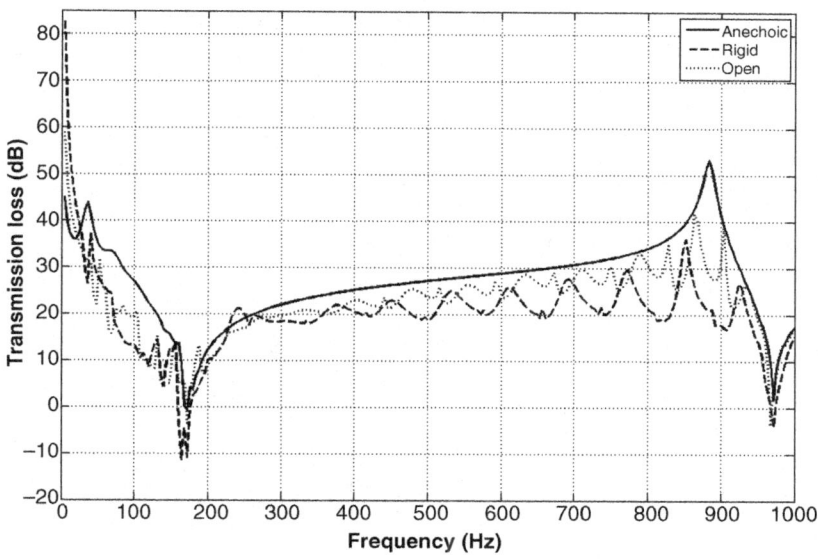

Figure 7.26 Effect of the termination on 1-D standing wave prediction of transverse transmission loss for square duct dimensions ($a = b = 0.203\,m$, $l = 2.1\,m$, $t = 1.22\,mm$, mild steel) (Reproduced with permission from [72])

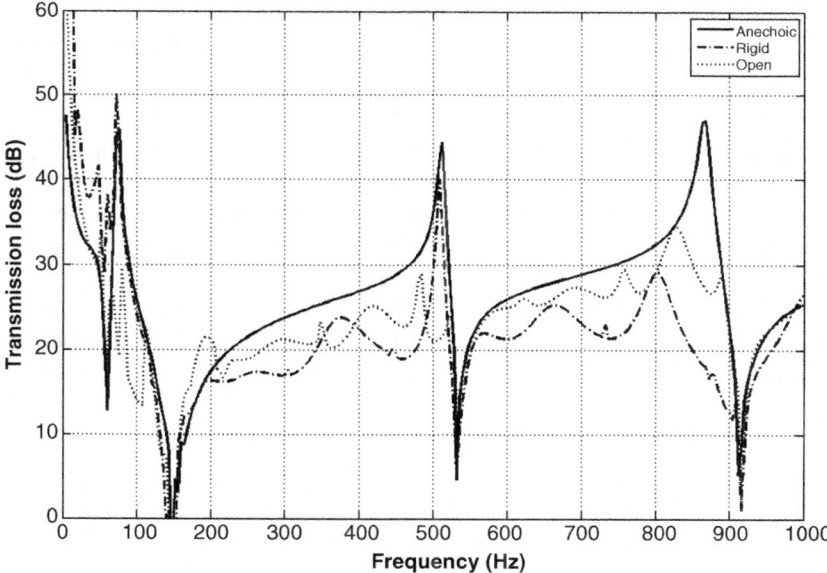

Figure 7.27 Effect of the termination on prediction of transverse transmission loss for rectangular duct dimensions ($a = 0.206\,m$, $b = 0.258\,m$, $l = 1.17\,m$, $t = 1.295\,m$, mild steel) (Reproduced with permission from [72])

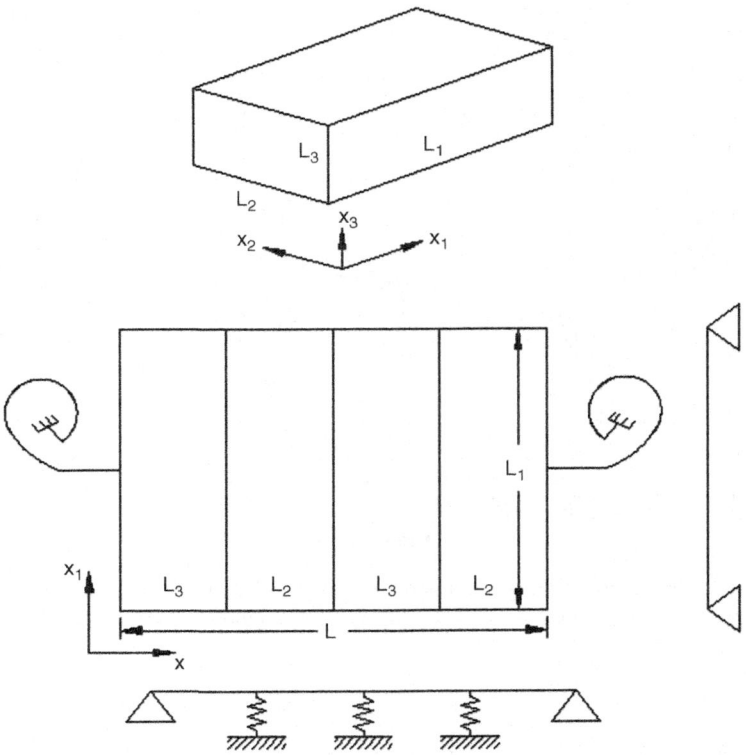

Figure 7.28 Simply supported rectangular duct and its unfolded plate representation (Reproduced with permission from [73])

pressure field and the normal duct wall vibration by using the solution of the governing differential equations in terms of Green's function. The resultant equations are arranged in terms of impedance and mobility, which results in a compact matrix formulation [74]. The radiated sound level is then calculated by means of an 'equivalent unfolded plate' model, illustrated in Figure 7.28.

Significantly, the transverse transmission loss of this three-dimensional model coincides very well with that of the one-dimensional model described above [72] for a square duct with anechoic termination, as shown in Figure 7.29. This would result in great convenience.

7.6.3 Breakout Noise from Elliptical Ducts

Cummings at al. [75] gave a theoretical treatment of sound transmission through the walls of distorted circular ducts for plane wave transmission within the duct. The transmission mechanism is essentially that of 'mode coupling', whereby higher structural modes in the duct walls are excited because of the wall distortion, by the internal sound field. Elliptical shells have continuous perimeteral shape which varies with angular coordinate (θ). Therefore, Munjal *et al.* [76] have extended the Cummings *et al.*'s approach [75] to elliptical ducts. They have validated their theory against finite element analysis, as shown in Figure 7.30.

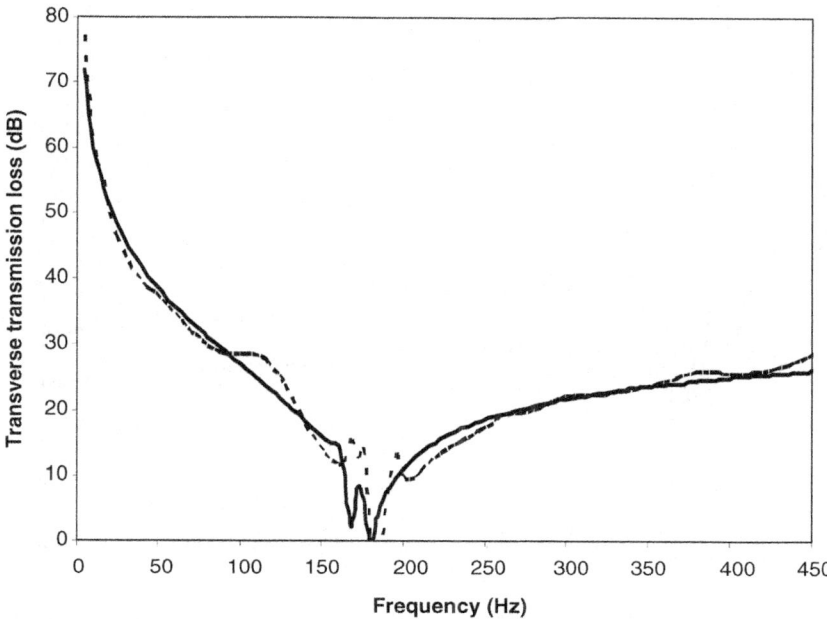

Figure 7.29 Comparison of transverse transmission loss prediction methods in square duct with anechoic termination, (——— one-dimensional model; - - - - three-dimensional model) (Reproduced with permission from [73])

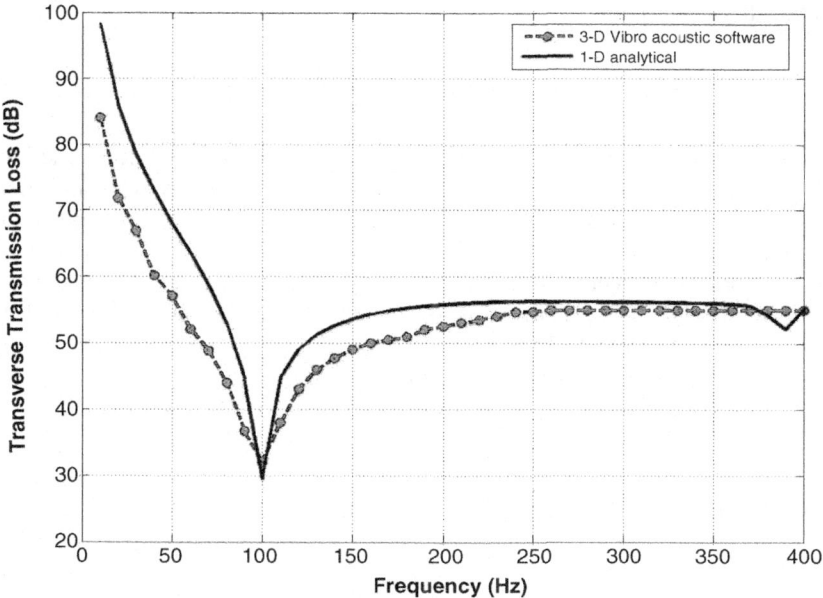

Figure 7.30 Comparison of analytical prediction for the elliptical duct with 3D vibro-acoustic software for anechoic termination (Reproduced with permission from [76])

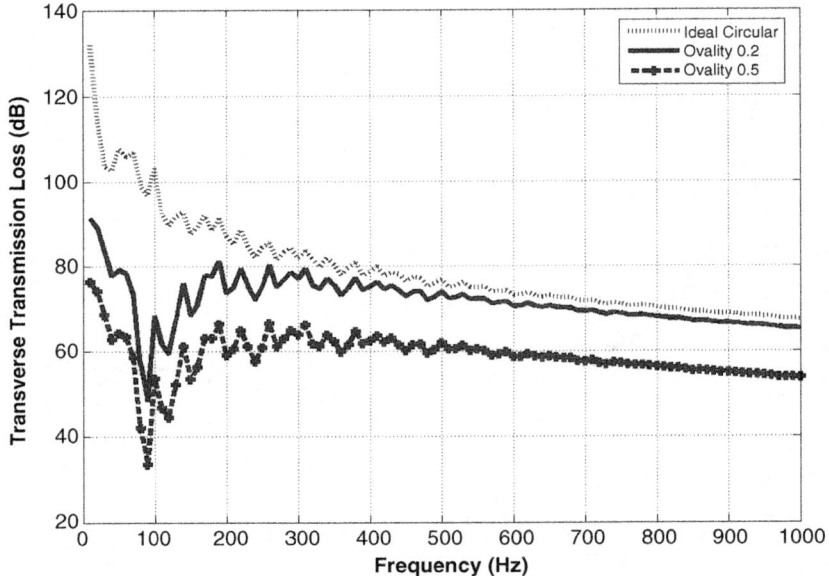

Figure 7.31 Effect of the ovality on the standing wave prediction of transverse transmission loss for elliptical duct with open termination (Reproduced with permission from [76])

If a and b are lengths of the major axis and minor axis of the ellipse, respectively, then ovality of the ellipse is defined as

$$b = a(1 - \varepsilon) \Rightarrow \varepsilon = \frac{a-b}{a} = 1 - \frac{b}{a} \qquad (7.171)$$

Effect of the duct ovality on the transverse transmission loss of an elliptical shell is shown in Figure 7.31. It can be observed that with an increase in the value of ovality there is an increase in the breakout noise. Increased deviation from circularity in general increases breakout noise, thereby decreasing TL_{tp}.

Finally, comparison between ideal circular duct, square duct, elliptical duct, and rectangular duct is shown in Figure 7.32. Here a square duct of dimensions $0.203\,m \times 0.203\,m$ is considered as reference configuration [72].

The equivalent dimension of ideal circular duct is,

$$r_{av} = \sqrt{((0.203 \times 0.203)/\pi)} = 0.1145\,m$$

The equivalent dimensions of elliptical duct are,

$$a = r_{av}/\sqrt{(1-\varepsilon)} = 0.1619\,m$$
$$b = a(1-\varepsilon) = 0.810\,m$$

whereas ovality $\varepsilon = 0.5$.

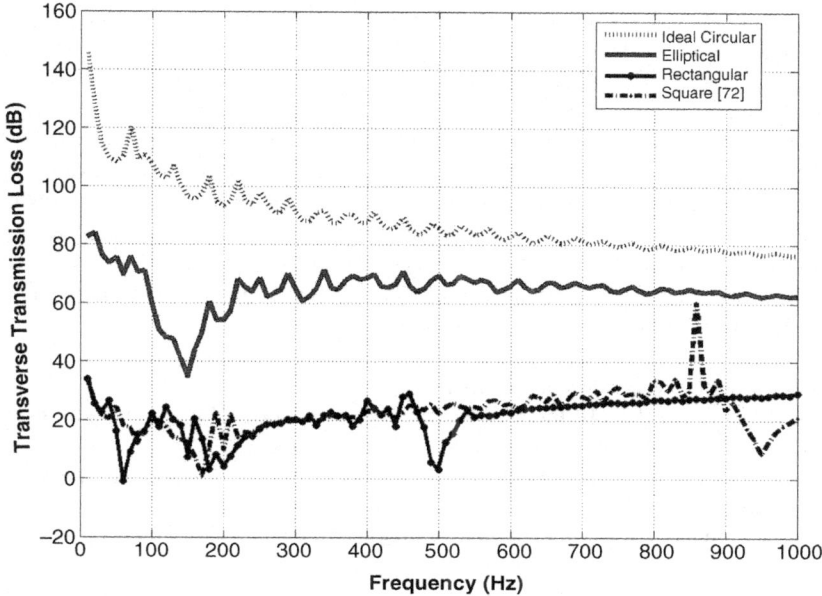

Figure 7.32 Comparison between ideal circular duct, square duct, elliptical duct and rectangular duct for open termination (Reproduced with permission from [76])

The equivalent dimensions of rectangular duct are,

$$a = 0.258\ m$$
$$b = 0.203 \times 0.203/0.258 = 0.01597\ m$$

It is assumed that all ducts are of 2.1 m in length. Obviously, the elliptical duct TL_{tp} lies in between the ideal circular duct and rectangular or square ducts. Increased deviation from circularity in general increases breakout noise, thereby decreasing TL_{tp}.

7.6.4 Breakout Noise from Mufflers

So far, we have discussed prediction of the breakout noise from ducts or shell. A muffler, however, consists of the end plates as well as the shell. Prediction of breakout noise or the transverse transmission loss (TTL), involves three steps; namely [59],

a. evaluation of acoustic pressure field excited by the source making use of the theory developed in the preceding chapters;
b. prediction of the normal velocity distribution of the shell and end-plates, making use of structural analysis; and
c. calculation of the breakout noise radiated by the surfaces exposed to the atmosphere making use of an appropriate acoustic radiation mechanism.

The foregoing three steps call for different modules of one of the commercially available finite element and/or boundary element softwares. These steps may be carried out in sequence in the so-called Uncoupled Analysis, or simultaneously by means of a Coupled Analysis. The uncoupled analysis is conceptually simpler but involves extensive book-keeping regarding meshing. The coupled analysis is computationally very expensive because it involves large random access memory as well as CPU time.

Narayana and Munjal [59] analyzed a typical simple expansion chamber muffler for its transverse transmission loss performance using finite element and boundary element methods considering the fluid-structure coupling for a given acoustic excitation. They predicted the breakout noise from the muffler shell together with end plates, and verified the same experimentally making use of the sound intensity measurements. They also conducted parametric studies, making use of a commercial FE/BE software [77]. These parametric studies resulted in the following design guidelines:

1. End plates are more critical than the shell of the same thickness.
2. Circular shells are relatively the best in this regard although logistic constraints under a passenger car may necessitate elliptic muffler configurations.
3. For a typical 20 cm diameter circular muffler, 2 mm thickness is adequate for the end plates as well as the shell.
4. Dish-type end plates are better than flat end plates as well as hemi-spherical ones.

References

1. Miles, J. (1944) The reflection of sound due to change in cross-section of a circular tube. *Journal of the Acoustical Society of America*, **16**(1), 14–19.
2. El-Sharkawy, A.I. and Nayfeh, A.H. (1978) Effect of an expansion chamber on the propagation of sound in circular ducts. *Journal of the Acoustical Society of America*, **63**(3), 667–674.
3. Jayaraman, K. (1984) A predictive model for symmetric mufflers, Nelson Acoustics Conference, Madison, WI.
4. Eriksson, L.J., Anderson, C.A., Hoops, R.H. and Jayaraman, K. (1983) Finite length effects on higher order mode propagation in silencers, 11th ICA, Paris, 329–332.
5. Ih, J.G. and Lee, B.H. (1985) Analysis of higher order mode effects in circular expansion chamber with mean flow. *Journal of the Acoustical Society of America*, **77**(4), 1377–1388.
6. Ih, J.G. and Lee, B.H. (1987) Theoretical prediction of the transmission loss for the circular reversing chamber mufflers. *Journal of Sound and Vibration*, **112**(2), 261–272.
7. Munjal, M.L. (1987) A simple numerical method for three-dimensional analysis of simple expansion chamber mufflers of rectangular as well as circular cross-section with stationary medium. *Journal of Sound and Vibration*, **116**(1), 71–88.
8. Zienkiewicz, O.C. (1971) *The Finite Element Methods in Engineering Sciences*, McGraw-Hill, London.
9. Desai, C.S. and Abel, J.F. (1972) *Introduction to the Finite Element Method: A Numerical Method for Engineering Analysis*, Affiliated East-West Press, New Delhi.
10. Gladwell, G.M.L. (1965) A finite element method for acoustics, Proceedings Fifth International Congress Acoustics Liege, **L33**.
11. Gladwell, G.M.L. and Zimmermann, G. (1965) On energy and complementary energy formulations of acoustics and structural vibration problems. *Journal of Sound and Vibration*, **3**(3), 233–241.
12. Gladwell, G.M.L. (1966) A variational formulation of damped acousto-structural vibration problems. *Journal of Sound and Vibration*, **42**(2), 177–186.
13. Gladwell, G.M.L. and Mason, V. (1971) Variational finite element calculation of the acoustic response of a rectangular panel. *Journal of Sound and Vibration*, **14**(1), 115–135.
14. Craggs, A. (1972) The use of simple three-dimensional acoustic finite elements for determining the natural modes and frequencies of complex shaped enclosure. *Journal of Sound and Vibration*, **23**(3), 331–339.
15. Craggs, A. (1973) An acoustic finite element approach for studying boundary flexibility and sound transmission between irregular enclosures. *Journal of Sound and Vibration*, **30**(3), 343–357.

16. Young, C.-I.J. and Crocker, M.J. (1975) Prediction of transmission loss in mufflers by the finite element method. *Journal of the Acoustical Society of America*, **57**(1), 144–148.
17. Young, C.-I.J. and Crocker, M.J. (1976) Acoustic analysis, testing and design of flow-reversing muffler chamber. *Journal of the Acoustical Society of America*, **60**(5), 1111–1118.
18. Gallagher, R.H. (ed.) (1975) *Finite Elements in Fluids*, John Wiley & Sons, London.
19. Craggs, A. (1976) A finite element method for damped acoustic systems: An application to evaluate the performance of reactive mufflers. *Journal of Sound and Vibration*, **48**(3), 377–392.
20. Kagawa, Y. and Omote, T. (1976) Finite-element simulation of acoustic filters of arbitrary profile with circular cross section. *Journal of the Acoustical Society of America*, **60**(5), 1003–1013.
21. Young, C.-I.J. and Crocker, M.J. (1977) Finite element analysis of complex muffler systems with or without wall vibrations. *Noise Control Engineering Journal*, **9**, 86–93.
22. Kagawa., Y., Yamabuchi, T. and Mori, A. (1977) Finite element simulation of an axi-symmetric acoustic transmission system with a sound absorbing wall. *Journal of Sound and Vibration*, **53**(3), 357–374.
23. Craggs, A. (1977) A finite element method for modeling dissipation mufflers with a locally reactive lining. *Journal of Sound and Vibration*, **54**(2), 285–296.
24. Abrahamson, A.L. (1977) A finite element algorithm for sound propagation in asixymmetric ducts containing mean flow, AIAA Paper No 77-1301.
25. Joppa, P.D. and Fyfe, I.M. (1978) A finite element analysis of the impedance properties of irregular shaped cavities. *Journal of Sound and Vibration*, **56**(1), 61–69.
26. Astley, R.J. and Eversman, W. (1978) A finite element method for transmission in non-uniform ducts without flow: Comparison with the method of weighted residuals. *Journal of Sound and Vibration*, **57**(3), 367–388.
27. Craggs, A. (1978) A finite element method for rigid porous absorbing materials. *Journal of Sound and Vibration*, **61**(1), 101–111.
28. Tag, I.A. and Lumsdaine, E. (1978) An efficient finite element technique for sound propagation in axisymmetric hard wall ducts carrying high subsonic Mach number flows, AIAA Paper No. 78-1154.
29. Sigman, R.K., Maijigi, R.K. and Zinn, B.T. (1978) Determination of turbofan inlet acoustics using finite elements. *Journal of AIAA*, **16**, 1139–1145.
30. Astley, R.J. and Eversman, W. (1979) A finite element formulation of the eigenvalue problem in lined ducts with flow. *Journal of Sound and Vibration*, **65**(1), 61–74.
31. Craggs, A. (1979) Coupling of finite element acoustic absorption models. *Journal of Sound and Vibration*, **66**(4), 605–613.
32. Maijigi, R.K., Sigman, R.K. and Zinn, B.T. (1979) Wave propagation in ducts using the finite element method, AIAA Paper No. 79-0659.
33. Astley, R.J. and Eversman, W. (1979) The application of finite element techniques to acoustic transmission in lined ducts with flow, AIAA Paper No. 79-0660.
34. Abrahamson, A.L. (1979) Acoustic duct linear optimization using finite elements, AIAA Paper No. 79-0662.
35. Tag, I.A. and Akin, J.E. (1979) Finite element solution of sound propagation in a variable are duct, AIAA Paper No. 79-0663.
36. Lester, H.C. and Parrott, T.L. (1979) Application of finite element methodology for computing grazing incidence wave experiment, AIAA Paper No. 79-0664.
37. Kagawa., Y., Yamabuchi, T. and Yoshikawa, T. (1980) Finite element approach to acoustic transmission radiation systems and application to horn and silencer design. *Journal of Sound and Vibration*, **69**(2), 207–228.
38. Baumeister, K.J. (1980) Numerical techniques in linear duct acoustics – a status report, NASA TM- 81553.
39. Ross, D.F. (1980) A finite element analysis of parallel-coupled acoustic systems. *Journal of Sound and Vibration*, **69**(4), 509–518.
40. Astley, R.J. and Eversman, W. (1980) The finite element duct eigenvalue problem: An improved formulation with Hermitian elements and no-flow condensation. *Journal of Sound and Vibration*, **69**(1), 13–25.
41. Astley, R.J. and Eversman, W. (1981) Acoustic transmission in non-uniform ducts with mean flow. Part II: The finite element method. *Journal of Sound and Vibration*, **74**(1), 103–121.
42. Astley, R.J. and Eversman, W. (1981) A note on the utility of a wave envelope approach in finite element duct transmission studies. *Journal of Sound and Vibration*, **76**(4), 595–601.
43. Ross, D.R. (1981) A finite element analysis of perforated component acoustic systems. *Journal of Sound and Vibration*, **79**(1), 133–143.

44. Peat, K.S. (1982) Evaluation of four-pole parameters for ducts with flow by the finite element method. *Journal of Sound and Vibration*, **84**(3), 389–395.
45. Craggs, A. (1982) A note on the theory and application of a simple pipe acoustic element. *Journal of Sound and Vibration*, **85**(2), 292–295.
46. Astley, R.J. and Eversman, W. (1983) Finite element formulation for acoustical radiation. *Journal of Sound and Vibration*, **88**(1), 47–64.
47. Mason, V. (1967) On the use of rectangular finite elements, Institute of Sound and Vibration Research, Report No. 161.
48. Craggs, A. (1971) The transient response of a coupled plate acoustic system using plate and acoustic finite elements. *Journal of Sound and Vibration*, **15**(4), 509–528.
49. Peat, K.S. (1983) A note on one-dimensional acoustic elements. *Journal of Sound and Vibration*, **88**(4), 572–575.
50. Tanaka, T., Fujikawa, T., Abe, T. and Utsuno, H. (1985) A method for the analytical prediction of insertion loss of a two-dimensional muffler model based on the transfer matrix derived from the boundary element method. *Transactions of the ASME*, **107**, 86–91.
51. Morse, P.M. and Feshbach, H. (1953) *Methods of Theoretical Physics*, McGraw-Hill, New York.
52. Chen, H.S. and Mei, C.C. (1975) Hybrid element for water waves, Proc. Modelling Techniques Conference (Modelling 1975) San Francisco, 1, 63–81.
53. Kim, Y.-H. and Kang, S.-W. (1993) Green's solution of the acoustic wave equation for a circular expansion chamber with arbitrary locations of inlet, outlet port, and termination impedance. *Journal of the Acoustical Society of America*, **94**(1), 473–490.
54. Ih, J.G. (1992) The reactive attenuation of rectangular plenum chamber. *Journal of Sound and Vibration*, **157**(1), 93–122.
55. Venkatesham, B., Tiwari, M. and Munjal, M.L. (2009) Transmission loss analysis of rectangular expansion chamber with arbitrary locations of inlet/outlet by means of Green's functions. *Journal of Sound and Vibration*, **323**, 1032–1044.
56. Morse, P.M. and Ingard, K.U. (1968) *Theoretical Acoustics*, McGraw Hill, New York.
57. Banerjee, S. and Jacobi, A.M. (2011) Analysis of sound attenuation in elliptical chamber mufflers by using Green's functions, Proceedings ASME 2011 International Mechanical Engineering Congress & Exposition IMECE 2011, Denver, Colorado, USA.
58. McMachlan, N.W. (1947) *Theory and Applications of Mathieu Functions*, Oxford University Press.
59. Narayana, T.S.S. and Munjal, M.L. (2007) Computational prediction and measurements of break-out noise of mufflers, SAE Paper No. 2007-26-040, SIAT-2007, Pune, India.
60. Munjal, M.L. and Thawani, P.T. (1996) Acoustic performance of hoses – A parametric study. *Noise Control Engineering Journal*, **44**(6), 274–280.
61. Washio, S. and Konishi, T. (1985) Research on wave phenomena in hydraulic lines (11th Report, Harmonic waves in coaxial double pipes). *Bull of JSME*, **28**, 1409–1415.
62. ASHRAE Handbook (1991) HVAC Applications, Chap. 42: Sound and Vibration Control.
63. Easwaran, V. and Munjal, M.L. (1995) A note on the effect of wall compliance on lowest order mode propagation in fluid-filled/submerged impedance tubes. *Journal of the Acoustical Society of America*, **97**, 3494–3501.
64. Fuller, C.R. (1986) Radiation of sound from an infinite cylindrical elastic shell by an internal monopole source. *Journal of Sound and Vibration*, **109**, 259–275.
65. Feng, L. (1994) Acoustic properties of fluid-filled elastic pipes. *Journal of Sound and Vibration*, **176**, 399–413.
66. Cummings, A. (2001) Sound transmission through duct walls. *Journal of Sound and Vibration*, **239**(4), 731–765.
67. Cummings, A. (1978) Low frequency acoustic transmission through the walls of rectangular ducts. *Journal of Sound and Vibration*, **61**, 327–345.
68. Cummings, A. (1979) Low frequency acoustic transmission through the walls of rectangular ducts: further comments. *Journal of Sound and Vibration*, **63**, 463–465.
69. Cummings, A. (1980) Low frequency acoustic radiation from duct walls. *Journal of Sound and Vibration*, **71**, 201–226.
70. Cummings, A., Chang, J.J. and Astley, R.J. (1984) Sound transmission at low frequencies through the walls of distorted circular ducts. *Journal of Sound and Vibration*, **97**, 261–286.
71. Astley, R.J., Cummings, A. and Sormaz, N. (1991) A finite element scheme for acoustical propagation in flexible walled ducts with bulk reacting liners and comparison with experiments. *Journal of Sound and Vibration*, **150**, 119–138.

72. Venkatesham, B., Pathak, A.G. and Munjal, M.L. (2007) A one-dimensional model for prediction of break-out noise from a finite rectangular duct with different acoustic boundary equations. *International Journal of Acoustics and Vibration*, **12**(3), 91–98.
73. Venkatesham, B., Tiwari, M. and Munjal, M.L. (2011) Prediction of breakout noise from a rectangular duct with compliant walls. *International Journal of Acoustics and Vibration*, **16**(4), 180–190.
74. Kim, S.M. and Brennan, M.J. (1999) A compact matrix formulation using the impedance and mobility approach for the analysis of structural-acoustic systems. *Journal of Sound and Vibration*, **223**(1), 97–113.
75. Cummings, A., Chang, J.J. and Astley, R.J. (1984) Sound transmission at low frequencies through the walls of distorted circular ducts. *Journal of Sound and Vibration*, **97**, 261–286.
76. Munjal, M.L., Gowtham, G.S.H., Venkatesham, B. and Reddy, M.H.K. (2010) Prediction of breakout noise from an elliptical duct of finite length. *Noise Control Engineering Journal*, **58**(3), 319–327.
77. SYSNOISE Rev 5.6. (2004) Technical Documentation, LMS International, Belgium.

8

Design of Mufflers

All the preceding chapters have dealt with analysis of a given system. In the process of the development of various evaluation procedures, a number of observations were made that can be put together to develop criteria for the design of mufflers for specific requirements. At the time of writing, the state of the art of muffler design does not allow synthesis of a unique muffler configuration on the design table (or work station) for given requirements. Nevertheless, the situation is not altogether hopeless; it is much better than it was 27 years ago when the original edition of the book appeared in print. One can synthesize a configuration, the dimensions of which can be optimized by means of a little experimentation and/or numerical computation.

8.1 Requirements of an Engine Exhaust Muffler

Generally an exhaust muffler is designed to satisfy some or all of the following requirements:

i. Adequate insertion loss: The exhaust muffler is designed so that muffled exhaust noise is at least 5 dB lower than the combustion induced engine body noise or other predominant sources of noise like transmission noise in earth-moving equipment. A frequency spectrum or speed order analysis of unmuffled exhaust noise is generally required for appreciation of the frequency range of interest, although it is well known that most of the noise is limited to the firing frequency and its first few harmonics.
ii. Back pressure: This represents the extra static pressure exerted by the muffler on the engine through restriction in the flow of exhaust gases. This needs to be kept to a minimum. However, for single-cylinder two-stroke-cycle engines, it is the instantaneous pressure exerted by large (usually nonlinear) waves that matters rather than the mean back pressure, which for a four-stroke-cycle engine would affect the brake power, volumetric efficiency, and hence the specific fuel consumption rate.
iii. Size: A large muffler would cause problems of accommodation, support (because of its weight), and, of course, excessive cost price.
iv. Durability: A uniform wall temperature is required to avoid thermal cracking of walls. The muffler must be made from a corrosion-resistive material.

v. Spark-arresting capability is also a requirement occasionally (particularly for agriculture use).
vi. The quality of the exhaust sound at idle and at curbside acceleration also matters (particularly for passenger cars).
vii. 'Breakout' noise from muffler shells must be minimized so that net transmission loss is nearly equal to the axial transmission loss.
viii. The muffler performance must not deteriorate with time.
ix. Flow-generated noise within muffler element and at the tail pipe exit should be sufficiently low, particularly for mufflers with large insertion loss. In fact, insertion loss of the HVAC system mufflers is often limited by the flow-generated noise.

8.2 Simple Expansion Chamber

We start the development of design guidelines with the most basic building block – a simple expansion chamber muffler, shown in Figure 8.1. Making use of the basic theory developed in Chapter 2 (Section 2.19) for stationary medium, we find that its axial transmission loss, TL, is given by

$$TL = 10 \log \left\{ 1 + \left(\frac{m - 1/m}{2} \sin kl \right)^2 \right\}. \tag{8.1}$$

It is plotted in Figure 8.2 for area expansion ratio, $m = 9, 16$ and 25, corresponding to the diameter ratio, $D/d = 3, 4$ and 5. For this graph, and indeed for most of the graphs in this chapter, the following data are used:

Length of the muffler paper, $l = 0.4\,m$.
Diameter of the exhaust pipe, tail pipe and intermediate pipes, $d = 40\,mm$
Default value of the shell diameter, $D = 4d = 160\,mm$.
Exhaust gas temperature $= 680\,°C = 953\,°K$.
Approximating exhaust gases as air with gas constant, $R = 287 m^2/(s^2\,°K)$ and $\gamma = 1.4$, we get
Exhaust gas density, $\rho_0 = \dfrac{p_0}{RT} = \dfrac{1.013 \times 10^5}{287 \times 953} = 0.37\,kg/m^3$.
Sound speed, $c_0 = (\gamma RT)^{1/2} = (1.4 \times 287 \times 953)^{1/2} = 619\,m/s$.

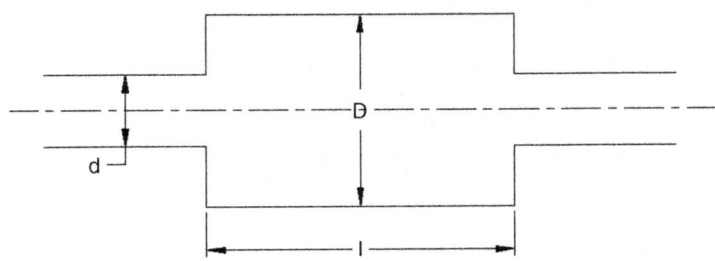

Figure 8.1 Schematic of a simple expansion chamber (SEC) muffler

Design of Mufflers

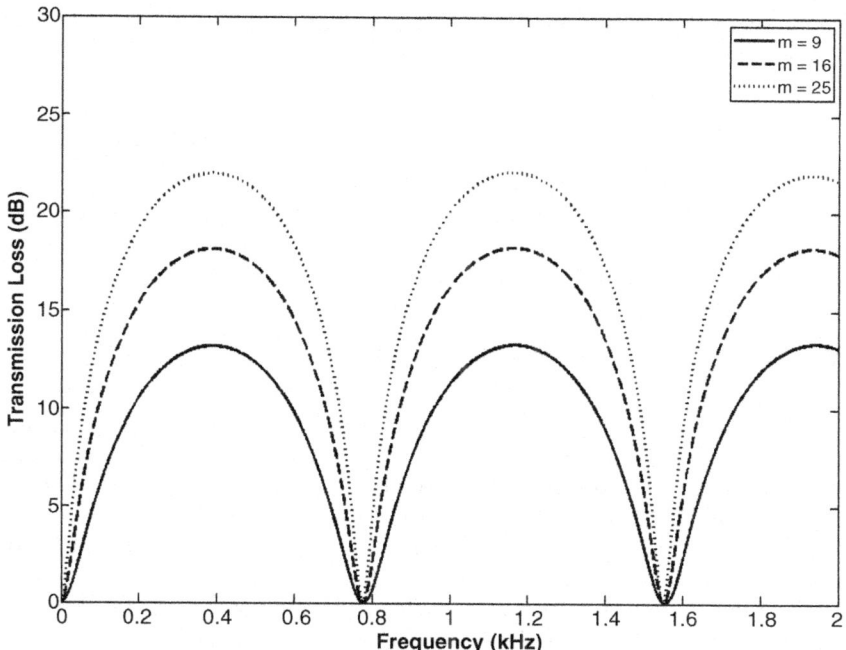

Figure 8.2 Effect of area ratio [m = (D/d)^2] on TL of a simple expansion chamber (SEC) muffler

Equation 8.1 suggests that for all values of m, TL would tend to zero at the Helmholtz number (or, nondimensional frequency parameter) kl, given by

$$kl = m\pi \Rightarrow \frac{2\pi f_n}{c_0} l = m\pi, \quad n = 0, 1, 2, 3, \ldots, \tag{8.2}$$

whence frequency of the nth trough, f_n, is given by

$$f_n = \frac{nc_0}{2l} = \frac{619n}{2 \times 0.4} = 774n, Hz.$$

It may be readily verified from Figure 8.2 that troughs occur periodically at 774 Hz and its integral multiples.

Equation 8.1 also indicates that the peaks would occur at $kl = (2n-1)\pi/2, n = 1, 2, 3, \ldots$; that is midway between the adjacent troughs. Specifically, making use the values of c_0 and l as above, the peaks would occur at odd multiples of 387 Hz.

Figure 8.2 as well as Equation 8.1 also indicates that an increase in area ratio increases TL at all frequencies, but does not raise the troughs; they still touch 0 dB.

In particular, peak value of TL is given approximately by

$$TL_{peak} \simeq 20 \log(m/2), \quad m \gg 1. \tag{8.3}$$

Relatively speaking, when area ratio is increased from m_1 to m_2, then

$$TL_{peak} \simeq 20 \log(m_2/m_1) \tag{8.4}$$

Equation 8.3 should be compared with Equation 2.106 for sudden area changes (expansion as well as contraction)

$$TL_c = TL_e = 10 \log \frac{(1+m)^2}{4m} \simeq 10 \log(m/4), \quad m \gg 1, \tag{8.5}$$

where subscripts c and e denote sudden contraction and sudden expansion, respectively.

For the pair of discontinuities that characterize a simple expansion chamber, TL would be double that given by Equation 8.5 at all frequencies:

$$TL_{e+c} = 20 \log(m/4) \tag{8.6}$$

This may be noted to be $20 \log 2 = 6\,dB$ less than the peak value given by Equation 8.3. Thus, the role of the shell of diameter D connecting the two discontinuities is to convert a frequency-independent value given by Equation 8.6 to near sinusoidal lobes with TL varying from 0 at the troughs to $TL_{e+c} + 6\,dB$ value at the peaks, as shown in Figure 8.2.

A heuristic study of the logarithmic addition of the exhaust noise at different harmonics indicates that zeros of the TL graph (rather, the IL graph) do big damage to the overall insertion loss of the muffler. A desirable TL or IL graph would have no sharp troughs even if this is achieved at the cost of the peaks. This is the main aim behind developing alternative muffler configurations discussed in the following sections.

8.3 Double-Tuned Extended-Tube Expansion Chamber

One ingenious way of raising the troughs of the TL graph of an expansion chamber of length l is to extend the inlet pipe into the chamber by $l/2$ and the outlet pipe by $l/4$ as shown in Figure 8.3 [1]. Comparing Equation 8.2 with Equation 2.145 indicates that an extended inlet of acoustic length $l_a = l_2 = l/2$ would tune out the troughs 1, 3, 5, 7, ... and an extended outlet of acoustic length $l_b = l_2 = l/4$ would tune out the troughs 2, 6, 10, 14, ... Thus a double-tuned extended-tube chamber (DTETC) with $l_a = l/2$ and $l_b = l/4$ would together tune out all troughs except those numbered 4, 8, 12..., as shown in

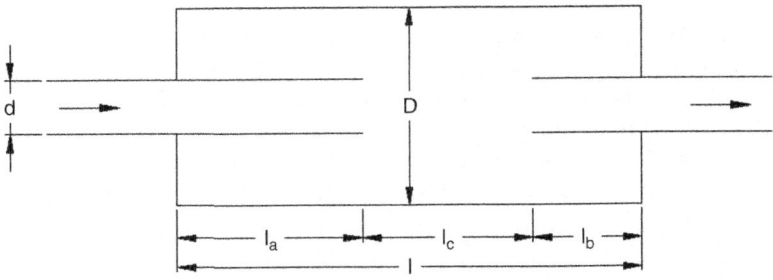

Figure 8.3 Schematic of an extended-tube expansion chamber (ETEC) muffler

Design of Mufflers

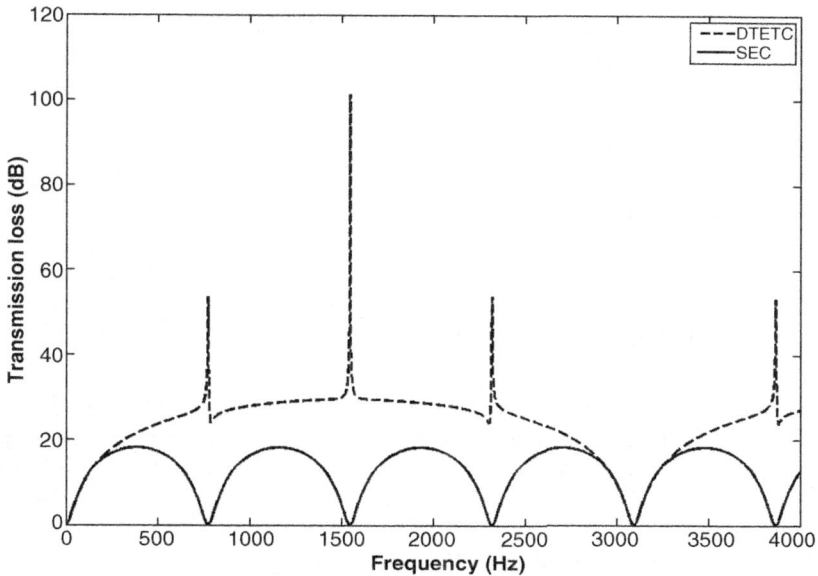

Figure 8.4 Effect of double-tuning on an extended-tube expansion chamber (ETEC) muffler $\left(l_a = 1/2 - \delta, l_b = 1/4 - \delta,\right)$

Figure 8.4. Interestingly, it may be noted from Figure 8.4 that the first four domes merge into a single dome that stands considerably higher than the domes of the simple expansion chamber (SEC) muffler.

There is, however, a small complication here. Acoustic length of a quarter-wave resonator exceeds the physical or geometric length of the annular resonator by an end correction, which as indicated earlier, is due to the generation of higher-order evanescent modes [2]. Therefore, the geometric or physical lengths of extensions should be given by [3]

$$l_{a,g} = l/2 - \delta \text{ and } l_{b,g} = l/4 - \delta \tag{8.7}$$

End correction δ is given by the following empirical expression [4]:

$$\frac{\delta}{d} = a_0 + a_1\left(\frac{D}{d}\right) + a_2\left(\frac{t_w}{d}\right) + a_3\left(\frac{D}{d}\right)^2 + a_4\left(\frac{D}{d}\frac{t_w}{d}\right) + a_5\left(\frac{t_w}{d}\right)^2 \tag{8.8}$$

Here, $a_0 = 0.005177$, $a_1 = 0.0909$, $a_2 = 0.537$, $a_3 = -0.008594$, $a_4 = 0.02616$, $a_5 = -5.425$; d and t_w are diameter and wall thickness of the inner inlet/outlet tube; and D is the (equivalent) shell diameter.

With the physical or geometric extension lengths given by Equations 8.7 and 8.8, we would be able to obtain the *TL* graph of Figure 8.4.

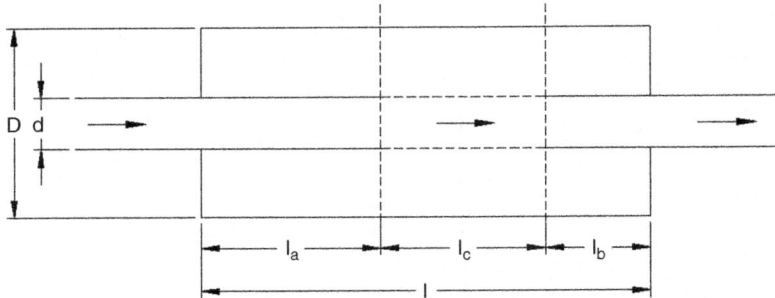

Figure 8.5 Schematic of a concentric tube resonator (CTR) muffler

However, the double-tuned extended inlet/outlet expansion chamber configuration of Figure 8.3 suffers from the weakness of free shear layer mean flow jets which generate aerodynamic noise and result in stagnation pressure loss (back-pressure) of the order of 1.5 H, where $H = 1/2\rho_o U^2$ is the hydrodynamic head or velocity head of the one-dimensional incompressible flow.

This disadvantage or weakness of DTETC can be avoided by means of a perforate bridge between the two extended pipes as shown in Figure 8.5. The theory of the resulting tuned concentric tube resonator (TCTR) is discussed below.

8.4 Tuned Concentric Tube Resonator

A perforate of large porosity (of the order of 20%) is known to be acoustically transparent, particularly at low or medium frequencies that interest us the most inasmuch as the exhaust noise predominates at the engine firing frequency and its first few multiples. Therefore, its TL performance is expected to be similar to that of the corresponding extended inlet/outlet muffler (without the perforate bridge). Acoustically, however, the configuration of Figure 8.5 differs from that of Figure 8.3 in two important ways as follows.

i. There would be no (or little) generation of the evanescent higher order modes, and therefore end corrections would be negligible. However, inertance of the holes or orifices constituting the perforate would result in what we propose to call 'differential lengths' [5].
ii. Analysis of the configuration of Figure 8.5 would follow Section 3.8.1 whereas that of Figure 8.3 follows Section 3.7.2.

A systematic investigation of the extended tube concentric-tube resonator of Figure 8.5 has resulted in the following formulae for tuning [5] corresponding to Equations 8.7 and 8.8:

$$l_a = l/2 - \Delta \text{ and } l_b = l/4 - \Delta \qquad (8.9)$$

$$\frac{\Delta}{d} = 0.6643 - 2.699\sigma + 4.522\sigma^2 \qquad (8.10)$$

Design of Mufflers

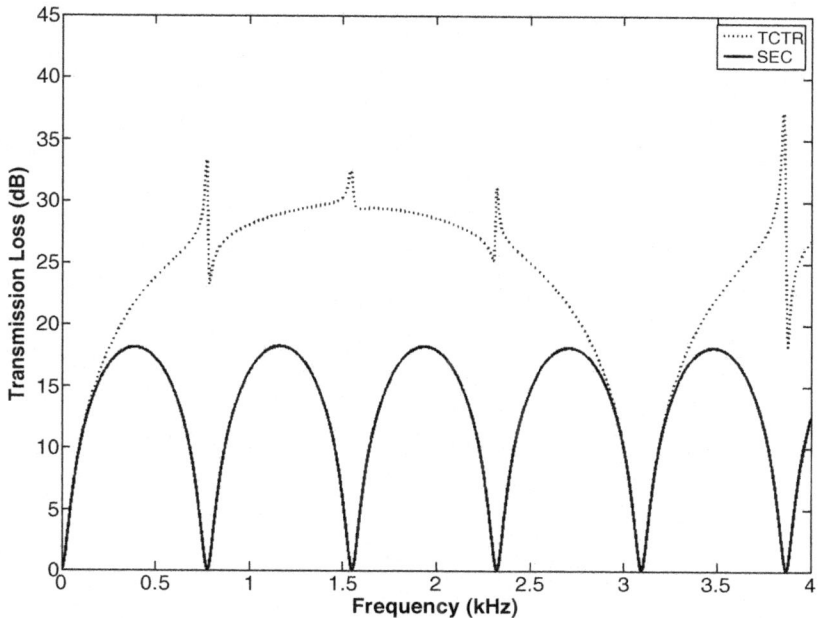

Figure 8.6 Effect of double-tuning with perforate bridge on TL of a tuned concentric tube resonator (TCTR). $(l_a = 1/2 - \Delta, l_b = 1/4 - \Delta)$ vis-à-vis the corresponding simple expansion chamber (SEC) muffler

where σ is porosity of the perforates (as a fraction). Equation 8.10 is applicable for σ ranging from 0.1 to 0.27.

Figure 8.6 shows the *TL*-performance of a tuned concentric tube resonator (TCTR), which can be seen to be similar to Figure 8.4. As indicated earlier, a TCTR has the following additional advantages:

i. There is no separation of boundary layer (or no free shear layer), and therefore there would be
   a. little aerodynamic noise
   b. little drop in stagnation pressure.
ii. There being no overhanging pipes inside the chamber, a TCTR is expected to be more robust and durable.

8.5 Plug Mufflers

The plug muffler shown in Figure 8.7 consists of a cross-flow expansion (Figure 3.11b) and cross-flow contraction (Figure 3.11c) with an annular uniform duct in-between. The plug forces the mean flow to move out from the pipe to the annular cavity and then reenter through the holes back into the pipe. In the process, it introduces a pair of inline flow-acoustic resistances into the system that dissipate the acoustic energy. Thus, a plug muffler is a reactive–cum-dissipative muffler. A transfer matrix based muffler program based on the

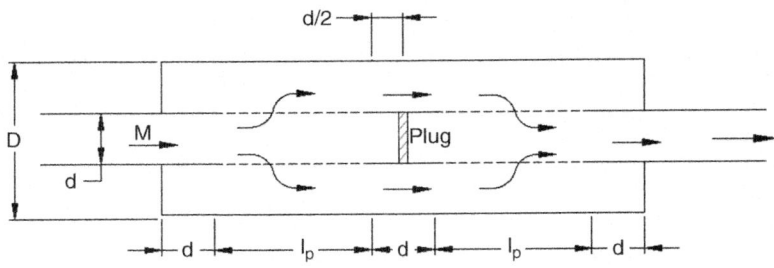

Figure 8.7 Schematic of a plug muffler

theory presented in Section 3.8 results in the TL graph shown in Figure 8.8 where open area ratio (OAR) has been used as a parameter. It is defined as [6]

$$\text{OAR} = \frac{\text{perforate area}}{\text{cross-sectional area of the pipe}} = \frac{\pi d \, l_p \sigma}{\frac{\pi}{4} d^2} = \frac{4 l_p \sigma}{d}, \quad (8.11)$$

where l_p and σ are the perforate length and porosity, respectively.

Use of a lower value of the OAR increases the mean flow velocity through the holes (or orifices), thereby increasing the flow-acoustic resistance and hence acoustic dissipation. This raises the *TL* curve at all frequencies as may be observed from Figure 8.8. Thus, plug muffler would appear to be an obvious choice for an efficient muffler configuration. However, plug muffler suffers from a tendency of very high stagnation pressure drop as the OAR is decreased

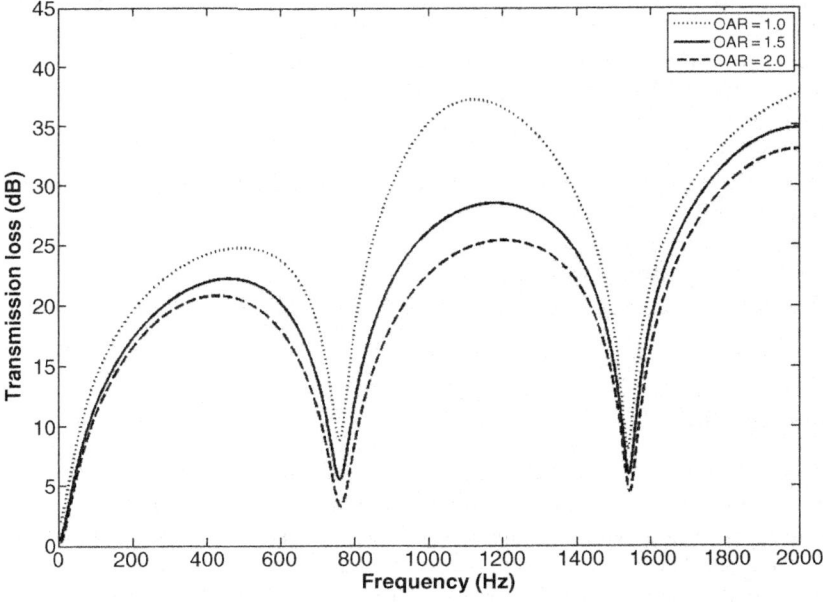

Figure 8.8 Effect of the open area ratio (OAR) on TL of the plug muffler of Figure 8.7 ($M = 0.15$, $l_p = 3.5d$)

to 1.0 or below [6], as will be clear later from the section on back pressure. On the other hand, as OAR is increased to 2.0 or more, the TL graph of plug muffler approaches that of a simple expansion chamber muffler, as is clear from Figure 8.8.

8.6 Side-Inlet Side-Outlet Mufflers

Often, logistic constraints make it necessary to adopt a side inlet and/or side outlet configuration. One such muffler is shown in Figure 8.9.

It may be noted that in such a muffler the axial dimension of the muffler (in the direction of the inlet/outlet) is much shorter than the transverse dimension (normal to the direction of the inlet/outlet). Therefore, the plane waves would move along the transverse direction. When the incoming waves in the exhaust pipe enter the muffler, they would encounter a quarter-wave resonator of length l_a. Then they move along the intermediate portion of length l_c towards the outlet junction where they would encounter another quarter-wave resonator of length l_b. Thus, the side-inlet side-outlet muffler of Figure 8.9 would behave like the extended-inlet extended-outlet muffler of Figure 8.3, and can be analyzed accordingly. The overall transfer matrix of the muffler can be obtained by successive multiplication of the transfer matrices of

a. uniform exhaust pipe
b. quarter-wave resonator of length l_a
c. uniform intermediate pipe section of length l_c
d. quarter-wave resonator of length l_b
e. uniform tail pipe.

Making use of the double-tuned extended-tube chamber concept, we can tune the side-inlet side-outlet muffler of Figure 8.9 by making

$$l_a = l/2, \ l_b = l/4, \ l_c = l/4. \tag{8.12}$$

End correction would not be applicable for this configuration [7].

The resulting TL graph is shown in Figure 8.10 for the mean flow Mach number of $M = 0.15$. The dissipation effect of the mean flow at the inlet junction and outlet junction is illustrated in Figure 8.11. This effect is typical of most reactive mufflers.

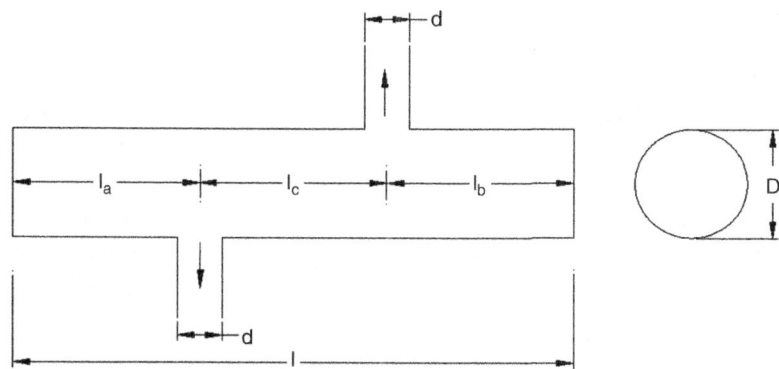

Figure 8.9 Schematic of a side-inlet side-outlet (SISO) chamber

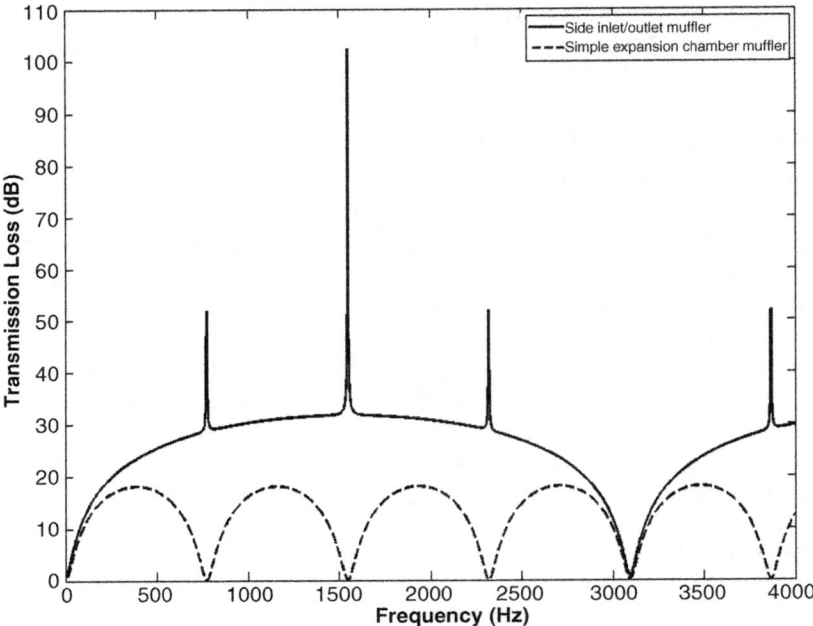

Figure 8.10 Effect of tuning on the side-inlet side-outlet chamber of Figure 8.9 ($l_a = l/2, l_b = l/4$)

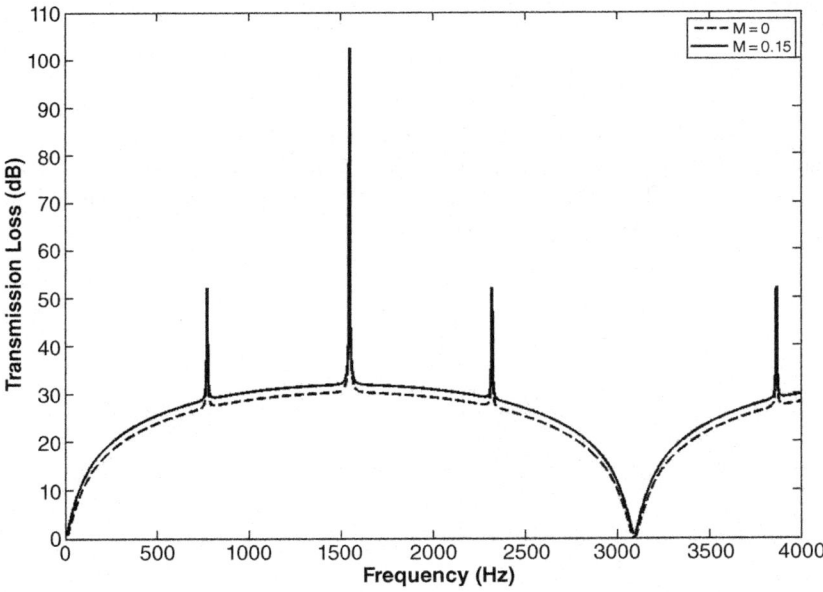

Figure 8.11 Effect of the mean flow Mach number M on TL of the side-inlet side-outlet muffler of Figure 8.9

8.7 Designing for Insertion Loss

The insertion loss (IL) curve is generally similar to the transmission loss (TL) curve. However, there is a significant difference. At very low frequencies ($\sim 50\, Hz$), the IL curve is characterized by a sharp dip at which the IL value is negative. At and around this frequency, the muffled SPL would exceed the unmuffled value! It is, therefore, important to understand the phenomenon and design the muffler so as to ensure that the dip occurs at a frequency that is much lower than the firing frequency of the engine. At such low frequency, we can model the muffler as a lumped-element acoustical filter with the exhaust pipe and tail pipe being represented by lumped inertances and the chamber cavity as a lumped compliance, as shown in Figure 8.12.

Assuming the engine exhaust system to be a constant velocity source ($Z_s \equiv Z_{n+1} \to \infty$), and neglecting the mean flow effect, Equation 3.51 would reduce to

$$IL = 20 \log|v_3/v_0| = 20 \log|v_2/v_1|. \tag{8.13}$$

At such low frequencies, radiation resistance (the real part of Z_0) would be negligible, and radiation reactance would only add the end correction of about $0.6 r_0$ to the tail pipe length. Therefore, we can safely neglect Z_o with respect to Z_1. Then it can readily be checked that $v_2/v_1 = 1 + Z_1/Z_2$, and therefore, Equation 8.13 yields

$$IL = 20 \log\left|1 + \frac{Z_1}{Z_2}\right| \tag{8.14}$$

Use of relations (2.49), (2.55) and (2.58) in Equation 8.14 yields

$$IL = 20 \log\left|1 - \frac{\omega^2 l_1}{S_1} \frac{V_2}{c_o^2}\right| \tag{8.15}$$

where V_2 is volume of the chamber.

It may be observed from Equation 8.15 that IL would start with 0 dB at zero Hz, then become negative and have a sharp dip at

$$f_{dip} = \frac{c_0}{2\pi}\left(\frac{S_1}{l_1 V_2}\right)^{1/2} \quad (Hz) \tag{8.16}$$

Beyond this frequency, IL would start increasing sharply, returning to zero and becoming positive as shown in Figure 8.13. So, we must maximize the tail pipe length and muffler volume and minimize the cross-sectional area of the tail pipe, so that f_{dip} given by Equation 8.16 is less than half of the firing frequency.

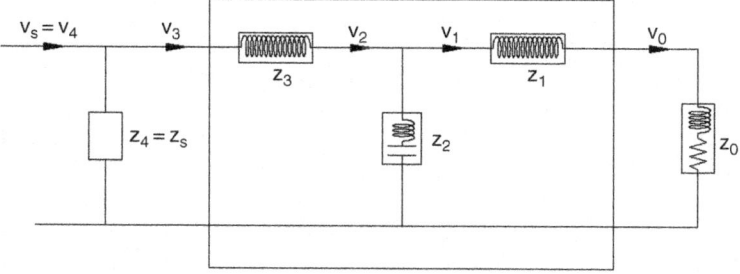

Figure 8.12 Lumped-element approximation of the simple expansion chamber of Figure 8.1 in order to capture the low frequency dip in the IL curve

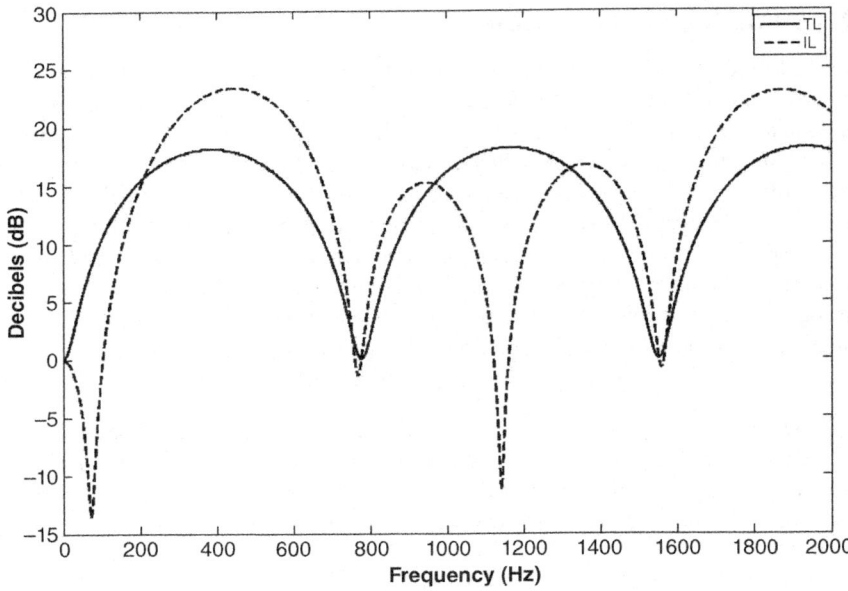

Figure 8.13 Comparison of the TL and IL spectra for the simple expansion chamber muffler of Figure 8.1 (Note the low frequency dip in the IL-graph)

In actual practice, dissipative effect of the mean flow at the inlet and outlet of the chamber would raise the IL troughs to near zero. Nevertheless, the rules for designing an exhaust muffler for insertion loss are as follows.

i. Make use of a narrower tail pipe. Make sure however that the mean flow Mach number in the tail pipe does not exceed 0.25, otherwise there may be too much jet noise and back pressure.
ii. Use large enough muffler volume. Table 8.1 indicates the approximate insertion loss of typical reactive mufflers used with reciprocating engines [8]. Typically, a muffler volume of the order of ten times the engine swept volume has been found to be adequate for insertion loss of about 20 dBA. This constitutes the 'medium' size muffler in Table 8.1.

While a narrower tail pipe would help to increase the value of IL at the firing frequency, a large muffler would increase IL at all frequencies.

For the two-chamber muffler shown in Figure 8.14, a similar investigation reveals that the intermediate pipe (numbered 3) as well as the tail pipe (numbered 1) should be made a little narrower than the exhaust pipe (numbered 5), and volumes of chambers 2 and 4 should be made as large as feasible. Figure 8.15 compares the insertion loss values of the muffler configuration of Figure 8.14 for $d_1 = d_3 = 30\,mm$ (retaining $d_5 = 40\,mm$) with those for $d_1 = d_3 = d_5 = 40\,mm$. In both these cases, $l_2 = l_4 = l/2 = 200\,mm$. Improvement in overall IL may be noted.

Here, and indeed throughout this chapter, the engine considered is a 1.2 liter capacity, turbocharged, three-cylinder, four-stroke cycle, diesel engine, running at 3750 RPM with air-fuel ratio of 20.35. The firing frequency works out to be $3750/60 \times 3/2 = 93.75\,Hz$. Therefore, the unmuffled SPL, insertion loss, and muffled SPL have been calculated for speed order of

Design of Mufflers

Table 8.1 Approximate insertion loss (dB) of typical reactive mufflers used with reciprocating engines[a] (Adapted from [8])

Octave band centre frequency (Hz)	Low pressure-drop muffler			High pressure-drop muffler		
	Small	Medium	Large	Small	Medium	Large
63	10	15	20	16	20	25
125	15	20	25	21	25	29
250	13	18	23	21	24	29
500	11	16	21	19	22	27
1000	10	15	20	17	20	25
2000	9	14	19	15	19	24
4000	8	13	18	14	18	23
8000	8	13	18	14	17	23

[a] Refer to manufacturers literature for more specific data.

1.5 and the firing frequency of 93.75 Hz, and integral multiples thereof. Overall insertion loss in dBA has been calculated by subtracting the total A-weighted muffled SPL from the total A-weighted unmuffled SPL.

It helps to make $l_2 \neq l_4$ and $l_{3a} \neq l_{3b}$. Figure 8.15 compares the IL performance of the unequal length configuration ($l_2 = 170\ mm, l_4 = 230\ mm, l_{3a} = l_4/2 - \delta, l_{3b} = l_2/2 - \delta$) with that of the equal length configuration ($l_2 = l_4 = l/2 = 200\ mm, l_{3a} = l_{3b} = 100\ mm$).

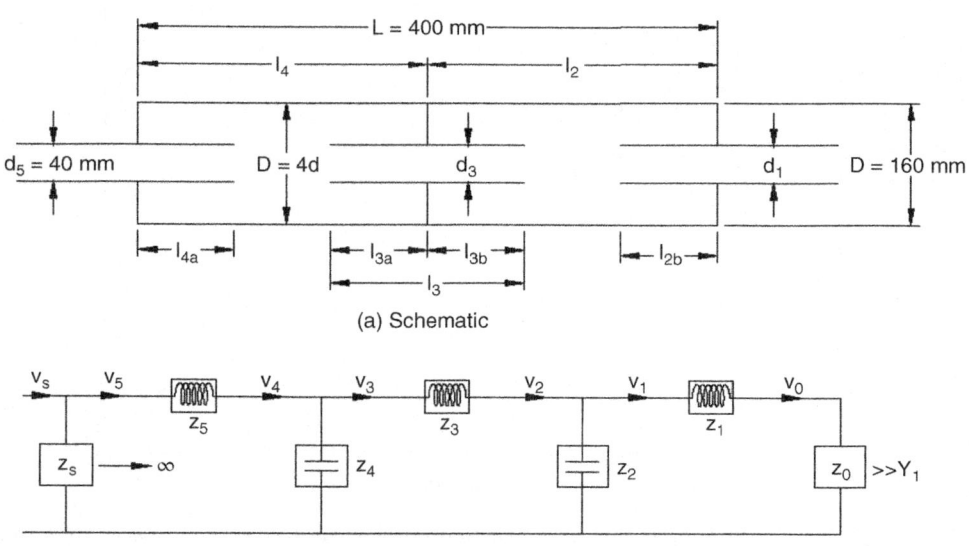

Figure 8.14 Two-chamber double-tuned extended-tube chamber muffler ($l_{4a} = l_4/4 - \delta_5, l_{3a} = l_4/2 - \delta_3, l_{3b} = l_2/2 - \delta_3, l_{2b} = l_2/4 - \delta_1, l_2 = l_4 = L/2 = 400/2 = 200\ mm, d_1 = d_3 = d_5 = 40\ mm, D = 160\ mm$)

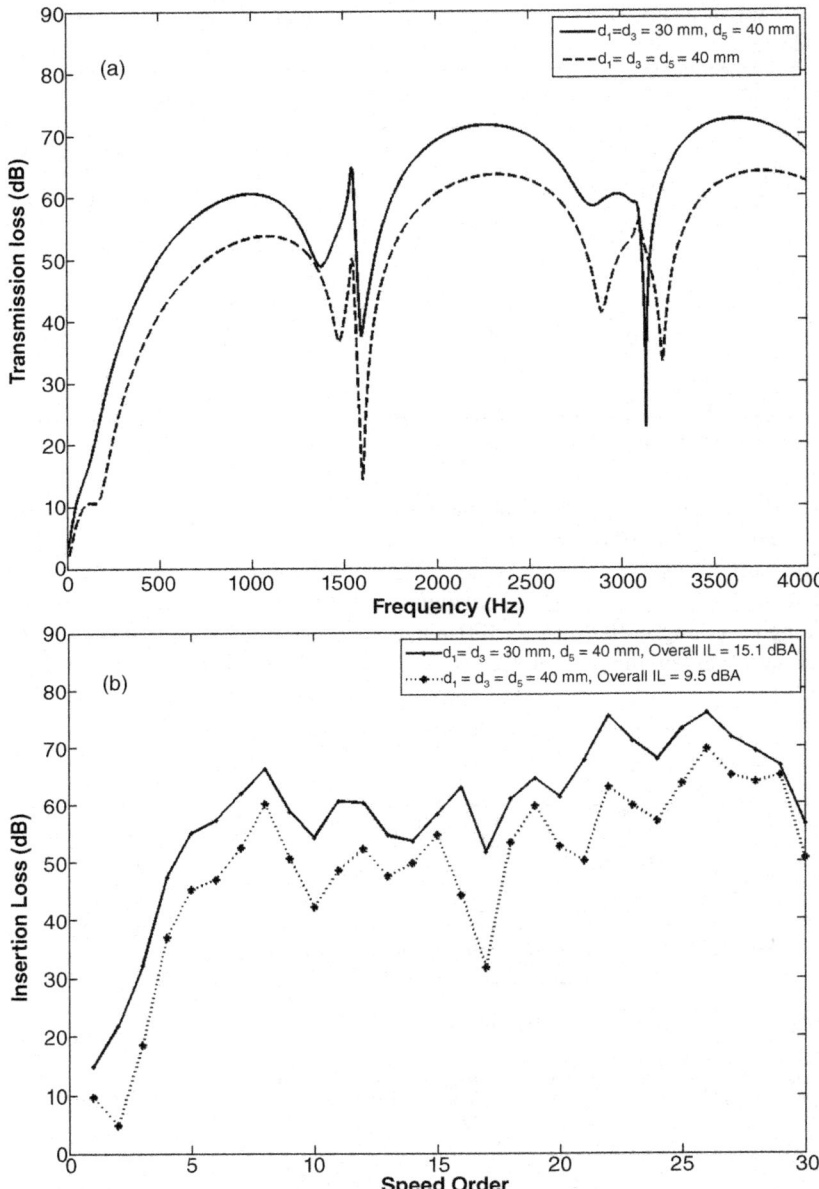

Figure 8.15 Effect of narrowing of the tail pipe and intermediate pipe of the two-chamber muffler of Figure 8.14 on (a) TL and (b) IL (Firing frequency of 93.75 Hz)

Making use of Equation 5.14 for $D = 160\ mm, d = 40\ mm, t_w = 1\ mm$, the end correction δ turns out to be 15.1 mm.

An overall lifting of the troughs (and lowering of the peaks) may be noted in Figure 8.16 for the unequal length configuration.

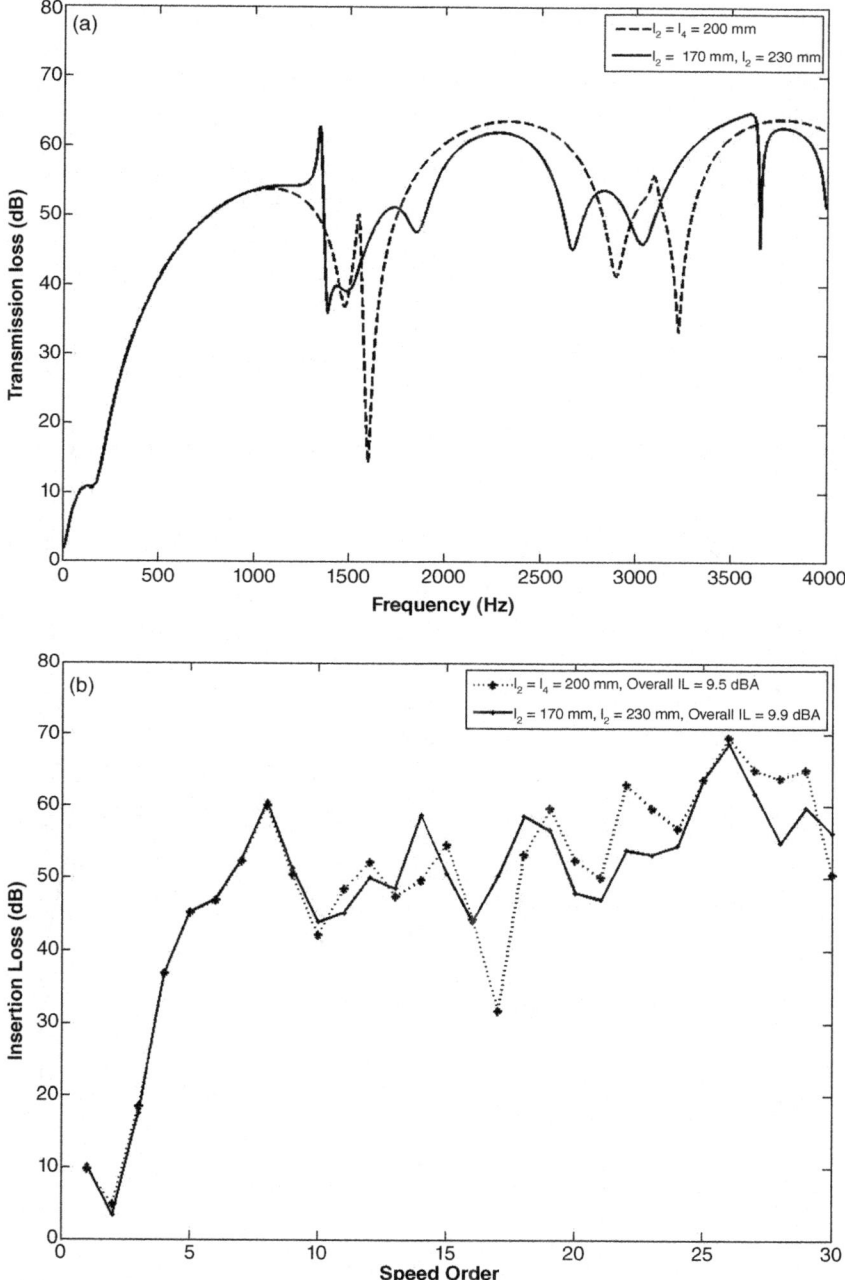

Figure 8.16 Effect of the unequal lengths of the two-chamber muffler of Figure 8.14 (for $d_1 = d_3 = d_5 = 40\,mm$) on (a) TL and (b) IL (Firing frequency of 93.75 Hz)

Incidentally, the IL-performance of the double-tuned two-chamber configuration of Figure 8.14 is compared with that of a single chamber configuration of Figure 8.1 in Figure 8.17 in order to investigate the effect of the number of chambers within the same overall length. It may be noted that the two-chamber configuration of Figure 8.14 yields

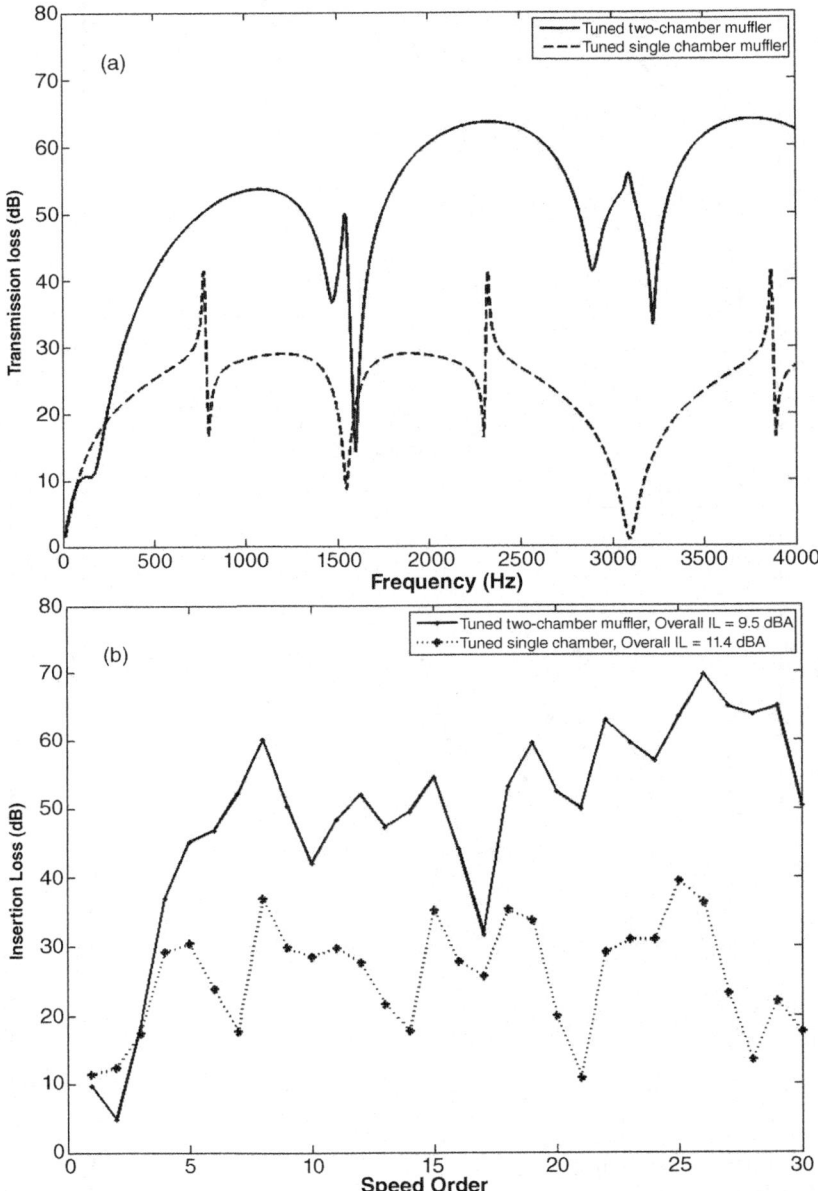

Figure 8.17 Effect of the number of chambers within the same overall muffler length on (a) TL and (b) IL (Firing frequency = 93.75 Hz)

higher TL and IL at most frequencies. However, its low-frequency (firing frequency) performance is poorer and therefore its overall IL is less than the single (longer) expansion chamber mufflers. Incidentally, this was first predicted way back in 1973 [9].

Similar effect may be noted for the corresponding tuned concentric tube resonator muffler configuration (Figure 8.14 with perforate bridges) in Figure 8.18.

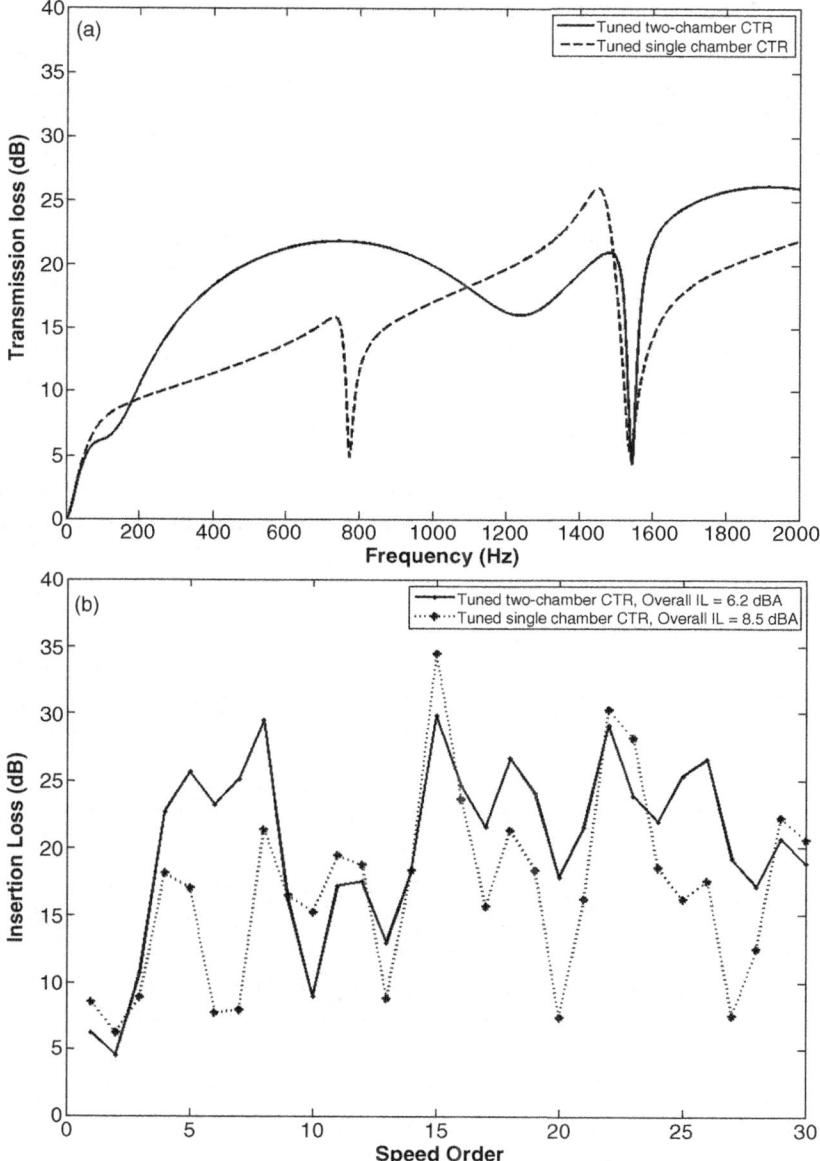

Figure 8.18 Effect of partitioning of the annular cavity of the tuned CTR on (a) TL and (b) IL (Firing frequency = 93.75 Hz)

However, if the constraint on the length of the muffler is removed (that is, the two chamber configuration could be longer than the corresponding single chamber configuration), then it would help to make use of two smaller chambers than a single longer chamber. But then, this would increase the overall length (and hence volume, weight and cost) of the muffler. In general, most automotive mufflers consist of two or three chambers.

8.8 Three-Pass Double-Reversal Chamber Mufflers

Flow-acoustic analysis of the three-pass double-reversal chamber muffler (Figure 8.19) has been discussed at length in Sections 4.3 and 4.4. Here we present results of some parametric studies in order to develop design guidelines for this multiply-connected configuration which combines a fairly wide-band TL (and IL) with minimal back pressure.

The parameters varied in this investigation [10] are:

Area expansion ratio for elliptical cross section, $m = \dfrac{D_1 D_2}{d^2} = 9, \underline{16}, 25$

The mean flow Mach number in the exhaust pipe, $M = 0.1, \underline{0.15}, 0.2$

Open area ratio of the perforate, $OAR = \dfrac{\pi d\, l_p \sigma}{(\pi/4)d^2} = 0.5, \underline{1.0}, 1.5$

Perforate length, $l_p/d = 3, \underline{4}, 5$

Length of the intermediate pipe, $l/d = 1.0, \underline{2.0}, 3.0$

End chamber length, $l_a/d = l_b/d = 3.0, \underline{4.0}, 5.0$.

Here, $d = d_2 = d_3 = d_4$. The default values of the parameters are underlined. The open ratio (OAR) is varied by varying porosity, σ. For doubly periodic set of holes in the perforate,

$$\sigma = \sigma_2 = \sigma_3 = \sigma_4 = \frac{(\pi/4)d_h^2}{C^2} \qquad (8.17)$$

where d_h is hole diameter, C is the center-to-center distance between the holes. Thus, porosity, in turn, is varied by varying the center-to-center distance, C.

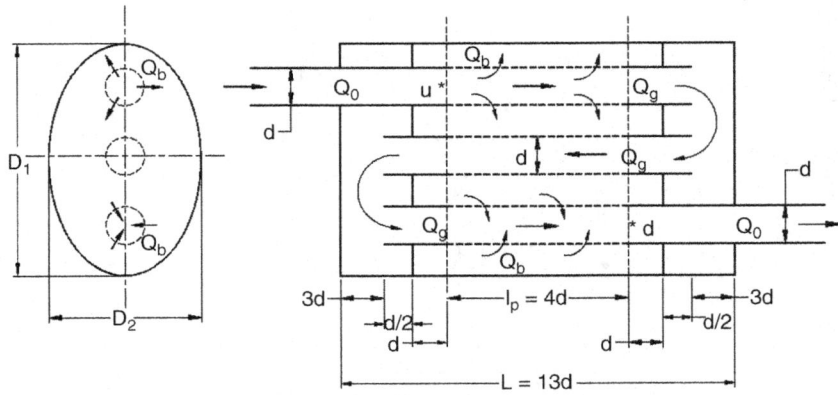

Figure 8.19 Schematic of a three-pass double-reversal chamber muffler

Design of Mufflers

Flow-distribution needed in the analysis is computed for each set of parameters as per Section 4.6. The end chambers are assumed to support axial plane wave propagation.

For the muffler configuration with default values,

length of the muffler shell, $L = 13\,d = 13 \times 40 = 520\,mm$
For $D_2/D_1 = 0.6$ and $D_1 D_2 = 16 d^2$, $D_1 = 5.164\,d = 206.6\,mm$
$$D_2 = 3.1\,d = 123.9\,mm$$
Volume of the muffler, $V = \pi/4\, D_1 D_2 L = 0.01045\,m^3 = 10.45$ litres.

For evaluation and comparison of the insertion loss values, an engine with the following parameters has been considered here:

Type of the engine: compression ignition, turbocharged
Engine capacity (or, swept volume) = 1.2 liters
Number of cylinders = 3
Number of strokes per cycle = 4
Air-fuel ratio, AFR = 20.35
Rated speed of the engine = 3750 RPM.

Internal impedance of the engine, $Z_s \equiv Y_0(0.707 - j0.707)$, where $Y = c/S$ is characteristic impedance of the exhaust pipe of diameter $d\,(S = \pi/4\, d^2)$.

Figure 8.20 confirms the well-known observation (see Figure 8.2) that an increase in the area expansion ratio m results in an increase in the TL as well as IL at most frequencies. In particular, a general increase in the value of the overall IL with increase in m is noteworthy. This method of increasing IL of a given muffler configuration can come handy, particularly for stationary gensets or compressor installations, where space constraint is not so critical as the mandatory requirement of reducing noise at the neighbor's premises.

Figure 8.21 shows that the effect of the mean flow Mach number for the muffler configuration of Figure 8.19 is not so marked as for the plug muffler configuration of Figure 8.7, shown in Figure 8.22 where the inline flow-acoustic resistance is directly proportional to the mean flow velocity or Mach number, and that results in acoustic absorption or dissipation, which in turn raises the TL and IL graphs considerably.

Figure 8.23 shows that increasing the open area ratio of the perforate (in all the three perforated tubes) in the middle chamber of the muffler configuration of Figure 8.19 by increasing the porosity is counter-productive. On the other hand, an increase in the length of the perforate while keeping the OAR at the default value of 1.0 (fixed) by reducing the porosity proportionally, increases the overall IL, as shown in Figure 8.24. Incidentally, for $l_p = 4d$ and OAR = 0.5, Equation 8.11 indicates that porosity σ equals $1/32$ or 0.0313 or 3.13%. This, then, is the optimal value of porosity.

Incidentally, a change in the porosity or OAR would change the flow distribution as per Equation 4.104 and thence the flow-acoustic impedance of the perforates in the intermediate pipes, and this has been duly incorporated in the analysis leading to the graphs of Figure 8.23.

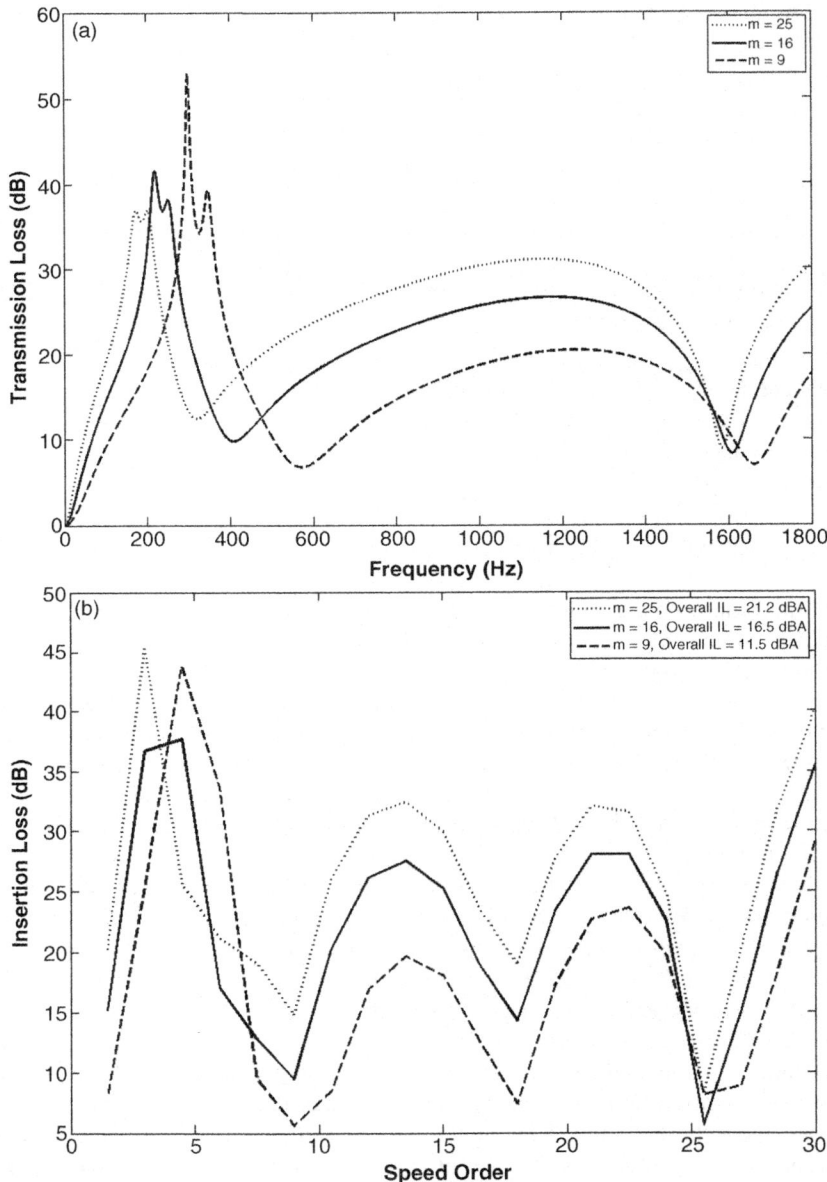

Figure 8.20 Effect of the area expansion ratio on the (a) TL and (b) IL of the muffler of Figure 8.19 (Firing frequency = 93.75 Hz)

Design of Mufflers

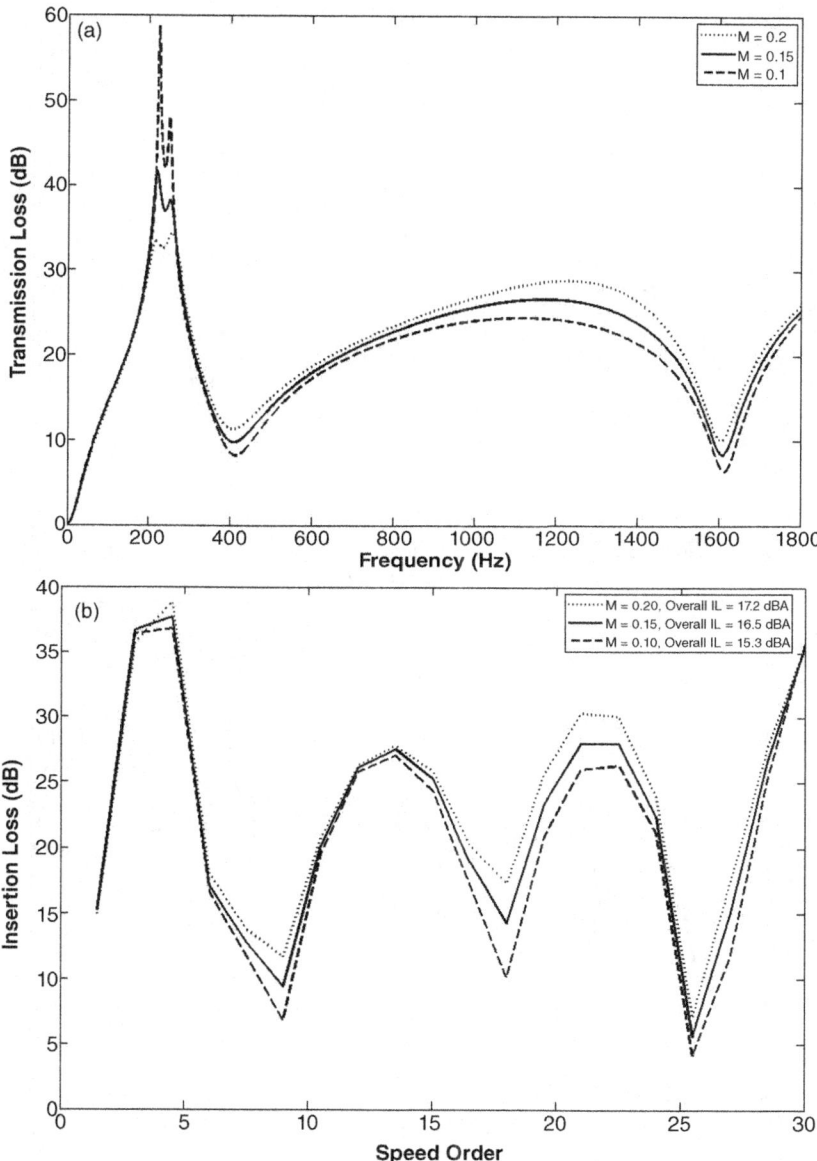

Figure 8.21 Effect of the mean flow mach number M on the TL and IL of the muffler configuration of Figure 8.19

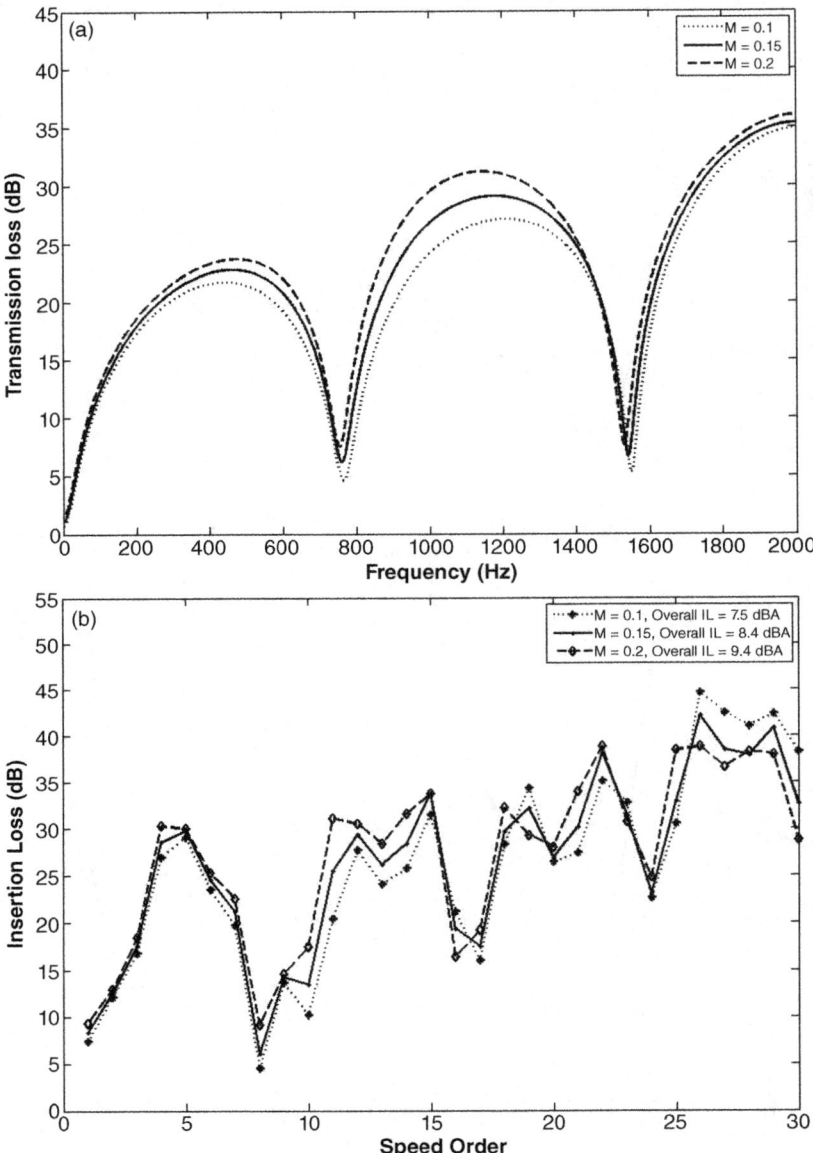

Figure 8.22 Effect of mean flow mach number M on (a) TL and (b) IL of the plug muffler of Figure 8.7 (Firing frequency = 93.75 Hz)

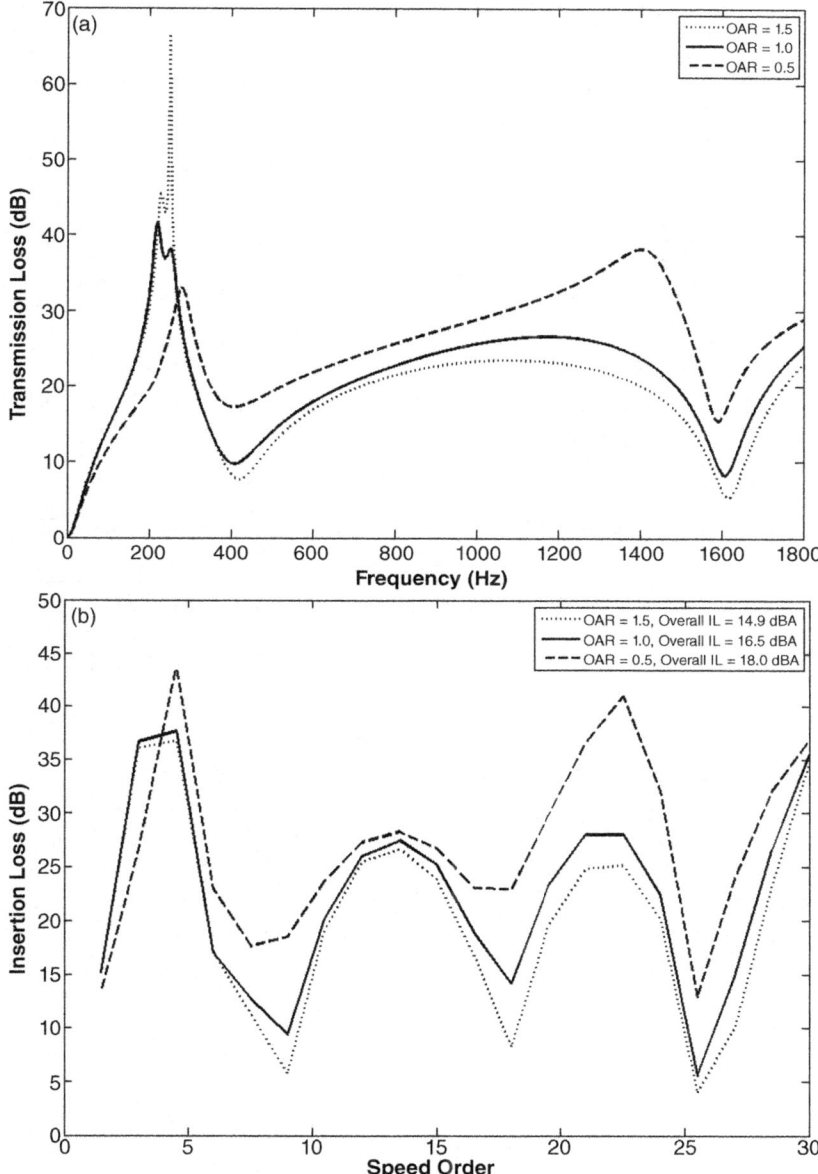

Figure 8.23 Effect of the open area ratio of the perforate on the (a) TL and (b) IL of the plug muffler of Figure 8.19 (Firing frequency = 93.75 Hz)

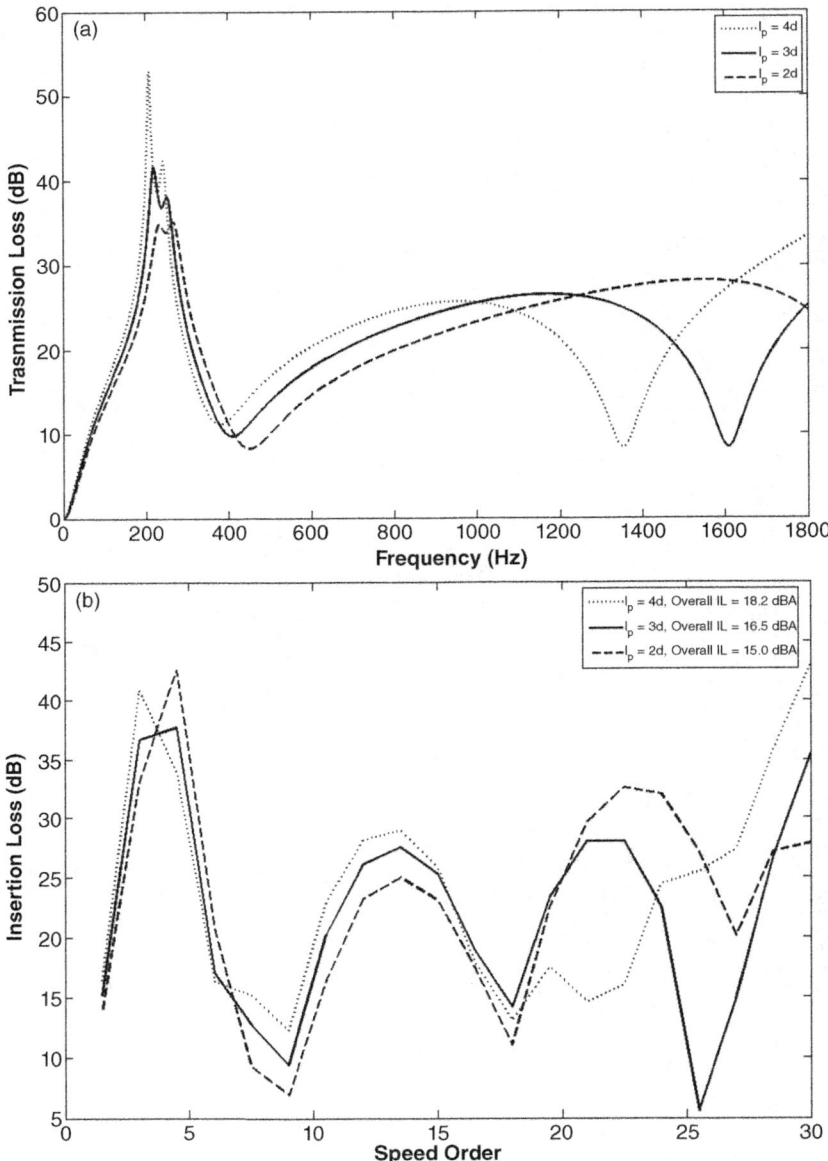

Figure 8.24 Effect of the length of the perforate l_p, keeping OAR fixed at 1.0, on (a) TL and (b) IL of the muffler configuration of Figure 8.19 (Firing frequency = 93.75 Hz)

Figure 8.25 shows that the effect of the extended pipe length (l_e) on the overall IL is negligible. Incidentally, $l_e = 0$ implies the perforated tubes being flush with the plates [11]. Therefore, l_e may be decided by the convenience or logistics of the welding process. Accordingly, the default value of l_e has been assumed here as $l = d$.

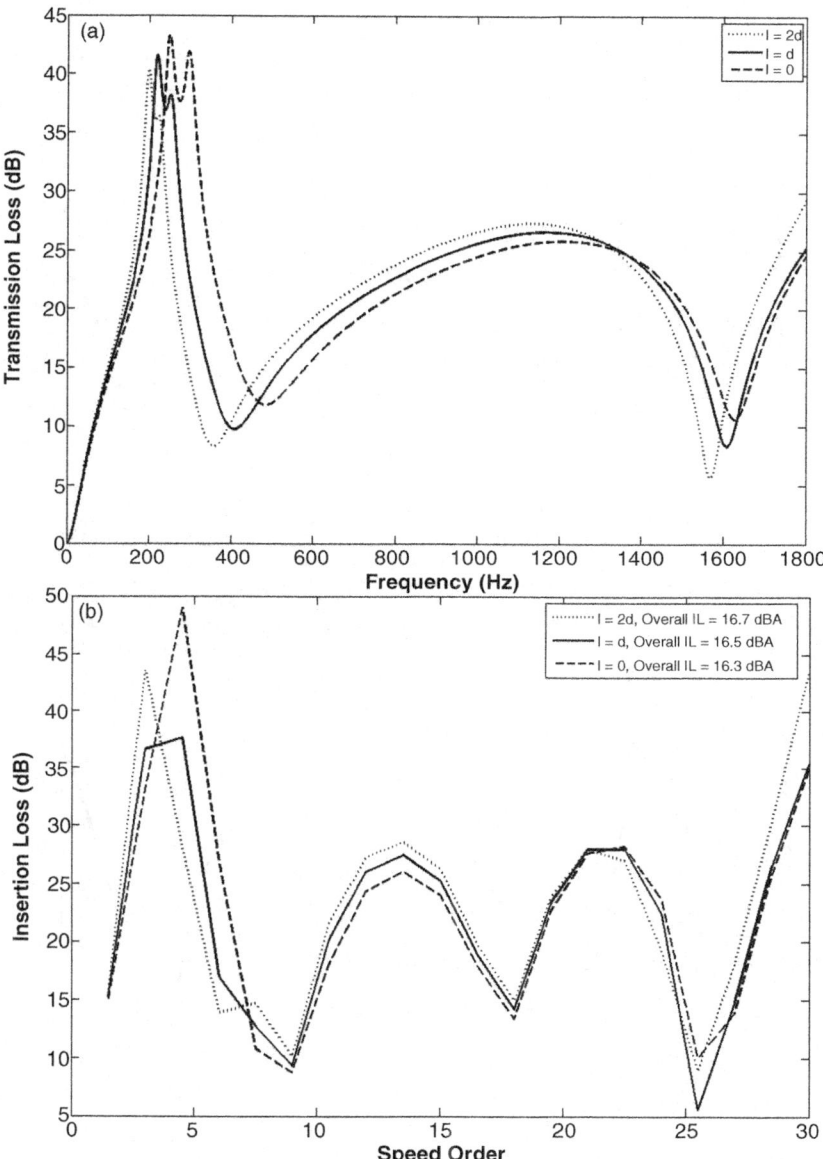

Figure 8.25 Effect of the extended pipe length l on (a) TL and (b) IL of the muffler configuration of Figure 8.19 (Firing frequency = 93.75 Hz)

Figure 8.26 shows that it is desirable to increase the length of the end chamber if space, weight and cost permit. Incidentally, a longer end chamber would also reduce the aerodynamic noise produced by jet impingement on the end plates of the muffler (see Figure 8.19).

Here, it must be pointed out that by increasing the perforate length and/or end-chamber length, we are increasing the total length and hence the volume of the muffler, and, as pointed

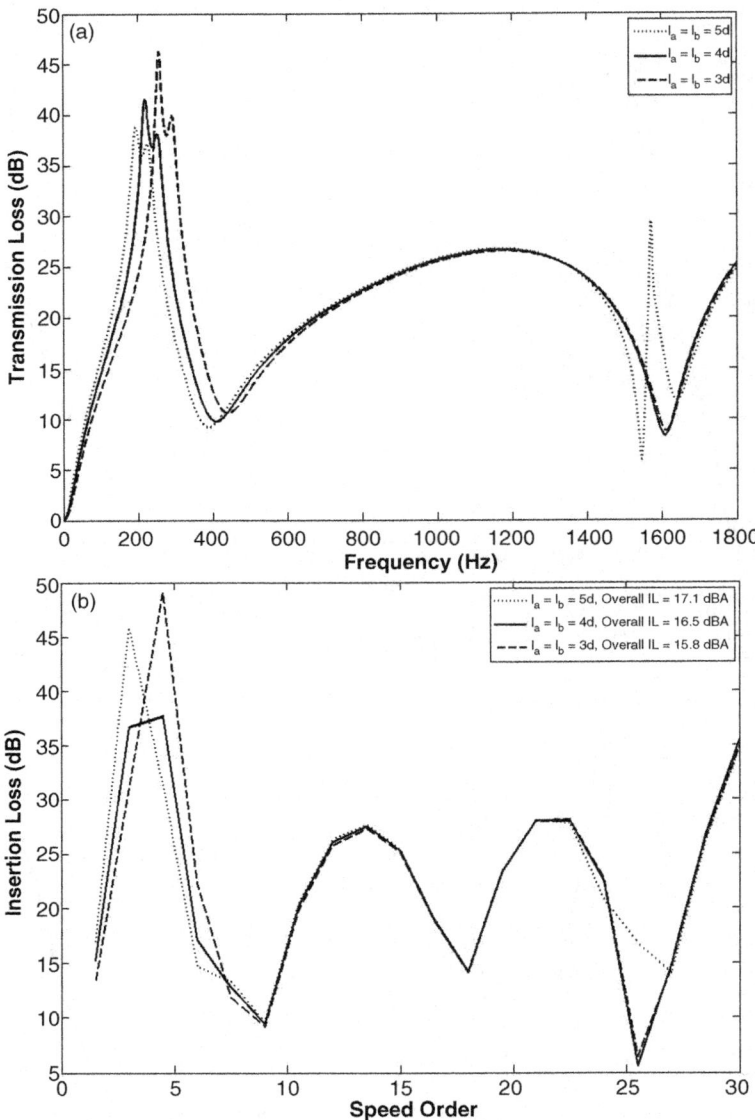

Figure 8.26 Effect of the end chamber length l_a, l_b on (a) TL and (b) IL of the muffler configuration of Figure 8.19 (Firing frequency = 93.75 Hz)

out earlier, an increase in the volume of the muffler would invariably result in increased insertion loss. Therefore, an increase in overall TL of the muffler shown by Figures 8.23 and 8.25 needs to be construed accordingly. However, the beneficial effect of the decrease in the open area ratio (or porosity) in Figure 8.23 is genuine inasmuch as it does not alter the overall length and hence volume of the muffler.

8.9 Perforated Baffle Muffler

The three-pass double-reversal muffler shown in Figure 8.19 and discussed at some length in the previous section makes use of two intermediate baffles which were assumed to be impervious. However, they could as well be perforated. Then there would be flow-acoustic interaction between the acoustic pressure field in the end chambers with that in the annular intermediate chamber, which in turn interacts with the acoustic field in all three perforated tubes. This configuration has been analyzed by Veerababu [10], making use of the Integrated Transfer Matrix Method [12] described earlier in Chapter 4. The results are shown in Figure 8.27 for

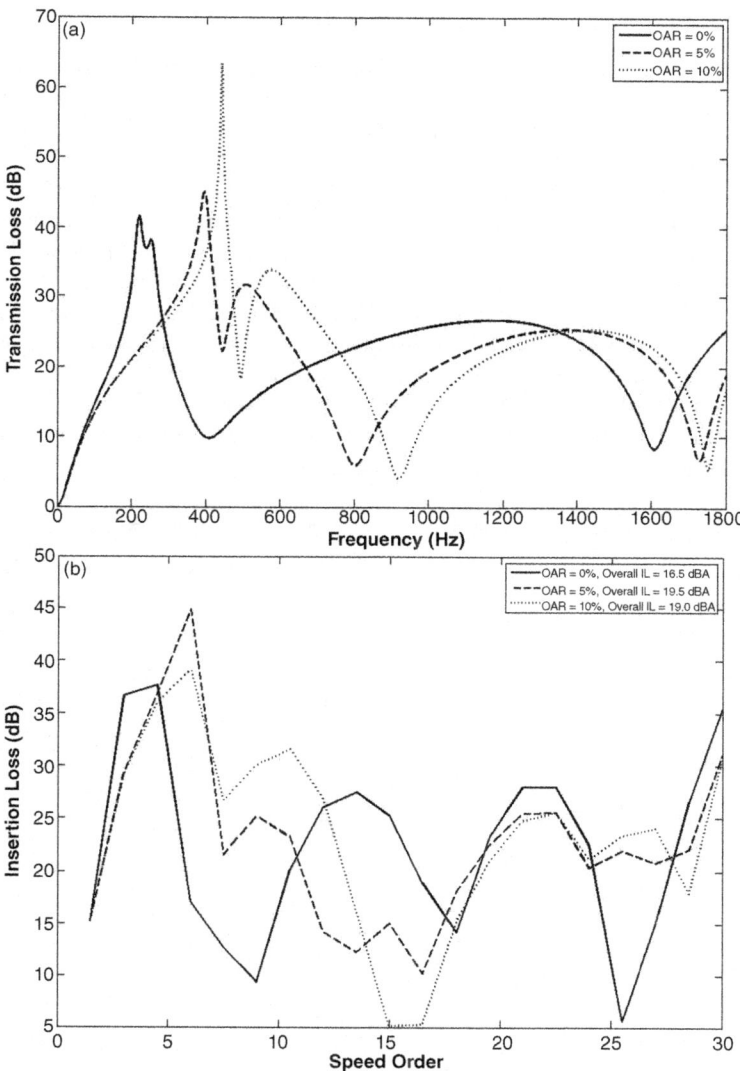

Figure 8.27 Effect of the porosity of perforated baffle plates on (a) TL and (b) IL muffler of Figure 8.19 (Adapted from [10])

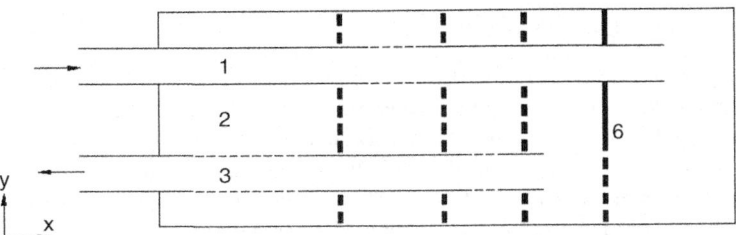

Figure 8.28 Schematic of a muffler with perforated elements, cross baffles and area discontinuities (Adapted from [13])

different values of porosity or open area ratio. It may be observed that perforation of the baffles adds considerably to the insertion loss, although increasing the porosity from 5% to 10% is of no use. Incidentally, for the symmetrical configuration of Figure 8.18, it may be noted that there will be no cross-flow across the perforated baffles.

The role of inline baffles placed parallel to the axial plane wave front can be exploited to ensure wideband TL right up to the low frequencies of the order of the engine firing frequency. In fact, perforated baffles have been used effectively by Elnady *et al.* [13] for the flow-reversal muffler configuration shown in Figure 8.28. They used ingenious four-port model and SIDLAB simulations [14] to predict TL of the muffler configuration of Figure 8.28 and validated the same against experiments [12]. The TL performance of the same, predicted by means of the Integrated Transfer Matrix method [13], is shown in Figure 8.29. It is noteworthy that the predicted TL exceeds 20 dB at all frequencies of 80 Hz onwards.

At the time of writing, however, there is no known methodology for rational design of mufflers with perforated baffles for the desired insertion loss.

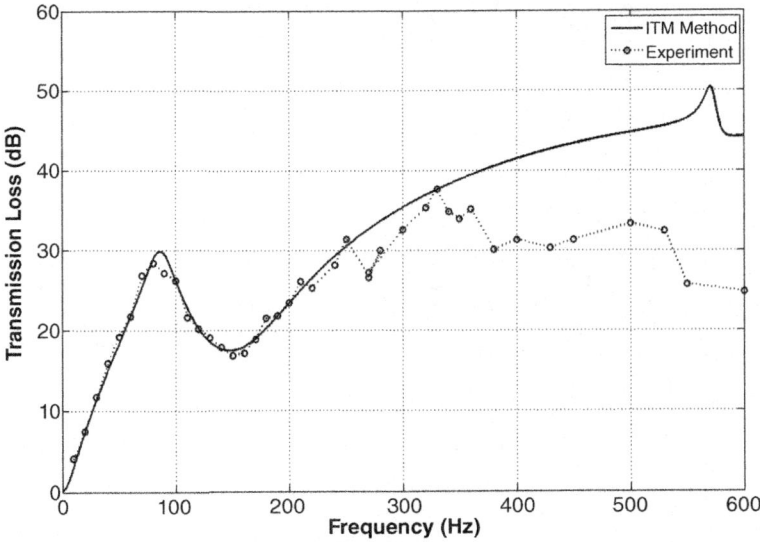

Figure 8.29 Validation of the ITM method with measurements [12,13] for the muffler of Figure 8.28 at $M = 0.15$ (Reproduced with permission from [12]; the experimental curve reproduced from [13])

8.10 Forked Dual Muffler System

Sometimes, under the vehicle bottom, sufficient space or room is not available for housing a single muffler of the required volume. One of the desirable alternatives is to use two identical mufflers of the same (desired) length, but with half the cross-sectional area of the respective exhaust pipe, tail pipe, intermediate pipes and the muffler shell. A schematic of such a forked dual muffler system is shown in Figure 8.30. The two branches are identified by subscripts 'a' and 'b', and the pipe internal diameter $d_a = d_b = d/\sqrt{2}$ so that the mean flow Mach number in each of the branches is equal to that in the upstream common exhaust pipe as well as in the equivalent single muffler configuration. In order to maintain the same area expansion ratio, the shell diameter, $D_a = D_b = D/\sqrt{2}$.

In general, making use of the procedure laid out in Chapters 2 and 3, the governing equations are given by

$$\begin{bmatrix} p_{u,a} \\ v_{u,a} \end{bmatrix} = \begin{bmatrix} T_{11,a} & T_{12,a} \\ T_{21,a} & T_{22,a} \end{bmatrix} \begin{bmatrix} p_{d,a} \\ v_{d,a} \end{bmatrix}, \qquad (8.18, 8.19)$$

$$\begin{bmatrix} p_{u,b} \\ v_{u,b} \end{bmatrix} = \begin{bmatrix} T_{11,b} & T_{12,b} \\ T_{21,b} & T_{22,b} \end{bmatrix} \begin{bmatrix} p_{d,b} \\ v_{d,b} \end{bmatrix}, \qquad (8.20, 8.21)$$

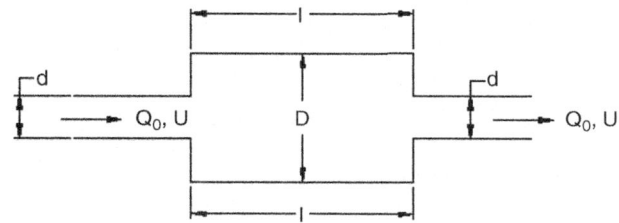

(a) Single muffler configuration ($Q_0 = SU$, $S = (\pi/4)d^2$)

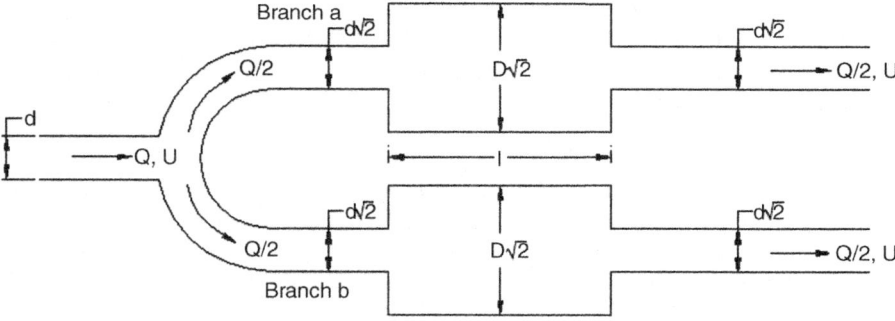

(b) Forked dual muffler and configuration

Figure 8.30 Schematic of the equivalent single and dual muffler configurations ($U_a = U_b = U$, $Q_a = Q_b = Q/2$)

$$p_u = p_{u,a} = p_{u,b}, \qquad (8.22, 8.23)$$

$$v_u = v_{u,a} + v_{u,b}, \qquad (8.24)$$

$$p_{d,a} = v_{d,a} Z_{0,a}, \qquad (8.25)$$

$$p_{d,b} = v_{d,b} Z_{0,b}, \qquad (8.26)$$

There are nine equations for ten unknowns (unknown variables), so we can obtain expressions for these nine variables in terms of the tenth variable, say v_u. Alternatively, let

$$v_u = 1. \qquad (8.27)$$

Then, the set of ten inhomogeneous equations may be written in the matrix form

$$\begin{bmatrix} 1 & 0 & -T_{11,a} & -T_{12,a} & 0 & 0 & 0 & 0 & 0 & 0 \\ 0 & 1 & -T_{21,a} & -T_{22,a} & 0 & 0 & 0 & 0 & 0 & 0 \\ 0 & 0 & 0 & 0 & 1 & 0 & -T_{11,b} & -T_{12,b} & 0 & 0 \\ 0 & 0 & 0 & 0 & 0 & 1 & -T_{21,b} & -T_{22,b} & 0 & 0 \\ -1 & 0 & 0 & 0 & 0 & 0 & 0 & 0 & 1 & 0 \\ 1 & 0 & 0 & 0 & -1 & 0 & 0 & 0 & 0 & 0 \\ 0 & -1 & 0 & 0 & 0 & -1 & 0 & 0 & 0 & 1 \\ 0 & 0 & 1 & -Z_{0,a} & 0 & 0 & 0 & 0 & 0 & 0 \\ 0 & 0 & 0 & 0 & 0 & 0 & 1 & -Z_{0,b} & 0 & 0 \\ 0 & 0 & 0 & 0 & 0 & 0 & 0 & 0 & 0 & 1 \end{bmatrix} \begin{bmatrix} p_{u,a} \\ v_{u,a} \\ p_{d,a} \\ v_{d,a} \\ p_{u,b} \\ v_{u,b} \\ p_{d,b} \\ v_{d,b} \\ p_u \\ v_u \end{bmatrix} = \begin{bmatrix} 0 \\ 0 \\ 0 \\ 0 \\ 0 \\ 0 \\ 0 \\ 0 \\ 0 \\ 1 \end{bmatrix}$$

(8.28)

This can be solved on computer in MATLAB® making use of Gaussian elimination with pivoting. However, an understanding of the flow-acoustic implication of the forked dual muffler configuration may be obtained as follows.

Let branch 'b' be identical to branch 'a' so that all variables and parameters subscripted 'b' would be equal to their 'a' counterparts. Retaining subscript 'a' only, Equations 8.18–8.27 reduce to the following equations:

$$\begin{bmatrix} p_{u,a} \\ v_{u,a} \end{bmatrix} = \begin{bmatrix} T_{11,a} & T_{12,a} \\ T_{21,a} & T_{22,a} \end{bmatrix} \begin{bmatrix} p_{d,a} \\ v_{d,a} \end{bmatrix}, \qquad (8.29)$$

$$p_u = p_{u,a}, \qquad (8.30)$$

$$1 = 2 v_{u,a}, \qquad (8.31)$$

$$p_{d,a} = v_{d,a} Z_{0,a}. \qquad (8.32)$$

Design of Mufflers

These can be solved systematically as follows:

$$p_u = p_{u,a} = (T_{11,a}Z_{0,a} + T_{12,a})v_{d,a},$$

$$v_{u,a} = (T_{21,a}Z_{0,a} + T_{22,a})v_{d,a} = \frac{1}{2} \Rightarrow v_{d,a} = \frac{1}{2(T_{21,a}Z_{0,a} + T_{22,a})}.$$

Thus,

$$p_u = \frac{(T_{11,a}Z_{0,a} + T_{12,a})}{2(T_{21,a}Z_{0,a} + T_{22,a})}. \tag{8.33}$$

Evaluation of TL involves complex amplitudes of the progressive waves A and B. These may be obtained from the corresponding standing wave variables p and v as follows.

$$\left.\begin{array}{l} p_u = A_u + B_u \\ v_u = \dfrac{A_u - B_u}{Y_u} \end{array}\right] \Rightarrow A_u = \frac{1}{2}(p_u + Y_u v_u) = \frac{1}{2}\left\{\frac{T_{11,a}Z_{0,a} + T_{12,a}}{T_{21,a}Z_{0,a} + T_{22,a}} + Y_{u,a}\right\}. \tag{8.34}$$

Similarly,

$$A_{d,a} = \frac{1}{2}(p_{d,a} + Y_{d,a}v_{d,a}) = \frac{v_{d,a}}{2}(Z_{d,a} + Y_{d,a}) = \frac{Z_{d,a} + Y_{d,a}}{4(T_{21,a}Z_{0,a} + T_{22a})}, \tag{8.35}$$

$$TL = 10\lg \frac{W_t}{W_{t,a} + W_{t,b}} = 10\lg \frac{\dfrac{1}{2\rho_0}|A_u|^2 Y_u}{2 \times \dfrac{1}{2\rho_0}|A_d|^2 Y_{d,a}} = 10\lg\left[\frac{|A_u|^2 Y_u}{2|A_d|^2 Y_{d,a}}\right]. \tag{8.36}$$

Let us now compare the forked configuration with the equivalent single configuration. Then, cross-sectional areas $S_{u,a} = S_{d,a} = \dfrac{S_0}{2} \Rightarrow Y_{u,a} = Y_{d,a} = 2Y_0$,

$$Z_{0,u} = Z_{0,d} = 2Z_0 = 2Y_0 \text{ (for anechoic termination)},$$

$$T_{11,a} = T_{11}, T_{12,a} = 2T_{12}, T_{21,a} = T_{21}/2, T_{22,a} = T_{22},$$

$$p_u = \frac{T_{11} \times 2Y_0 + 2T_{12}}{2\left(\dfrac{T_{21}}{2} \times 2Y_0 + T_{22}\right)} = \frac{T_{11}Y_0 + T_{12}}{T_{21}Y_0 + T_{22}},$$

$$A_u = \frac{1}{2}(p_u + Y_u v_u) = \frac{1}{2}\left(\frac{T_{11}Y_0 + T_{12}}{T_{21}Y_0 + T_{22}} + Y_0\right),$$

$$A_d = \frac{2Y_0 + 2Y_0}{4\left(\dfrac{T_{21}}{2} \times 2Y_0 + T_{22}\right)} = \frac{Y_0}{2(T_{21}Y_0 + T_{22})}.$$

$$TL_{dual} = 10 \lg \left[\frac{1}{4}\left|\frac{A_u}{A_d}\right|^2\right] = 10 \lg \left|\frac{T_{11}Y_0 + T_{12}}{Y_0} + T_{21}Y_0 + T_{22}\right|$$

$$= 10 \lg \left[\frac{1}{4}\left|T_{11} + \frac{T_{12}}{Y_0} + T_{21}Y_0 + T_{22}\right|^2\right] \quad (8.37)$$

$$= 20 \lg \left[\frac{1}{2}|T_{11} + T_{12}/Y_0 + T_{21}Y_0 + T_{22}|\right] = TL_{single}$$

Thus, TL of the forked dual configuration would be the same as that of the corresponding single muffler configuration.

Let us now calculate and compare the back-pressure (or, stagnation pressure drop) of the two muffler configurations. As the mean flow velocity, U, in each branch is the same as at the upstream point 'u', and the area ratio $\left((D/d)^2\right)$ of each of the parallel mufflers is the same as for the equivalent single muffler, it follows that the overall loss coefficient K and hence

$$\Delta p = K \cdot \left(\frac{1}{2}\rho_0 U^2\right) \quad (8.38)$$

for each of the branch mufflers would be equal to that of the equivalent single muffler.

Incidentally, it is instructive to analyze Δp in terms of the lumped flow-resistance network, shown in Figure 8.31.

$$\Delta p = K \cdot \frac{1}{2}\rho_0 U^2 = Q^2 R, Q = SU, \quad (8.39)$$

whence

$$R = \frac{\rho_0}{2S^2} K. \quad (8.40)$$

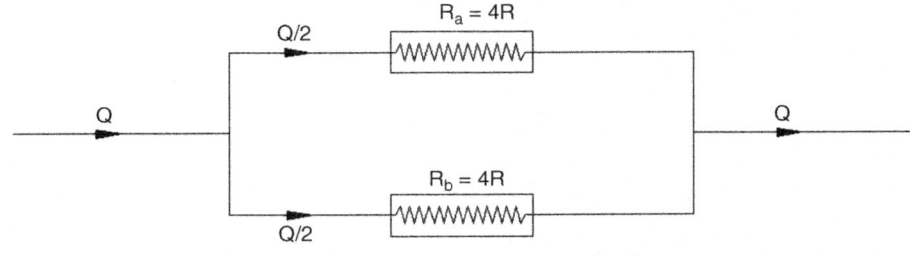

Figure 8.31 Flow-resistance networks for the two muffler configurations of Figure 8.30 for evaluation of back-pressure

Referring to Figure 8.30,

$$S_a = S_b = \frac{S}{2}.$$

Therefore,

$$R_a = R_b = 4R, \tag{8.41}$$

where R is the flow-resistance of the equivalent single muffler configuration.

Use of Equations 4.91 with $R_1 = R_a = 4R$ and $R_2 = R_b = 4R$ yields $R_{eq} = R$, and

$$\Delta p_{dual} = Q^2 R_{eq} = Q^2 R = \Delta p_{single}. \tag{8.42}$$

This confirms the observations made above.

In conclusion, for the special case of two identical parallel mufflers of total volume equal to that of the single equivalent muffler shown in Figure 8.30, there would be no change in the overall transmission loss (TL) and back pressure (Δp).

8.11 Design of Short Elliptical and Circular Chambers

The axially short elliptical (or circular) cylindrical chambers are often used as a flow-reversing end chambers in multi-pass perforated duct system as illustrated in Figure 8.32(a) and (b). This work concerns with obtaining optimal locations of the end ports to formulate guidelines

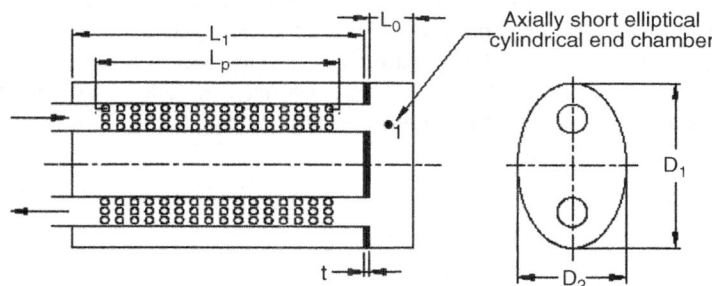

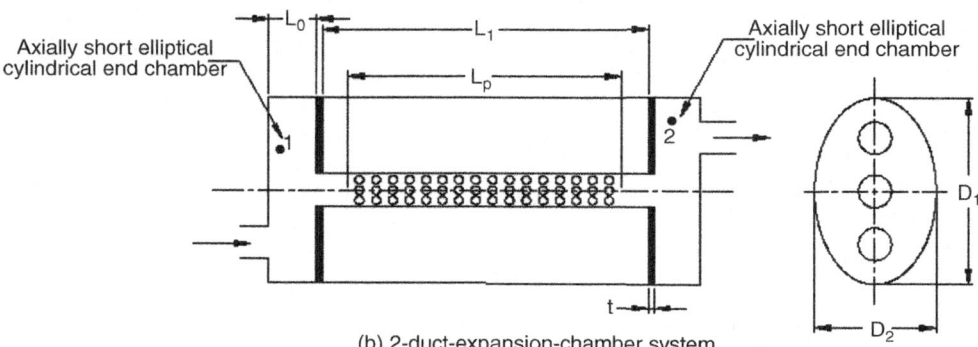

Figure 8.32 A schematic diagram illustrating the use of an axially short elliptical cylindrical end chamber in multi-pass perforated duct muffler system (Reproduced with permission from [26])

for designing the short end chambers based on the 3D modal summation and uniform piston-driven model [15,16].

8.11.1 Short Chamber Configurations

Figure 8.32(a) depicts a two-port flow-reversal end chamber comprising of an elliptical cylindrical chamber having two perforated pipes (of circular cross-section), one of the ends of which is connected to an axially short elliptical cylindrical chamber which facilitates the reversal of the flow direction. Therefore, the two ports of the axially short elliptical cylindrical chamber are located on the same end face. Similarly, Figure 8.32(b) depicts an expansion chamber type of configuration with the middle portion being the perforated pipe in an elliptical cylindrical chamber which in turn is connected to an axially short elliptical cylindrical chamber at both ends of the perforated pipe.

The length of the middle elliptical cylindrical chamber is denoted by L_1 while the length of the perforated pipe is symbolically denoted by L_p, whereas the thickness of the end plate partitioning the perforated chamber and the short chamber is denoted by t. The major and minor axis of the elliptical cross-section is given by D_1 and D_2, respectively. Figure 8.23(a) and (b) depict elliptical cylindrical chamber configurations having an end-inlet and end-outlet port [17]. An expansion chamber type of configuration is shown in Figure 8.33(a) whilst a flow-reversal end chamber configuration is shown in Figure 8.33(b). The chamber length is given by L. The end ports E1 and E2 may in general be located at arbitrary locations given by ξ_{E1}, η_{E1} and ξ_{E2}, η_{E2} on the elliptical cross-section, respectively, although in Figure 8.33(a) and (b), the locations of the end ports are expressed in terms of the radial distance δ_{E1}, δ_{E2} from the center and angular location θ_{E1}, θ_{E2} measured with respect to $\eta = 0$ or $\theta = 0$ axis. The offset distance δ and angular location θ may, in general, be obtained from the elliptical coordinates by means of the following relationships:

$$\delta = h\sqrt{\cosh^2\xi - \sin^2\eta}, \quad (8.43)$$

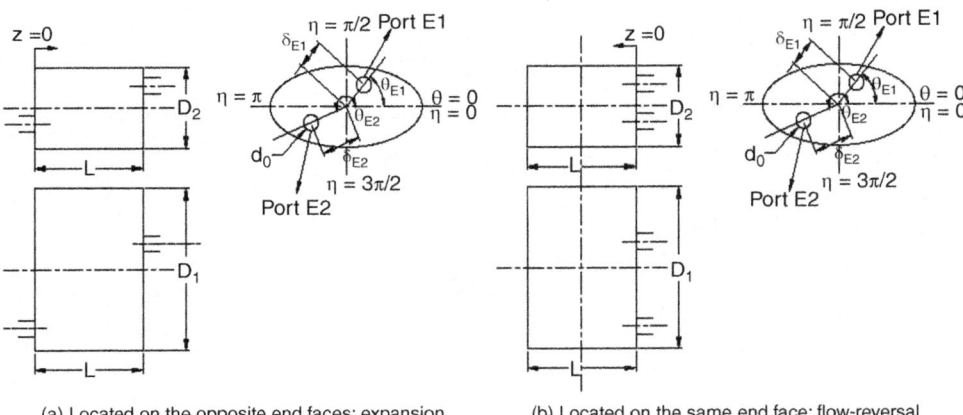

(a) Located on the opposite end faces: expansion chamber configuration.

(b) Located on the same end face: flow-reversal chamber configuration.

Figure 8.33 A two-port elliptical cylindrical chamber muffler having an end-inlet and end-outlet (Reproduced with permission from [17])

and

$$\theta = \cos^{-1}\left\{\frac{\cosh\xi\cos\eta}{\sqrt{\cosh^2\xi - \sin^2\eta}}\right\} = \sin^{-1}\left\{\frac{\sinh\xi\sin\eta}{\sqrt{\cosh^2\xi - \sin^2\eta}}\right\}, \qquad (8.44)$$

where ξ and η are the elliptical coordinates and h is the semi-interfocal distance.

8.11.2 Theory

The two-port muffler is characterized by means of an impedance [**Z**] matrix shown hereunder [18,19]

$$\left\{\begin{matrix}p_1\\p_2\end{matrix}\right\} = \begin{bmatrix}Z_{11} & Z_{12}\\Z_{21} & Z_{22}\end{bmatrix}_{2\times 2}\left\{\begin{matrix}v_1\\v_2\end{matrix}\right\}. \qquad (8.45)$$

In Equation 8.45, p_1 and p_2 are the acoustic pressures whereas v_1 and v_2 are the acoustic mass velocities (considered positive 'looking into' the system) at the ports 1 and 2, respectively. A harmonic time-dependence is assumed so that $p_i = p_i e^{j\omega t}$ and $v_i = v_i e^{j\omega t}$, $(i = 1, 2)$, where $j = \sqrt{-1}$ and ω is the excitation angular frequency. The acoustic pressure response $p(\xi_r, \eta_r, z_r | \xi_\theta, \eta_\theta, z_\theta)$ due to a point source located at an arbitrary point either on the end face of the elliptical cylindrical chamber may be expressed by means of the following Green's function [18,20].

$$p(\xi_r,\eta_r,z_r|\xi_\theta,\eta_\theta,z_\theta) = G(\xi_r,\eta_r,z_r|\xi_\theta,\eta_\theta,z_\theta) = G(R_r|R_\theta) = \rho_0 Q_0 (jk_0 c_0)$$

$$\times\left\{\begin{matrix}\sum\limits_{P=0,1,2,\ldots}^{\infty}\sum\limits_{m=0,1,2,\ldots}^{\infty}\sum\limits_{n=1,2,\ldots}^{\infty}\dfrac{Ce_m(\xi_r,q_{m,n})ce_m(\eta_r,q_{m,n})\cos\left(\dfrac{P\pi z_r}{L}\right)Ce_m(\xi_\theta,q_{m,n})ce_m(\eta_\theta,q_{m,n})\cos\left(\dfrac{P\pi z_\theta}{L}\right)}{\left\{(P\pi/L)^2 + 4q_{m,n}/h^2 - k_0^2\right\}N_{m,n,P}} +\\ \sum\limits_{P=0,1,2,\ldots}^{\infty}\sum\limits_{m=1,2,\ldots}^{\infty}\sum\limits_{n=1,2,\ldots}^{\infty}\dfrac{Se_m(\xi_r,\bar{q}_{m,n})se_m(\eta_r,\bar{q}_{m,n})\cos\left(\dfrac{P\pi z_r}{L}\right)Se_m(\xi_\theta,\bar{q}_{m,n})se_m(\eta_\theta,\bar{q}_{m,n})\cos\left(\dfrac{P\pi z_\theta}{L}\right)}{\left\{(P\pi/L)^2 + 4\bar{q}_{m,n}/h^2 - k_0^2\right\}\overline{N}_{m,n,P}}\end{matrix}\right\}e^{j\omega t},$$

(8.46)

where $(\xi_\theta, \eta_\theta, z_\theta)$ and (ξ_r, η_r, z_r) are the coordinates of the center of source and receiver ports, respectively, ρ_0, c_0 and Q_0 are the ambient density, sound speed and the volume flow rate, respectively and $k_0 = \omega/c_0$ is the excitation wavenumber. In Equation 8.46, $Ce_m(\xi, q_{m,n})$ and $ce_m(\eta, q_{m,n})$ are the even radial and angular Mathieu functions, respectively, of the first kind of order $m = 0, 1, 2 \ldots \infty$, whilst $Se_m(\xi, \bar{q}_{m,n})$ and $se_m(\eta, \bar{q}_{m,n})$ are the odd radial and angular Mathieu functions, respectively, of the first kind of order $m = 1, 2 \ldots \infty$ [21–23]. $q_{m,n}$ and $\bar{q}_{m,n}$ are parametric zeros or n^{th} root of derivative of even and odd type of Mathieu function, respectively, of order m, corresponding to the rigid wall condition in transverse direction. Furthermore, $N_{m,n,P}$ and $\overline{N}_{m,n,P}$ are integrals of the square of the product of a particular set of mode shape functions (m, n, P) defined over the volume of elliptical cylindrical chamber [18–20]. The uniform piston-driven model is employed to obtain the acoustic pressure response function (or equivalently, the $Z_{Ek\ Ei}$ matrix parameter) due to excitation at the end port Ei and response at the end port Ej shown hereunder.

$$Z_{EkEi} = \frac{p_{piston}(\xi_{Ek},\eta_{Ek},l_{Ek}|\xi_{Ei},\eta_{Ei},l_{Ei})}{\rho_0 Q_0}$$

$$= \frac{1}{S_{Ek}}\iint\limits_{S_{Ek}}\frac{1}{S_{Ei}}\left\{\iint\limits_{S_{Ei}}\frac{G(\xi_{Ek},\eta_{Ek},l_{Ek}|\xi_{Ei},\eta_{Ei},l_{Ei})}{\rho_0 Q_0}h_\xi(\xi_{Ei},\eta_{Ei})h_\eta(\xi_{Ei},\eta_{Ei})d\xi_{Ei}d\eta_{Ei}\right\}h_\xi(\xi_{Ek},\eta_{Ek})h_\eta(\xi_{Ek},\eta_{Ek})d\xi_{Ek}d\eta_{Ek},$$

(8.47)

where h_ξ and h_η are the scale factors in the elliptical cylindrical coordinate system, whilst S_{E1} and S_{E2} are the cross-sectional area of the end ports E1 and E2, respectively. The Z_{E2E1} matrix parameter may be obtained by substituting $i=1$ and $k=2$ in Equation 8.47, whereas Z_{E1E1} and Z_{E2E2} matrix parameters are obtained by substituting $i=k=1$ and $i=k=2$, respectively in Equation 8.47. Also, $Z_{E1E2}=Z_{E2E1}$ due to acoustic reciprocity. The TL for a single-inlet and single-outlet (SISO) system may be expressed in terms of the [**Z**] matrix parameters (obtained in Equation 8.45 as follows.

$$TL = 10\log_{10}\left(\frac{1}{4Y_{E1}Y_{E2}}\left|\frac{(Z_{E1E1}+Y_{E1})(Z_{E2E2}+Y_{E2})-Z_{E1E2}Z_{E2E1}}{Z_{E2E1}}\right|^2\right),\qquad(8.48)$$

where Y_{E1} and Y_{E2} are the characteristic impedances of the end ports E1 and E2.

8.11.3 Results and Discussion

In the ensuing results, the major and minor axis of the elliptical cylindrical chamber are given by $D_1 = 0.25\,\text{m}, D_2 = 0.15\,\text{m}$ whilst the axial length of the chamber $L = 0.05\,\text{m}$, therefore $L/D_1 = 0.2$, implying the chamber is axially short. Furthermore, the diameter d_0 of the end ports E1 and E2 is equal to 0.035 m and the sound speed c_0 is taken as 343.14 m·s^{-1}.

Figure 8.34 depicts the TL performance of the configuration shown in Figure 8.33(b) for the following locations of the end ports E1 and E2: $\delta_{E1} = 51.5\,\text{mm}, \theta_{E1} = 0$ and $\delta_{E2} = 39.0\,\text{mm}, \theta_{E2} = \pi/2$. It is noted that location of the end port E1 corresponds to the pressure node of the (2, 1) even mode denoted by $\eta_{(2,1)e}$, whilst the location of the end port E2 corresponds to the pressure node of the (0, 2) even mode denoted by $\xi_{(0,2)e}$. (The vertical solid lines in Figure 8.34 indicate the cut-on frequencies of the different higher-order modes of the elliptical chamber).

An excellent agreement is observed between the 3D analytical approaches based on the uniform piston-driven model and the 3D FEA predictions (obtained using the commercial package, SYSNOISE) which validates the 3D analytical uniform piston-driven model. A broadband attenuation is observed throughout the frequency range of interest, precisely, up to the cut-on frequency of the (4, 1) even mode. This broadband attenuation is due to the offset locations of the end ports E1 and E2 which may be understood by the following discussion.

The reason for the occurrence of the peaks at the (1, 1) even and (3, 1) even modes may be explained on the basis of the Green's function solution obtained in Equation 8.46. It can be analytically shown that the Z_{E1E1} parameter due to the port E1 tends to infinity at the (1, 1) and (3, 1) even modes, whereas Z_{E2E2} and Z_{E2E1} (or Z_{E1E2}) due to the port E2 are finite at the said frequencies. Thus, while the numerator in Equation 8.48 tends to infinity about (and

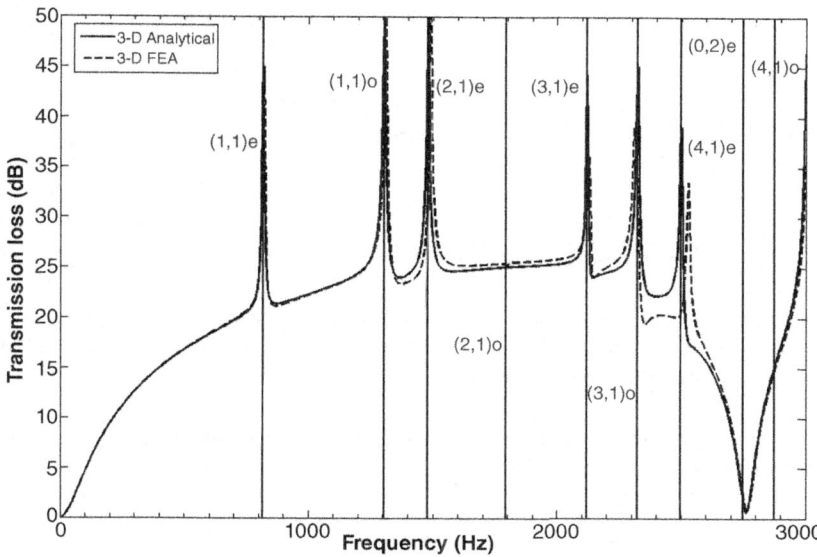

Figure 8.34 Comparison of the TL performance for the muffler configuration shown in Figure 8.33b based on 3D uniform piston-driven model and 3D FEA (SYSNOISE simulations) having the following dimensions: $D_1 = 0.25$ m, $D_2 = 0.15$ m, $L = 0.05$ m, $d_0 = 0.035$ m, $\delta_{E1} = 51.5$ mm, $\theta_{E1} = 0$, $\delta_{E2} = 39.0$ mm and $\theta_{E2} = \pi/2$ (Reproduced with permission from [17])

strictly speaking, at) the cut-on frequencies of the (1, 1) and (3, 1) even modes, the denominator is a finite quantity, implying that the argument of the logarithm in Equation 8.48 tends to infinity, thereby explaining the occurrence of resonance peaks.

It may similarly be observed that while the Z_{E2E2} parameter due to the port E2 tends to infinity at the cut-on frequencies of the (1, 1) and (3, 1) odd modes, the Z_{E1E1} and Z_{E2E1} (or Z_{E1E2}) parameters due to the port E1 are finite in the neighborhood and also at the cut-on frequencies of the above mentioned modes, thereby explaining the occurrence of a well defined resonance peak in the TL spectrum at the above mentioned frequencies.

The occurrence of the resonance peak at the cut-on frequency of the (2, 1) even mode is due to the location of the center of port E1 on the pressure node of the (2, 1) even mode [24]. This is explained by noticing that the contribution of the (2, 1) even modal term in the Z_{E1E1} and Z_{E2E1} (or Z_{E1E2}) parameters is zero, thence, these parameters are finite at the cut-on frequency of the (2, 1) even mode, whilst the Z_{E2E2} parameter tends to infinity, thus, signifying that the TL tends to infinity at the cut-on frequency of the (2, 1) even mode, thereby accounting for the resonance peak. The resonance peak at the (0, 2) even mode is similarly explained by noting that Z_{E2E2} and Z_{E2E1} (or Z_{E1E2}) parameters are finite at the (0, 2) even mode due to the location of the centre of port 2 on the pressure node of (0, 2) even mode, while the Z_{E1E1} parameter tends to infinity.

The breakdown of the broadband attenuation at the cut-on frequency of the (4, 1) even mode may be explained by noticing that since neither of the two end ports is located on the center of the pressure node of the (4, 1) even mode, all of the [**Z**] matrix parameters tend to infinity at the (4, 1) even mode so that the TL expression shown in Equation 8.48 simplifies as follows.

$$\text{TL} \approx 10 \log_{10}\left(\frac{1}{4Y_0^2}\left|\frac{Y_0(Z_{E1E1} + Z_{E2E2})}{Z_{E2E1}}\right|^2\right) \approx 10 \log_{10}\left(\left|\frac{Z_{E1E1}}{Z_{E2E1}}\right|^2\right) \approx 10 \log_{10}(1) = 0, \tag{8.49}$$

where $Y_0 = Y_{E1} = Y_{E2}$ for ports of equal diameters, thereby explaining the trough at the (4, 1) even mode.

Based on the foregoing discussion, the following guidelines are recommended for the optimal location of the end ports E1 and E2 for designing an 'axially short' end-inlet and end-outlet elliptical cylindrical chamber in order to obtain a broadband attenuation pattern.

1. The center of the end port E1 must be located on the pressure node of the (2, 1) even mode $\eta_{(2,1)e}$, and in terms of the radial offset distance, it is expressed as $\delta_{E1} = x_{(2,1)e}, \theta_{E1} = 0$, where

$$x_{(2,1)e} = \frac{D_1}{2} e \cos\left(\eta_{(2,1)e}\right) \cosh\left(\xi_{(2,1)e}\right) = \frac{D_1}{2} e \cos\left(\eta_{(2,1)e}\right), \tag{8.50}$$

that is on the semi-interfocal length [17].

2. The center of the end port E2 must be located on the pressure node of the (0, 2) even mode $\xi_{(0,2)e}$, and in terms of the radial offset distance, it is expressed as $\delta_{E2} = y_{(0,2)e}, \theta_{E2} = \pi/2$, where

$$y_{(0,2)e} = \frac{D_1}{2} e \sin\left(\eta_{(0,2)e}\right) \sinh\left(\xi_{(0,2)e}\right) = \frac{D_1}{2} e \sinh\left(\xi_{(0,2)e}\right), \tag{8.51}$$

that is on the semi-minor axis [17].

Here, $e = \sqrt{1 - (D_2/D_1)^2}$ denotes the eccentricity of the elliptical section. The optimal offset distances of end ports E1 and E2 are tabulated in Table 8.2 for different values of the aspect ratio (D_2/D_1) [17].

It is noted that the pressure node of the (2, 0) circumferential mode of the circular cylindrical chamber is located at the center. Therefore, the offset-distance $\delta_{(2,0)}$ of the end port E1 is zero indicating that this port should be centerd. The pressure node of the (0, 1) radial mode of the circular cylindrical chamber is located on the circle of radius $r = 0.6276R_0$, where R_0 is the radius of the circular section. (The diameter D_0 of the circular cross-section is taken to be equal to the major-axis D_1 of the elliptical cross-section.) Hence, the end port E2 should be offset at $\delta_{E2} = 0.6276R_0$ from the center [25].

A least squares interpolating polynomial is developed from the discrete values of the pressure node of the (2, 1) even mode $\eta_{(2,1)e}$ and the nondimensional offset distance $y_{(0,2)e}/(0.5D_1)$ from Table 8.2 to obtain

$$\begin{aligned}\eta_{(2,1)e} = {}& 1.1681 \times 10^{-5}\beta_1^{10} + 2.6073 \times 10^{-5}\beta_1^9 - 9.3584 \times 10^{-5}\beta_1^8 - 0.0003483\beta_1^7 \\ & - 9.7421 \times 10^{-5}\beta_1^6 + 0.001466\beta_1^5 + 0.0037612\beta_1^4 + 0.0015273\beta_1^3 \\ & - 0.02432\beta_1^2 - 0.11191\beta_1 + 0.928,\end{aligned}$$

$$\tag{8.52}$$

Design of Mufflers

Table 8.2 Pressure nodes of the (2, 1) even mode and the (0, 2) even mode of the elliptical cross-section and the corresponding non-dimensional offset distances of the end port E1 $\left(\delta_{E1} = x_{(2,1)e}, \theta_{E1} = 0\right)$ and the end port E2 $\left(\delta_{E2} = y_{(2,1)e}, \theta_{E1} = \pi/2\right)$ in the end-inlet and end-outlet elliptical cylindrical chamber muffler configuration (shown in Figure 8.33a and b) for different values of the aspect ratio (D_2/D_1) (Data adapted from [17])

$\dfrac{D_2}{D_1}$	Pressure node of (2, 1) even mode $\left\{\xi_{(2,1)e} = 0, y_{(2,1)e} = 0\right\}$		Pressure node of (0, 2) even mode $\left\{x_{(0,2)e} = 0\right\}$	
	$\eta_{(2,1)e}$	$\dfrac{x_{(2,1)e}}{(D_1/2)}$	$\xi_{(0,2)e}$	$\dfrac{y_{(0,2)e}}{(D_1/2)}$
0.50	1.0594	0.4238	0.2914	0.2559
0.55	1.0457	0.4187	0.3332	0.2835
0.60	1.0297	0.4121	0.3807	0.3120
0.65	1.0110	0.4036	0.4357	0.3417
0.70	0.9891	0.3924	0.5010	0.3729
0.75	0.9637	0.3774	0.5810	0.4063
0.80	0.9343	0.3566	0.6829	0.4423
0.85	0.9010	0.3271	0.8198	0.4819
0.90	0.8641	0.2831	1.0198	0.5257
0.95	0.8249	0.2119	1.3693	0.5743
0.96	0.8169	0.1917	1.4822	0.5846
0.97	0.8090	0.1678	1.6277	0.5951
0.98	0.8011	0.1385	1.8323	0.6057
0.99	0.7932	0.0990	2.1810	0.6166
0.999	0.7862	0.0316	3.3344	0.6265
1 (Circular Chamber)		$\dfrac{\delta_{(2,0)}}{(D_0/2)} = 0$		$\dfrac{\delta_{(0,1)}}{(D_0/2)} = 0.6276$

and

$$\frac{y_{(0,2)e}}{(D_1/2)} = 1.6078 \times 10^{-5}\beta_2^{10} + 6.8916 \times 10^{-5}\beta_2^9 + 7.3693 \times 10^{-5}\beta_2^8 - 9.0283 \times 10^{-5}\beta_2^7$$
$$- 0.00045173\beta_2^6 - 0.0009785\beta_2^5 - 0.00035942\beta_2^4 + 0.0057184\beta_2^3$$
$$+ 0.023465\beta_2^2 + 0.13715\beta_2 + 0.45908,$$

(8.53)

where

$$\beta_1 = \frac{(D_2/D_1) - 0.80993}{0.17395} \quad \text{and} \quad \beta_2 = \frac{(D_2/D_1) - 0.82181}{0.17464}. \quad (8.54, 8.55)$$

The coefficients of Equations 8.52–8.55 are correct to five significant decimal places and may be used to obtain the pressure node of the (2, 1) even mode $\eta_{(2,1)e}$ and the nondimensional offset distance $y_{(0,2)e}/(0.5D_1)$ for any arbitrary values of the aspect ratio $0.5 \leq (D_2/D_1) \leq 1$. Equation 8.52 (along with Equation 8.54) is then used to compute the nondimensional offset distance $x_{(2,1)e}/(0.5D_1)$ of the end port E1, whilst Equations 8.53 and

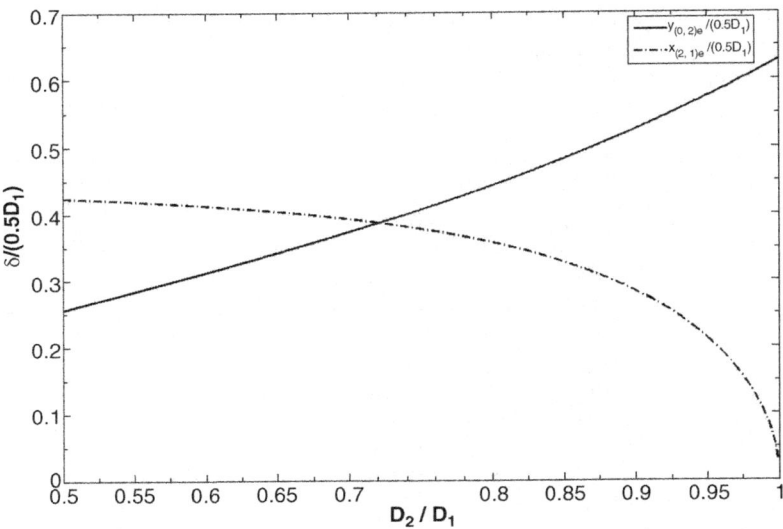

Figure 8.35 Variation of the non-dimensional offset distance $x_{(2,1)e}/(0.5D_1)$ of the end-port E1 and $y_{(0,2)e}/(0.5D_1)$ of the end-port E2 with the aspect ratio (D_2/D_1) (Reproduced with permission from [26])

8.55 may be used to compute $\xi_{(0,2)e}$. Figure 8.35 shows the variation of the nondimensional offset distances $x_{(2,1)e}/(0.5D_1)$ and $y_{(0,2)e}/(0.5D_1)$ of the end ports E1 and E2, respectively.

It is observed from Figure 8.35 that $x_{(2,1)e}/(0.5D_1)$ decreases slowly with an increase in aspect-ratio near $0.5 \leq D_2/D_1 \leq 0.7$ and thereafter, rapidly decreases and approaches zero as $D_2/D_1 \to 0$. Furthermore, $y_{(0,2)e}/(0.5D_1)$ increases almost linearly with an increase in D_2/D_1 and tends to 0.6276 as $D_2/D_1 \to 1$. Therefore, as noted earlier, in the limit of the elliptical cross-section tending to the circular cross-section, the end port E1 should be centerd, therefore, $\delta_{E1} = 0$ whilst the end port E2 should be offset at $\delta_{E2} = 0.6276R_0$ from the center to obtain a broadband attenuation in the TL graph as may be appreciated from Figure 8.36.

Figure 8.36 depicts the TL graph of an axially short flow-reversal circular cylindrical end chamber (similar to the configuration shown in Figure 8.33(b)) of dimensions $D_0 = 0.25$ m, $L = 0.05$ m whilst the end ports E1 and E2 are located at $\delta_{E1} = 0$ and $\delta_{E2} = 0.6276R_0 = 0.07845$ m. The diameters of both the end ports are taken to be equal to 0.04 m. (The vertical solid lines in Figure 8.36 indicate the cut-on frequencies of the different higher-order modes of the circular cylindrical chamber.) The TL graph in Figure 8.36 exhibits resonance peaks at the cut-on frequencies of all the circumferential modes, that is $(m, 0)$ modes, where $m = 1, 2, 3, \ldots$ due to the centerd location of end port E1, whilst the resonance peak observed at the cut-on frequency of the $(0, 1)$ radial mode is attributed to location of the center of end port E2 on the pressure node of the $(0, 1)$ radial mode. These features result in a broadband attenuation pattern up to the cut-on frequency of the $(0, 2)$ radial mode wherein a trough is observed leading to the breakdown of the broadband attenuation [A. Mimani, Personal communication, 2013].

From the foregoing discussion, it is evident that the conditions pertaining to the optimal locations of the end ports E1 and E2 in the context of obtaining a broadband attenuation TL graph as well as the nondimensional frequency range of broadband attenuation are

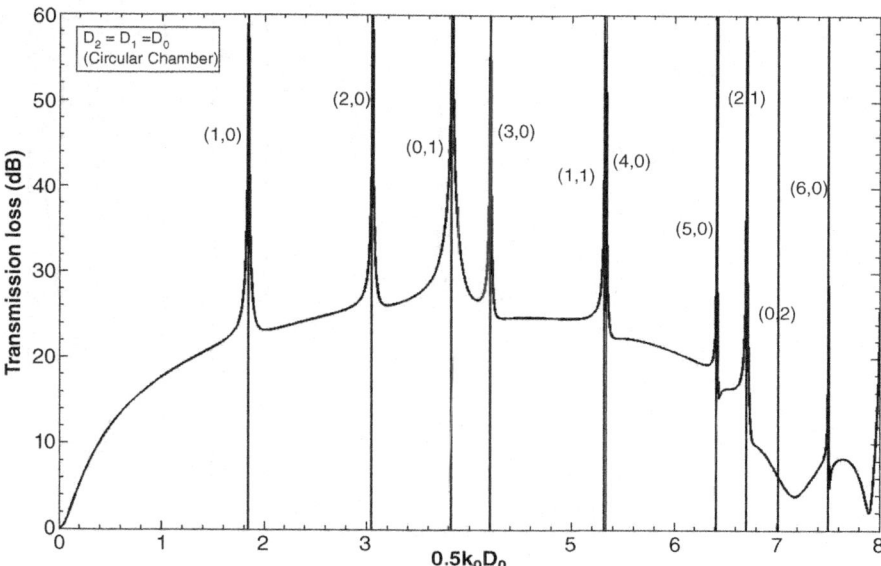

Figure 8.36 TL performance for the axially short circular cylindrical chamber configuration having the following dimensions: $D_0 = 0.25$ m, $L = 0.05$ m, $\delta_{E1} = 0$, $\delta_{E2} = 0.6276 R_0 = 0.07845$ m, $d_0 = 0.04$ m obtained using the 3D uniform piston-driven model (Reproduced with permission from [26])

significantly different for an axially short elliptical and a circular cylindrical chamber. For elliptical chambers with low aspect ratio or high eccentricity, the (2, 1) even mode is significantly lower than the (0, 2) even mode, therefore, the end port E1 must be located at $\delta_{E1} = x_{(2,1)e}$, $\theta_{E1} = 0$ to nullify the deteriorating influence of the (2, 1) even mode. Then, the other end port E2 must be offset on the minor-axis at the pressure node of the (0, 2) even mode, that is at $\delta_{E2} = y_{(0,2)e}$, $\theta_{E2} = \pi/2$ to nullify the trough due to resonance at the (0, 2) even mode. A broadband attenuation up to the cut-on frequency of the (4, 1) even mode (about 2747 Hz) is observed for elliptical chambers of low aspect-ratio ($D_2/D_1 = 0.6$) as compared to the broadband attenuation limit of (0, 2) radial mode (about 3065 Hz) for circular cylindrical chambers (of diameter D_0 equal to the major-axis D_1 of the elliptical chambers) as shown in Figure 8.36 [17,26]. The gradual transition in the theoretical broadband attenuation limit (α) as well as the corresponding maximum value of the ratio

$$\frac{L}{D_1} \leq \frac{\pi}{2\alpha}, \tag{8.56}$$

deciding the shortness limit needs further investigation.

8.11.4 Acoustical Equivalence of Short Expansion and Flow-reversal End Chambers

The acoustical equivalence of the axially short end-inlet and end-outlet elliptical cylindrical expansion chamber and the flow-reversal end chamber configurations shown in Figure 8.33(a) and (b) for identical locations of the end ports is shown by means of the 3D analytical uniform

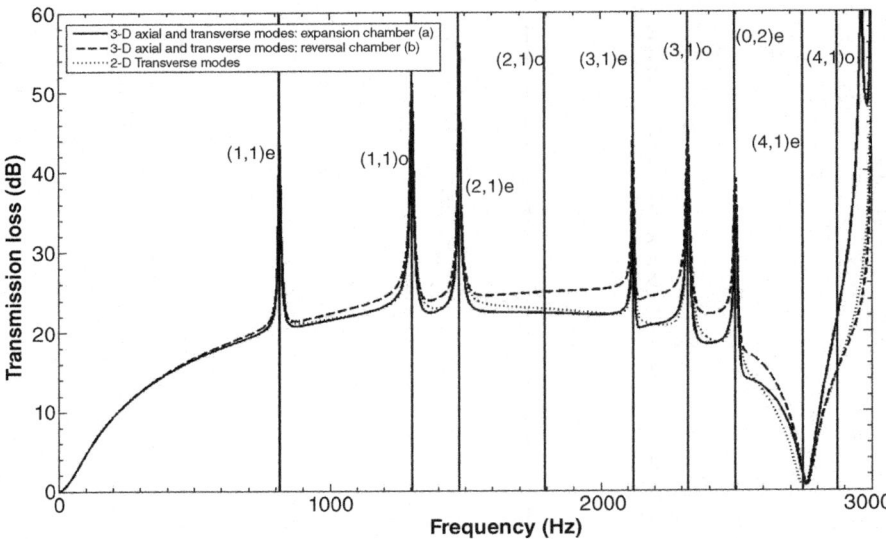

Figure 8.37 Comparison of the TL performance for the configurations shown in Figure 8.33(a) and (b) based on 2D approach considering only the transverse modes and 3D approach considering both axial and transverse modes having the following dimensions: $D_1 = 0.25$ m, $D_2 = 0.15$ m, $L = 0.05$ m, $d_0 = 0.035$ m and (a) $\delta_{E1} = 0, \delta_{E2} = 51.5$ mm, $\theta_{E2} = 0$, and (b) $\delta_{E1} = 51.5$ mm, $\theta_{E1} = 0$, $\delta_{E2} = 39.0$ mm, $\theta_{E2} = \pi/2$ (Reproduced with permission from [17])

piston-driven model. Figure 8.37 compares the TL characteristics obtained by considering the complete 3D acoustical field (comprising of the axial and transverse modes) with the 2D transverse model [A. Mimani, Personal communication, 2013].

This 2D transverse model is based upon altogether neglecting the axial modes ($P = 1, 2, 3, \ldots, \infty$) in the Green's function solution obtained in Equation 8.46, retaining only $P = 0$, implying that the variation of the acoustic field along the axial direction is completely ignored owing to the short chamber length. An excellent agreement is observed between the 2D transverse modes approach and the 3D approach for expansion and reversal chamber configurations for the frequency range up to the (4, 1) even mode, thereby demonstrating that the design guidelines obtained for short flow-reversal end chambers in the foregoing discussion are equally valid for short expansion chamber configurations [17,26].

8.12 Back-Pressure Considerations

The most important detrimental effect in a muffler with good insertion loss is the back pressure that it would exert on the engine. The back pressure is due to loss in stagnation pressure in various tubular elements and across various junctions. When this back pressure is low enough (less than 0.1 bar), it simply represents a corresponding loss in the brake mean effective pressure (BMEP) of the engine. However, higher back pressure would result in relatively much greater loss in BMEP owing to a sharp fall in the volumetric efficiency of the engine. The loss in stagnation pressure across various muffler elements is generally described in terms of the (hydro) dynamic head H of mean flow in the exhaust pipe:

$$H \equiv \frac{1}{2}\rho_0 U^2 = \gamma \frac{p_0}{2} M^2 \simeq 0.7 M^2 \quad \text{(bars)} \tag{8.57}$$

where $M = U/c_0$ is the mean flow Mach number in the exhaust pipe.

Defining the area ratio n at any area discontinuity as

$$n = \frac{\text{cross-sectional area of the thinner pipe}}{\text{cross-sectional area of the wider pipe}}, \tag{8.58}$$

the loss in stagnation pressure at various types of junctions has been measured from the steady flow experiments to be as follows:

Sudden expansion and extended inlet [27]: $(1-n)^2 H$.
Sudden expansion and extended outlet [27]: $\{(1-n)/2\}H$.
Flow-reversal cum expansion [28]: $H(n < 0.25)$.
Flow-reversal cum contraction [28]: $0.5H(n < 0.25)$.
Free expansion at the tail pipe end: $0.4H$.

The pressure (or stagnation pressure) drop across a tube (or pipe) due to boundary layer is given by

$$\nabla p_0 = f\left(\frac{1}{2}\rho_0 U^2\right)\frac{l}{d}, \tag{8.59}$$

where l is the length of the tube,

d is the internal diameter (or hydraulic diameter),

$\frac{1}{2}\rho_0 U^2$ is the dynamic head H,

f is the Froude's friction factor; for G.I. pipes, it is given by

$$f = 0.0072 + 0.612/R_e^{0.35}, \quad R_e < 4 \times 10^5 \tag{8.60}$$

R_e is the Reynold's number, $U d \rho_0 / \mu$, and
μ is the coefficient of dynamic viscosity.

Generally, for typical exhaust muffler systems, stagnation pressure drop across pipes (the main ones being exhaust pipe and tail pipe) is small enough to be ignored at the design stage. In fact, more often than not, lengths of exhaust pipe and tail pipe are decided by external factors like the desirable locations of the 'muffler proper' under (or on) the vehicle, and the muffler designer has to worry about the 'muffler proper' only.

The stagnation pressure drops across perforated elements are given by the empirical expressions [29]:

Plug muffler (Figure 8.7):

$$y = 5.6e^{-0.23x} + 67.3e^{-3.05}. \tag{8.61}$$

Three-duct cross-flow chamber (Figure 8.5):

$$y = 4.2e^{-0.06x} + 16.7e^{-2.03x}. \tag{8.62}$$

Concentric tube resonator (Figure 3.10a):

$$y = 0.06x. \tag{8.63}$$

Here,

$$y \equiv \Delta p / H, \qquad H \equiv \frac{1}{2}\rho U^2 \tag{8.64}$$

$$x \equiv \frac{\text{total area of the holes constituting the perforate}}{\text{cross-sectional area of the perforated pipe}} \equiv OAR \tag{8.65}$$

$$= \frac{\pi dl_p \sigma}{\frac{\pi}{4}d^2} = \frac{n_h \left(\frac{\pi}{4}d_h^2\right)}{\frac{\pi}{4}d^2} \tag{8.66}$$

where l_p is length of the perforate, σ is porosity, n_h is the total number of holes constituting the perforate, d_h is the hole diameter (or equivalent diameter), and d is diameter of the perforated pipe.

These general expressions have been found to be independent of the mean flow velocity U (for Mach number $M \leq 0.2$), and have a very weak dependence on the expansion ratio d_2/d_1 and lengths l_1 and l_3 (for porosity $\sigma \leq 0.1$). The length l_2 was varied to obtain the desired variation in the open area ratio x.

The stagnation pressure drop across an inline perforated baffle (see Figure 8.18) is given by [13].

$$\Delta p \simeq 1.56 H \tag{8.67}$$

where H is the dynamic head or velocity head in the baffle holes or orifices.

Equations 8.61–8.63 indicate that the through-flow perforated elements (cross-flow expansion and cross-flow contraction) introduce much larger drops in stagnation pressure than the corresponding simple expansion chambers when the open area of the perforated section A_{op} is less than the cross-sectional area of the pipe A_{cs}. But this is exactly what is required for high transmission loss, TL. Thus, there is a direct conflict between requirements of high TL and low back pressure in the case of plug mufflers. The only way out is to decrease dynamic head H.

But then, this calls for a decrease in mean-flow velocity U, which in turn calls for an increase in diameters of all tubes (the mean-flow flux being constant). This would increase the size, and hence weight and cost, of the muffler. The designer has to compromise between

a. insertion loss.
b. back pressure (and hence engine performance), and
c. size (and hence weight and cost).

In this context, choice of a double-tuned concentric-tube resonator (Figure 8.5) gains merit. For such a resonator, back pressure and size are minimum and TL can be improved by tuning the extensions.

Many of these design considerations are summarized in [30].

8.13 Practical Considerations

A well-designed muffler would have adequate insertion loss over the frequency range of interest; minimum or optimum restriction; moderate volume, weight, and cost; high durability; good styling and tonal quality; and would be easy to manufacture and maintain. There may also be specific constraints on the geometry and packaging of the silencing system.

It has been observed in practice that total muffler volume V_m is proportional to the total piston displacement and that insertion loss, IL in dB(A), generally increases with the ratio V_m/V_p. Inversely, for an extra dB(A) of IL, the ratio V_m/V_p may have to be increased by about 0.38. Typical values of this volume ratio for about 20 dB(A) of insertion loss may vary from about 3.5 for low-specific-output engines to about 12.5 for high-specific-output engines [31]. Approximate values of insertion loss of typical reactive mufflers used with reciprocating engines are listed in Table 8.1.

Durability is affected by several factors, such as material of fabrication, vibration, thermal expansion, protective coating on the exposed surface, and the way the muffler is mounted.

Material used in the fabrication of mufflers is generally mild steel or aluminized steel for low exhaust gas temperature (up to about 500 °C), type-409 stainless steel (the so-called muffler stainless steel) for temperatures of up to 700 °C, and true stainless steels (such as type 321) for still higher temperatures [31].

A surface coating is necessary for weather protection as well as aesthetic appeal. Normally, a suitable paint is used. However, for a high-quality surface finish, one may have to resort to a ceramic coating.

The thickness and type of the shell have a bearing on shell noise as well as durability. Standard 22-gauge plate is about the minimum requirement. For larger shells or for reduction in shell noise, one may use increased body thickness, double-wrapped bodies, and insulated body (two layers of metal with a layer of acoustic material in between), in that order, for increasing reduction in shell noise. Double-wrapped bodies may be a good compromise inasmuch as they can effectively reduce shell noise with minimum material usage.

For easy removal, the muffler is connected to the manifold with a slotted tube, threaded connection, or mounting flange with bolts. Noise leakages may occur if all openings are not properly sealed and if the clamps are not properly installed. This extra noise is often confused with shell noise.

Maintenance of exhaust systems (and, for that matter, intake systems) may require replacement of muffler, clamps, rain caps, and so forth. In smaller economy countries where labor is comparatively cheap, it is often more economical to open, repair, and remount the repaired muffler. In such cases, the muffler configuration has to be designed in such a way that its constituent parts can be easily separated and reassembled.

If an engine has to be used in an area where there is grass or other vegetation that is likely to be ignited by a hot spark from the engine, one must incorporate (into the muffler) spark-arresting units that make use of centrifugal action to spin these hot sparks into a collection chamber.

For exhaust gas pollution control one may also have to incorporate a catalytic material, such as platinum deposited on a ceramic base such as alumina, to oxidize excess hydrocarbons and carbon monoxide to water vapor and carbon dioxide. The process of oxidation requires additional oxygen, which may be provided by means of an aspirating ventury upstream of the catalytic convertor element.

Typical configuration of an aspirating muffler, a spark-arresting muffler, a catalytic muffler, and combination mufflers are shown in Figure 8.38. These have been adapted from [31].

There are several acoustical complications resulting from the extra elements or features. An aspirating ventury provides some incidental acoustic absorption but, if not properly designed, may also generate considerable aerodynamic sound. Spark-arresting mufflers introduce spinning modes that may have considerable and not yet predictable influence on the insertion

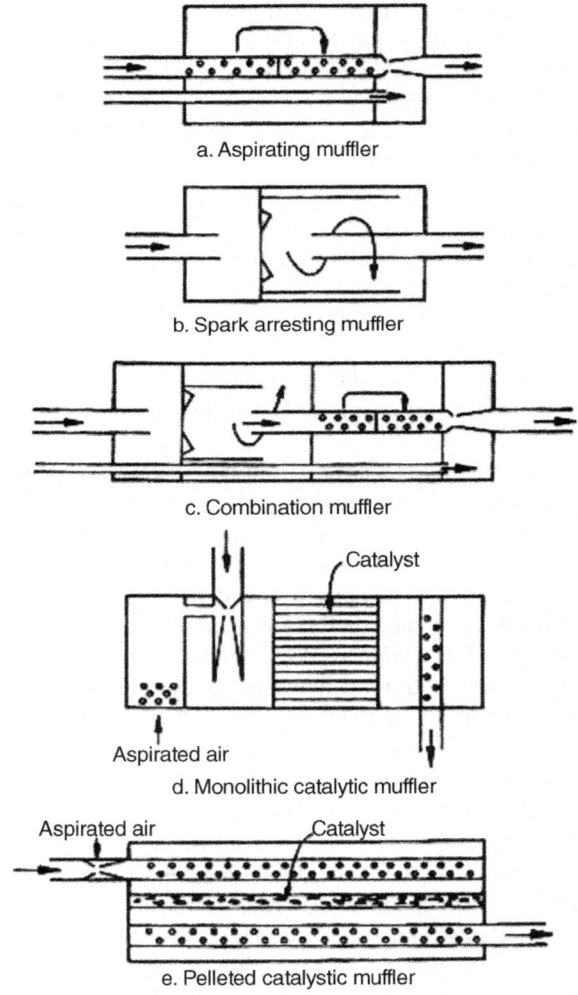

Figure 8.38 Typical configurations of special muffler (Reproduced with permission from [31])

loss of the muffler. Similarly, the oxidation process being exothermic, catalytic devices generate very high temperatures. These may not only require special materials in the construction of the exhaust muffler, but could also generate additional sound and alter the acoustic propagation because of localized addition of heat into the system.

At the time of writing, there are no techniques available in the published literature for prediction of the acoustical performance of these special types of mufflers, and therefore design continues to be one of trial and error.

8.14 Design of Mufflers for Ventilation Systems

The principle of mismatch of acoustic impedance used in the reciprocating machinery does not hold for fans or blowers (whether forced-draft type or suction type) employed in ventilation systems. Besides, back-pressure considerations are much stronger here, that is, much less back pressure can be tolerated here. This rules out mufflers of the type used on reciprocating engines. Instead, resort is made to dissipation of acoustic energy into heat by lining the air ducts with acoustically absorptive material. The theory of lined ducts and parallel-baffle mufflers has been dealt with at length in Chapter 6.

The air-conditioning or ventilation system ducts are generally rectangular and are simply lined at two or all four walls. The latter, of course, have nearly twice as much TL and allow much less acoustical radiation from duct walls. The intake systems of large industrial fans have parallel-baffle mufflers as shown in Figure 6.13, the design of which has been discussed in Chapter 6 (see Figures 6.14–6.17).

Normally fiberglass-type, highly porous materials are used. One may, however, also use less porous materials like ceramic absorbers, which have the advantages of much better weatherability, mechanical strength, and refractory properties. Such absorbers provide much lower absorption at low frequencies. This may be made up by providing an entrapped air gap at the back of the wall lining and in the middle of the intermediate baffles. The effect of the air gap is shown in Figure 8.39, and a typical silencer of this type is shown in Figure 8.40.

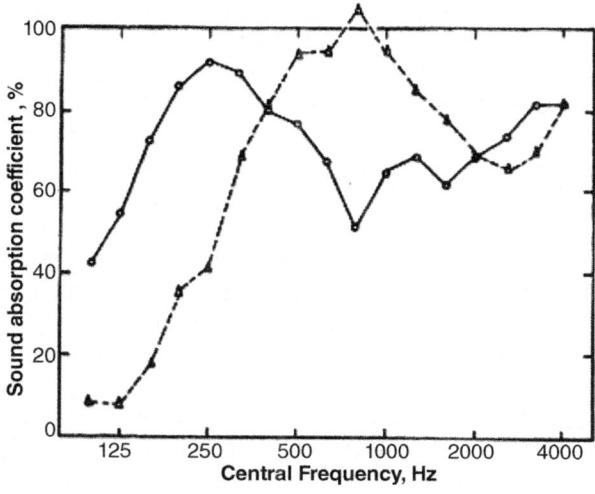

Figure 8.39 Effect of air gap behind porous ceramic absorbers of 20 mm thickness. ___, Air space 200 mm; ------, air space 50 mm

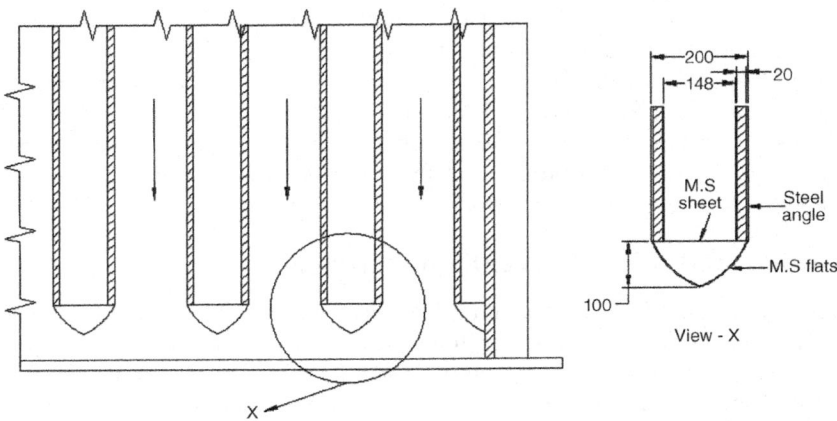

Figure 8.40 Intake silencer for a large industrial fan making use of porous ceramic absorbers

Back pressure on the fan may be calculated by making use of the formulae of Section 8.12, in particular Equation 8.59. However, Froude's friction factor for acoustically lined ducts may be two to three times that for G.I. pipes, given by Equation 8.60, because of extra roughness.

Compared to engine exhaust mufflers, air ducts of ventilation systems and industrial fans radiate considerably more noise through their walls. An approximate evaluation of this can be made by means of the methods described in Chapter 7 (Section 7.6).

This secondary emission of noise may be minimized by

a. acoustic lining on the inside,
b. acoustic lagging on the outside, or
c. using double-wall sandwich construction for the duct walls.

The first method, if it can be used, is, of course, the most effective.

It is generally a good practice to provide a receiver or plenum chamber between the fan (or fans) and the duct system. Many times it is a physical necessity when a single fan has to serve many parallel ducts. Incidentally, however, it provides additional transmission loss in the process, combining principles of reflective muffling with absorptive silencing if it is lined on the inside with absorptive material. The attenuation due to a plenum chamber (for a supposedly progressive wave) is given by Equation 6.101. When a plenum chamber is a part of the system (as, of course, it invariably is), its dissipative function can be modeled as a lumped resistance, given by Equation 6.103, and volume represented by a lumped compliance, given by Equation 6.102. The two, of course, cannot be separated!

Another element in ventilation system is a bend (generally a right-angle one). As pointed out in Section 6.11, bends should be acoustically lined, and typical lined bends give an insertion loss of 1 dB at lower frequencies to 10 dB at higher frequencies. For better insertion loss at lower frequencies, the lining thickness may be increased. Thus, ends can be acoustically as useful as they are functionally necessary. Of course, 180° bends would cause a large pressure drop and therefore should be avoided. In fact, back pressure would be smaller if the bends were made smoother by decreasing sharpness (acuteness) of the angle or by making use of guide vanes. These guide vanes could again be made of a hard but porous material (like ceramic absorber) for better acoustic absorption.

Finally, in the acoustic design of a ventilation system, one must not lose sight of the flow-generated noise given by Equation 6.104, which sets a limit on the maximum insertion loss that can be obtained. Symbolically,

$$\text{maximum insertion loss} = (\text{sound power level upstream of the muffler}) \\ - (\text{flow generated sound power level}). \tag{8.68}$$

Thus, it is no use increasing the length of the dissipative section beyond the limit set by Equation 8.68 without increasing the cross section of the flow passages at the same time.

An equally important, but analytically much less tractable, limiting factor is the flanking transmission. The walls of the duct at the upstream end are readily excited into vibrations. This vibration is transmitted to the lower end. The result is not only a loss of effective transmission loss but also secondary radiation of sound to the atmosphere. The remedy lies in use of double-leaf sandwich-type walls or, where possible, use of a lossy material like plywood for the walls of the duct.

8.15 Active Sound Attenuation

This book has dealt with the theory and design of passive acoustic filters and mufflers. Both the reflective as well as dissipative mufflers have generally poor performance at low frequencies, and for a good wideband response, they would be very large and expensive. These limitations of passive mufflers have given rise to the idea of active attenuation, which consists in sensing the undesired noise in the exhaust pipe and reintroducing an inverted signal through a loudspeaker. The basic idea is more than 75 years old [32] and depends on the fact that electrical signals move much more rapidly than acoustic waves.

Although the basic idea is very simple and attractive, there are a variety of complications and problems, namely [33,34]:

a. The pickup microphone and loudspeaker do not have a flat frequency response over the entire frequency range of interest; the loudspeaker is particularly weak inasmuch as its low-frequency response is generally very poor [35].
b. The loudspeaker radiates energy upstream as well as downstream with the noise to be attenuated. This results in a contamination of the input to the microphone.
c. Reflection at the downstream termination (generally, the radiation end) of the exhaust pipe result in standing waves and the associated problems of feedback [26].
d. A typical engine exhaust system produces a signal that varies with time because of small (but unavoidable) variations in speed. This calls for adaptive signal processing [36–38].

The solution to all these problems calls for very sophisticated hardware. One typical system [34], shown in Figure 8.41, involves

a. two microphones and one loudspeaker, with amplifiers,
b. a powerful, high-speed microprocessor,
c. two A/D and one D/A convertors,
d. three low-pass filters (LPF), and
e. an analog interface board with two input channels and one output channel.

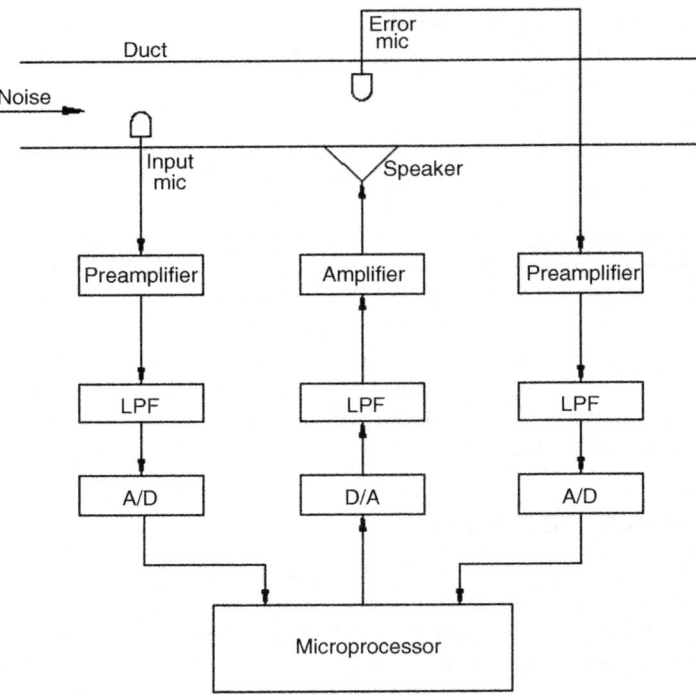

Figure 8.41 Block diagram of a typical active noise control system (Reproduced with permission from [34])

Design of efficient hardware (input microphones, microprocessor, loudspeakers, etc.) would be facilitated by a priori knowledge of the transfer function that would be generated by an adaptive digital filter at steady stage for total cancellation. An analytical one-dimensional standing-wave model of a linear active noise control system in a duct (see Figure 8.42) revealed that at the junction of the auxiliary (or secondary) source and the duct, impedance is zero [39]. In other words, a tuned auxiliary source acts so as to produce an active short circuit in the duct. Tuning is achieved by means of an adaptive controller that uses the error microphone signal for the process of adaptation [40] in order to generate the required transfer function, $H(\omega)$ shown in Figure 8.42a. A frequency-domain analysis [39,40] has revealed that at this actively produced zero-impedance junction, the volume velocity produced by the primary source is equal and opposite to that produced by the tuned auxiliary sources. This investigation [39,41], where electroacoustic analogies have been used following Small [42], has indicated that for a perfectly tuned system, the primary and secondary sources present to each other zero impedance (acoustical short circuit) at the junction of the auxiliary source; the downstream load impedance Z_0 is short-circuited. Referring to Figure 8.43, the use of transfer matrix relationships (derived in Chapter 2) yields [43]

$$v_p = \frac{v_{spi}}{\cos kl_i + j\dfrac{Y}{Z_{spi}} \sin kl_i} \tag{8.69}$$

Design of Mufflers

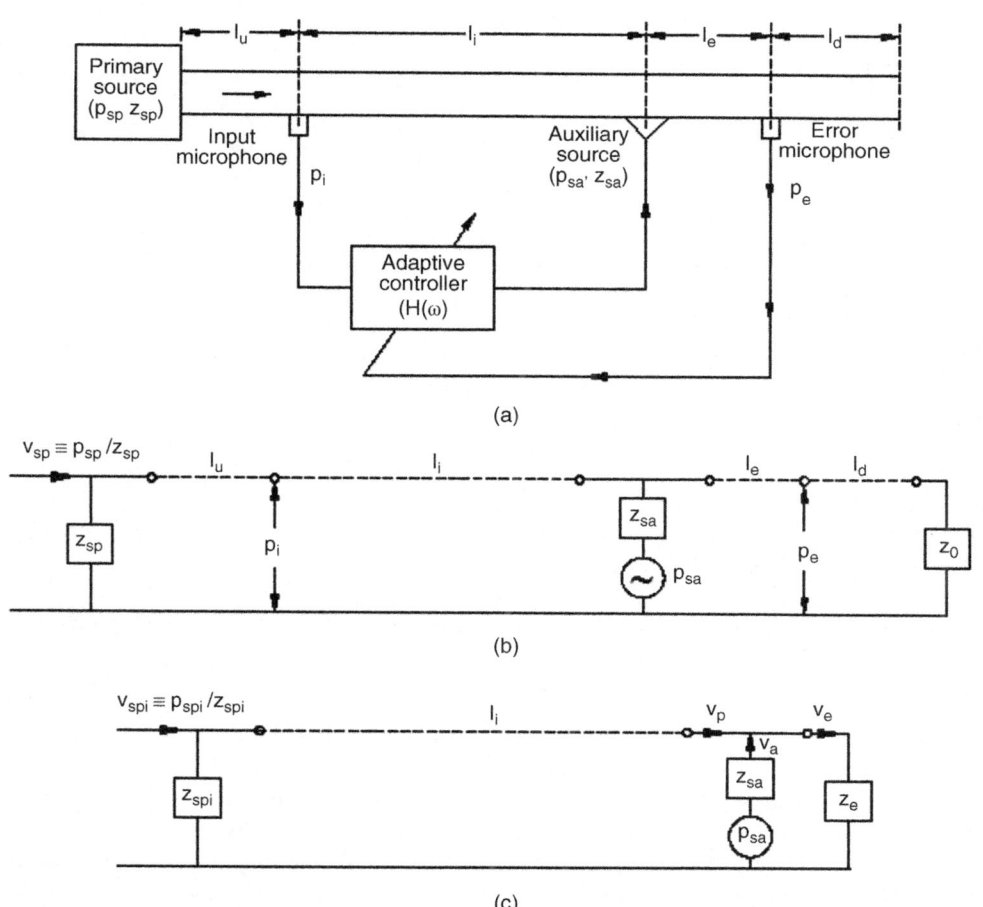

Figure 8.42 (a) Schematic of an active noise control system in a duct with single-speaker auxiliary source, (b) its electrical analogous circuit using direct electroacoustic analogies; (c) the interactive portion of the electrical analogous circuit (Adapted from [43])

and

$$v_a = v_{sa} = p_{sa}/Z_{sa} \tag{8.70}$$

where Z_{sa} is the acoustic impedance of the auxiliary speaker, when inactive, as seen from the main duct side.

For v_e (and hence p_e) in Figure 8.42(c) to be zero, we must ensure that

$$v_a = -v_p. \tag{8.71}$$

Substituting from Equation 8.69 and 8.70 yields

$$v_{sa} = \frac{v_{spi}}{\cos kl_i + j\dfrac{Y}{Z_{spi}}\sin kl_i} \tag{8.72}$$

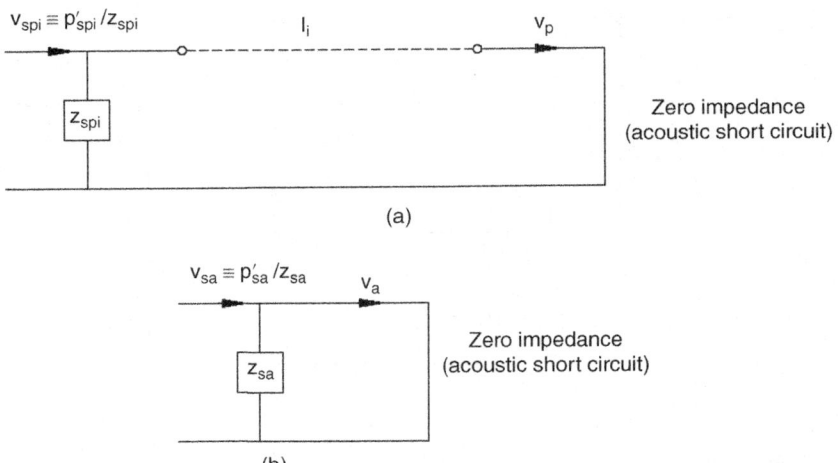

Figure 8.43 Electrical analogous circuits for the tuned system. Contribution of (a) primary source, and (b) auxiliary source (Adapted from [43])

Thus, p_{sa}, the acoustical equivalent of the open-circuit voltage of the power amplifier, is given by [43]

$$p_{sa} = -\frac{Z_{sa}}{Z_{spi}} \left\{ \frac{p_{spi}}{\cos kl_i + j\, Y/Z_{spi} \sin kl_i} \right\}. \tag{8.73}$$

Equation 8.71 indicates that the auxiliary source, often a loudspeaker, needs to be designed for the required volume velocity and not power, because the steady-state acoustical power output of the auxiliary source as well as the primary source of an ideally tuned system would be zero. The two sources then act together as an acoustical dipole in as much as they cancel each other's velocity output at the junction rather than power output which is zero. In other words, the two sources unload each other in that the resistive part of the acoustical load impedance faced by both of them is zero.

This explains the real mechanism of an active noise control system in a duct; there is no cancellation of acoustic fluxes as was surmised heretofore.

There are, however, several practical complications. The plane wave (or, one-dimensional) theory holds at the frequencies lower than the cut-off frequency of the duct (or, the cut-on frequency of the first azimuthal higher-order mode). This poses a problem for the intake as well as exhaust ducts of large industrial fans used with boilers of thermal power stations working on superheated steam. This can be addressed by radial as well as azimuthal partitioning of the large round duct, as shown in Figure 8.44. The cut-on frequencies of such portioned ducts have been calculated and tabulated by Munjal [44].

Most ducts carry turbulent mean flow that is a potential source of error for input microphones used in active noise control systems. This is addressed by means of antiturbulance probe tubes that are used for measurement of acoustic pressure in ducts with turbulent mean flow for use in sound power measurement of ducted fans as well as in active noise control systems for low-frequency noise flow machinery. A typical tube would have a thin longitudinal slit covered with a number of layers of cloth, with a small microphone at the

Design of Mufflers

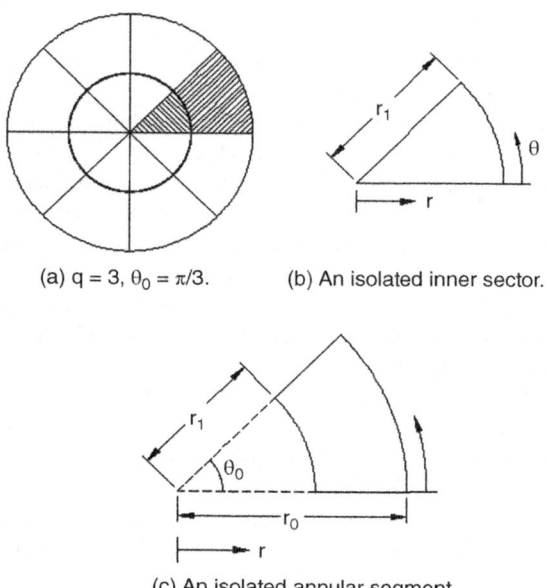

(a) $q = 3$, $\theta_0 = \pi/3$.   (b) An isolated inner sector.

(c) An isolated annular segment.

Figure 8.44 Radial as well as azimuthal partitioning of a round duct (Adapted from [43])

downstream end, as shown in the schematic diagram of Figure 8.45. As the convection velocity of turbulence is much smaller than the sound speed, the probe tube averages out the turbulent pressure fluctuates. However, the pressure picked up by the microphone inside the probe tube is in error inasmuch as it is not the same as outside. This error has been referred to by Neise [45] as the 'Acoustic Sensitivity' of the probe, and 'Frequency Response' by Wang and Crocker [46]. This has been investigated analytically by Munjal and Eriksson, leading to useful design guidelines [47].

The microprocessor would make use of fast and accurate algorithms, the development of which continues to be a challenge [49–51]. Furthermore, its performance at high frequencies

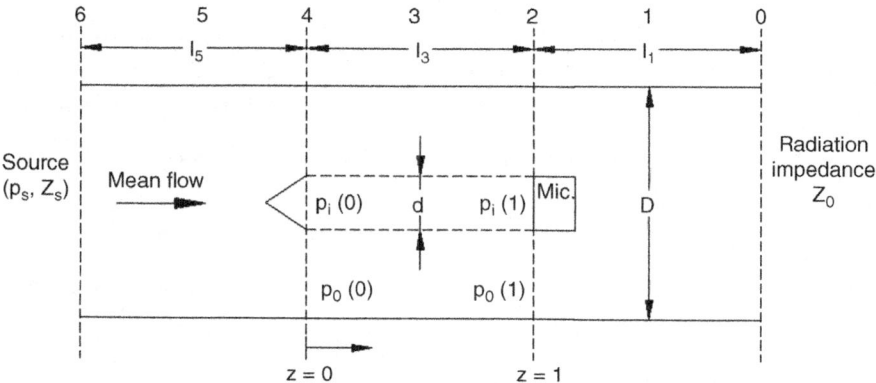

Figure 8.45 Schematic diagram of the acoustical system with the microphone located at the downstream end of the probe tube ($l = l_3$) (Adapted from [47])

is limited by several factors including (a) the existence of three-dimensional waves and (b) the need for an increased sampling frequency that requires increased speed (better than cycle time of 200 ns) from the digital microprocessor.

An active noise control system has a very important advantage over passive mufflers. It does not exert any back pressure on the system. This special feature, coupled with its low frequency attenuation advantage, makes it particularly suitable for large industrial fans, the primary noise of which is limited to the blade passing frequency and its first couple of multiples. Because of its very little back pressure, not only the running cost (electrical consumption) of the fan is very low, we can use a smaller power motor, and thereby save on the initial cost of the motor as well. Thus the cost of the active noise control system (microphones, loudspeakers, adaptive microprocessor, etc.) is recovered in no time in the case of noise control of large fan installations.

The active noise control system is particularly effective for suppressing the discrete frequency components at low frequencies. This makes it ideally suited to quieten large bladed machinery. It not only reduces their low frequency noise drastically but also improves the resultant sound quality.

There is, however, a major limitation. The active noise control system is not particularly suited for global noise attenuation in a room or the passenger cabin of an automobile or an aeroplane. In such a 3D environment we have to use tiny loudspeakers embedded in the headrest of each seat, because the active cancellation is limited to an area or a volume described by a radius of a quarter wavelength [52]. This feature makes active noise control costly, yet it is often used for noise control in the passenger cabin of a propeller aircraft or a luxury helicopter.

In view of this, a high-performance muffler system of the future may be a hybrid muffler-combination of a passive muffler (reactive and/or resistive) to reduce high frequency sound and active attenuation to reduce low-frequency sound (see Figure 8.46). Meanwhile, for

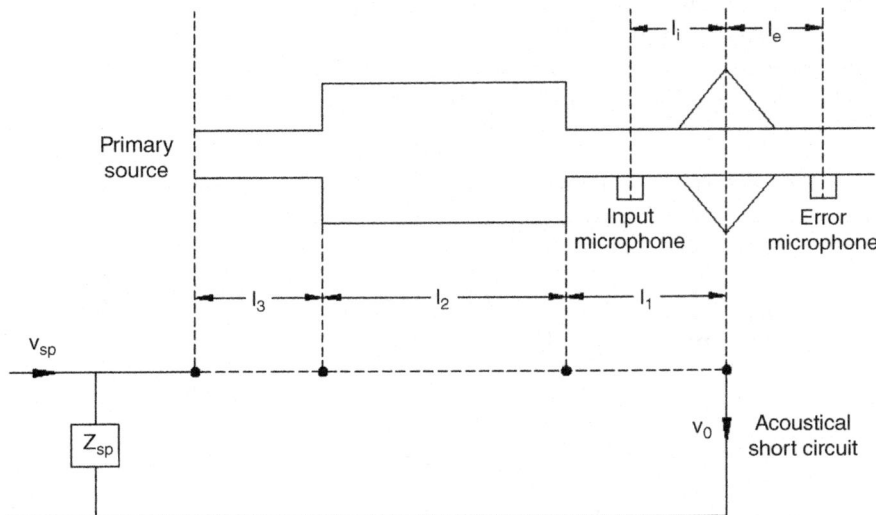

Figure 8.46 Schematic diagram and the electrical analogous circuit of a hybrid system with a simple expansion chamber as its passive component (Adapted from [48])

active attenuation to be economically viable, much work needs to be done on the development of better and cheaper microprocessors and loudspeakers (secondary or auxiliary sources).

References

1. Munjal, M.L., Galaitsis, A.G. and Ver, I.L. (2006) *Passive Silencers, Noise and Vibration Control Engineering* (eds I.L. Ver and L.L. Beranek), John Wiley & Sons, Inc., New York.
2. Karal, F.C. (1953) The analogous impedance for discontinuities and constrictions of circular cross-section. *Journal of the Acoustical Society of America*, **25** (2), 327–334.
3. Munjal, M.L. and Gowri, S. (2009) Theory and design of tuned extended tube chambers and concentric tube resonators. *Journal of the Acoustical Society of India*, **36** (2), 53–71.
4. Chaitanya, P. and Munjal, M.L. (2011) Effect of wall thickness on the end corrections of the extended inlet and outlet of a double-tuned expansion chamber. *Applied Acoustics*, **72** (1), 65–70.
5. Chaitanya, P. and Munjal, M.L. (2011) Tuning of the extended concentric tube resonators, SAE International Symposium on International Automotive Technology (SIAT 2011) 2011-26-0070, ARAI, Pune India.
6. Munjal, M.L., Krishnan, S. and Reddy, M.M. (1993) Flow-acoustic performance of the perforated elements with application to design. *Noise Control Engineering Journal*, **40** (1), 159–167.
7. Munjal, M.L. (1997) Plane wave analysis of side inlet/outlet chamber mufflers with mean flow. *Applied Acoustics*, **52** (2), 165–175.
8. Joint Department of the Army, Airforce and Navy, USA, Power Plant Acoustics (1983) Technical manual TM 5-805-9 AFT 88-20 NAVFAC Dhf 3.14.
9. Munjal, M.L., Narasimhan, M.V. and Sreenath, A.V. (1973) A rational approach to the synthesis of one-dimensional acoustic filters. *Journal of Sound and Vibration*, **29** (2), 263–280.
10. Veerababu, D. (June 2013) Flow-acoustic analysis and design of multiply-connected automotive mufflers, M.E. thesis, Department of Mechanical Engineering, Indian Institute of Science, Bangalore, India.
11. Munjal, M.L. (1997) Analysis of a flush-tube three-pass perforated element muffler by means of transfer matrices. *International Journal of Acoustics and Vibration*, **2** (2), 59–62.
12. Vijayasree, N.K. and Munjal, M.L. (2012) On an integrated transfer matrix method for multiply connected mufflers. *Journal of Sound and Vibration*, **331**, 1926–1938.
13. Elnady, T., Abom, M. and Allam, S. (2010) Modeling perforates in mufflers using two-ports. *ASME Journal of Vibration and Acoustics*, **132**, 1–11.
14. Elnady, T. and Abom, M. (2006) SIDLAB: New 1-D sound propagation simulation software for complex duct networks, Proceedings of the Thirteenth International Congress on Sound and Vibration (ICSV13) Vienna.
15. Ih, J.-G. (1992) The reactive attenuation of rectangular plenum chambers. *Journal of Sound and Vibration*, **157** (1), 93–122.
16. Venkatesham, B., Tiwari, M. and Munjal, M.L. (2009) Transmission loss analysis of rectangular expansion chamber with arbitrary location of inlet/outlet by means of Green's function. *Journal of Sound and Vibration*, **323** (3–5), 1032–1044.
17. Mimani, A. (2012) 1-D and 3-D analysis of multi-port muffler configurations with emphasis on elliptical cylindrical chamber, Doctor of Philosophy Thesis, Department of Mechanical Engineering, Indian Institute of Science, Bangalore, India, http://hdl.handle.net/2005/1931.
18. Mimani, A. and Munjal, M.L. (2012) Acoustical analysis of a general network of multi-port elements – An impedance matrix approach. *International Journal of Acoustics and Vibration*, **17** (1), 23–46.
19. Mimani, A. and Munjal, M.L. (2012) 3-D acoustic analysis of elliptical chamber mufflers having an end inlet and a side outlet: An impedance matrix approach. *Wave Motion*, **49** (2), 271–295.
20. Mimani, A. and Munjal, M.L. (2012) Acoustical behaviour of single inlet and multiple outlet elliptical cylindrical chamber muffler. *Noise Control Engineering Journal*, **60** (5), 605–626.
21. Mclachlan, N.W. (1947) *Theory and Application of Mathieu Functions*, Oxford University Press, London.
22. Stamnes, J.K. and Spjelkavik, B. (1995) New method for computing eigenfunctions (Mathieu functions) for scattering by elliptical cylinders. *Pure and Applied Optics*, **4** (3), 251–262.
23. A. Mimani and M.L. Munjal (2011) 3-D acoustic analysis of spherical chambers having single inlet and multiple outlet: An impedance matrix approach, *International Journal of Applied Mechanics*, **3** (4), 685–710.
24. Denia, F.D., Albelda, J., Fuenmayor, F.J. and Torregrosa, A.J. (2001) Acoustic behaviour of elliptical chamber mufflers. *Journal of Sound and Vibration*, **241** (3), 401–421.

25. Selamet, A. and Ji, Z.L. (1998) Acoustic attenuation performance of circular flow-reversing chambers. *Journal of the Acoustical Society of America*, **104** (5), 2867–2877.
26. Mimani, A. and M.L. Munjal (2012) On the role of higher-order evanescent modes in end-offset inlet and end-centered outlet elliptical flow-reversal chamber mufflers, *International Journal of Acoustics and Vibration*, **17** (3), 139–154.
27. Panicker, V.B. and Munjal, M.L. (1981) Aeroacoustic analysis of straight-through mufflers with simple and extended-tube expansion chambers. *Journal of the Indian Institute of Science*, **63(A)**, 1–19.
28. Panicker, V.B. and Munjal, M.L. (1981) Aeroacoustic analysis of straight-through mufflers with flow reversals. *Journal of the Indian Institute of Science*, **63(A)**, 21–38.
29. Munjal, M.L., Krishnan, S. and Reddy, M.M. (1993) Flow-acoustic performance for perforated element mufflers with application to design. *Noise Control Engineering Journal*, **40** (1), 159–167.
30. Eriksson, L.J. and Thawani, P.T. (1985) Theory and practice in exhaust system design, SAE Vehicle Noise and Vibration Conference, Traverse City, MI.
31. Eriksson, L.J. (1982) Chapter 5, Silencers, in *Noise Control in Internal Combustion Engine* (ed. D.E. Baxa), John Wiley & Sons, Inc., New York.
32. Lueg, P. (June 6 (1936)) Process of silencing sound oscillations, U. S. Patent #2,043,416.
33. Warnaka, G.E. (1982) Active attenuation of noise – the state of the art. *Noise Control Engineering Journal*, **18** (3), 100–110.
34. Eriksson, L.J. (1985) Active sound attenuation using adaptive signal processing techniques, Ph.D. Thesis, University of Wisconsin, Madison.
35. Small, R.H. (1971) Direct-radiator loudspeaker analysis. *IEEE Transactions on Audio and Electroacoustics*, **19**, 269–281.
36. Gritton, C.W.K. and Lin, D.W. (1984) Echo cancellation algorithms. *IEEE ASSP Magazine*, **1** (2), 30–38.
37. Ross, C.F. (1982) An adaptive digital filter for broadband active sound control. *Journal of Sound and Vibration*, **80**, 381–388.
38. Burgess, J.C. (1981) Active adaptive sound control in a duct: A computer simulation. *Journal of the Acoustical Society of America*, **70** (3), 715–726.
39. Munjal, M.L. and Eriksson, L.J. (1988) An analytical one-dimensional, standing wave model of a linear active noise control system in a duct. *Journal of the Acoustical Society of America*, **84**, 1086–1093.
40. Eriksson, L.J. (1991) Development of the filtered U-algorithm for active noise control. *Journal of the Acoustical Society of America*, **89**, 257–265.
41. Munjal, M.L. and Eriksson, L.J. (1989) Analysis of linear one-dimensional noise control system by means of block diagrams and transfer functions. *Journal of Sound and Vibration*, **129** (2), 443–455.
42. Small, R.H. (1973) Closed-box loudspeaker systems. *Journal of the Audio Engineering Society*, **21**, 11.
43. Munjal, M.L. (1995) Analysis of an active noise control system with several loudspeakers spread along the axis of the duct. *Journal of the Indian Institute of Science*, **75**, 651–666.
44. Munjal, M.L. (1988) On the cut-on frequencies of a large round duct with azimuthal as well as radial partitions. *Journal of the Acoustical Society of America*, **84** (5), 1936–1939.
45. Neise, W. (1975) Theoretical and experimental investigations of microphone probes for sound measurements in turbulent flow. *Journal of Sound and Vibration*, **39**, 371–400.
46. John, S. and Wang and Malcolm, J. (1974) Crocker, Tubular windscreen design for microphones for in-duct fan sound power measurements. *Journal of the Acoustical Society of America*, **55** (3), 568–575.
47. Munjal, M.L. and Eriksson, L.J. (1989) An exact one-dimensional analysis of the acoustic sensitivity of the anti-turbulence probe tube in a duct. *Journal of the Acoustical Society of America*, **85** (2), 582–587.
48. Munjal, M.L. and Eriksson, L.J. (1989) Analysis of a hybrid noise control system for a duct. *Journal of the Acoustical Society of America*, **86** (1), 832–834.
49. Honig, M.L. and Messerschmitt, D.G. (1984) *Adaptive Filters: Structures, Algorithms and Applications*, Kluwer, Boston.
50. Widrow, B. and Stearns, S.D. (1985) *Adaptive Signal Processing*, Prentice-Hall, Englewood Cliffs, NJ.
51. Ljung, L. and Soderstrom, T. (1983) *Theory and Practice of Recursive Identification*, MIT Press, Cambridge, MA.
52. Hansen, C.H. (2001) *Understanding Active Noise Cancellation*, E& FN Spon Press, London.

Appendix A

Bessel Functions and Some of Their Properties

Bessel functions or cylinder functions appear frequently in the text, for analysis of waves in tubes of circular cross section. Here we list for ready reference some of their more important properties.

The second-order linear differential equation

$$\frac{d^2R}{dr^2} + \frac{1}{r}\frac{dR}{dr} + \left(1 - \frac{n^2}{r^2}\right)R = 0 \tag{A.1}$$

is called Bessel's equation of order n and has the following general solution:

$$R = C_1 J_n(r) + C_2 N_n(r). \tag{A.2}$$

$J_n(r)$ is known as the Bessel function of the first kind of order n, and $N_n(r)$ is known as Neumann function as well as Bessel function of the second kind of order n. For a nonnegative integer value of n, $J_n(r)$ is defined, for arbitrary finite r, by the series

$$J_n(r) = \sum_{k=0}^{\infty} \frac{(-1)^k (r/2)^{n+2k}}{k!(n+k)!}, \tag{A.3}$$

and satisfies the relations

$$\frac{d}{dr}[r^n J_n(r)] = r^n J_{n-1}(r), \quad \frac{d}{dr}[r^{-n} J_n(r)] = -r^{-n} J_{n+1}(r), \tag{A.4}$$

$$J_{n-1}(r) + J_{n+1}(r) = \frac{2n}{r} J_n(r), \quad n = 1, 2 \ldots, \tag{A.5}$$

$$J_{n-1}(r) - J_{n+1}(r) = 2J'_n(r), \quad n = 1, 2 \ldots, \tag{A.6}$$

For $n = 0$, Equation A.6 should be replaced by

$$J'_0(r) = -J_1(r). \tag{A.7}$$

Also

$$J_{-n}(r) = (-1)^n J_n(r). \tag{A.8}$$

$N_n(r)$ is called the Bessel function of the second kind of order n. It is defined by the series

$$N_n(r) = \frac{2}{\pi} J_n(r) \log \frac{r}{2} - \frac{1}{\pi} \sum_{k=0}^{n-1} \frac{(n-k-1)!}{k!} \left(\frac{r}{2}\right)^{2k-n} \\ - \frac{1}{\pi} \sum_{k=0}^{\infty} \frac{(-1)^k (r/2)^{n+2k}}{k!(n+k)!} [\psi(k+1) + \psi(k+n+1)], \tag{A.9}$$

where

$$\psi(m+1) = -\gamma + 1 + \frac{1}{2} + \ldots + \frac{1}{m}, \quad \psi(1) = -\gamma, \tag{A.10}$$
$$\gamma = 0.5772156$$

is Euler's constant, and in the case $n = 0$, the second term in Equation A.9 should be set equal to zero.

Relations (A.4)–(A.8) are also satisfied by $N_n(r)$. Asymptotically,

$$N_0(r) \simeq \frac{2}{\pi} \log r \quad N_0(r) \simeq \frac{-(n-1)!}{\pi} \left(\frac{r}{2}\right)^{-n}, \quad r \to 0, \quad n = 1, 2, \ldots. \tag{A.11}$$

The Bessel function of the first kind of arbitrary order ν is defined as

$$J_\nu(r) \equiv \sum_{k=0}^{\infty} \frac{(-1)^k (r/2)^{\nu+2k}}{\Gamma(k+1)\Gamma(k+\nu+1)}, \quad \arg|r| < \pi, \tag{A.12}$$

and satisfies the relations (A.4)–(A.6), with n being replaced by ν.

The general solution to Equation A.1 for a noninteger order is

$$R(r) = C_1 J_\nu(r) + C_2 J_{-\nu}(r), \quad \nu \neq \pm 1, \pm 2, \ldots. \tag{A.13}$$

Bessel functions of the first and second kind of noninteger order ν are related as

$$N_\nu(r) = \frac{J_\nu(r) \cos \nu\pi - J_{-\nu}(r)}{\sin \nu\pi}. \tag{A.14}$$

By making changes of variables, it can be shown that the general solution to the equation

$$\frac{d^2 R}{dr^2} + \frac{1}{r} \frac{dR}{dr} + \left(k_r^2 - \frac{n^2}{r^2}\right) R = 0 \tag{A.15}$$

is
$$R(r) = C_1 J_n(k_r r) + C_2 N_n(k_r r). \tag{A.16}$$

Bessel functions of the third kind, or Hankel functions, are defined as

$$H_\nu^{(1)}(r) = J_\nu(r) + j N_\nu(r) \tag{A.17}$$

and

$$H_\nu^{(2)}(r) = J_\nu(r) - j N_\nu(r). \tag{A.18}$$

We can write the general solution of Equation A.1 as

$$\begin{aligned} R(r) &= A_1 J_\nu(r) + A_2 H_\nu^{(1)}(r) \\ &= B_1 J_\nu(r) + B_2 H_\nu^{(2)}(r) \\ &= C_1 H_\nu^{(1)}(r) + C_2 H_\nu^{(2)}(r), \end{aligned} \tag{A.19}$$

where $A_1, \ldots, C_2$ are arbitrary constants, as in the form (A.13).
Hankel functions also satisfy relations (A.4)–(A.6). Additionally,

$$H_{-\nu}^{(1)}(r) = e^{\nu \pi j} H_\nu^{(1)}(r) \tag{A.20}$$

$$H_{-\nu}^{(2)}(r) = e^{-\nu \pi j} H_\nu^{(2)}(r). \tag{A.21}$$

A real function $f(r)$ defined in the interval $(0, r_0)$ can be written as the Fourier Bessel series

$$f(r) = \sum_{m=1}^{\infty} C_m J_\nu \left(x_{\nu m} \frac{r}{r_0} \right), \quad 0 < r < r_0, \quad \nu \gg -\frac{1}{2}, \tag{A.22}$$

where

$$C_m = \frac{2}{r_0^2 J_{\nu+1}^2(x_{\nu m})} \int_0^{r_0} r f(r) J_\nu \left\{ x_{\nu m} \frac{r}{r_0} \right\} dr, \quad m = 1, 2, \ldots, \tag{A.23}$$

and

$$0 < x_{\nu 1} < x_{\nu 2} < \ldots < x_{\nu m} < \ldots, \tag{A.24}$$

are the positive roots of the equation $J_\nu'(x) = 0$.

This representation makes use of the following orthogonal property of the Bessel functions:

$$\int_0^{r_0} r J_\nu\left(x_{\nu m}\frac{r}{r_0}\right) J_\nu\left(x_{\nu n}\frac{r}{r_0}\right) dr = 0 \quad \text{if } m \neq n \tag{A.25}$$

$$\begin{aligned}\int_0^{r_0} r J_\nu^2\left(x_{\nu m}\frac{r}{r_0}\right) dr &= \frac{r_0^2}{2} J_\nu^2(x_{\nu n})\left\{\frac{x_{\nu n}^2 - \nu^2}{x_{\nu n}^2}\right\} \\ &= \frac{r_0^2}{2} J_\nu^2(x_{\nu n}). \\ &= 0 \text{ if } \nu = n = 0\end{aligned} \tag{A.26}$$

Appendix B

Entropy Changes in Adiabatic Flows

B.1 Stagnation Pressure and Entropy

Adiabatic flow is characterized by zero heat transfer. For this kind of flow, the first law of thermodynamics yields (for a unit mass)

$$p\nu + C_\nu T + \frac{u^2}{2} = constant \tag{B.1}$$

or

$$pd\nu + \nu dp + C_\nu dT + udu = 0$$

or

$$\nu dp + udu = -(pd\nu + C_\nu dT). \tag{B.2}$$

On making use of the second law of thermodynamics ($pd\nu + C_\nu dT = Tds$), for the right-hand side, one gets

$$\nu dp + udu = -Tds$$

or

$$\frac{dp}{\rho} + udu = -Tds. \tag{B.3}$$

Integrating this equation for incompressible flow yields

$$\frac{p_{0.2} - p_{0.1}}{\rho_0} + \frac{1}{2}\left(U_2^2 - U_1^2\right) = T_0(s_{0.1} - s_{0.2})$$

Acoustics of Ducts and Mufflers, Second Edition. M. L. Munjal.
© 2014 John Wiley & Sons, Ltd. Published 2014 by John Wiley & Sons, Ltd.

or

$$s_{0.2} - s_{0.1} = \frac{1}{\rho_0 T_o}(p_{s.1} - p_{s.2}), \tag{B.4}$$

where

s connotes entropy,
p_s connotes stagnation pressure, $p_0 + \frac{1}{2}\rho_0 U^2$, and
ν connotes specific volume, $1/\rho$.

Relation (B.4) is applicable to all area discontinuities across which the flow is adiabatic but not isentropic, inasmuch as there is a drop in stagnation pressure. Equation (B.2) throws some light on the physics of the situation; a change in the stagnation pressure (left-hand side) generates heat that in turn raises temperature, and this internal adjustment of energy is irreversible.

Across area discontinuities shown in Figures 2.11 and 2.13, the stagnation pressure drops and therefore entropy increases. Referring to Figure 2.13 in particular, assuming the upstream entropy level to be zero, the entropy downstream can be written as

$$s_{0.1} = \frac{1}{\rho_0 T_0}(p_{s,3} - p_{s,1}) = \frac{R}{p_0}(p_{s,3} - p_{s,1}). \tag{B.5}$$

The corresponding relation for aeroacoustic perturbations would obviously be

$$s_1 = \frac{R}{p_0}(p_{c,3} - p_{c,1}). \tag{B.6}$$

It may be recalled that for incompressible flow, changes in ρ_0, T_0, and hence c_0, consequent to change in flow velocity, can be neglected in the foregoing expressions for changes in pressure.

B.2 Pressure, Density, and Entropy

$$\text{Equation of state}: \quad \frac{p}{\rho} = RT \tag{B.7}$$

or

$$\frac{dp}{p} - \frac{d\rho}{\rho} = \frac{dT}{T}. \tag{B.8}$$

According to the second law of thermodynamics,

$$p\,dv + C_\nu dT = T\,ds \tag{B.9}$$

or

$$-\frac{p}{\rho^2}d\rho + C_\nu dT = T\,ds \tag{B.10}$$

or

$$-\frac{p}{\rho T}\frac{dp}{\rho} + C_\nu \frac{dT}{T} = ds. \qquad (B.11)$$

On making use of Equations B.7 and B.11 becomes

$$-R\frac{d\rho}{\rho} + C_\nu \frac{dT}{T} = ds$$

or

$$\frac{dT}{T} = \frac{ds}{C_\nu} + \frac{R}{C_\nu}\frac{d\rho}{\rho}$$

or

$$\frac{dT}{T} = \frac{ds}{C_\nu} + (\gamma - 1)\frac{d\rho}{\rho}. \qquad (B.12)$$

Substituting for dT/T from Equation B.12 in Equation B.8 yields

$$\frac{dp}{p} = -\frac{d\rho}{\rho} = \frac{ds}{C_\nu} + (\gamma - 1)\frac{d\rho}{\rho}$$

or

$$\frac{dp}{p} - \gamma\frac{d\rho}{\rho} = \frac{ds}{C_\nu}. \qquad (B.13)$$

In terms of acoustic perturbations, this relation becomes

$$\frac{p}{p_0} - \gamma\frac{\rho}{\rho_0} = \frac{s}{C_\nu}. \qquad (B.14)$$

Again, on applying Equation B.14 to the area discontinuities of Figure 2.13, and assuming that the entropy upstream is zero, one gets from Equation B.14

$$\frac{p_3}{p_{0.3}} - \gamma\frac{\rho_3}{\rho_{0.3}} = 0 \quad \text{and} \quad \frac{p_1}{p_{0.1}} - \gamma\frac{\rho_1}{\rho_{0.1}} = \frac{s_1}{C_\nu} \qquad (B.15)$$

or

$$\rho_3 = \frac{p_{0.3}}{\gamma p_{0.3}}p_3 \quad \text{and} \quad \rho_1 = \frac{\rho_{0.1}}{\gamma p_{0.1}}p_1 - \frac{s_1}{C_\nu}\frac{\rho_{0.1}}{\gamma}$$

or

$$\rho_3 = \frac{p_3}{c_{0.3}^2} \quad \text{and} \quad \rho_1 = \frac{p_1 - s_1 p_{0.1}/C_v}{c_{0.1}^2} \qquad (B.16)$$

For incompressible flow, these relations further reduce to

$$\rho_3 = \frac{p_3}{c_0^2} \quad \text{and} \quad \rho_1 = \frac{p_1 - s_1 p_0/C_v}{c_0^2}, \qquad (B.17)$$

where s_1 is given in terms of decrease in aeroacoustic pressure by relation (B.6).

Appendix C

Nomenclature

Because every symbol has been described at the place of its first appearance, this appendix includes only the symbols that appear often in the text.

A	Area; Complex amplitude of the forward pressure wave; A four-pole parameter
a	Local sound speed
B	Complex amplitude of the reflected pressure wave
b	Breadth; Width: x Dimension of a rectangular duct
C	A constant; Compliance
Ce_m	Modified radial Mathieu function of order m
ce_m	Radial Mathieu function of order m
C_p	Specific heat at constant pressure
C_v	Specific heat at constant volume
D	Diameter
d	Diameter
F	Froude's friction factor; Firing frequency (Hz)
f	Frequency (Hz); Friction factor
H	Dynamic head or velocity head for incompressible flow
h	Height; y-Dimension of a rectangular duct; half the interfocal length; half the flow passage width
I	Acoustic intensity
IL	Insertion loss
J	Bessel function of the first kind
j	Iota, $(-1)^{1/2}$
K	Loss factor
k	Wave number
L	Length; Reference length
LD	Level difference
L_I	Sound intensity level
L_p	Sound pressure level (SPL)
L_w	Sound power level (SWL)

Acoustics of Ducts and Mufflers, Second Edition. M. L. Munjal.
© 2014 John Wiley & Sons, Ltd. Published 2014 by John Wiley & Sons, Ltd.

l	Length
L_W	Sound power level
Δl	End correction; Peturbation in length
M	Mean-flow Mach number; Inertance
m	The number of nodes in the x direction or diametral direction
n	The number of nodes in the y direction or azimuthal direction; the number of elements in the muffler
P	Complex amplitude of acoustic pressure
p	Pressure; Acoustic pressure
q	Rate of heat transfer into the medium per unit mass;
R	Reflection coefficient, $|R|e^{j\theta}$; Lumped flow resistance
r	Radius coordinate
r_0	Radius of a circular pipe
(r, ϕ, z)	Cylindrical coordinates
S	Area of cross section of a pipe; Area of a surface; Scattering matrix
Se_m	Modified angular Mathieu function of order m
se_m	Angular Mathieu function of order m
SPL	Sound pressure level
T	Temperature; Time period; Transfer matrix
TL	Transmission loss
t	Time variable
U	Mean-flow velocity averaged over the cross section
u	Velocity; Acoustic particle velocity
V	Complex amplitude of acoustic mass velocity; Volume
ν	Acoustic mass velocity, $\rho_0 Su$
W	Acoustic power flux
(x, y, z)	Cartesian coordinates of a point
Y	Characteristic impedance for waves in a pipe, p/ν
Z_0	Characteristic impedance of the medium; Radiation impedance
Z	Acoustic impedance of an element; Impedance matrix
α	Acoustic pressure attenuation constant
$\overline{\alpha}$	Acoustic power absorption coefficient
β	Complex propagation constant
γ	Ratio of the specific heats
δ	Variation of
Δ	Vector differential operator; Variation of
μ	Coefficient of dynamic viscosity of the medium
ω	Radian frequency
ϕ	Velocity potential; Azimuthal coordinate
ρ	Density of the medium
θ	Phase of the reflection coefficient
ξ	Particle displacement; $F/2d$; Radial elliptical coordinate
η	Angular elliptical coordinate
ζ	Acoustic impedance at a point

Subscripts

0	Ambient; Atmospheric; In ideal fluid; Unperturbed Value
1	In the positive (forward) direction; Tail pipe
2	In the negative (rearward) direction
a	Axial
c	Convective; Incorporating the convective effect of mean flow; Aeroacoustic; Cavity Cylinder
cs	Cross-sectional area of the perforated tube
e	Element; Entering
ec	End correction
eq	Equivalent
ex	Exhaust; Outgoing gases
h	Hole
in	Inlet; Incoming gases
n	Exhaust pipe; Neck of Helmholtz resonator
n	Net
m, n	Corresponding to the (m, n) mode
$n+1$	Acoustic source
op	Open area of the perforate
ref	Reference state (hypothetical)
rms	Root mean square value
s	Stagnation; Source
t	Transverse
w	At the wall
x	Partial derivative with respect to x; x Component
y	Partial derivative with respect to y; y Component
z	Partial derivative with respect to z; z Component

Superscripts

$-$	(overbar) Time average
$'$	Perturbation: Space derivative
$+$	In the positive (forward) direction
$-$	(minus) In the negative (rearward) direction

Index

1-D analysis, 11
2-D transverse model, 362
3-D analysis, 11
3D wave equation, 5, 24
Absorption coefficient, 250
Absorptive boundary conditions, 292
Absorptive duct, 234
Absorptive layer, 252
Absorptive lining, 184, 185, 233
Absorptive material, 1, 23, 27, 233, 252, 261
Absorptive mufflers, 233
Acoustic analysis, 101
Acoustic diode, 211
Acoustic driver, 211
Acoustic element, 41
Acoustic energy flux, 104, 105
Acoustic filter, 5, 41, 53, 91, 97
Acoustic impedance, 44, 48, 199, 221
Acoustic impedance of perforates, 126
Acoustic intensity, 41, 102, 104
Acoustic lagging, 368
Acoustic lining, 1, 246, 368
Acoustic perturbation, 382, 383
Acoustic power flux, 42, 111
Acoustic pressure, 43, 297
Acoustic radiation, 48
Acoustic radiation mechanism, 315
Acoustic resistance, 129
Acoustic sensitivity, 373
Acoustic source characteristics, 217
Acoustic theory, 101
Acoustic transmission line, 41
Acoustical dipole, 372

Acoustical equivalence, 361
Acoustical load impedance, 372
Acoustical short circuit, 370
Acoustically absorptive duct, 243, 249
Acoustically absorptive layer, 23
Acoustically absorptive material, 27, 69
Acoustically lined duct, 24, 234, 244
Acoustically long, 70
Acoustically long tube, 51
Active attenuation, 369, 374
Active boundary, 291
Active noise control system, 370, 372, 374
Active short circuit, 370
Adaptive controller, 370
Adaptive digital filter, 370
Adaptive signal processing, 369
Adiabatic flow, 381
Adjoint system, 283, 284
Adjoint system equations, 284
Aeroacoustic elements, 112
Aeroacoustic filter, 41
Aeroacoustic mass velocity, 106
Aeroacoustic perturbations, 382
Aeroacoustic power flux, 108
Aeroacoustic pressure, 106, 108, 113
Aeroacoustic state variables, 106, 111, 115
Aeroacoustic variables, 112, 114
Aerodynamic noise, 326, 327, 345
Air fuel ratio, 226–228
Air-conditioning system, 1
Algebraic algorithm, 88, 91
Ambient density, 106
Analogous circuit, 89, 170

Index

Analog-to-digital converter, 193
Anechoic boundary condition, 270
Anechoic environment, 58
Anechoic termination, 46, 55, 57, 60, 61, 197, 201, 260, 310, 312
Angular elliptical coordinate, 34
Angular matieu function, 35
Annular cavity, 70, 71, 94
Antisymmetric modes, 26
Antiturbulance probe tube, 372
Area discontinuities, 1, 61, 115, 118, 273, 382, 383
Area discontinuity, 61
Aspect ratio, 37, 358, 359
Aspirating muffler, 366
Asymmetric expansion chamber, 267
Attenuation constant, 14, 22, 23, 191, 200, 234, 245, 250, 261
Autocorrelation, 197
Automotive exhaust muffler, 34, 158
Automotive exhaust system, 234
Automotive muffler, 34, 149, 151
Auto-spectral density, 194, 197, 221
Auto-spectrum, 221
Auxiliary sources, 372, 375
Auxiliary speaker, 371
Axial attenuation, 27
Axial mass velocity, 242
Axial particle velocity, 238, 242
Axial transmission loss, 322
Axial wave model, 162
Axisymmetric modes, 10
Axisymmetric ring element, 280
Axisymmetric wave propagation, 12
Axisymmetry, 12
Azimuthal higher-order mode, 372
Azimuthal mode, 29
Azimuthal partitioning, 372

Back-pressure, 97, 150, 169, 175, 321, 326, 329, 332, 362, 364, 365, 367, 368, 374
Backward-moving wave, 4
Banded matrix approach, 275
Band-structured matrices, 276
Basic equations, 12
Basis function, 286, 377
Bessel function, 9, 32, 163, 301
Bessel's equation, 9, 12, 377
Bias flow, 163, 171
Bias flow Mach number, 163

Binder material, 233
Blade passing frequency, 310, 374
Bladed machinery, 374
Blow-down, 211
Boundary conditions, 33, 154, 156, 236, 289, 291, 292
Boundary integral equation techniques, 290
Boundary-layer friction, 21
Branch element, 68, 89
Branch impedance, 69, 70
Branch lumped element, 77
Branch resonator, 68, 161
Break mean effective pressure, 362
Break-in and break-out of noise, 310
Breakout noise, 306, 309, 310, 314–316, 322
Bulk reacting lining, 29, 32, 251, 252
Bulk reaction, 29
Bulk reaction model, 255, 257

Calibration methods, 201
Canonical form, 130, 133, 153
Cartesian coordinates, 5, 24, 35, 301
Cascaded-element muffler, 137, 147
Catalytic converter, 366
Catalytic muffler, 366
Cavity compliance, 69
Center-to-center distance, 164, 338
Ceramic absorber, 368
Characteristic impedance, 4, 13, 15, 18, 23, 33, 43, 46, 52, 63, 84, 104, 113, 126, 148, 177, 184, 194, 199, 225, 226, 246, 252, 297, 339, 356
Characteristic polynomial, 122
Choked sonic condition, 211
Choked sonic flow, 211
Circular cylindrical chambers, 301
Circular duct, 8, 10, 19, 27, 29, 32, 38
Circular function product, 88
Circular inlet/outlet ports, 301
Circular tube, 2, 5, 11, 20
Circumferential modes, 360
Classical acoustic variables, 106
Classical method, 71
Classical state variables, 137, 221
Closed-end cavity, 68
Closed-end termination, 98
Coefficient matrix, 129, 130, 134
Coefficient of discharge, 175
Coefficient of thermal conductivity, 15
Coefficient of viscothermal friction, 15
Collocation method, 285

Combination muffler, 176, 184, 234, 366
Compact matrix formulation, 312
Compatibility condition, 6, 31, 32
Compatibility equations, 269
Compatibility requirements, 277
Compliance, 52
Compliant walls, 23, 29, 239, 251
Compliant-wall tubes, 306
Compression-ignition engine, 97
Computation time, 290
Concentric hole-cavity resonator, 70, 71
Concentric tube resonator, 124, 128, 129, 284, 364
Condenser microphone, 212
Confocal ellipses, 34
Confocal hyperbolae, 34
Conical tube, 62, 78, 79
Conservation of energy, 102
Constant pressure source, 54, 57, 85
Constant velocity source, 331
Continuity equation, 2, 21, 151, 152
Convected 3D wave equation, 18
Convected wave, 107, 234, 239
Convective effect of mean flow, 18, 20, 22, 37, 105, 126, 133, 239, 245, 275, 284, 285
Convective four-pole parameters, 217
Convective one-dimensional wave equation, 17
Convective source impedance, 112, 217
Convective state variables, 114, 126, 137, 217
Convective variables, 114, 218
Cooling jacket, 212
Correction factor, 201, 202
Coupled analysis, 316
CPU time, 275
Cross baffle, 178
Cross mode, 37
Cross-correlation, 197
Cross-flow contraction, 125, 166, 327, 364
Cross-flow expansion, 135, 327, 364
Cross-flow expansion element, 124, 164
Cross-spectral density, 194, 195
Cross-talk, 310
Curvature factor, 307
Cut-off frequency, 8, 11, 19, 20, 34, 38, 65, 372
Cut-on frequency, 37, 268, 270, 357, 360, 372
Cyclic passive systems, 214
Cylindrical polar coordinates, 5, 27, 301, 302

Damped wave equation, 16
Damping matrix, 284

Data processing system, 202
Data-processing errors, 208
Degrees of freedom, 277
Design guidelines, 338
Diametral mode, 10
Differential length, 326
Differential matrix form, 152
Diffractive effects of mean flow, 259
Direct analogies, 51
Direct least-squares method, 224
Direct least-squares multi-load method, 218
Direct measurement, 211, 212
Discharge coefficient, 128
Discrete Fourier transform, 225
Discrete frequency components, 374
Discrete frequency method, 216
Discretization errors, 200
Displacement volume, 228
Dissipative duct, 1, 23, 27, 245
Dissipative effect of mean flow, 112
Dissipative element, 262
Dissipative muffler, 1
Distorted circular duct, 312
Distributed element, 76, 77, 88, 89, 126
Distributed parameter approach, 126
Doppler effect, 23
Double-leaf sandwich type walls, 369
Double-reversal chamber, 162, 173
Double-tuned extended inlet/outlet expansion chamber, 326
Double-tuned extended-tube chamber, 324, 329
Double-wall sandwich construction, 368
Double-wrapped bodies, 365
Doubly-periodic set of holes, 338
Downstream convection, 259
Dual-channel FFT analyzer, 204
Ducted fans, 372
Dynamic head, 163, 363, 364
Dynamic shape factor, 248
Dynamical equilibrium, 2, 4, 5, 11, 12, 15, 16, 18, 21, 24
Dynamical filter, 91

Eccentricity, 37
Eccentricity of the elliptical section, 358
Effect of mean flow, 259
Eigen equation, 25, 26, 29
Eigenanalysis, 177, 185
Eigenfrequency, 301, 304
Eigenfunction, 301, 304

Eigenfunction expansion, 290
Eigenmatrix, 122
Eigenvalue problem, 135
Eigenvalues, 26, 130, 131
Electrical analogous circuit, 225
Electrical analogy, 225
Electrical circuit analogy, 163
Electrical circuit representation, 52, 58
Electrical filter, 41
Electrical network theory, 48
Electrical resistance, 163
Electrical source, 210
Electroacoustic analogies, 370
Electroacoustic driver, 211
Electromotive force, 52
Electropneumatic driver, 211, 212, 228
Electropneumatic exciter, 193
Elements with curved-line boundaries, 276
Elliptical chamber, 36, 65, 276, 290
Elliptical coordinates, 303
Elliptical cross-section, 154
Elliptical cylindrical chamber, 356
Elliptical cylindrical coordinates, 34, 35
Elliptical duct, 37, 38, 162
Elliptical shell, 2
End cavity, 158
End chamber, 154, 159–162, 182, 345
End correction, 47, 51, 159, 160, 325, 329, 331, 334
End plates, 49, 316
Energy analysis, 60
Energy flux, 102, 104
Energy functional, 285, 292
Engine exhaust source, 218
Engine exhaust system, 331
Engine manifold, 217
Engine swept volume, 332
Ensemble averaging, 203
Entropy, 2, 382, 383
Entropy fluctuation, 103, 116, 118
Entropy fluctuation hypothesis, 119
Entropy gradients, 103
Equalization errors, 208
Equivalent acoustic impedance, 75
Equivalent circuit, 60, 68–70, 74, 83, 167
Equivalent cylindrical source, 310
Equivalent flow resistance, 167, 173
Equivalent impedance, 44, 46, 53, 261
Equivalent resistance, 167–169, 171, 173
Equivalent unfolded plate, 312

Error function, 218, 221
Even-even, 35, 36
Even-membership combinations, 90, 92
Even-odd, 36
Exhaust muffler, 11, 20, 21, 23, 41, 43, 53, 98, 111
Exhaust noise, 1, 11
Exhaust pipe, 53, 56, 86, 93, 98, 104, 217
Exhaust process, 97
Exhaust source, 210, 211
Exhaust source impedance, 212
Exhaust system, 97, 197
Exhaust system of CI engine, 227
Exhaust valve, 97
Expansion chamber, 91, 93, 94
Expansion ratio, 91, 94
Expansion stroke, 97
Experimental corroboration, 267
Experimental setup, 192
Exponential tube, 64, 79
Extended inlet, 115, 117, 118, 363
Extended outlet, 115, 117, 118, 363
Extended-inlet extended-outlet muffler, 329
Extended-inlet junction, 161
Extended-tube elements, 114
Extended-tube expansion chamber, 91, 94
Extended-tube resonator, 67–69, 91, 94, 118
External measurements, 218
External set of nodes, 291

Fan noise, 1
Far field, 105
Far-field solution, 290
Fiber-based porous sound-absorbing materials, 247
Fibonocci series, 89
Fibrous materials, 1
Field function, 276
Field variables, 277, 281, 286, 292
Finite amplitude wave effects, 101
Finite element analysis, 276
Finite element equations, 289
Finite element formulation, 286
Finite element method, 76, 276
Finite element solution, 290
Finite elements, 276, 286
Finite Fourier transform, 195
Finite volume analysis, 229
Finite volume approach, 225
Finite volume method, 224

Finite wave analysis, 101
Firing frequency, 98, 100, 211, 226, 233, 337
Firing mode, 211
First law of thermodynamics, 381
First-order equations, 122
Flanking transmission, 369
Flare tube, 78, 203
Flexible membrane, 284
Flexible walls, 292
Flow distribution, 175
Flow resistance, 163, 164
Flow resistance network, 173
Flow resistivity, 29, 33, 247, 255, 257
Flow shear, 104
Flow-acoustic impedance, 339
Flow-acoustic interaction, 347
Flow-acoustic resistance, 328, 339
Flow-generated noise, 262, 263, 322, 369
Flow-reversal end chamber, 34, 154, 158
Flow-reversal muffler, 151
Flush tube, 154
Forked dual configuration, 351, 352
Forward progressive wave, 13
Forward wave, 3, 104, 107
Forward-moving wave, 3
Fourier Bessel series, 379
Fourier transform, 194
Four-load method, 218, 224
Four-pole form, 81
Four-pole parameters, 74–76, 81, 126, 148, 185, 203, 204, 208, 276, 288, 289, 292, 297–299, 304
Four-stroke-cycle engine, 97, 98, 100
Free field, 109
Free shear layer, 326, 327
Free-space radiation, 108
Frequency domain, 43, 101, 225
Frequency-domain analysis, 51, 370
Frobenius polynomial function, 160
Frobenius solution, 65
Froude's friction factor, 21, 113, 191, 193, 198, 207
Fully developed turbulent flow, 102
Fundamental frequency, 98

Gain factor, 201
Galerkin approach, 285
Galerkin formulation, 286
Galerkin method, 285, 286
Gaussian elimination, 180, 183

Gaussian form, 179
Generalized algorithm, 177, 178
Generalized decoupling approach, 120
Generalized variables, 122
Geometric invariance, 277
Geometric isotropy, 277
Glass fibers, 1
Global finite element shape function, 286
Global matrix, 291, 292
Global noise attenuation, 374
Granular materials, 248
Grazig flow resistance, 169
Grazing flow, 127, 163, 171
Grazing flow Mach number, 128, 163
Grazing mean flow, 234
Grazing-flow impedance, 127
Green's function, 292, 293, 295, 296, 298, 302, 304, 312, 352
Green's function solution, 356
Green's theorem, 287
Guide vanes, 368

Hamilton's principle, 281, 285
Hankel function, 308, 379
Harmonic time dependence, 5, 130, 152, 252, 292, 352
Heat conduction, 12, 15
Heat transfer, 381
Heaviside function, 296
Helmholtz equation, 34, 63, 160, 295, 303, 307
Helmholtz number, 87, 109
Helmholtz resonator, 71, 299
Hermite polynomials, 290
Herschel-Quincke tube, 147
Heuristic study, 91
Hexahedral element, 280
High specific-output engines, 365
Higher order acoustic modes, 118
Higher order mode excitation, 119
Higher order mode hypothesis, 119
Higher-order elements, 277, 290
Higher-order evanescent modes, 325
Higher-order mode effects, 210
Higher-order model, 280
Higher-order modes, 7, 10, 12, 18, 19, 202, 272, 273, 275, 360
Hole-cavity combination, 94
Hole-cavity resonator, 91
Hose wall, 306, 308
Hot-wire anemometer, 127

HVAC ducts, 292
HVAC system, 292, 310, 322
Hybrid approach, 224
Hybrid muffler, 374
Hypothetical apex, 62, 78
Hypothetical pistons, 267

Impedance analysis, 212
Impedance approach, 51
Impedance mismatch, 1, 68
Impedance tube, 105, 187, 188, 190, 191, 193, 215, 216
Impedance tube method, 201, 215
Impedance tube technology, 56, 58, 202
Impervious layer, 257
Incident power, 61, 108
Incident wave, 45, 197
Incompressible flow, 115, 163, 167, 170, 384
Incompressible moving medium, 234
Indicial equation, 66
Indirect measurement, 211
In-duct measurements, 229
Inertance, 52
Inertive impedance, 59, 119
Infinite flange, 46
Infinite series, 267
Inhomogeneous wave equation, 292
Inlet junction, 162
Inline aeroacoustic resistance, 118
Inline baffle, 348
In-line lumped element, 77, 89
Inner loop, 168
Insertion loss, 53, 57, 58, 71, 72, 82, 84, 94, 111, 112, 183, 187, 213, 261–263, 321, 322, 331, 332, 365, 369
Instantaneous pressure, 321
Insulated body, 365
Intake source characterization, 225
Intake system of CI engine, 227
Integrated transfer matrix, 175, 181, 347
Intensity level, 42
Inter element compatibility, 277
Interacting ducts, 131, 151, 169
Interacting perforated elements, 184
Interfocal distance, 34
Intermediate baffle plate, 159
Intermediate baffles, 347
Internal combustion engine, 1, 11, 23, 97, 102, 197, 211
Internal impedance, 210, 339

Interpolation functions, 277, 279
Interpolation model, 277
Isentropicity, 2, 17, 62, 152
Isentropicity relations, 116
Isotropic field model, 277

Jet impingement, 345
Jet noise, 109, 262, 332

Kirchoff's first law, 167, 170
Kirchoff's second law, 168, 170

Lagrange polynomials, 290
Lagrangian, 285
Lagrangian functional, 281
Laplacian, 5, 8, 27, 34
Least attenuated mode, 241, 243
Least squares method, 285
Level difference, 53, 56–58, 71, 183, 187, 189
Library function, 135
Line element, 278
Line source model, 310
Linear interpolation model, 279
Linearized equation, 2
Linearized theory, 101
Lined bend, 261
Lined duct, 244
Lined wall, 2
Load impedance, 214
Load resistance, 48, 51
Loading termination, 208
Local coordinate system, 277
Local reaction, 29
Local reaction model, 255
Locally reacting, 23
Locally reacting lining, 24, 33, 251–253
Long-strand fibers, 234
Loss coefficient, 115, 163, 164
Loss factor, 118
Loudspeaker, 211, 372, 375
Low frequency dip, 331
Low specific-output engines, 365
Lowest order mode, 30, 261
Low-pass acoustic filter, 71
Low-pass filter, 59, 60
Lumped compliance, 69
Lumped element, 58, 88, 89
Lumped element acoustical filter, 331
Lumped element approximation, 331
Lumped flow-resistance network, 352

Index

Lumped impedance, 51
Lumped inertance, 50, 69
Lumped in-line element, 78
Lumped in-line inertance, 59
Lumped resistance network, 175
Lumped shunt compliance, 59
Lumped shunt element, 78
Lumped-element network theory, 169

Manifold, 100
Mass continuity, 2, 5, 12, 16–18, 62, 116, 120, 130, 131, 151
Mass flux, 103
Mass flux perturbation, 105
Mass velocity, 4, 5, 7, 12, 15, 18, 23, 42, 43, 45, 104, 113, 297
Matching surface, 290
Mathieu differential equation, 35
Mathieu function, 35, 36, 303
MATLAB® notation, 156, 158
MATLAB® programming, 158
Matrix calculus theory, 129
Matrix equation, 85
Matrizant approach, 131
Mean back pressure, 321
Mean flow distribution, 173, 175
Mean flow Mach number, 102, 286, 339, 363
Mean-flow convection, 23, 239
Mean-flow velocity profile, 104
Measurement channels, 201
Memory requirements, 290
Method of characteristics, 224
Method of direct measurement, 228
Microphone spacing, 201
Microphone switching, 201
Midstream Mach number, 102
Mimani tables, 38
Mismatch of characteristic impedances, 61
Modal expansion method, 292
Modal matrix, 130
Mode coupling, 312
Mode shape function, 294
Modified Mathieu differential equation, 35
Modified Mathieu function, 35–37, 303
Momentum balance, 2, 116, 120, 130, 151
Momentum equation, 2, 7, 17, 27, 28, 117, 131, 133, 151, 152, 240
Monopole, 108
Motored engine, 214
Muffler proper, 53, 55, 58, 288

Multi-cylinder engine, 100, 224, 226
Multi-load least-squares method, 217
Multi-load method, 225
Multiply-connected configuration, 338
Multiply-connected elements, 148

Natural coordinate system, 277, 286
Naturally aspirated diesel engine, 227
Naturally aspirated SI engine, 228
Naturally-aspirated engine, 100
Navier-Stokes equations, 12
Neck inertance, 69
Neck tube, 69
Negative work, 97
Net transmission loss, 322
Neumann factor, 302
Neumann function, 9, 32, 377
Newton-Raphson iteration scheme, 32, 237, 241
Newton's second law of motion, 49
Nodal degrees of freedom, 286
Noise reduction, 53, 56, 58, 183, 187
Nondimenstional offset distance, 358–360
Non-equalization errors, 200
Nonlinearity effects, 212
Non-overlapping perforated ducts, 176
Normal impedance of lined wall, 244
Normal wall impedance, 25
Normalization factors, 294, 301
Normalized frequency, 258
Normalized impedance, 196, 221
Normalized pressure drop, 164, 166
Normalized source impedance, 213
Normalized state variables, 153
Normalized wall admittance, 307
Norton theorem, 53
Numerical computation, 267

Odd-even, 36
Odd-odd, 36
Offset inlet and outlet configuration, 292
One-dimensional elements, 276
One-dimensional model, 312
One-dimensionality, 202
Open area ratio, 164, 339, 348
Open tube, 47
Open-circuit voltage, 210
Open-end, 47, 49
Open-ended termination, 201
Order-of-magnitude considerations, 22

Order-of-magnitude relations, 13, 21
Orifice discharge coefficient, 163
Orifice impedance, 126, 128
Orifice interaction effects, 163
Orifice plate meter, 193
Orthogonal expansion, 267
Orthogonal property, 380
Orthogonality, 301
Orthonormal function, 294
Orthonormal set of eigenfunction, 301
Outer loop, 168
Outlet junction, 162
Output impedance, 210
Ovality, 314
Overall exhaust noise level, 98
Overall insertion loss, 333
Overall performance of a muffler, 288
Overall pressure drop, 167
Overall transfer matrix, 86, 148, 162, 183, 288

Parallel baffle muffler, 27, 30, 234, 255, 257, 258, 261, 263, 367
Parallel paths, 169
Parametric zero, 37
Particle velocity, 4, 7, 12, 16, 20, 24, 29, 42, 297
Particle velocity components, 235
Partitioned matrix approach, 158
Partitioning of the global matrix, 291
Passive muffler, 369, 374
Passive subsystem, 44, 52
Passive termination, 188, 199
Perforate admittance, 284
Perforate bridge, 326
Perforate impedance, 121, 126, 129, 163
Perforated baffle, 348
Perforated element mufflers, 119, 126
Perforated elements, 125, 131
Perforated impedance expressions, 133
Perforated plate, 252
Perforates with cross-flow, 127
Perforates with grazing flow, 129
Permeable membrane, 284
Phase angle, 191
Phase factor, 201
Phase front, 63
Phase mismatch, 201
Phase surface, 2
Phase velocity, 3

Pienings empirical formula, 250
Piezoelectric pressure transducer, 212
Plane wave, 2, 6–9, 12, 15, 18, 19, 23, 30, 38, 42, 46
Plane wave propagation, 9, 41, 62, 129
Plane wave propagation criterion, 208
Plane wave theory, 208, 210
Plenum chamber, 261, 262, 292, 368
Plug muffler, 284, 327, 328, 339, 363
Poisson's ratio, 247
Polynomial model, 277
Porosity, 164, 247, 252
Porous ceramic absorber tiles, 249
Porous rigid tiles, 248
Positional subscripts, 89
Post-processing time, 300
Power level, 42
Prandtl number, 15, 248
Pre-computation effort, 290, 292
Pressure attenuation constant, 250
Pressure function, 277
Pressure minimum, 190, 192, 202
Pressure model, 277
Pressure nodal lines, 241
Pressure node, 47, 48, 191
Pressure pattern, 277
Pressure representation, 53
Pressure-volume diagram, 97
Primary nodes, 277
Primary source, 372
Principal variable, 122
Principle of linear superposition, 295, 298
Probe tube, 187, 192, 373
Probe-tube method, 187, 188, 193
Product matrix, 83, 84, 112
Progressive plane wave, 234
Progressive wave, 3, 14, 43, 46
Progressive wave approach, 310
Progressive wave components, 60
Propagating solution, 16
Propagation constant, 4, 245, 261
Propagation of higher-order modes, 258
Protective layer, 33, 251, 252
Pseudo-random source, 204
Pumping losses, 97

q-parameter, 37, 38
Quadratic interpolation function, 279
Quadrilateral element, 279
Quarter-wave resonator, 69, 70, 160, 329

Index

Radial elliptical coordinate, 37
Radial mode, 37
Radial particle velocity, 121
Radial velocity, 9
Radial wave number, 9
Radiation end, 48, 72
Radiation impedance, 43, 46, 48, 51, 52, 54, 58, 69, 83, 110, 187, 226, 261, 308
Radiation load, 112
Radiation reactance, 48, 51, 331
Radiation resistance, 48, 51, 70, 331
Random excitation, 203
Random-excitation method, 212
Random-noise generator, 193
Random-noise signal, 193
Rated load, 211
Rated speed of the engine, 339
Reactive muffler, 1, 61
Reactive-cum-dissipative muffler, 327
Reciprocity relation, 293
Rectangular duct, 2, 5, 8, 10, 18, 24, 29, 30
Rectangular plenum, 301
Refined multi-load method, 218, 221
Reflected power, 61
Reflected wave, 46, 104, 107, 197
Reflection coefficient, 45, 47–49, 61, 68, 69, 104, 105, 107, 108, 110, 187, 188, 191, 196, 199–203
Reflective muffler, 1, 61, 233
Regenerated sound, 262
Reliability inequality, 208
Residual, 286
Residual method, 276, 285, 286
Resistances in parallel, 167
Resonance frequency, 68, 70
Resonator cavity, 69
Response coordinates vector, 294
Resultant transfer matrix, 83
Reversal cum contraction, 115
Reversal cum expansion, 115
Reversal-contraction, 117
Reversal-expansion, 117
Reversal-expansion element, 159
Reversing chamber mufflers, 267
Reynold's number, 21, 102, 126
Riemann variables, 224
Rigid end plate, 69, 71
Rigid termination, 188, 201
Rigid unlined duct, 239
Rigid-end cavity, 117

Rigid-pipe annulus, 177
Rigid-walled duct, 6
Rigid-walled pipe, 244
Rigid-walled tube, 104
Rotating machinery, 43
Runners, 100

Sampling frequency, 374
Second law of thermodynamics, 381, 382
Secondary emission of noise, 368
Secondary radiation of sound, 369
Second-order equation, 122
Second-order terms, 106, 116
Section reversal method, 201
Segmentation approach, 126
Segmentation procedure, 119
Shape factor, 248
Shape functions, 277, 279, 280
Shell noise, 306, 365
Shunt element, 88
Shunt impedance, 70
Side-inlet side-outlet muffler, 329
Sideways diffraction, 259
Signal enhancement technique, 201, 203
Signal-to-noise ratio, 201, 211, 212
Simple area change, 61
Simple area discontinuities, 60, 83, 118
Simple expansion chamber, 85, 94, 100, 167, 257, 322, 325, 329, 331
Singular points, 65
Sinusoidal signals, 43
Small-amplitude waves, 12
Sound power measurement, 372
Sound pressure level, 42
Sound speed, 3, 102
Sound wave, 109
Source, 72, 108
Source characteristics, 211, 214, 217, 224, 226, 228
Source coordinates vector, 294
Source impedance, 58, 83, 93, 212, 221, 224
Source pressure, 221, 226
Source strength, 210, 217, 218, 221
Source strength level, 226
Space-average Mach number, 102
Spark arresting muffler, 366
Spark-ignition engine, 97
Spatial isotropy, 277
Specific fuel consumption, 97, 321
Specific transmission loss, 257

Spectral analyzer, 201
Spectral density, 197
Speed order, 100, 225, 226
Speed order analysis, 321
Spherical wave propagation, 105
Spherical waves, 63
Spinning mode, 267
Splitter silencer, 263
Stagnation enthalpy, 103, 106
Stagnation enthalpy perturbation, 105
Stagnation pressure drop, 106, 115, 163, 164, 166, 167, 169, 175, 326, 328, 364, 382
Stagnation pressure probe, 108
Standing wave, 3, 60, 238, 241
Standing wave effects, 217
Standing wave method, 212
Standing wave model, 370
Standing wave parameters, 190
Standing wave pattern, 190
Standing wave pressure, 198
Standing wave relations, 76
Standing wave solution, 14, 15
Standing wave variables, 56, 351
State variables, 43, 70, 74, 177, 204, 295, 298
State vector, 74, 77, 82, 129
Static pressure probe, 108
Stationary frame of reference, 17
Straight-edge elements, 276
Structural factor, 247
Successive iteration method, 241
Sudden area changes, 203
Sudden area discontinuity, 61, 118, 203
Sudden contraction, 61, 78, 86, 115, 118
Sudden expansion, 61, 78, 86, 115, 118, 167, 363
Sudden-contraction element, 159
Swept volume, 228
Symmetric expansion chamber, 267
Symmetric inlet-outlet chamber, 270
Symmetric modes, 26
Synthesis criteria, 94
System matrix, 292

Tail pipe, 47, 53, 56, 57, 86, 93
Tail-pipe end, 48
Terminal reflection coefficient, 200
Test element, 204, 208
Theory of filters, 43
Thermal insulation, 29
Thermodynamic cycle, 97
Thermo-kinetic energy flux, 105

Thevenin theorem, 53
Three-dimensional analysis, 268
Three-dimensional effects, 11, 208, 270, 276, 310
Three-dimensional elements, 276
Three-dimensional model, 312
Three-dimensional wave, 2, 288, 374
Three-duct cross-flow chamber, 364
Three-duct element, 135
Three-pass double-reversal chamber, 338
Three-pass muffler, 151, 154
Threshold pressure, 226
Through-flow perforated elements, 364
Time-domain analysis, 101
Time-domain ensemble averaging, 204
Time-invariant, 218
Top dead center, 97
Tortuosity factor, 248, 249
Total back-pressure, 173
Total pressure drop, 169, 171
Tracking frequency multiplier, 100
Transfer function, 197, 199, 201, 204, 206, 208, 370
Transfer function approach, 212
Transfer function method, 197, 202
Transfer matrix, 74, 76, 77, 80–82, 162
Transfer matrix based muffler program, 143, 228, 300, 327
Transfer matrix method, 83
Transfer matrix relation, 75, 77–79, 81, 83, 114, 118, 123–125, 130, 136, 241, 243, 244, 270
Transfer matrix representation, 203
Transformation matrix, 114
Transient testing method, 202, 203
Transmission loss, 53, 55, 57, 58, 61, 71, 83, 84, 180, 183, 187, 197, 249, 250, 263, 270, 298, 300, 331
Transmission matrix, 74
Transmission tube, 193
Transmission wave number, 7
Transverse impedance, 23
Transverse modes, 362
Transverse plane wave, 162, 182
Transverse pressure distribution, 10
Transverse radiation impedance, 307
Transverse transmission loss, 308, 310, 312, 314, 315
Transverse wave number, 244
Trial solution, 286
True transfer function, 202
Tube attenuation, 190

Index

Tubular cavity, 69
Tuned concentric tube resonator, 326, 327, 337, 365
Tuned frequency, 71
Turbocharged diesel engine, 226, 227
Turbocharged engine, 100, 224
Turbulent flow friction, 22
Turbulent incompressible mean flow, 113
Two source-location method, 203, 204, 207–209
Two-channel Fourier analyzer, 193
Two-dimensional elements, 276, 290
Two-dimensional waves, 252, 257
Two-duct element, 123
Two-load method, 203, 207–209, 225
Two-microphone method, 127, 188, 202
Two-row array of integers, 89
Two-stroke-cycle engine, 98, 100

Uncoupled analysis, 316
Uniform duct, 159, 162
Uniform pipe, 177, 182
Uniform piston-driven model, 356
Uniform tube, 77, 78, 130, 131, 203
Unit step function, 296

Valve throat, 211
Variational formulation, 285
Variational methods, 281, 292
Variational principle, 276, 281
Velocity antinode, 47
Velocity head, 364
Velocity of wave propagation, 3, 14, 16, 109
Velocity potential, 267, 286, 295–297, 302
Velocity ratio, 55, 72, 74, 82–84, 88, 91, 94, 111, 112
Velocity representation, 53
Ventilation ducts, 1

Ventilation system, 369
Vibration isolator, 41
Viscothermal attenuation constant, 198
Viscothermal dissipation, 21
Viscothermal friction, 21
Volume flow rate, 163
Volume velocity, 4, 7, 42, 372
Volumetric efficiency, 97
Vortex waves, 108

Wall friction, 21, 169
Wall friction effect, 244
Wall friction losses, 171
Wall friction resistance, 169
Wall impedance, 239
Wall lining, 31, 32
Wall-boundary condition, 240
Wall-mounted microphones, 202
Wave coupling phenomena, 310
Wave envelope, 290
Wave equation, 3, 4, 9, 17, 24, 41, 62, 131
Wave front, 2, 42
Wave number, 4, 14, 22, 24, 29, 33, 43, 130, 163, 177, 184, 189, 190, 226, 246, 248, 252, 301
Wave operator, 3
Wave propagation speed, 52
Waveguide, 202
Wavelength, 10
Webster equation, 62
Weighted equations, 286
Weighting factor, 218
Weighting function, 286

Yielding of the walls, 104
Yielding walls, 27
Y-Z combination, 90, 92, 94